T0206330

Hindustan Publishing Corporation

The Hindustan Publishing Corporation series publishes high-quality content in the domain of mathematical sciences and applications with primary focus on mathematics, statistics, and computer sciences under the publishing co-operation with Hindustan Publishing Corporation (India). Editorial board comprises of active scientists from across the world. The series will publish textbooks, monographs, and lecture notes. Literature in this series will appeal to a wide audience of students, researchers, educators, and professionals across mathematics, statistics and computer science disciplines.

More information about this series at http://www.springer.com/series/16628

Andrei Bourchtein · Ludmila Bourchtein

Complex Analysis

Andrei Bourchtein
Institute of Physics and Mathematics
Universidade Federal de Pelotas
Pelotas, Brazil

Ludmila Bourchtein
Institute of Physics and Mathematics
Universidade Federal de Pelotas
Pelotas, Brazil

ISSN 2730-6569 ISSN 2730-6577 (electronic)
Hindustan Publishing Corporation
ISBN 978-981-15-9221-8 ISBN 978-981-15-9219-5 (eBook)
https://doi.org/10.1007/978-981-15-9219-5

This work is a co-publication with Hindustan Publishing Corporation, New Delhi, licensed for sale in all countries in electronic form, in print form only outside of India. Sold and distributed in print within India by Hindustan Publishing Corporation, 4805 Bharat Ram Road, 110002 New Delhi, India. ISBN: 9788170750994 © Hindustan Publishing Corporation 2021.

Jointly published with Hindustan Publishing Corporation
The print edition is not for sale in India. Customers from India please order the print book from: Hindustan Publishing Corporation.
ISBN of the Co-Publisher's edition: 9788170750994

Mathematics Subject Classification: 30-01, 30Axx, 30Bxx, 30C20, 30C35, 33-01, 33B10, 33E05

This Springer imprint is published by the registered company Springer Nature Singapore Pte Ltd.
The registered company address is: 152 Beach Road, #21-01/04 Gateway East, Singapore 189721, Singapore

To Haim and Maria with the warmest memories;

To Maxim and Natalia with love and inspiration;

To Valentina for her affectionate and cheerful heart;

To Victoria for her wonderful ocean of imagination

Preface

This textbook is aimed at university students specializing in mathematics and natural sciences and at all the readers interested in studying complex analysis. The mathematical level of the exposition corresponds to advanced undergraduate courses of mathematical analysis and first graduate introduction to the discipline. The book covers all the traditional topics of complex analysis, starting from the very beginning—the definition and properties of complex numbers, and study of the "pre-differentiable" properties of functions similar to those in real analysis—and ending with properties of conformal mappings, including the proofs of the fundamental results of conformal transformations.

As the subject is developed, the exposition becomes gradually more complicated. Within each major topic, the exposition is as a rule inductive and starts with rather simple definitions and/or examples, but generally it becomes more compressed and sophisticated as the course progresses. The only exception is the last chapter which deals primarily with theoretical concepts and proofs of the major theorems on conformal mappings. In this last part, we consider almost exclusively theoretical questions. Even the examples provided are of theoretical rather than practical importance.

This book is designed for the reader who has a good working knowledge of real analysis or, at least, advanced calculus. No additional prior knowledge is required. All the preliminary topics related to complex variables and required for the development of the complex function theory are covered in the initial part of the text and, when necessary, at the starting point of each topic. This makes the book self-sufficient and the reader independent of any other texts on complex variables. As a consequence, this work can be used both as a textbook for advanced undergraduate/graduate courses and as a source for self-study of selected (or all) topics presented in the text.

The book is divided into five chapters. Each chapter is comprised of different sections, which are numbered separately within each chapter, bearing both the number of chapter and the sequential number of section inside the chapter. For instance, Sect. 3.2 means the second section in the third chapter. The formulas and figures are numbered sequentially within each chapter regardless of the section where they are found. In this way, Formula (5.4) is the reference to the fourth formula in the fifth chapter, while Figure 2.17 indicates the seventeenth figure in the second chapter. The theorems, propositions, lemmas, definitions, and examples are numbered independently within each section.

Chapter 1 deals with the basic concepts and preliminary results of complex analysis: it starts with the properties of complex numbers and their geometric interpretation, then proceeds to description of essential properties of sets and curves, and, in the main part, it focuses on the properties of complex functions analogous to those studied in real analysis. Among such properties, the following are analyzed in detail: continuity of a function, behavior of functions depending on a parameter, including sequences and series, definition and properties of line integrals, and integrals depending on a parameter.

Chapter 2 starts with differentiability—the first topic that separates radically the real and complex analyses. The comparison between real- and complex-differentiability is performed in great detail, including different illustrative examples. The following part of this chapter addresses the concepts of analytic and holomorphic functions and the relationship between them. In the course of this study, some background results of fundamental importance are established and discussed—Cauchy's and Goursat's theorems on vanishing of line integral of an analytic function, the Cauchy integral formula, and the equivalence between analyticity and holomorphy in a domain (the main criterion for analyticity). In the subsequent sections, we consider the relationship between harmonic functions and real as well as imaginary parts of an analytic function, and also introduce an indefinite integral of an analytic function. Different properties of analytic functions are studied in Sects. 2.7 and 2.8, including Morera's theorem (the converse to Cauchy's theorem), limit properties of families of analytic functions, and the theorem on uniqueness of analytic function determined in an arbitrary small disk inside the domain of analyticity. The remaining two sections are devoted to the procedure of analytic continuation and construction of single-valued and multi-valued analytic functions on extended domains, with initial notion of the Riemann surface.

In Chap. 3, we study systematically singular points, their classification, expansion in the Laurent series, the concept of residues, and their properties and applications to evaluation of integrals. First, we show that there exists at least one singular point on the boundary of the disk of convergence and analyze the relationship between the behavior of the power series on the circle of convergence and the number of the singular boundary points. In Sects. 3.3 and 3.4, we consider the expansion of functions in the Laurent series and classification of singular points. The next four sections deal with a deeper investigation of different types of single-valued isolated singular points: removable singularities, poles, and essential singularities. In Sect. 3.9, we introduce the concept of residues and prove the fundamental theorem on residues, which paves the way for developing the methods of calculation of different types of line integrals. Many of these techniques are illustrated in Sects. 3.10–3.12, including application of the residues for evaluation of improper integrals. In the last two sections, we consider the argument principle, which expresses the relationship between the numbers of zeros and poles in a domain and a line integral along the boundary of this domain, and Rouché's theorem, which has a number of applications, including the evaluation of the number of zeros and fixed points of a holomorphic function and also the use as an important tool in the proofs of other statements such as the fundamental theorem of algebra and Hurwitz's theorem.

Chapter 4 starts with the geometric interpretation of derivative and notion of conformal mapping. The fundamental principles of conformal mappings are formulated and used for investigation of elementary functions in the main part of this chapter, but their proofs are postponed to Chap. 5. The question of symmetric points is considered as one more preliminary point applied later to study of elementary functions. All the remaining (and main) part of Chap. 4 is devoted to a detailed discussion of the properties (including conformal ones) of a large number of analytic functions which are of importance both in theory and in applications. Some elementary and also a bit more sophisticated functions are given to work out as exercises at the end of this chapter, but almost all the essential elementary functions are analyzed in detail in the text.

Finally, in Chap. 5, we provide the proofs of fundamental results on conformal mappings. Unlike other parts of the text, the subjects of this chapter are almost exclusively theoretical, including some specific problems, which have theoretical significance rather than representing only illustrative examples. This is due to the fact that introductory examples and essential geometric properties of conformal mappings were already discussed in detail in Chap. 4, and numerous examples of application of the principles of conformal mappings to study of elementary functions were provided in the same chapter. Another reason is an attempt to focus attention on the logical reasoning of the proofs, some of which are rather intricate, at least for an inexperienced reader. This is especially true for the proof of the Riemann mapping theorem on possibility to map conformally any simply connected domain with at least two boundary points onto the open unit disk. In sequel, we consider the Riemann–Schwarz symmetry principle as one of the important forms of application of analytic continuation. The two final sections are devoted to construction of conformal mappings between the domains of theoretical and practical importance—half-planes, rectangles, and polygons.

The book contains a large number of problems and exercises, which should make it suitable for both classroom use and self-study. Many standard exercises are included in each section to develop basic techniques and test understanding of concepts. Other problems are more theoretically oriented and illustrate more intricate points of the theory or provide counterexamples to false propositions which seem to be natural at first glance. Some harder exercises of striking theoretical interest are also included as examples or applications of theoretical results, but they may be omitted in courses addressed to less advanced students. Many different additional problems are proposed as homework tasks at the end of each chapter. There are about 400 problems proposed in all. Their level ranges from straightforward, but not overly simple, exercises to problems of considerable difficulty, but of comparable interest. The most involved and challenging problems are marked with an asterisk.

The presented text has the following features:

1. Completeness: the text covers all the traditional topics of complex analysis at the advanced undergraduate/graduate level, starting with the definition of complex numbers and ending with the fundamental principles of conformal mappings.
2. Self-sufficiency: all the background topics related to complex variables are covered in the text, and, consequently, this work can be used as both a textbook and a source for self-study.
3. Exercises: there are a large number of problems and exercises, solved and proposed, which should make the book suitable for both classroom use and self-study.
4. Visualization: there are about 130 figures providing systematic illustration of the presented material to make the exposition clearer and develop geometric intuition.
5. Generality: our intention was to present all the results in a more general form while avoiding major complications of their proofs.
6. Accessibility: all the topics are covered in a rigorous mathematical manner while keeping the exposition at a level acceptable for advanced undergraduate courses.

The theory of functions of a complex variable is one of the most significant and original mathematical creations. This branch of mathematics dominated the nineteenth century to the extent that this specific subject was often referred to as the theory of functions, reflecting its leading position in a general theory. The theory of functions (that is, complex analysis) has been called the mathematical joy of the nineteenth century and has also been acclaimed as one of the most harmonious theories in the abstract sciences. During the twentieth century, the complex analysis has achieved many elegant, fundamental, and even surprising results in a short span of time. We may join R. Dedekind who wrote: "The splendid creations of this theory have excited the admiration of mathematicians mainly because they have enriched our science in an almost unparalleled way with an abundance of new ideas and opened up heretofore wholly unknown fields to research." We hope that this text will help the reader to appreciate the intrinsic harmony and power of this subject.

Pelotas, Brazil

Andrei Bourchtein
Ludmila Bourchtein

Contents

About the Authors

Andrei Bourchtein is Professor at the Institute of Physics and Mathematics, Federal University of Pelotas, Brazil. He received his Ph.D. in Mathematics and Physics from the Hydrometeorological Center of Russia. He began his academic and research career as an Associate Professor at Mathematics Institute, Far East State University, Russia, and a Research Scientist at the Hydrometeorological Institute, Russia. In 1995, he joined the Federal University of Pelotas, Brazil, as Associate Professor at the Institute of Physics and Mathematics before being promoted to Full Professor. An author of more than 100 refereed articles and 7 books, Prof. Bourchtein's research interests include real and complex analysis, numerical analysis, computational fluid dynamics and numerical weather prediction. During his research career, he was awarded a number of grants from Brazilian science foundations and scientific societies, including the International Mathematical Union (IMU) and the International Council for Industrial and Applied Mathematics (ICIAM).

Ludmila Bourchtein is Professor Emeritus and Senior Research Scientist at the Institute of Physics and Mathematics, Federal University of Pelotas, Brazil. She received her Ph.D. in Mathematics from Saint Petersburg State University, Russia. During the span of 34 years at the Mathematics Institute, Far East State University, Russia, she held different positions, from Assistant Professor, Associate Professor, to the rank of Full Professor. In the last 15 years, she served as Associate and Full Professor at the Institute of Physics and Mathematics, Federal University of Pelotas, where she was conferred with the title of Emeritus Professor in 2019. An author of more than 80 referred articles and 5 books, her research interests include real and complex analysis, conformal mappings and numerical analysis. During her research career, she was awarded a number of grants of Russian and Brazilian science foundations and scientific societies, including the International Mathematical Union (IMU) and the International Council for Industrial and Applied Mathematics (ICIAM).

About the Authors

Chapter 1
Introduction

We begin this chapter with a quick review of the algebraic and analytic properties of complex numbers followed by their geometric interpretation on the Cartesian plane and through the stereographic projection in the three-dimensional space (Sects. 1.1–1.2). In Sect. 1.3, we consider basic topological notions of sets, focusing on properties of sets and curves on a plane.

Using these background concepts, in the subsequent part of this chapter (Sects. 1.4–1.9), we consider the properties of complex functions similar to those of real-valued functions. First, we consider the notion of limit and different types of continuity of complex functions. In Sect. 1.5, we discuss how to define a function on the boundary of a domain in the case when the boundary contains folds. Section 1.6 is concerned with functions depending on a parameter (including as a special case the sequences and series of functions) and different types of their convergence—pointwise, uniform, and normal. The last type is specific for complex functions, which is important for different theoretical results, such as analyticity of a limit function, univalence of a limit function, the Riemann mapping theorem, etc. In Sect. 1.7, we study power series (the basis of analytic functions) and present the main results of their convergence: Abel's lemma, the existence of the disk of convergence and the Cauchy–Hadamard theorem on the radius of convergence. The last two sections deal with the properties of the two types of integrals of complex functions—the line integral and the integral depending on a parameter.

1.1 Complex Numbers

Definition A *complex number* z is an ordered pair (x, y) of real numbers x and y.

The complex number $(x, 0)$ is identified with the real number x: $x = (x, 0)$. In particular, $0 = (0, 0)$ and $1 = (1, 0)$.

Two complex numbers $z_1 = (x_1, y_1)$ and $z_2 = (x_2, y_2)$ are *equal* if and only if $x_1 = x_2$ and $y_1 = y_2$

A. Bourchtein and L. Bourchtein, *Complex Analysis*, Hindustan Publishing Corporation,
https://doi.org/10.1007/978-981-15-9219-5_1

$$z_1 = z_2 \Leftrightarrow \begin{cases} x_1 = x_2 \\ y_1 = y_2 \end{cases} .$$

Addition of two complex numbers $z_1 = (x_1, y_1)$ and $z_2 = (x_2, y_2)$ is defined in a component-wise manner

$$z_1 + z_2 = (x_1 + x_2, y_1 + y_2),$$

and *multiplication* is defined by the formula

$$z_1 \cdot z_2 = (x_1, y_1) \cdot (x_2, y_2) = (x_1 x_2 - y_1 y_2,\ x_1 y_2 + y_1 x_2) .$$

The last definition implies the following rule of *multiplication* of a complex number $z = (x, y)$ *by a real constant* a:

$$az = (a, 0) \cdot (x, y) = (ax - 0, ay + 0) = (ax, ay) .$$

The set of all the complex numbers together with the defined property of equality and operations of addition and multiplication is denoted by $\mathbb{C}$.

Remark 1.1 Hereinafter the natural indexing of complex numbers is used: $z_k = (x_k, y_k)$, $k = 1, 2, \ldots$.

Remark 1.2 The definition of the complex numbers as the ordered pairs of real numbers, their equality and the operations of addition and multiplication by a real number are the same as for the elements of $\mathbb{R}^2$. However, multiplication of complex numbers differs $\mathbb{C}$ from $\mathbb{R}^2$, since in the latter there is no similar property. The implications of this operation will be seen further in many topics, and especially in the properties of complex-valued functions related to differentiability.

It is easy to check that the above *operations with complex numbers* satisfy the following properties:

(1)
$$z_1, z_2 \in \mathbb{C} \Rightarrow z_1 + z_2 \in \mathbb{C},\ z_1 \cdot z_2 \in \mathbb{C};$$

(2) *commutative property* of addition and multiplication

$$z_1 + z_2 = z_2 + z_1;\ z_1 z_2 = z_2 z_1;$$

(3) *associative property* of addition and multiplication

$$z_1 + (z_2 + z_3) = (z_1 + z_2) + z_3;\ z_1 \cdot (z_2 \cdot z_3) = (z_1 \cdot z_2) \cdot z_3;$$

(4) *distributive property*

$$z_1 \cdot (z_2 + z_3) = z_1 \cdot z_2 + z_1 \cdot z_3;$$

(5) *additive zero and inverse*

$$z + 0 = z; \;\; z + (-z) = 0;$$

(6) *multiplicative unit and inverse*

$$1 \cdot z = z; \;\; z \cdot \frac{1}{z} = 1, \forall z \neq 0.$$

The proofs of these properties follow immediately from the corresponding properties of the real numbers. For instance, for commutative property one has

$$z_1 + z_2 = (x_1 + x_2,\, y_1 + y_2) = (x_2 + x_1,\, y_2 + y_1) = z_2 + z_1,$$

$$z_1 z_2 = (x_1 x_2 - y_1 y_2,\;\; x_1 y_2 + y_1 x_2) = (x_2 x_1 - y_2 y_1,\;\; x_2 y_1 + y_2 x_1) = z_2 z_1.$$

For multiplicative unit and inverse one obtains

$$1 \cdot z = (1, 0) \cdot (x, y) = (1 \cdot x, 1 \cdot y) = z,$$

$$z \cdot \frac{1}{z} = (x, y) \cdot \left(\frac{x}{x^2+y^2}, \frac{-y}{x^2+y^2}\right) = \left(x\frac{x}{x^2+y^2} - y\frac{-y}{x^2+y^2}, x\frac{-y}{x^2+y^2} + y\frac{x}{x^2+y^2}\right) = (1, 0) = 1,$$

where the inverse element $\frac{1}{z} = \left(\frac{x}{x^2+y^2}, \frac{-y}{x^2+y^2}\right)$ is defined for any $z = (x, y) \neq (0, 0)$. The remaining properties are verified in the same elementary way.

These properties reveal that the set $\mathbb{C}$ is a field. However, contrary to the field $\mathbb{R}$, the set $\mathbb{C}$ is not an ordered field.

The last two properties indicate that the numbers $0 = (0, 0)$ and $1 = (1, 0)$ play a special role: the former acts as zero additive element and the latter is the multiplicative unit, also called the *real unit*. Besides, there is one more special number called the *imaginary unit* $i = (0, 1)$. According to the properties of arithmetic operations, for this number one has $i^2 = (0,\ 1)\,(0,\ 1) = (-1,\ 0) = -1\,(1,\ 0) = -1$. Then any complex number z can be uniquely represented in the form:

$$z = (x, y) = (x,\ 0) + (0, y) = x\,(1,\ 0) + y\,(0,\ 1) = x + iy.$$

The representation $z = x + iy$, called the *algebraic form of a complex number*, is more common and convenient than the coordinate form $z = (x, y)$.

If $z = x + iy$, then $x = \operatorname{Re} z$ is called the *real part* of z and $y = \operatorname{Im} z$—the *imaginary part* of z. The number $\bar{z} = x - iy$ is called the *complex conjugate* of z. The product $z \cdot \bar{z} = x^2 + y^2$ is the squared *distance from the origin* to z (or $\bar{z}$).

The following two examples illustrate the algebraic operations with complex numbers.

Example 1.1 Find the value of the numerical expression

$$\left(\frac{2+i^5}{1+i^{19}}\right)^2.$$

Recall that $i^1 = i$, $i^2 = -1$, $i^3 = -i$, $i^4 = 1$, $i^5 = i$, ..., and then $i^5 = i^4 \cdot i = i$, $i^{19} = i^{16} \cdot i^3 = \left(i^4\right)^4 \cdot i^3 = -i$. Consequently,

$$\left(\frac{2+i^5}{1+i^{19}}\right)^2 = \left(\frac{2+i}{1-i}\right)^2 = \frac{4+4i-1}{1-2i-1} = \frac{3+4i}{-2i} = -2+\frac{3}{2}i.$$

Example 1.2 Show that for any complex numbers z_1 and z_2 one has

$$|z_1\bar{z}_2+1|^2+|z_1-z_2|^2 = \left(1+|z_1|^2\right)\left(1+|z_2|^2\right).$$

Transform the left-hand side of the given expression using the identity $z \cdot \bar{z} = |z|^2$

$$|z_1\bar{z}_2+1|^2+|z_1-z_2|^2 = (z_1\bar{z}_2+1)(\bar{z}_1z_2+1)+(z_1-z_2)(\bar{z}_1-\bar{z}_2) = |z_1|^2|z_2|^2+z_1\bar{z}_2+\bar{z}_1z_2+1$$

$$+|z_1|^2-z_1\bar{z}_2-\bar{z}_1z_2+|z_2|^2 = |z_1|^2|z_2|^2+|z_1|^2+|z_2|^2+1 = \left(1+|z_1|^2\right)\left(1+|z_2|^2\right).$$

Let us now consider a *geometric interpretation* of complex numbers. Since a complex number is the ordered pair of real numbers, each complex number $z = (x, y) = x+iy$ can be associated to the corresponding point with coordinates (x, y) in Cartesian plane, and conversely, each point with the coordinates (x, y) is related to the unique complex number $z = x+iy$. Therefore, like in $\mathbb{R}^2$, it is possible to establish a one-to-one correspondence between all the complex numbers and the points of the Cartesian plane. Accordingly, the set of all the complex numbers $\mathbb{C}$ is frequently called the *complex plane* (finite). The *distance between two points* $z_1 = (x_1, y_1)$ and $z_2 = (x_2, y_2)$ in $\mathbb{C}$ is defined by the standard formula of $\mathbb{R}^2$: $|z_2 - z_1| = \sqrt{(x_2-x_1)^2+(y_2-y_1)^2}$, which can also be expressed through the complex product: $|z_2-z_1| = \sqrt{(z_2-z_1)\cdot(\overline{z_2-z_1})}$.

If one adds to $\mathbb{C}$ the point ∞, then the obtained set $\overline{\mathbb{C}} = \mathbb{C} \cup \infty$ is called the *extended complex plane*. Notice that the point ∞ is supposed to be unique in the complex plane: whatever direction is taken to move infinitely away from the origin, one will arrive at the same point ∞.

One can also introduce the polar coordinates r and φ in the Cartesian plane, where (as usual) r is the distance of z from the origin (that is, $r = \sqrt{x^2+y^2}$) and φ is the angle between the vector z and positive direction of the x-axis (that is, $\tan\varphi = \frac{y}{x}$, $x \neq 0$). Conversely, the Cartesian coordinates are found from the polar ones by the formulas $x = r\cos\varphi$, $y = r\sin\varphi$ (see Fig. 1.1). Recall that φ is not defined uniquely—adding 2π to any φ results in the same point on the plane (it follows immediately from the last formulas). In order to establish a one-to-one correspondence between all the pairs (r, φ) of polar coordinates and all the complex numbers (pairs (x, y) or points of complex plane), except for the origin where φ is not defined, we need to restrict the range of φ to an interval of the length 2π with one endpoint included and another excluded. For instance, frequently used intervals of variation of φ are $[0, 2\pi)$, or $(-\pi, \pi]$, or $[-\frac{\pi}{2}, \frac{3\pi}{2})$. The following representation

$$z = x + iy = r\cos\varphi + ir\sin\varphi = r(\cos\varphi + i\sin\varphi)$$

is called the *trigonometric form* of a complex number. In this respect, the number r is called the *modulus* or *the absolute value* of z and is denoted by $r = |z|$, while φ is called the *argument* of z and is denoted by $\varphi = \arg z$. (One more time, pay attention that the argument of a complex number is generally not defined in a unique way, i.e., $\arg z$ is a multi-valued function, except when the variation of φ is restricted to the corresponding interval).

Let us elaborate a bit more on the expression for the argument φ—although it can be defined by the relation $\tan\varphi = \frac{y}{x}$ when $x \neq 0$, the formula for φ itself depends on the chosen interval of variation of the angle, and also the values for the case $x = 0$ should be specified. If we confine φ to the interval $\left[-\frac{\pi}{2}, \frac{3\pi}{2}\right)$, then the next expression follows (see Fig. 1.1)

$$r = \sqrt{x^2 + y^2}\,,\ \varphi = \begin{cases} \arctan\dfrac{y}{x}\,, & x > 0; \\ \arctan\dfrac{y}{x} + \pi\,, & x < 0; \\ \dfrac{\pi}{2}\,, & x = 0,\ y > 0; \\ -\dfrac{\pi}{2}\,, & x = 0,\ y < 0. \end{cases}$$

Notice that φ can also be found from the trigonometric equations $\cos\varphi = \frac{x}{x^2+y^2}$, $\sin\varphi = \frac{y}{x^2+y^2}$, using an appropriate restriction on the variation of the angle.

Applying Euler's formula

$$e^{it} = \cos t + i\sin t, \forall t \in \mathbb{R},$$

to the trigonometric form of a complex number, one obtains the *exponential form*

$$z = x + iy = r(\cos\varphi + i\sin\varphi) = re^{i\varphi}.$$

Usually, the algebraic or exponential form are used in solution of problems; the trigonometric form itself is not a convenient tool and it is commonly used only to transform the algebraic form into exponential and vice versa.

Consider now finding the natural *powers* and *roots* of complex numbers. The first task is elementary if the exponential form is used: given a complex number z $(z \neq 0)$, the nth power $w = z^n$, $n \in \mathbb{N}$ is found by formula $w = z^n = (re^{i\varphi})^n = r^n e^{in\varphi}$. To

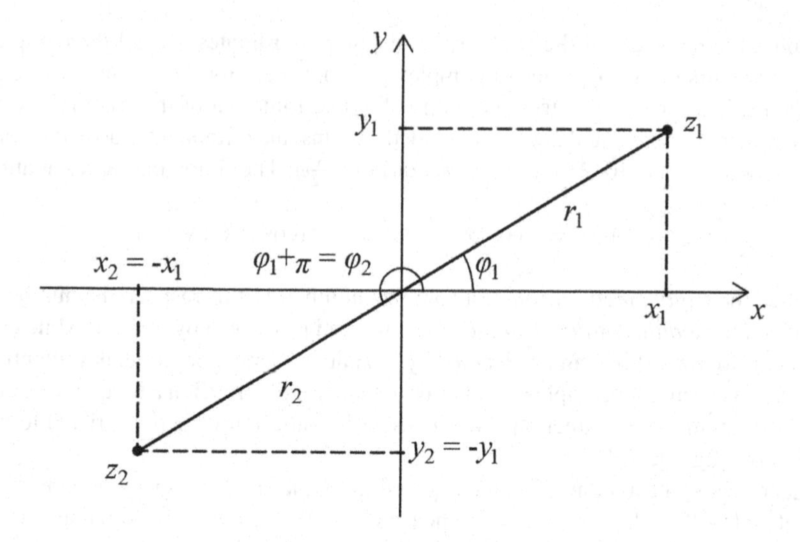

Fig. 1.1 Polar coordinates

transfer this representation to the algebraic form $w = u + iv$, one can apply the trigonometric form of complex numbers: $z^n = r^n e^{in\varphi} = r^n(\cos n\varphi + i\sin n\varphi)$, that is, $u = r^n \cos n\varphi$ and $v = r^n \sin n\varphi$. Additionally, $0^n = 0$ by definition.

For nth root $\sqrt[n]{z}$, $n \in \mathbb{N}$, we should find all $w \in \mathbb{C}$ such that $w^n = z$. Using the exponential form for both numbers—$z = re^{i\varphi}$ and $w = \rho e^{i\theta}$, where ρ and θ are to be found—we rewrite the equation $w^n = z$ in the form $\rho^n e^{in\theta} = re^{i\varphi}$. Since the two complex numbers are equal, it means that their absolute values coincide and the arguments can differ by angle $2k\pi$, $\forall k \in \mathbb{Z}$

$$\begin{cases} \rho^n = r; \\ n\theta = \varphi + 2k\pi \ , \ \forall k \in \mathbb{Z}. \end{cases}$$

Therefore

$$\begin{cases} \rho = \sqrt[n]{r}; \\ \theta_k = \dfrac{\varphi}{n} + \dfrac{2k\pi}{n} \ , \ \forall k \in \mathbb{Z}. \end{cases}$$

Notice that in the first line of the last formula, the root is arithmetic and not complex (due to positivity of r and ρ), which means that ρ is uniquely determined.

At first sight, there is an infinite number of the roots due to infinity of the values of θ_k, but it is not so. Let us see how many different points in the complex plane the last formula determines, verifying location of these points for different values of k. For $k = 0$, one has

$$k = 0\,,\ \theta_0 = \frac{\varphi}{n}\,,\ w_0 = \sqrt[n]{r}e^{i\frac{\varphi}{n}};$$

then, for $k = 1$, one gets

$$k = 1\,,\ \theta_1 = \frac{\varphi}{n} + \frac{2\pi}{n}\,,\ w_1 = \sqrt[n]{r}e^{i\left(\frac{\varphi}{n}+\frac{2\pi}{n}\right)}.$$

Therefore, the point w_1 can be obtained by rotating w_0 about the origin by the angle $\frac{2\pi}{n}$. For the next group of points (roots), one has the following results:

$$k = 2\,,\ \theta_2 = \frac{\varphi}{n} + \frac{2\pi}{n}\cdot 2 = \theta_1 + \frac{2\pi}{n}\,,\ w_2 = \sqrt[n]{r}e^{i\theta_2};$$

$$\vdots$$

$$k = n-1\,,\ \theta_{n-1} = \frac{\varphi}{n} + \frac{2\pi}{n}\cdot(n-1)\,,\ w_{n-1} = \sqrt[n]{r}e^{i\theta_{n-1}};$$

$$k = n\,,\ \theta_n = \frac{\varphi}{n} + \frac{2\pi}{n}\cdot n = \frac{\varphi}{n} + 2\pi\,,\ w_n = \sqrt[n]{r}e^{i\left(\frac{\varphi}{n}+2\pi\right)} = \sqrt[n]{r}e^{i\frac{\varphi}{n}} = w_0.$$

Hence, the point w_n coincides with the point w_0 (it can be obtained by rotating w_0 about the origin by the angle $\frac{2\pi}{n}\cdot n = 2\pi$). For the following values of k, one obtains $w_{n+1} = w_1,\ w_{n+2} = w_2,\ \ldots$, that is, each of the roots with $k > n-1$ corresponds to one of the roots with k between 0 and $n-1$. Similarly, for negative values of k, one has $w_{-1} = w_{n-1}, w_{-2} = w_{n-2}, \ldots$, that is, the roots with negative k also repeat those with k between 0 and $n-1$. Thus, we have exactly n different complex numbers representing nth roots of a complex number z ($z \neq 0$):

$$w_k = \sqrt[n]{r}e^{i\left(\frac{\varphi}{n}+\frac{2\pi}{n}k\right)}\ ,\quad k = 0, 1,\ \ldots\ , n-1.$$

Finally, for $z = 0$ one has the only root $\sqrt[n]{0} = 0$.

The following example illustrates the application of this theory:

Example Find all the roots of the number $\sqrt[5]{-\sqrt{3}+i}$.

First of all, represent the number $-\sqrt{3}+i$ in the exponential form: $-\sqrt{3}+i = re^{i\varphi}$, where $r = \sqrt{\left(-\sqrt{3}\right)^2 + 1^2} = 2$, $\tan\varphi = -\frac{1}{\sqrt{3}}$. Since $-\sqrt{3}=\text{Re}\left(-\sqrt{3}+i\right) < 0$, the angle φ is in the second quadrant: $\varphi = -\arctan\frac{1}{\sqrt{3}} + \pi = \pi - \frac{\pi}{6} = \frac{5}{6}\pi$. Therefore, the five different roots of the given number are

$$z_0 = \sqrt[5]{2}e^{i\frac{\pi}{6}} = \sqrt[5]{2}\left(\cos\frac{\pi}{6} + i\,\sin\frac{\pi}{6}\right) = \sqrt[5]{2}\left(\frac{\sqrt{3}}{2} + \frac{i}{2}\right),$$

$$z_1 = \sqrt[5]{2}e^{i\left(\frac{\pi}{6}+\frac{2\pi}{5}\right)} = \sqrt[5]{2}e^{\frac{17}{30}\pi i} = \sqrt[5]{2}\left(\cos\frac{17}{30}\pi + i\,\sin\frac{17}{30}\pi\right),$$

$$z_2 = \sqrt[5]{2} e^{i\left(\frac{\pi}{6}+\frac{4\pi}{5}\right)} = \sqrt[5]{2} e^{\frac{29}{30}\pi i} = \sqrt[5]{2}\left(\cos\frac{29}{30}\pi + i\,\sin\frac{29}{30}\pi\right),$$

$$z_3 = \sqrt[5]{2} e^{i\left(\frac{\pi}{6}+\frac{6\pi}{5}\right)} = \sqrt[5]{2} e^{\frac{41}{30}\pi i} = \sqrt[5]{2}\left(\cos\frac{41}{30}\pi + i\,\sin\frac{41}{30}\pi\right),$$

$$z_4 = \sqrt[5]{2} e^{i\left(\frac{\pi}{6}+\frac{8\pi}{5}\right)} = \sqrt[5]{2} e^{\frac{53}{30}\pi i} = \sqrt[5]{2}\left(\cos\frac{53}{30}\pi + i\,\sin\frac{53}{30}\pi\right).$$

1.2 Stereographic Projection of Complex Numbers

In the previous section, we have discussed the geometric interpretation of complex numbers, establishing a one-to-one correspondence between complex numbers in $\mathbb{C}$ and points in the Cartesian plane. Consider now another geometric interpretation of complex numbers, in this case including ∞, called the *stereographic projection*. To construct this projection, let us introduce the two Cartesian coordinate systems: the first one is spatial (three-dimensional) system $\xi\eta\zeta$ and the second is plane (bi-dimensional) xy. These two systems are related in such a way that the x-axis coincides with the ξ-axis and the y-axis coincides with the η-axis. In the spatial system, let us consider the sphere of the radius $\frac{1}{2}$ centered at $\left(0, 0, \frac{1}{2}\right)$. Evidently, this sphere is tangent to the plane $\xi\eta$ (and also to xy) at the origin O, which will be called the south pole of the sphere. The point $P\,(0, 0, 1)$ will be called the north pole.

To establish a one-to-one correspondence between the points of the plane xy (that is, complex numbers $z = x + iy$) and the points of the sphere, let us pick an arbitrary point z on the plane and trace the straight line connecting this point z and the north pole P. This line intersects the sphere (at P) and is not tangent to the sphere (otherwise, it would lie in the plane that contains P and is parallel to the plane xy, so it would not cross the plane xy). Therefore, this line crosses the sphere at one more point that we denote by $A\,(z)$. In this way, we show that each point z of complex plane is related to the unique point $A\,(z)$ of the sphere (see Fig. 1.2).

Conversely, by choosing any point $A \neq P$ on the sphere, we can trace the straight line through the points A and P. Since this line is not parallel to the plane xy (otherwise, such a line would have the only common point P with the sphere), it crosses the plane xy at some point z that corresponds to A (see Fig. 1.2). Evidently, the points z and A form the same pair as obtained in the previous step. Thus, we establish a one-to-one correspondence between all the points z of the complex (finite) plane $\mathbb{C}$ and all the points A of the sphere with the hole at the north pole P: $z \leftrightarrow A$. Notice that when a point in the plane moves infinitely away from the origin ($|z| \to +\infty$), the corresponding point on the sphere approaches P ($A\,(z) \to P$) and vice versa (see Fig. 1.2). For this reason, it is natural to associate the point P with the point $z = \infty$: $A\,(\infty) = P$. With this complementary relation, we obtain a one-to-one correspondence between all the points of the extended complex plane $\overline{\mathbb{C}}$ and all the points of the sphere. Let us study a relation between the (Cartesian) coordinates of

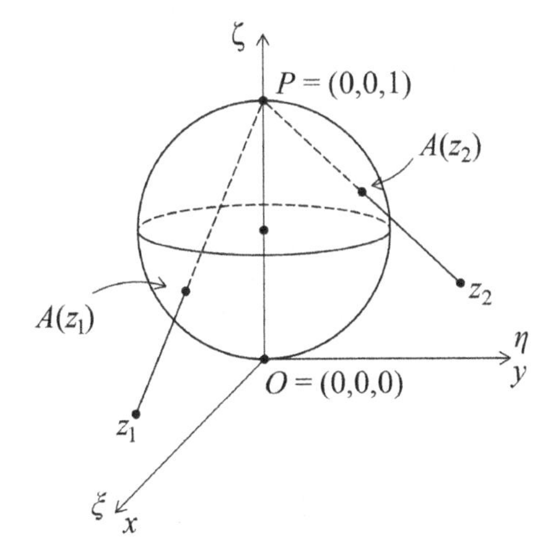

Fig. 1.2 Stereographic projection of the complex plane

the corresponding points $z = x + iy$ and $A(z) = (\xi, \eta, \zeta)$. The point $A(z)$ belongs to the straigh line connecting $z = (x, y, 0)$ and $P = (0, 0, 1)$, and consequently, the coordinates of $A(z)$ satisfy the equations of this line

$$\frac{\xi - 0}{x - 0} = \frac{\eta - 0}{y - 0} = \frac{\zeta - 1}{0 - 1} = t,$$

where $t \in \mathbb{R}$ is a parameter of the straight line. Then

$$\xi = tx, \ \eta = ty, \ \zeta = 1 - t. \tag{1.1}$$

Also, the point $A(z)$ lies on the sphere, and consequently, its coordinates satisfy the sphere equation

$$(\xi - 0)^2 + (\eta - 0)^2 + \left(\zeta - \frac{1}{2}\right)^2 = \frac{1}{4}.$$

Substituting the expressions (1.1) of ξ, η and ζ in the last equation, one has

$$t^2\left(x^2 + y^2\right) + t^2 - t + \frac{1}{4} = \frac{1}{4}$$

or

$$t\left(x^2 + y^2 + 1\right) = 1,$$

that is,

$$t = \frac{1}{1 + |z|^2}.$$

Bringing this expression for t in (1.1), we obtain the desired relation between the coordinates (x, y) and (ξ, η, ζ)

$$\xi = \frac{x}{1+|z|^2}, \eta = \frac{y}{1+|z|^2}, \zeta = \frac{|z|^2}{1+|z|^2}. \tag{1.2}$$

The next task is to find the (Euclidean) distance ρ between the sphere points $A(z_1) = (\xi_1, \eta_1, \zeta_1)$ and $A(z_2) = (\xi_2, \eta_2, \zeta_2)$, which correspond to the complex plane points $z_1 = x_1 + iy_1$ and $z_2 = x_2 + iy_2$. Using the derived relations (1.2), one gets:

$$\rho(A(z_1), A(z_2)) = \sqrt{(\xi_1 - \xi_2)^2 + (\eta_1 - \eta_2)^2 + (\zeta_1 - \zeta_2)^2}$$

$$= \sqrt{\left(\frac{x_1}{1+|z_1|^2} - \frac{x_2}{1+|z_2|^2}\right)^2 + \left(\frac{y_1}{1+|z_1|^2} - \frac{y_2}{1+|z_2|^2}\right)^2 + \left(\frac{|z_1|^2}{1+|z_1|^2} - \frac{|z_2|^2}{1+|z_2|^2}\right)^2}$$

$$= \sqrt{\frac{|z_1|^2(1+|z_1|^2)}{(1+|z_1|^2)^2} + \frac{|z_2|^2(1+|z_2|^2)}{(1+|z_2|^2)^2} - \frac{2|z_1|^2|z_2|^2}{(1+|z_1|^2)(1+|z_2|^2)} - \frac{2x_1x_2 + 2y_1y_2}{(1+|z_1|^2)(1+|z_2|^2)}}$$

$$= \sqrt{\frac{|z_1|^2 + 2|z_1|^2|z_2|^2 + |z_2|^2 - 2|z_1|^2|z_2|^2 - 2x_1x_2 - 2y_1y_2}{(1+|z_1|^2)(1+|z_2|^2)}}$$

$$= \sqrt{\frac{x_1^2+x_2^2-2x_1x_2+y_1^2+y_2^2-2y_1y_2}{(1+|z_1|^2)(1+|z_2|^2)}} = \sqrt{\frac{(x_1-x_2)^2+(y_1-y_2)^2}{(1+|z_1|^2)(1+|z_2|^2)}} = \frac{|z_1 - z_2|}{\sqrt{1+|z_1|^2} \cdot \sqrt{1+|z_2|^2}}.$$

Thus, the distance sought is

$$\rho(A(z_1), A(z_2)) = \frac{|z_1 - z_2|}{\sqrt{1+|z_1|^2} \cdot \sqrt{1+|z_2|^2}}. \tag{1.3}$$

First, notice that $z_1 \to z_2$ implies $\rho(A(z_1), A(z_2)) \to 0$, that is, $A(z_1) \to A(z_2)$. Second, if one of the points approaches infinity, for example, $z_1 \to \infty$, then taking the limit in (1.3) (and using z instead of z_2), one has:

$$\lim_{z_1\to\infty} \rho(A(z_1), A(z)) = \lim_{z_1\to\infty} \frac{|z_1 - z|}{\sqrt{1+|z_1|^2} \cdot \sqrt{1+|z|^2}}$$

$$= \lim_{z_1\to\infty} \frac{|z_1| \cdot \left|1 - \frac{z}{z_1}\right|}{|z_1| \cdot \sqrt{1+\frac{1}{|z_1|^2}} \cdot \sqrt{1+|z|^2}} = \frac{1}{\sqrt{1+|z|^2}}.$$

On the other hand, the same result can be obtained by calculating directly the distance between $A(z)$ and P (the latter corresponding to ∞):

$$\rho(A(z), P) = \sqrt{(\xi - 0)^2 + (\eta - 0)^2 + (\zeta - 1)^2}$$

$$= \sqrt{\frac{x^2}{\left(1+|z|^2\right)^2} + \frac{y^2}{\left(1+|z|^2\right)^2} + \frac{1}{\left(1+|z|^2\right)^2}} = \sqrt{\frac{1+|z|^2}{\left(1+|z|^2\right)^2}} = \frac{1}{\sqrt{1+|z|^2}}.$$

This justifies the imposed earlier correspondence between ∞ of $\overline{\mathbb{C}}$ and P of the sphere.

1.3 Sets. Curves

In this section, we present metrics and analytical concepts that will be frequently used in the next sections and chapters. The majority of them have roots and are studied in the courses of Topology and Real Analysis. Indeed, since $\mathbb{C}$ and $\mathbb{R}^2$ have identical elements (points) and the same Euclidean metric (distance) $|z_2 - z_1| = \sqrt{(x_2 - x_1)^2 + (y_2 - y_1)^2}$, it implies that the metric structure of the sets in $\mathbb{C}$ is the same as in $\mathbb{R}^2$, including the properties of boundedness, openness, convexity, connectedness, etc. At the same time, there are some properties inherent to Complex Analysis.

Definitions Let $z_0 \in \mathbb{C}$, $r, R \in \mathbb{R}$, $r, R > 0$. An open disk with the radius r centered at z_0 is called *r-neighborhood* of z_0 and denoted as follows: $U_r(z_0) = \{|z - z_0| < r\}$. When the radius is not important we also write $U(z_0)$.

An exterior of the disk of the radius R is a *neighborhood of the infinite point* denoted by $U_R(\infty) = \{|z| > R\}$.

A r-neighborhood of z_0 without the central point z_0 is called *deleted r-neighborhood* of z_0:

$$\overset{0}{U}_r(z_0) = \{0 < |z - z_0| < r\}.$$

A neighborhood of ∞ without the point ∞ is a *deleted neighborhood of infinite point*: $\overset{0}{U}_R(\infty) = \{R < |z| < +\infty\}$.

Let E be a (non-empty) set in $\overline{\mathbb{C}}$. The point z_0 (finite or infinite) is an *accumulation point* (or a *limit point*) of a set E, if any deleted neighborhood of z_0 contains points of E, that is, for $\forall U(z_0) \subset \overline{\mathbb{C}}$ there exists $z \neq z_0$ such that $z \in U(z_0) \cap E$. Evidently, this condition implies that there is an infinite number of points of E in such a neighborhood: otherwise it would be possible to decrease the radius of the neighborhood in such a way that no point of E lies inside this neighborhood, except, possibly, z_0 itself.

A point z_0 is an *interior point* of a set E if there exists a neighborhood of z_0, which is entirely contained in E. Or, in symbols:

$\exists r: \; U_r(z_0) = \{|z - z_0| < r\} \subset E$, if z_0 is a finite point;
$\exists R: \; U_R(\infty) = \{|z| > R\} \subset E$, if $z_0 = \infty$.

A point z_0 is an *exterior point* of a set E if there exists a neighborhood of z_0, which contains no point of E, that is, $\exists r: \; U_r(z_0) \cap E = \varnothing$.

If there exists a neighborhood of z_0 such that the only point of this neighborhood that belongs to E is the central point z_0, then z_0 is an *isolated point* of E. In symbols, it can be written as follows: $\exists r: \; U_r(z_0) \cap E = z_0$.

If any neighborhood of z_0 contains both points of E and out of E, then z_0 is called a *boundary point* of E.

The set of all the boundary points of E is called a *boundary* of E.

A set $E \subset \overline{\mathbb{C}}$ is *open* if all its points are interior.

A set $E \subset \overline{\mathbb{C}}$ is *closed* if it contains all its limit points.

Denote by E' the set of all the limit points of a set E. The set $\overline{E} = E \cup E'$ is called the *closure* of E.

A set $E \subset \mathbb{C}$ is *bounded* if there exists a real number $R > 0$ such that $E \subset \{|z| < R\}$, that is, if E is contained in a disk of the radius R. Otherwise, a set is *unbounded* .

A set $E \subset \mathbb{C}$ is *compact* if it is bounded and closed. In any finite-dimensional space with Euclidean metric, in particular, in $\mathbb{C}$, this definition is equivalent to the following characterization: each sequence in $E \subset \mathbb{C}$ has a convergent subsequence whose limit is in E.

A subset A of a set E is *dense* in E if its closure contains E.

A *curve* *(arc)* Γ on the complex plane is the set of the points defined by the equation

$$z(t) = x(t) + iy(t), t \in [a, b], \tag{1.4}$$

where t is a real parameter of the curve and $x(t)$, $y(t)$ are given functions on the interval $[a, b]$. The point $A = z(a)$ is called the initial point and $B = z(b)$ the final point.

A curve Γ is a *continuous curve* if both functions $x(t)$ and $y(t)$ in (1.4) are continuous on the interval $[a, b]$.

Two functions $z(t)$, $t \in [a, b]$ and $z(\theta)$, $\theta \in [\alpha, \beta]$ define the same curve Γ on the plane if there exists a function $\theta = \theta(t)$, continuous and strictly monotone on $[a, b]$, such that $\theta(a) = \alpha$, $\theta(b) = \beta$ and $z(\theta(t)) = z(t)$, $\forall t \in [a, b]$.

A curve Γ: $z(t) = x(t) + iy(t), t \in [a, b]$ is a *simple curve* (or a *Jordan's curve*) if for $\forall t_1, \; t_2 \in [a, b], t_1 \neq t_2$, it follows that $z(t_1) \neq z(t_2)$ (that is, no point of the curve is repeated when the parameter t varies on $[a, b]$).

A curve Γ is *simple and closed* if for $\forall t_1, \; t_2 \in [a, b]$ $(t_1 \neq a, \; t_2 \neq b), t_1 \neq t_2$, it follows that $z(t_1) \neq z(t_2)$, and additionally $z(a) = z(b)$.

Two examples of the definition of curves are given below.

Example 1.1 Classify the curve defined in the parametric form $z = 3\cos t + 2i \sin t$, $t \in [0, 2\pi]$ and give its sketch.

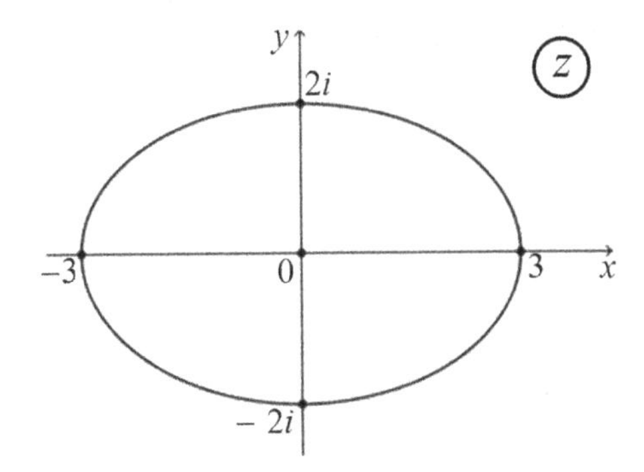

Fig. 1.3 The ellipse defined by the equation $z = 3\cos t + 2i\ \sin t$, $t \in [0, 2\pi]$

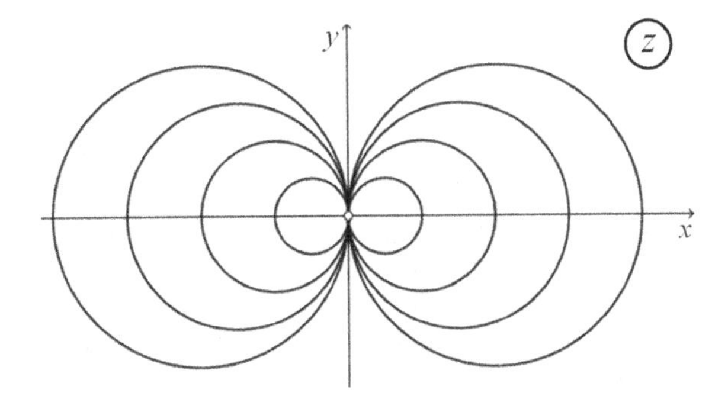

Fig. 1.4 Family of the curves $\mathrm{Re}\,\frac{1}{z} = c$, $\forall c \in \mathbb{R}$

Denoting as usual $z = x + iy$, we can rewrite the equation in the form $\begin{cases} x = 3\cos t \\ y = 2\ \sin t \end{cases}$, or, eliminating the parameter, $\frac{x^2}{9} + \frac{y^2}{4} = 1$. Therefore, we have the ellipse with half-axes $a = 3$ and $b = 2$. Since $t \in [0, 2\pi]$, all the points of the ellipse are included (see Fig. 1.3).

Example 1.2 Construct the family of the curves $\mathrm{Re}\,\frac{1}{z} = c$, $\forall c \in \mathbb{R}$.

First notice that the expression $\frac{1}{z}$ is defined for $\forall z \neq 0$. Since $\mathrm{Re}\,\frac{1}{z} = \mathrm{Re}\,\frac{x-iy}{x^2+y^2} = \frac{x}{x^2+y^2}$, the original equation can be written as follows:

$$\frac{x}{x^2 + y^2} = c. \tag{1.5}$$

If $c \neq 0$ then we get $x^2 + y^2 = \frac{x}{c}$ or $\left(x - \frac{1}{2c}\right)^2 + y^2 = \frac{1}{4c^2}$, which is a set of circles. If $c = 0$, then the solution of (1.5) is $x = 0$, that is, the imaginary axis, which we can interpret as the circle of an infinite radius. Notice that all the found circles (including the imaginary axis) have a hole at the point $z = 0$, since the left-hand side of the inequality is not defined at the origin (see Fig. 1.4).

A curve Γ is *rectifiable* if the set of the lengths of polygonal lines inscribed in Γ is bounded above. In this case, the *curve length* is defined as the supremum (the

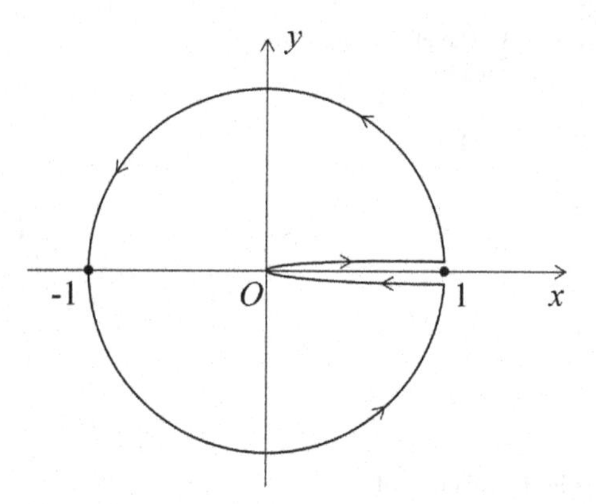

Fig. 1.5 A piecewise smooth curve with a fold (1.6)

least upper bound) of the set of the lengths of polygonal lines, that is, $l_\Gamma = \sup_{\forall p} l_p$, where l_Γ is the curve length, p is a polygonal line and l_p is its length. (Recall that a *polygonal line*, or *polygonal chain*, is a curve consisting of a finite sequence of straight-line segments such that the final point of the current segment coincides with the initial point of the next segment.)

A curve Γ is *differentiable* if the functions $x(t)$ and $y(t)$ in (1.4) are differentiable on $[a, b]$. In this case, the derivative of $z(t)$ is defined following the definition of the derivative of vector-function:

$$z'(t) = \lim_{\Delta t \to 0} \frac{\Delta z(t)}{\Delta t} = \lim_{\Delta t \to 0} \frac{z(t+\Delta t) - z(t)}{\Delta t} = \lim_{\Delta t \to 0} \left(\frac{x(t+\Delta t) - x(t)}{\Delta t} + i \frac{y(t+\Delta t) - y(t)}{\Delta t} \right)$$

$$= \lim_{\Delta t \to 0} \left(\frac{\Delta x(t)}{\Delta t} + i \frac{\Delta y(t)}{\Delta t} \right) = x'(t) + iy'(t)$$

and $z'(t)$ represents the tangent vector to the curve Γ.

A curve Γ is called *smooth* if the functions $x(t)$ and $y(t)$ in (1.4) are continuously differentiable in $[a, b]$ and $x'(t)^2 + y'(t)^2 \neq 0$, $\forall t \in [a, b]$. In this case, $z'(t) \neq 0$, $\forall t \in [a, b]$, or, in geometrical language, the tangent line is defined at each point of the curve and its inclination varies continuously when moving along the curve.

A curve Γ is *piecewise smooth* if it can be divided into a finite chain of smooth arcs. In this case, there exist one-sided tangents (possibly not equal) at the contact points.

A curve Γ is *piecewise smooth with folds* if it can be divided in a finite chain of smooth simple arcs in such a way that different arcs either have no common points (except, possibly, for their endpoints) or completely coincide (as the set of points on the complex plane).

Example Consider the curve defined in the following way:

$$z(t)=\begin{cases} e^{it}, & 0\le t\le 2\pi;\\ \cos t, & 2\pi<t<\dfrac{5}{2}\pi;\\ \dfrac{2t}{\pi}-5, & \dfrac{5\pi}{2}\le t\le 3\pi. \end{cases} \qquad (1.6)$$

Notice that point z moves along the interval $[0, 1]$ on the x-axis twice in opposite directions, which means that this curve is piecewise smooth with a fold. See the illustration of the curve in Fig. 1.5.

Let Γ be a continuous simple and closed curve. The curve is said to be *positively oriented* if the parameterization traverses it in a counterclockwise direction. In this case, the plane domain bounded by Γ always lies on the left as the point $z(t)$ traverses Γ. Otherwise, the curve is *negatively oriented*.

Given two sets $E_1, E_2 \subset \overline{\mathbb{C}}$, the *distance* between E_1 and E_2 is defined as

$$\rho(E_1, E_2)=\inf_{\forall z_1\in E_1, \forall z_2\in E_2} |z_1-z_2|\,.$$

Two sets A and B, $A\neq\varnothing$, $B\neq\varnothing$ are said to be *separated*, if $A\cap\overline{B}=\varnothing$ and $\overline{A}\cap B=\varnothing$.

A set E is called *connected* if E is not a union of two non-empty separated sets.

A set E is called *linearly connected* if any two of its points can be joined by a continuous curve, which lies entirely in E.

A set E is said to be *convex* if it contains all line segments between its points.

Example Find the set of points defined by the inequality

$$\frac{1}{2}<\operatorname{Re}\frac{1}{z}\le 1$$

and sketch it in the complex plane.

Expressing the middle part of the inequality in coordinates x, y: $\frac{1}{z}=\frac{1}{x+iy}=\frac{x-iy}{x^2+y^2}$, $z\neq 0$, and $\operatorname{Re}\frac{1}{z}=\frac{x}{x^2+y^2}$, we obtain

$$\frac{1}{2}<\frac{x}{x^2+y^2}\le 1. \qquad (1.7)$$

It follows from the right-hand inequality that $x^2+y^2\ge x$, $(x, y)\neq(0,0)$, and consequently $(x-1/2)^2+y^2\ge 1/4$, $(x, y)\neq(0,0)$. The left-hand inequality can be expressed in the form $x^2+y^2<2x$, $(x, y)\neq(0,0)$ or, equivalently, $(x-1)^2+y^2<1$. Therefore, the double inequality (1.7) is equivalent to the following system of inequalities:

$$\begin{cases} (x-1/2)^2+y^2\ge 1/4\\ (x-1)^2+y^2<1 \end{cases}.$$

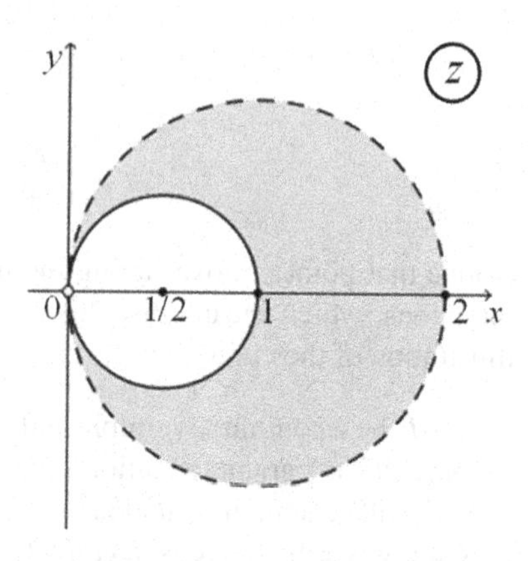

Fig. 1.6 Set of the points defined by the inequality $\frac{1}{2} < \text{Re}\,\frac{1}{z} \le 1$

Notice that in the last system the condition $(x, y) \neq (0, 0)$ need not be added to the first inequality, because the second one excludes this point. Geometrically, we have the set of points between the two circles $(x - 1/2)^2 + y^2 = 1/4$ and $(x - 1)^2 + y^2 = 1$, where the first (inner) circle belongs to this set and the second (outer) does not (see Fig. 1.6). This is a linearly connected set, whose boundary consists of the inner and outer circles. It is neither open (since it contains the points of the inner circle) nor closed (since it does not contain the points of the outer circle).

Let us return to the theoretical part and continue with the definitions of sets and their characteristics intimately related to the study of complex functions.

A linearly connected open set is called *open domain* or simply *domain*.

Remark The last definition is very important because the domain of complex functions (the set in $\overline{\mathbb{C}}$ where a complex function is defined) considered in this text is usually a linearly connected open subset of $\overline{\mathbb{C}}$, that is, a domain in the sense of the last definition.

A crosscut (cut) of a domain D is a simple Jordan arc related to the domain D in one of the two ways: either it consists of only the boundary points of D, which are interior for the closure of D, or it lies entirely in the domain D except for the end points, which belong to the boundary of D. In the former case, the cut is called a fold and in the latter—an auxiliary cut (or simply cut).

Let D be a domain. The set of all the boundary points of D is the *boundary of the domain D* denoted by ∂D. Notice that ∂D is a closed set. The boundary ∂D can be divided into a finite or infinite number of separated sets, each of which is a connected closed set called a *component* (*boundary component*) of ∂D. In particular, a boundary can consist of only one boundary component. Any boundary component is represented by a closed curve, or a closed curve with folds, or a separated fold or, finally, an isolated point. If a boundary component contains the only (isolated) point, then it is said to be *singular*. Any other type of a boundary component is

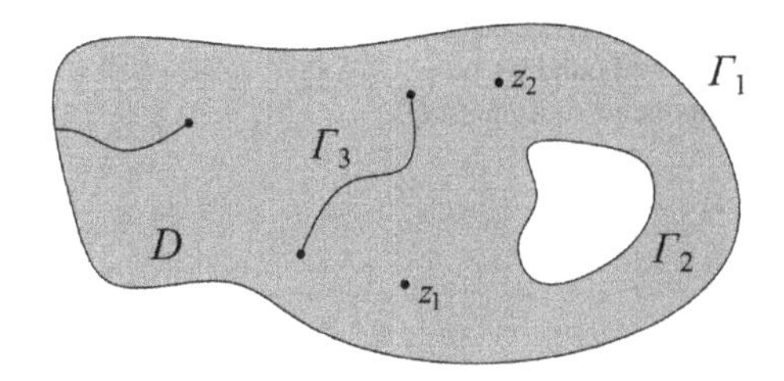

Fig. 1.7 An example of 5-connected domain with different boundary components

non-singular. A fold can be both a separated boundary component or a part of a boundary component.

The number of the boundary components determines the *degree of connectedness* of a domain in $\overline{\mathbb{C}}$: if a domain possesses n boundary components, it is called *n-connected (finitely connected)*, more specifically, if $n = 1$, then a domain is *simply connected* , and if $n > 1$, then a domain is *multiply connected*; if the number of components is infinite, then a domain is called *infinitely connected*. Figure 1.7 illustrates a 5-connected domain D, whose boundary consists of the following five components: the closed curve with a fold Γ_1, the closed curve (without a fold) Γ_2, the fold Γ_3, the isolated point z_1 and the isolated point z_2.

A domain D together with its boundary ∂D is called a *closed domain* and is denoted by $\overline{D}$: $\overline{D} = D \cup \partial D$. Notice that for a domain the concepts of a closed domain and a *closure of a domain* coincide.

A domain D is said to be *finite* if $\infty \notin D$. Otherwise a domain is *infinite*.

In this text, we will usually consider finitely connected (open) domains whose boundary components are either piecewise smooth curves with folds or singular points.

If ∂D is the boundary of a domain D, the *positive orientation* (*positive traversal*) of ∂D is such that D always lies on the left under the traversal of ∂D, whatever boundary component is traversed. Notice that the positive traversal of a closed curve Γ may not coincide with the positive traversal of the same curve considered as a boundary component of some domain.

Example Consider the domain D shown in Fig. 1.8, which is bounded by the curves Γ_1 and Γ_2. The indicated traversal of Γ_2 is the positive traversal of the boundary component of D, but at the same time it is the negative traversal for the curve Γ_2 itself.

1.4 Functions. Limit of a Function. Continuous Functions

Definition 1.1 Given a set $E \subset \overline{\mathbb{C}}$, let us say that a rule f defines a *complex-valued function* $w = f(z)$ on E, if a single complex number w is associated with each

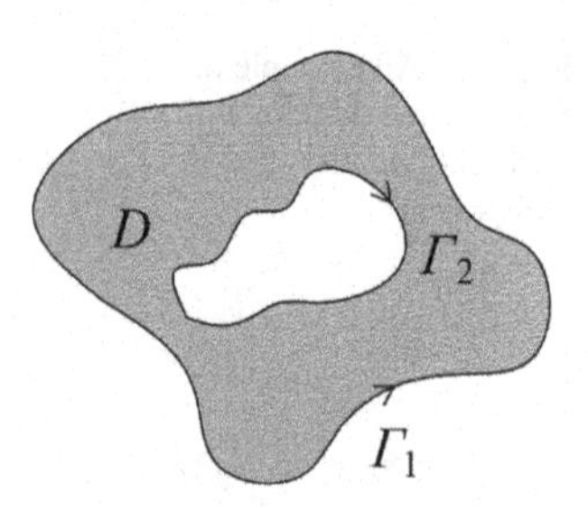

Fig. 1.8 Positive traversal along the boundary components of domain D

complex point z of E. Together with the term *complex-valued function* we will equivalently use the terms *complex function* or *function* for short.

Remark 1.3 This definition of a complex function is equivalent to the definition of two real-valued functions of two real variables x and y: $w = f(z) = u(x, y) + iv(x, y)$, $z = x + iy$.

Remark 1.4 Strictly speaking, one should distinguish between the function f and its value $f(z)$ at the point z, but it is customary to use $f(z)$ to denote both the function and its values, and we will follow this simplified notation in this text.

Unlike Real Analysis, the theory of complex functions requires consideration of multi-valued functions, which leads to the generalization of Definition 1.1 in the following form:

Definition 1.1′. A *complex function* $f(z)$ defined on $E \subset \overline{\mathbb{C}}$ is a rule according to which a (non-empty) set of complex numbers w is associated with each point z of E. If each of these sets consists of a single element w for every z, then $f(z)$ is said to be *single-valued*, otherwise $f(z)$ is said to be *multi-valued*.

Remark 1.5 Usually, we will use the term *function* without any explanation of its type in the case of a single-valued function. Otherwise, we will use the term *multi-valued function* or will specify what single-valued part (branch) of a multi-valued function is considered.

Let us illustrate the concept of single-valued and multi-valued functions.

Example 1.1 The function $f(z) = |z|$ is a single-valued function defined on $\mathbb{C}$, since a single number $w = \sqrt{x^2 + y^2}$ is associated with every complex point $z = x + iy \in \mathbb{C}$. Other examples of single-valued functions defined on $\mathbb{C}$ (already considered in Sect. 1.1, albeit without a formal definition of function) are $f(z) = \operatorname{Re} z$, $f(z) = \operatorname{Im} z$ and $f(z) = z^n$, $n \in \mathbb{N}$.

Example 1.2 In Sect. 1.1, it was shown that $\sqrt[n]{z}$, $n \in \mathbb{N}$ has exactly n different values for any $z \in E = \mathbb{C}\backslash\{0\}$. In terms of functions, it means that $f(z) = \sqrt[n]{z}$, $n \in \mathbb{N}, n > 1$ is a multi-valued (n-valued) function, which to each element $z = re^{i\varphi}$, $r \neq 0$ associates a set of n different complex numbers found by the formula $w_k = \sqrt[n]{r}e^{i\left(\frac{\varphi}{n} + \frac{2\pi}{n}k\right)}$, $k = 0, 1, \ldots, n-1$ (see Sect. 1.1).

Example 1.3 The function $f(z) = \arg z$ defined on the set E of all non-zero points of the complex plane is multi-valued. Moreover, with each point $z \in E$ it associates the countably infinite set of points $\arg z = \varphi + 2n\pi, \forall n \in \mathbb{Z}$, where the range of φ is restricted to $[-\frac{\pi}{2}, \frac{3\pi}{2})$ (see Sect. 1.1).

Let us define some elementary properties of functions similar to those in Real Analysis.

Definition 1.2 A function $f(z)$ is called *injective (univalent) on a set* E, if for $\forall z_1, z_2 \in E,\ z_1 \neq z_2$ it follows that $f(z_1) \neq f(z_2)$, that is, different points of a set E in the z-plane are carried to different points in the w-plane.

Definition 1.3 If a function $w = f(z)$ is defined on a set E in the z-plane, its *image* $G = f(E)$ is the set G of all the points in the w-plane such that for $\forall w_0 \in G$ there exists at least one point $z_0 \in E$ for which $f(z_0) = w_0$.

Definition 1.4 If a function $w = f(z)$ defined on E with the image $G = f(E)$ is injective on E, then $f(z)$ is said to be a *bijective function* (or *one-to-one correspondence*) between E and G.

Remark It is important to notice that in the last definition, according to our convention, the term *function* means a *single-valued function*. An *injective multi-valued function* doesn't establish a one-to-one correspondence. For example, the function $w = \sqrt{z}$ maps the ring $D = \{1 < |z| < 4\}$ onto the ring $G = f(D) = \{1 < |w| < 2\}$ and it is univalent on D, since $w_1 = w_2$ means $\sqrt{z_1} = \sqrt{z_2}$, which implies $z_1 = w_1^2 = w_2^2 = z_2$. However, it was shown in Sect. 1.1 that the two different points $w_1 = \sqrt{r}e^{i\frac{\varphi}{2}}$ and $w_2 = \sqrt{r}e^{i(\frac{\varphi}{2}+\pi)} = -\sqrt{r}e^{i\frac{\varphi}{2}}$ correspond to every $z = re^{i\varphi} \neq 0$, in particular, to every $z \in D$. Since both w_1 and w_2 belong to G, this means that the two-valued function $w = \sqrt{z}$ is not bijective on D.

Now we define one of the fundamental concepts—the limit of a function.

Definition 1.5 Let $w = f(z)$ be a function defined on a set $E \subset \overline{\mathbb{C}}$ and z_0 be a limit point of E. We say that the function $f(z)$ has a *limit* A as z approaches z_0, which is denoted by $\lim\limits_{z\to z_0} f(z) = A$, if for any $\varepsilon > 0$ there exists $r > 0$ such that for all $z \in \overset{0}{U}_r(z_0) \cap E$ it follows that $f(z) \in U_\varepsilon(A)$.

This definition of a limit, formulated using the concept of neighborhood, has a general (topological) form and is applicable to any kind of points —no matter if z_0 and A are finite or infinite.

If the points z_0 and A are finite, we can rewrite the definition of a limit $\lim\limits_{z\to z_0} f(z) = A$ specifying the definition of neighborhoods in the terms of inequalities

$$\forall \varepsilon > 0\ \exists r > 0 : \ \forall z \in E \cap \{0 < |z - z_0| < r\} \Rightarrow |f(z) - A| < \varepsilon.$$

For finite z_0 and infinite A, that is $\lim_{z\to z_0} f(z) = \infty$, the specification of the definition goes as follows:

$$\forall\varepsilon > 0\, \exists r > 0 : \forall z \in E \cap \{0 < |z - z_0| < r\} \Rightarrow |f(z)| > \varepsilon.$$

For infinite z_0 and finite A one has $\lim_{z\to\infty} f(z) = A$

$$\forall\varepsilon > 0\, \exists r > 0 : \forall z \in E \cap \{r < |z| < +\infty\} \Rightarrow |f(z) - A| < \varepsilon.$$

Finally, for infinite z_0 and infinite A one obtains $\lim_{z\to\infty} f(z) = \infty$

$$\forall\varepsilon > 0\, \exists r > 0 : \forall z \in E \cap \{r < |z| < +\infty\} \Rightarrow |f(z)| > \varepsilon.$$

One can prove the following elementary result.

Theorem *The existence of a finite limit of a complex function* $f(z) = u(x, y) + iv(x, y)$ *(i.e.,* $\lim_{z\to z_0} f(z) = A$*, where* $A = B + iC$*) is equivalent to the existence of the limits of the two real functions* $u(x, y)$ *and* $v(x, y)$ *of the real variables* x, y *(real and imaginary parts of* $f(z)$*), that is*

$$\exists \lim_{z\to z_0} f(z) = A \Leftrightarrow \begin{cases} \exists \lim\limits_{\substack{x\to x_0 \\ y\to y_0}} u(x, y) = B \\ \exists \lim\limits_{\substack{x\to x_0 \\ y\to y_0}} v(x, y) = C \end{cases}.$$

Proof In fact, suppose that there exists a limit of a complex function $\lim_{z\to z_0} f(z) = A$, that is

$$\forall\varepsilon > 0\ \exists r > 0 : \forall z \in E \cap \{0 < |z - z_0| < r\} \Rightarrow |f(z) - A| < \varepsilon.$$

The following inequalities are true for any $z \in E$, $z = x + iy$, $x, y \in \mathbb{R}$:

$$\left.\begin{matrix} |x - x_0| \\ |y - y_0| \end{matrix}\right\} \le \sqrt{(x - x_0)^2 + (y - y_0)^2} = |z - z_0| \le |x - x_0| + |y - y_0|.$$

Therefore, if one pick any point of E that satisfies the conditions $0 < |x - x_0| < \frac{r}{2}$, $0 < |y - y_0| < \frac{r}{2}$, then one has $0 < |z - z_0| < r$, which implies that $|f(z) - A| < \varepsilon$.

Since the following evaluation is true

$$\left.\begin{matrix} |u(x, y) - B| \\ |v(x, y) - C| \end{matrix}\right\} \le \sqrt{(u - B)^2 + (v - C)^2} = |f(z) - A| \le |u(x, y) - B| + |v(x, y) - C|,$$

where $A = B + iC$, then one obtains

$$|u(x, y) - B| \le |f(z) - A| < \varepsilon , \ |v(x, y) - C| \le |f(z) - A| < \varepsilon ,$$

that is,

$$\lim_{\substack{x \to x_0 \\ y \to y_0}} u(x, y) = B , \quad \lim_{\substack{x \to x_0 \\ y \to y_0}} v(x, y) = C .$$

The converse statement is also easily proved. □

Since the definition of the limit of a complex function resembles that of a real function and its existence is equivalent to the existence of the limits of its real and imaginary parts, all the algebraic properties of the limits of real functions (studied in Calculus and Real Analysis courses) can be translated without any modification to the limits of a complex function.

Definition 1.6 Let $w = f(z)$ be a function defined on a set $E \subset \mathbb{C}$ and $z_0 \in E$ be a limit point of E. We say that the function $f(z)$ is *continuous* at z_0 if there exists a finite limit equal to $f(z_0)$: $\lim_{z \to z_0} f(z) = f(z_0)$.

Remark Actually, a bit more general definition of continuity, corresponding to that adopted in Real Analysis, goes as follows: a function $w = f(z)$ defined on $E \subset \mathbb{C}$ is *continuous* at $z_0 \in E$ if for $\forall \varepsilon > 0$ there exists $\delta > 0$ such that whenever $z \in E$ and $|z - z_0| < \delta$ it follows that $|f(z) - f(z_0)| < \varepsilon$. The minor difference between the two definitions is that the first one requires z_0 to be a limit point of E, while the second admits both limit and isolated points z_0. Since isolated points are, in certain sense, too far from other points of the domain to contribute to any interesting phenomena, we will consider the definition of continuity by the limit (Definition 1.6), as it is usually assumed in Complex Analysis courses.

A function $f(z)$ is *continuous on a set* E, if it is continuous at each point of E. The $\epsilon - \delta$ version of this definition can be stated as follows: for each fixed $z_0 \in E$ it is provided that

$$\forall \varepsilon > 0 \ \exists \delta = \delta(\varepsilon, z_0) > 0 \ : \ \forall z \in E, |z - z_0| < \delta \Rightarrow |f(z) - f(z_0)| < \varepsilon. \quad (1.8)$$

This type of continuity on a set is also referred to as *pointwise continuity* in the cases when other types of continuity are considered concurrently.

Since *the continuity of a complex function* is defined by the concept of the limit, this definition is *equivalent to the continuity of the real and imaginary parts of a complex function*. Consequently, all the algebraic properties of real continuous function are valid for complex continuous functions.

Let us give two examples of verification of the continuity of a function.

Example 1.1 Let us investigate the continuity of the function $f(z) = \frac{\operatorname{Re} z}{z}$.

Notice that this function is defined in the entire finite complex plane $\mathbb{C}_z$ except at the point $z = 0$. Since the functions $\operatorname{Re} z$ and z are continuous in the entire plane $\mathbb{C}_z$, the function $f(z)$ is continuous in $\mathbb{C}_z \setminus \{0\}$. Let us check if it is possible to remove the discontinuity of $f(z)$ at $z = 0$. To do this, choose different paths of approaching $z = 0$. If, for instance, we take $z = x \to 0 \quad (y = 0)$, then $\lim\limits_{z=x\to 0} f(z) = \lim\limits_{x\to 0} \frac{x}{x} = 1$. On the other hand, along the path $z = iy$ (that is, $x = 0$), we have $\lim\limits_{z=iy\to 0} f(z) = \lim\limits_{y\to 0} \frac{0}{iy} = 0$. Since the two partial limits are different, it is impossible to remove the discontinuity of $f(z)$ at the origin.

Example 1.2 In this example, we analyze how the continuity (and limits) of a multi-valued function should be considered.

Let us take the function $f(z) = \sqrt{z}$. In Sect. 1.1, it was shown that $\sqrt[n]{z}$ has exactly n different values for any $\forall z \neq 0$, that is, $w = \sqrt{z}$ has two different (opposite in sign) values for $\forall z = re^{i\varphi} \neq 0$: $w_1 = \sqrt{z} = \sqrt{r}e^{i\varphi/2}$ and $w_2 = \sqrt{z} = \sqrt{r}e^{i(\varphi/2+\pi)} = -\sqrt{r}e^{i\varphi/2}$. It is easy to note that this two-valued nature of the function $f(z) = \sqrt{z}$ prevents the function from being continuous at any point, unless we take special measures to isolate its single-value part. Indeed, let us pick for simplicity a point $z_0 = 1$. Approaching this point along the real axis, we can choose the path $z = 1 - \delta$ at the left and $z = (1+\delta)e^{i2\pi}$ at the right, $\delta > 0$. In the first case, we obtain $\lim\limits_{z\to z_0} \sqrt{z} = \lim\limits_{\delta\to 0} \sqrt{1-\delta} = 1$, while for the second path we get $\lim\limits_{z\to z_0} \sqrt{z} = \lim\limits_{\delta\to 0} -\sqrt{1+\delta} = -1$. Of course, we can interchange the parameterization of the paths, using $z = (1-\delta)e^{i2\pi}$ for the first path and $z = 1 + \delta$ for the second one, with the opposite results. The same is true for any other point $z \neq 0$. Hence, we can see that the definition of the continuity at $z_0 \neq 0$ is not satisfied, and the reason of this behavior is that $w = \sqrt{z}$ is a two-valued function.

Therefore, to arrive at a continuous function, we should choose only one value of the root. It can be done by restricting the range of the variation of the argument to avoid the loops around the origin (called *the branch point* for the root function). It is seen that the origin is the unique point of the domain of $\sqrt{z}$ such that any loop around it returns geometrically to the initial point, but increases the argument by 2π, which leads to the change of the value of the square root from w_1 to w_2 (or vice versa), generating in this way different values of $\sqrt{z}$ that correspond to the same point z. The largest interval we can use is of the length 2π, for instance, $[0, 2\pi]$: $0 \leq \arg z = \varphi \leq 2\pi$. In this case, the formula of Sect. 1.1 for the square root gives the only value: $\sqrt{z} = \sqrt{r}e^{i\varphi/2} = \sqrt{r}\left(\cos\frac{\varphi}{2} + i\sin\frac{\varphi}{2}\right)$ and to avoid different values at $\varphi = 0$ and $\varphi = 2\pi$, we consider additionally that the plane is cut along the positive part of the real axis and the two edges of this cut (the upper γ_+ and the lower γ_-) are not connected (see Fig. 1.9). This cut (called the *branch cut*) is required because on its upper edge the angle $\varphi = 0$ and the corresponding values of the function are $\sqrt{z} = \sqrt{r}e^{i0} = \sqrt{r}$, while on the lower edge $\varphi = 2\pi$ and $\sqrt{z} = \sqrt{r}e^{i\pi} = -\sqrt{r}$.

On such a slit plane the function $w = \sqrt{z}$ is single-valued and continuous at each point. In fact, denoting $w = \sqrt{z} = u + iv$, we have $u(r,\varphi) = \sqrt{r}\cos\frac{\varphi}{2}$ and $v(r,\varphi) = \sqrt{r}\sin\frac{\varphi}{2}$. Since all the involved functions—$\sqrt{r}$, $\cos\frac{\varphi}{2}$, $\sin\frac{\varphi}{2}$—are continuous, the real $u(r,\varphi)$ and imaginary $v(r,\varphi)$ parts are continuous functions on the slit plane, and consequently, $w = \sqrt{z}$ is continuous on the slit plane. The only restriction

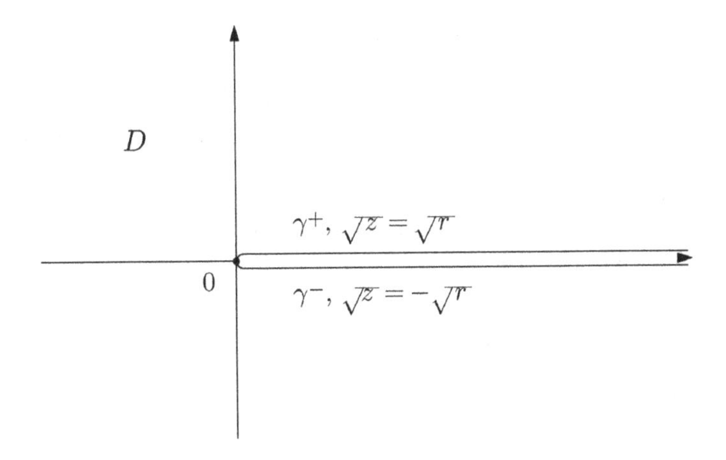

Fig. 1.9 Complex plain with the branch cut

with respect to the continuity of $w = \sqrt{z}$ on the slit plane is that the points on the upper edge can be approached only by the points of the first quadrant and the points on the lower edge—by the points of the fourth quadrant.

Notice that the interval of variation of φ can be different, for instance, $[-\pi, \pi]$, but its largest length is 2π and it requires the insertion of a corresponding cut. Such a cut can be a ray or a curve connecting the origin (the branch point of the root function) with ∞. A detailed analysis of multi-valued functions, branch points, and cuts will be provided in Chap. 4.

Definition 1.7 A function $w = f(z)$ is said to be *uniformly continuous* on a set E if

$$\forall \varepsilon > 0 \ \exists \delta = \delta(\varepsilon) > 0 : \ \forall z, z' \in E, \left|z - z'\right| < \delta \Rightarrow \left|f(z) - f\left(z'\right)\right| < \varepsilon. \quad (1.9)$$

Notice that δ in the last definition does not depend on the choice of a point in E, and therefore for each ε one can find a single value of δ for all the points of E.

Evidently, the uniform continuity on a set implies the pointwise continuity on the same set (just like in Real Analysis). The converse statement is generally not true, but for some special types of a set E the converse holds. The most well-known statement of this kind is the Cantor Theorem.

Cantor Theorem *If a function $w = f(z)$ is continuous on a compact set, then it is uniformly continuous on this set.*

We will also consider another notion of continuity, which has no counterpart in Real Analysis. To this end, let us first introduce a new measure.

Definition 1.8 The *distance between two points* $z_1, z_2 \in D$ *inside a domain* D is the infimum (the greatest lower bound) of the set of lengths of polygonal lines connecting the points z_1 and z_2 and lying entirely in D. We will use the following notation: $\rho_D(z_1, z_2) = \inf\limits_{\forall p \in D} l_p$, where p is a polygonal line and l_p is its length.

It follows directly from this definition that $\rho_D(z_1, z_2) \geq |z_1 - z_2|$ (see Fig. 1.10). In particular case, when the straight line segment between the points z_1 and z_2 is entirely contained in D, one has $\rho_D(z_1, z_2) = |z_1 - z_2|$.

Definition 1.9 A function $w = f(z)$ is called *continuous on a domain D up to its boundary* if

$$\forall \varepsilon > 0 \ \exists \delta = \delta(\varepsilon) > 0 : \ \forall z_1, z_2 \in D, \rho_D(z_1, z_2) < \delta \Rightarrow |f(z_1) - f(z_2)| < \varepsilon. \tag{1.10}$$

Let us analyze how this new type of continuity is related to other known types. The following proposition is true:

Theorem

$$\begin{array}{ccc} Continuity\ in\ D & \Leftarrow & Uniform\ Continuity \\ \Uparrow & & \Downarrow \\ & Continuity\ up\ to\ the\ boundary & \end{array}$$

Proof (1) First, we show that the uniform continuity on D implies the continuity up to the boundary.

Let $f(z)$ be uniformly continuous on D. According to the definition, one has

$$\forall \varepsilon > 0 \ \exists \delta = \delta(\varepsilon) > 0 : \ \forall z, z' \in D, \ |z - z'| < \delta \ \Rightarrow \ |f(z) - f(z')| < \varepsilon.$$

Taking any two points $z_1, z_2 \in D$ such that $\rho_D(z_1, z_2) < \delta$ (where δ is the same as in the definition of uniform continuity) and noting that $|z_1 - z_2| \leq \rho_D(z_1, z_2)$, one obtains $|z_1 - z_2| < \delta$. Then, the condition of the uniform continuity implies that $|f(z_1) - f(z_2)| < \varepsilon$. Thus, the definition of the continuity of $f(z)$ up to the boundary on D is satisfied.

(2) Now we prove that the continuity on D up to its boundary guarantees the continuity on D, which means the continuity at each point of D.

Let z_0 be an arbitrary point of D. Since D is an open set, z_0 is an interior point of D, that is, there exists a neighborhood of z_0 contained in D: $\exists \sigma > 0$ such that $\{|z - z_0| < \sigma\} \subset D$. By hypothesis, $f(z)$ is continuous on D up to its boundary, that is, for $\forall \varepsilon > 0$ there exists $\delta = \delta(\varepsilon) > 0$ such that whenever $z_1, z_2 \in D$, $\rho_D(z_1, z_2) < \delta$ it follows that $|f(z_1) - f(z_2)| < \varepsilon$. Let us consider a r-neighborhood of z_0, $|z - z_0| < r$, where the radius r is chosen in a special way: $r = \min\{\sigma, \delta\}$. Since $r \leq \sigma$, this neighborhood is contained in the disk $|z - z_0| < \sigma$, and consequently, it is contained in D. Then, for any point $z \in U_r(z_0)$, the segment between the points z and z_0 is entirely contained in D, implying that $|z - z_0| = \rho_D(z, z_0)$, $\forall z \in U_r(z_0)$. Therefore, for $\forall z \in U_r(z_0)$ one has $\rho_D(z, z_0) =$

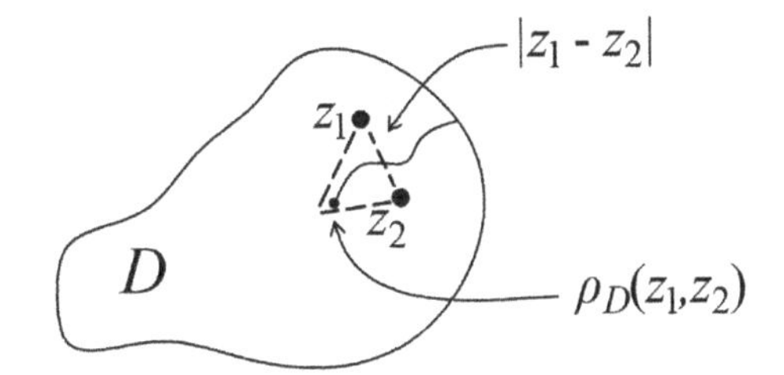

Fig. 1.10 The difference between $|z_1 - z_2|$ and $\rho_D(z_1, z_2)$

$|z - z_0| < r \leq \delta$. According to the definition of the continuity up to the boundary, this ensures that $|f(z) - f(z_0)| < \varepsilon$. Summarizing, we have shown that for $\forall\varepsilon > 0$ there exists $r > 0$ such that whenever $\forall z \in U_r(z_0)$ it follows that $|f(z) - f(z_0)| < \varepsilon$. Hence, the definition of the continuity of $f(z)$ holds at an arbitrary point $z_0 \in D$. □

Notice that if the boundary of a domain D can be represented as a finite number of piecewise smooth curves (without folds), the usual proximity between points and the proximity in the domain coincide: $\rho_D(z_1, z_2) = |z_1 - z_2|$. For this reason, for domains with such a boundary, the uniform continuity and the continuity up to the boundary are two equivalent concepts. However, this is not true for domains whose boundary contains folds. In the latter case, it may happen that $|z_1 - z_2| \to 0$, but $\rho_D(z_1, z_2) \not\to 0$ (see Fig. 1.10).

1.5 Definition of a Function on the Domain Boundary

Let $f(z)$ be a continuous function on a domain D up to its boundary.

1st case. If the boundary of a domain D is composed of a finite number of piecewise smooth curves without folds, then, as it was discussed in the previous section, the continuity up to the boundary is equivalent to the uniform continuity on D. In this case, the value of $f(z)$ at each point z_0 of the boundary ∂D is defined to be the limit of $f(z)$ as z approaches z_0, that is,

$$f(z_0) = \lim_{z \to z_0,\, z \in D} f(z)\ ,\ \forall z_0 \in \partial D.$$

To justify this definition we should check that the above limit exists and is finite.

First, we show that the limit $\lim\limits_{z \to z_0,\, z \in D} f(z)$ exists. Assume, by contradiction, that the limit does not exist. Then, one can choose at least two different sequences of points $\{z'_n\}_{n=1}^{\infty}$ and $\{z''_n\}_{n=1}^{\infty}$, $z'_n, z''_n \in D$, both convergent to z_0, which result in two different tendencies of the function: $f(z'_n) \to A$, while $f(z''_n) \to B$, where $A \neq B$. Since $f(z)$ is uniformly continuous on D, one has

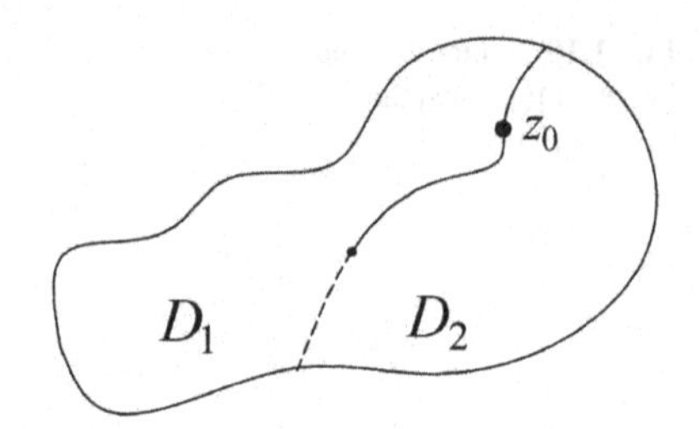

Fig. 1.11 Additional cuts in domain D to avoid folds on the boundaries of subdomains

$$\forall\varepsilon > 0\ \exists\delta = \delta(\varepsilon) > 0 : \forall z', z'' \in D, \left|z' - z''\right| < \delta \Rightarrow \left|f(z') - f(z'')\right| < \varepsilon\,. \tag{1.11}$$

In the last definition, one can choose $\varepsilon < \frac{|A-B|}{2}$. Consider now $\frac{\delta}{2}$-neighborhood of the point z_0. Since $z'_n \to z_0$ and $z''_n \to z_0$, one can find the index n_0 such that for all $n \geq n_0$ the points z'_n and z''_n belong to $\frac{\delta}{2}$-neighborhood of z_0. Then, for $n \geq n_0$, one has

$$\left|z'_n - z''_n\right| = \left|z'_n - z_0 + z_0 - z''_n\right| \leq \left|z'_n - z_0\right| + \left|z''_n - z_0\right| < \frac{\delta}{2} + \frac{\delta}{2} = \delta.$$

Consequently, $\left|f(z'_n) - f(z''_n)\right| < \varepsilon < \frac{|A-B|}{2}$ for $\forall n \geq n_0$. Taking the limit as $n \to \infty$ in the last inequality, one arrives at the false statement: $|A - B| \leq \frac{|A-B|}{2}$. This means that the initial assumption that the limit does not exist is not true. Therefore, the limit $\lim\limits_{\substack{z\to z_0 \\ z\in D}} f(z)$ exists.

Second, we prove that this limit is finite. Applying again the proof by contradiction, let us suppose that the limit is infinite: $\lim\limits_{z\to z_0,\, z\in D} f(z) = \infty$. Since $f(z)$ is uniformly continuous on D, the definition (1.11) is satisfied. Consider $\frac{\delta}{2}$-neighborhood of z_0 and fix any point $z^* \in D$ such that $|z^* - z_0| < \frac{\delta}{2}$. Then, for $\forall z \in D \cap \left\{|z - z_0| < \frac{\delta}{2}\right\}$ one gets $|z - z^*| \leq |z - z_0| + |z^* - z_0| < \frac{\delta}{2} + \frac{\delta}{2} = \delta$, which implies (due to the uniform continuity) that $|f(z) - f(z^*)| < \varepsilon$. On the other hand, according to the initial supposition

$$\left|f(z) - f(z^*)\right| \geq |f(z)| - \left|f(z^*)\right| \underset{z\to z_0,\, z\in D}{\longrightarrow} \infty\,,$$

that is, we arrive at a contradiction.

Hence, for $\forall z_0 \in \partial D$ the limit $\lim\limits_{z\to z_0,\, z\in D} f(z)$ exists and is finite. This limit is naturally denoted by $f(z_0)$.

2nd case (general). Let us turn to the general case of a domain D bounded by a finite number of piecewise smooth curves with folds.

We start by dividing a domain D in a finite number of subdomains D_k such that each of D_k is bounded by a finite number of piecewise smooth curves without folds. For this, we introduce additional crosscuts in the domain D made by piecewise smooth curves in such a way that the boundaries of D_k have no folds (see Fig. 1.11).

Then, each new domain D_k is of the type considered in the 1st case, and consequently, for $\forall z_0^{(k)} \in \partial D_k$ there exists the finite limit $\lim\limits_{z \to z_0^{(k)},\, z \in D_k} f(z) = f\left(z_0^{(k)}\right)$. If a point $z_0^{(k)}$ belongs only to the boundary of one of the subdomains D_k, then the function $f(z)$ is uniquely defined at the point $z_0^{(k)}$.

The situation is more intricate if $z_0 \in \partial D$ is a boundary point of different subdomains D_k. Let z_0 be a boundary point of domains D_k and D_j: $z_0 = z_0^{(k)} \in \partial D_k$ and $z_0 = z_0^{(j)} \in \partial D_j$, $k \neq j$. Then, according to the 1st case, there exist two limits:

$$\exists \lim_{z \to z_0 = z_0^{(k)},\, z \in D_k} f(z) = f\left(z_0^{(k)}\right) \text{ and } \exists \lim_{z \to z_0 = z_0^{(j)},\, z \in D_j} f(z) = f(z_0^{(j)}).$$

Notice that generally $f\left(z_0^k\right) \neq f\left(z_0^j\right)$. In this case, the function $f(z)$ can be defined at $z_0 \in \partial D$ as a multi-valued function. In fact, if $z_0 \in \partial D$ is considered as a point of the complex plane, then at this point the function has different values. On the other hand, if z_0 is treated as a point of the curve ∂D, then (in the case when z_0 belongs to a fold) this single point of plane represents distinct points of ∂D, which correspond to different values of a real parameter in the analytic form (1.4) of the curve definition. Therefore, at the points of the boundary ∂D, the function $f(z)$ is defined as a single-valued function.

1.6 Functions Depending on a Parameter. Sequences and Series

Definition 1.1 Let function $f(z, w)$ of two complex variables be defined on a set $E_1 \times E_2 \subset \mathbb{C}^2$ $(z \in E_1, w \in E_2)$ and let w_0 be a limit point of E_2. Assume that for any fixed $z \in E_1$ there exists a finite limit $\lim\limits_{w \to w_0} f(z, w) = \varphi(z)$. Evidently, this limit depends on the choice of the point $z \in E_1$, and consequently, we obtain a function $\varphi(z)$ defined on E_1, which is called the *limit function*. In the $\varepsilon - \delta$ notation it can be written as follows:

$$\forall z \in E_1\ \forall \varepsilon > 0\ \exists \delta = \delta(\varepsilon, z) > 0 : \forall w \in E_2, 0 < |w - w_0| < \delta \Rightarrow |f(z, w) - \varphi(z)| < \varepsilon$$

(notice that δ depends on the choice of both ε and z). In this case, it is said that $f(z, w)$ *converges on* E_1 *to a limit function* $\varphi(z)$ as w approaches w_0 and the following notation is used:

$$f(z, w) \underset{w \to w_0}{\to} \varphi(z), \ \forall z \in E_1 \text{ or } f(z, w) \overset{E_1}{\underset{w \to w_0}{\to}} \varphi(z).$$

To differ this kind of convergence from other types, it is also called *pointwise convergence*.

Definition 1.2 Let $f(z, w)$ be defined on $E_1 \times E_2 \subset \mathbb{C}^2$ and w_0 be a limit point of E_2. If
(1) for $\forall z \in E_1$ there exists a finite limit $\lim\limits_{w \to w_0} f(z, w) = \varphi(z)$ (that is, $f(z)$ is pointwise convergent on E_1),
(2) for $\forall \varepsilon > 0$ there exists $\delta = \delta(\varepsilon) > 0$ such that for $\forall w \in E_2$, $0 < |w - w_0| < \delta$ it follows that $|f(z, w) - \varphi(z)| < \varepsilon$ simultaneously for $\forall z \in E_1$ (here δ depends only on ε and does not depend on points of E_1),
then we will say that the function $f(z, w)$ *converges uniformly on* E_1 *to a limit function* $\varphi(z)$ as w approaches w_0 and use the following notation:

$$f(z, w) \underset{w \to w_0}{\overset{E_1}{\rightrightarrows}} \varphi(z) .$$

Definition 1.2′ Let $f(z, w)$ be defined on $E_1 \times E_2 \subset \mathbb{C}^2$ and w_0 be a limit point of E_2. If
(1′) for $\forall z \in E_1$ there exists a finite limit $\lim\limits_{w \to w_0} f(z, w) = \varphi(z)$ (that is, $f(z)$ is pointwise convergent on E_1),
(2′) there exists $\varepsilon_0 > 0$ such that for $\forall \delta > 0$ $\exists w_\delta \in E_2$, $0 < |w_\delta - w_0| < \delta$ and $\exists z_\delta \in E_1$ such that $|f(z_\delta, w_\delta) - \varphi(z_\delta)| \geq \varepsilon_0$,
then we say that $f(z, w)$ *converges non-uniformly on* E_1 *to function* $\varphi(z)$ as w approaches w_0. The corresponding notation is as follows:

$$f(z, w) \underset{w \to w_0}{\overset{E_1}{\rightarrow}} \varphi(z) \text{, but } f(z, w) \underset{w \to w_0}{\overset{E_1}{\not\rightrightarrows}} \varphi(z) .$$

Evidently, the uniform convergence implies the pointwise convergence (see the first condition in Definition 1.2), but the converse is not true. Notice that the uniform convergence guarantees that for any $\varepsilon > 0$ can be found a single value of δ simultaneously for all the points of E_1. To the contrary, if the convergence is non-uniform, then for an arbitrary $\varepsilon > 0$, it is possible to find δ for each fixed point $z \in E_1$, but this δ depends on the choice of z and there is no unique value of δ suitable for all the points of E_1 at the same time.

Definition 1.2″ Let $f(z, w)$ be defined on $E_1 \times E_2 \subset \mathbb{C}^2$ and w_0 be a limit point of E_2. The function $f(z, w)$ *converges normally on* E_1 as w approaches w_0 if it converges uniformly on every compact subset of E_1. Sometimes this kind of convergence is called *uniform convergence inside* E_1. The following notation is used:

$$f(z, w) \underset{w \to w_0}{\overset{normal}{\rightrightarrows}} \varphi(z) ,$$

where $\varphi(z)$ is a limit function.

Let us consider the convergence of sequences of complex numbers and complex functions.

Definition 1.3 A sequence of complex numbers C_n converges to a complex number C if for $\forall\varepsilon > 0\ \exists N(\varepsilon) \in \mathbb{N}$ such that for $\forall n > N(\varepsilon)$ it follows that $|C_n - C| < \varepsilon$. The corresponding notation is $\lim\limits_{n\to\infty} C_n = C$.

Since a sequence of complex numbers is a specific case of a complex function with domain $\mathbb{N}$, the result on the equivalence of the limit of a complex function and the limits of its real and imaginary parts (see Sect. 1.4) is applied to this case. Denoting $C_n = A_n + iB_n$ and $C = A + iB$, we restate that result in the following form:

$$\exists \lim_{n\to\infty} C_n = C \;\Leftrightarrow\; \exists \lim_{n\to\infty} A_n = A\ ,\ \ \exists \lim_{n\to\infty} B_n = B\ ;\ C = A + iB. \tag{1.12}$$

Hence, *the convergence of a complex sequence is equivalent to the convergence of the two corresponding real sequences.*

Definition 1.4 Let a sequence of functions $\{f_n(z)\}_{n=1}^{\infty}$ be defined on a set $E \subset \mathbb{C}$. We will say that this sequence *converges pointwise on a set* E if it converges at each point of E, that is, the numerical sequence $f_n(z)$ converges for any fixed $z \in E$. The $\varepsilon - N$ version goes as follows:

$$\forall z \in E,\ \forall\varepsilon > 0\ \exists N = N(\varepsilon, z)\ :\ \forall n > N(\varepsilon, z) \Rightarrow\ |f_n(z) - \varphi(z)| < \varepsilon$$

(notice that N depends on both ε and z). The function $\varphi(z)$ is referred to as the limit function and the following notation is used:

$$f_n(z) \underset{n\to\infty}{\overset{E}{\rightarrow}} \varphi(z)\,.$$

Definition 1.5 Let a sequence of functions $\{f_n(z)\}_{n=1}^{\infty}$ be defined on a set $E \subset \mathbb{C}$. We will say that this sequence *converges uniformly on* E if
(1) for $\forall z \in E$ there exists a finite limit $\lim\limits_{n\to\infty} f_n(z) = \varphi(z)$, that is, the sequence is pointwise convergent on E;
(2) for $\forall\varepsilon > 0$ there exists $N = N_\varepsilon$ such that for $\forall n > N_\varepsilon \Rightarrow |f_n(z) - \varphi(z)| < \varepsilon$ simultaneously for $\forall z \in E$.
In this case we write

$$f_n(z) \underset{n\to\infty}{\overset{E}{\rightrightarrows}} \varphi(z)\,.$$

Definition 1.5′. A sequence of functions $\{f_n(z)\}_{n=1}^{\infty}$ defined on $E \subset \mathbb{C}$ *converges non-uniformly on a set* E if
(1′) for $\forall z \in E$ there exists a finite limit $\lim\limits_{n\to\infty} f_n(z) = \varphi(z)$, that is, the sequence is pointwise convergent on E;
(2′) there exists $\varepsilon_0 > 0$ such that for $\forall N$ there exist $n_N > N$ and $z_N \in E$ such that $\left|f_{n_N}(z_N) - \varphi(z_N)\right| \geq \varepsilon_0$.

Definition 1.5″. A sequence of functions $\{f_n(z)\}_{n=1}^{\infty}$ defined on a set $E \subset \mathbb{C}$ *converges normally on* E if it converges uniformly on every compact subset of E. This kind of convergence is also called *uniform convergence inside E*. The following notation is used

$$f_n(z) \underset{n\to\infty}{\overset{normal}{\rightrightarrows}} \varphi(z)\,,$$

where $\varphi(z)$ is a limit function.

Notice that Definitions 1.4, 1.5, 1.5′, and 1.5″ are particular cases of Definitions 1.1, 1.2, 1.2′, and 1.2″, respectively, where the set E_2 is specified to be the set of natural numbers $\mathbb{N}$ whose only limit point is ∞.

Let us turn our attention to a series of numbers and functions. First, consider a series of complex numbers

$$\text{(C)}\ \sum_{n=1}^{\infty} c_n,\ c_n = a_n + ib_n\ (a_n, b_n \in \mathbb{R})\,.$$

The two series of real numbers intimately related to the series (C) are the series composed of the real and imaginary parts of c_n

$$\text{(A)}\ \sum_{n=1}^{\infty} a_n \quad \text{and} \quad \text{(B)}\ \sum_{n=1}^{\infty} b_n.$$

The *partial sums* of series (A), (B), and (C) will be denoted by $A_n,\ B_n$ and C_n, respectively

$$C_n = \sum_{k=1}^{n} c_k = \sum_{k=1}^{n} (a_k + ib_k) = \sum_{k=1}^{n} a_k + i\sum_{k=1}^{n} b_k = A_n + iB_n.$$

Definition 1.6 Series (C) is called *convergent* if there exists a finite limit of the sequence of its partial sums $\exists \lim\limits_{n\to\infty} C_n = C$.

As it was discussed above, the existence of the limit C of a complex sequence is equivalent to the existence of the limits A and B of its real and imaginary parts, and also $C = A + iB$, that is, the relation (1.12) is satisfied. Hence, *the convergence of a complex series is equivalent to the convergence of the two real series* composed of the real and imaginary parts of the elements of the series (C).

Definition 1.7 A series $\sum_{n=1}^{\infty} c_n$ is called *absolutely convergent* if the series of the absolute values $\sum_{n=1}^{\infty} |c_n|$ converges.

Like in Real Analysis, the absolute convergence implies the convergence of a series, but the converse is not true (the proof is straightforward).

Example Analyze if the numerical series is convergent or divergent: $\sum_{n=1}^{+\infty} \frac{i^n}{2^n}$.

Since $\left|\frac{i^n}{2^n}\right| = \frac{1}{2^n}$ and the real numerical series $\sum_{n=1}^{+\infty} \frac{1}{2^n}$ converges, the original series converges absolutely.

Consider now a series of functions

$$\sum_{n=1}^{\infty} u_n(z), z \in E \subset \mathbb{C}.$$

Denote by $f_n(z)$ and $\alpha_n(z)$ the nth *partial sum* and the nth *residual* of this series: $f_n(z) = \sum_{k=1}^{n} u_k(z)$, $\alpha_n(z) = \sum_{k=n+1}^{\infty} u_k(z)$.

Definition 1.8 Let $\sum_{n=1}^{\infty} u_n(z)$ be a series of functions defined on a set $E \subset \mathbb{C}$. The series is said to be *pointwise convergent on E* if the sequence of its partial sums $f_n(z)$ converges pointwise on E, which means that for each fixed $z \in E$ the following property holds: for $\forall \varepsilon > 0$ there exists $N = N(\varepsilon, z)$ such that $|f_n(z) - \varphi(z)| < \varepsilon$ whenever $n > N(\varepsilon, z)$. (Notice that N depends on both ε and z.)

The same definition can be equivalently restated as the *pointwise convergence of the residuals* $\alpha_n(z)$ *on E to* 0: for any fixed $z \in E$ one has that for $\forall \varepsilon > 0$ there exists $N = N(\varepsilon, z)$ such that $|\alpha_n(z)| = \left|\sum_{k=n+1}^{\infty} u_k(z)\right| < \varepsilon$ whenever $n > N(\varepsilon, z)$.

Definition 1.9 Let $\sum_{n=1}^{\infty} u_n(z)$ be a series of functions defined on a set $E \subset \mathbb{C}$. The series is said to be *uniformly convergent on E* if the sequence of its partial sums $f_n(z)$ converges uniformly on E, which means that
(1) $\exists \lim_{n\to\infty} f_n(z) = \varphi(z), \forall z \in E$, that is, there exists the sum of the given series: $\varphi(z) = \sum_{n=1}^{\infty} u_n(z)$,
(2) $\forall \varepsilon > 0\ \exists N = N_\varepsilon : \forall n > N_\varepsilon, \forall z \in E \Rightarrow |f_n(z) - \varphi(z)| < \varepsilon$.

Equivalently, the same definition can be formulated in the terms of the residuals. The series is said to be *uniformly convergent on a set E* if $\alpha_n(z)$ converges uniformly on E to 0, or in the detailed form:
(1) $\exists \lim_{n\to\infty} \alpha_n(z) = 0,\ \forall z \in E$,
(2) $\forall \varepsilon > 0\ \exists N = N_\varepsilon : \forall n > N_\varepsilon, \forall z \in E \Rightarrow |\alpha_n(z)| = \left|\sum_{k=n+1}^{\infty} u_k(z)\right| < \varepsilon$.

Definition 1.9′. Let $\sum_{n=1}^{\infty} u_n(z)$ be a series of functions defined on a set $E \subset \mathbb{C}$. If (1′) there exists a sum of the series $\varphi(z) = \sum_{n=1}^{\infty} u_n(z)$, that is, the series converges at each point of E,
(2′) $\exists \varepsilon_0 > 0 : \forall N\ \exists n_N > N$ and $\exists z_N \in E$ such that $\left|f_{n_N}(z_N) - \varphi(z_N)\right| \geq \varepsilon_0$
then the *series converges pointwise but non-uniformly on a set E*.

The same definition can be given also using the residuals. The *series converges pointwise but non-uniformly on a set E* if
(1′) $\exists \lim_{n\to\infty} \alpha_n(z) = 0,\ \forall z \in E$,
(2′) $\exists \varepsilon_0 > 0 : \forall N\ \exists n_N > N$ and $\exists z_N \in E$ such that $\left|\alpha_{n_N}(z_N)\right| \geq \varepsilon_0$.

Definition 1.9″. A series $\sum_{n=1}^{\infty} u_n(z)$ of functions defined on a set $E \subset \mathbb{C}$ *converges normally on E* if it converges uniformly on every compact subset of E. This kind of convergence is also called *uniform convergence inside E*.

Definitions 1.8, 1.9, 1.9′, and 1.9″ are actually Definitions 1.4, 1.5, 1.5′, and 1.5″ reiterated for the case of series, since by definition, the convergence of a series means the convergence of the sequence of its partial sums.

One of the first and useful tests for the uniform convergence is the Cauchy criterion, which is similar to that studied in Real Analysis.

Theorem (Cauchy criterion for uniform convergence of a series) *A series* $\sum_{n=1}^{\infty} u_n(z)$ *is uniformly convergent on* $E \subset \mathbb{C}$ *if and only if the following Cauchy condition holds: for* $\forall \varepsilon > 0\ \exists N_\varepsilon$ *such that for* $\forall n > N_\varepsilon$ *and* $\forall p \in \mathbb{N}$ *it follows that* $\left|\sum_{k=n+1}^{n+p} u_k(z)\right| < \varepsilon$ *simultaneously for* $\forall z \in E$.

Proof *Necessity*. According to the definition, a series of functions converges uniformly if the sequence of its partial sums converges uniformly, that is, $\forall \varepsilon > 0$ (take $\frac{\varepsilon}{2}$ for convenience) $\exists N_\varepsilon$ such that for $\forall n > N_\varepsilon$ and for $\forall z \in E$ it follows that $|f_n(z) - \varphi(z)| < \frac{\varepsilon}{2}$, where $\varphi(z)$ is the limit function—$\varphi(z) = \lim_{n\to\infty} f_n(z)$. Then, for $\forall p \in \mathbb{N}$, one has $(n+p) > N_\varepsilon$, and therefore, for $\forall z \in E$ the following inequality holds: $\left|f_{n+p}(z) - \varphi(z)\right| < \frac{\varepsilon}{2}$. Consequently, for $\forall z \in E$, it follows that

$$\left|\sum_{k=n+1}^{n+p} u_k(z)\right| = \left|\sum_{k=1}^{n+p} u_k(z) - \sum_{k=1}^{n} u_k(z)\right| = \left|f_{n+p}(z) - f_n(z)\right|$$

$$= \left|f_{n+p}(z) - \varphi(z) + \varphi(z) - f_n(z)\right| \le \left|f_{n+p}(z) - \varphi(z)\right| + |f_n(z) - \varphi(z)| < \frac{\varepsilon}{2} + \frac{\varepsilon}{2} = \varepsilon.$$

In this way, the necessity is proved.

Sufficiency. First of all, we prove that the series is convergent on E. Take an arbitrary $z_0 \in E$ and fix this point. Then the Cauchy condition of the Theorem turns into the Cauchy condition for the numerical series $\sum_{n=1}^{\infty} u_n(z_0)$ and the last series converges according to the Cauchy criterion for numerical series. Since z_0 is an arbitrary point in E, it means that the series $\sum_{n=1}^{\infty} u_n(z)$ converges on E.

Now we show that the convergence is uniform. Rewrite the Cauchy condition of the Theorem in the form

$$\forall \varepsilon > 0\ \exists N_\varepsilon : \forall n > N_\varepsilon,\ \forall p \in \mathbb{N} \text{ and } \forall z \in E \Rightarrow \left|\sum_{k=n+1}^{n+p} u_k(z)\right| = \left|f_{n+p}(z) - f_n(z)\right| < \frac{\varepsilon}{2} \tag{1.13}$$

(take $\frac{\varepsilon}{2}$ for convenience). Since this inequality holds for $\forall p \in \mathbb{N}$ and it was already proved that the series is convergent, that is, $\lim_{p\to\infty} f_{n+p}(z) = \varphi(z)$, we can pass to the limit as $p \to \infty$ in the inequality (1.13) and obtain the following result (recall that in the limit form a strict inequality is transformed into a non-strict one):

$$|\varphi(z) - f_n(z)| \le \frac{\varepsilon}{2} < \varepsilon.$$

The last evaluation holds for all $z \in E$ simultaneously, which means that the series $\sum_{n=1}^{\infty} u_n(z)$ is uniformly convergent on E. □

We consider below a simple result, which is the most used test for uniform convergence of series of complex functions. A similar theorem for real functions is also a ubiquitous tool in the analysis of uniform convergence.

Theorem (Weierstrass M-test) *Let $\sum_{n=1}^{\infty} u_n(z)$ be a series of functions defined on a set E. If*
(1) for $\forall n \in \mathbb{N}$ and for $\forall z \in E$ one has $|u_n(z)| \le a_n$ (a_n is a positive real number for each index n),
(2) the positive series of numbers $\sum_{n=1}^{\infty} a_n$ is convergent,
then the series $\sum_{n=1}^{\infty} u_n(z)$ converges uniformly (and absolutely) on E.

Proof The convergence of the positive series $\sum_{n=1}^{\infty} a_n$ implies that (according to the Cauchy criterion for convergence) for $\forall \varepsilon > 0\ \exists N_\varepsilon$ such that for $\forall n > N_\varepsilon$ and for $\forall p \in \mathbb{N}$ one has $\left|\sum_{k=n+1}^{n+p} a_k\right| = \sum_{k=n+1}^{n+p} a_k < \varepsilon$. Applying the last inequality, the properties of absolute values and the first condition of the Theorem, we arrive at the following evaluation:

$$\left|\sum_{k=n+1}^{n+p} u_k(z)\right| \le \sum_{k=n+1}^{n+p} |u_k(z)| \le \sum_{k=n+1}^{n+p} a_k < \varepsilon.$$

Therefore, by the Cauchy criterion for uniform convergence, the series $\sum_{n=1}^{\infty} u_n(z)$ converges uniformly and absolutely on E. □

We finalize this section by proving one more important theorem.

Theorem (Continuity of the limit function) *Let $f(z, w)$ be a function defined on $E_1 \times E_2$, and w_0 be a limit point of E_2. If*
(1) $f(z, w)$ is continuous with respect to z on E_1 (for each fixed value of $w \in E_2$),
(2) $f(z, w) \underset{w \to w_0}{\overset{E_1}{\rightrightarrows}} \varphi(z)$,
then the limit function $\varphi(z)$ is continuous on E_1.

Proof To show that $\varphi(z)$ is continuous at an arbitrary point $z_0 \in E_1$, we evaluate the difference $|\varphi(z) - \varphi(z_0)|$

$$|\varphi(z) - \varphi(z_0)| = |\varphi(z) - f(z, w) + f(z, w) - f(z_0, w) + f(z_0, w) - \varphi(z_0)|$$

$$\le |f(z, w) - \varphi(z)| + |f(z, w) - f(z_0, w)| + |f(z_0, w) - \varphi(z_0)| . \quad (1.14)$$

The first and third summands in the last sum can be evaluated by using the condition of the uniform convergence: for $\forall \varepsilon > 0$ (choose for convenience $\frac{\varepsilon}{3}$) $\exists \sigma_\varepsilon > 0$ (the number σ_ε depends only on ε and does not depend on the points of E_1) such that for $\forall w \in E_2, 0 < |w - w_0| < \sigma_\varepsilon$ and simultaneously for $\forall z \in E_1$, it follows that $|f(z, w) - \varphi(z)| < \frac{\varepsilon}{3}$, and in particular, $|f(z_0, w) - \varphi(z_0)| < \frac{\varepsilon}{3}$.

The second summand in (1.14) can be estimated through the continuity of $f(z, w)$ with respect to z. In fact, take an arbitrary w in a deleted σ_ε-neighborhood of the point

w_0 and fix this number, that is: $\forall w \in E_2, 0 < |w - w_0| < \sigma_\varepsilon$. Then $f(z, w)$ (with w fixed) is continuous in E_1, and in particular, it is continuous at z_0, which means that for $\forall \varepsilon > 0$ (again take for convenience $\frac{\varepsilon}{3}$) $\exists \delta = \delta(\varepsilon) > 0$ such that $\forall z \in E_1$, $|z - z_0| < \delta$ the following inequality holds $|f(z, w) - f(z_0, w)| < \frac{\varepsilon}{3}$.

Joining all the inequalities, we obtain that for $\forall \varepsilon > 0$ $\exists \delta(\varepsilon) > 0$ such that $\forall z \in E_1, |z - z_0| < \delta$ it follows that $|\varphi(z) - \varphi(z_0)| < \frac{\varepsilon}{3} + \frac{\varepsilon}{3} + \frac{\varepsilon}{3} = \varepsilon$, that is, we arrive at the definition of the continuity of $\varphi(z)$ at the point z_0. Since z_0 is an arbitrary point of E_1, the function $\varphi(z)$ is continuous on E_1. □

Corollary 1.1 *Let $\{f_n(z)\}_{n=1}^{\infty}$ be a sequence of functions defined on $E \subset \mathbb{C}$. If*
(1) each function $f_n(z)$ is continuous on E (for $\forall n \in \mathbb{N}$),
(2) $f_n(z) \underset{n\to\infty}{\overset{E}{\rightrightarrows}} \varphi(z)$,
then $\varphi(z)$ is continuous on E.

This Corollary is a particular case of the Theorem with $E_2 = \mathbb{N}$.

Corollary 1.2 *Let $\sum_{n=1}^{\infty} u_n(z)$ be a series of functions defined on $E \subset \mathbb{C}$. If*
(1) each function $u_n(z)$ is continuous on E (for $\forall n \in \mathbb{N}$),
(2) the series is uniformly convergent on E,
then the sum of the series is a continuous function on E.

Since any type of convergence of a series is defined through the corresponding convergence of the sequence of partial sums, the last Corollary is merely a restatement of Corollary 1.1 in the terms of series.

1.7 Power Series

Definition A *power series* is a particular case of a series of functions, which is given in the following form : $\sum_{n=0}^{\infty} c_n (z - z_0)^n$, where the constants c_n are called the coefficients and z_0—the central point of the series.

Notice that any power series $\sum_{n=0}^{+\infty} c_n (z - z_0)^n$ is convergent at the central point z_0: $\sum_{n=0}^{+\infty} c_n (z_0 - z_0)^n = c_0$.

The following result is fundamental for determining the set of convergence of a power series:

Abel's Lemma *If a power series $\sum_{n=0}^{\infty} c_n (z - z_0)^n$ converges at a point $z_1 \neq z_0$, then it converges absolutely on the open disk $|z - z_0| < |z_1 - z_0|$. If a power series diverges at some point z_2, then it diverges on the set $|z - z_0| > |z_2 - z_0|$. (See illustration in Fig. 1.12.)*

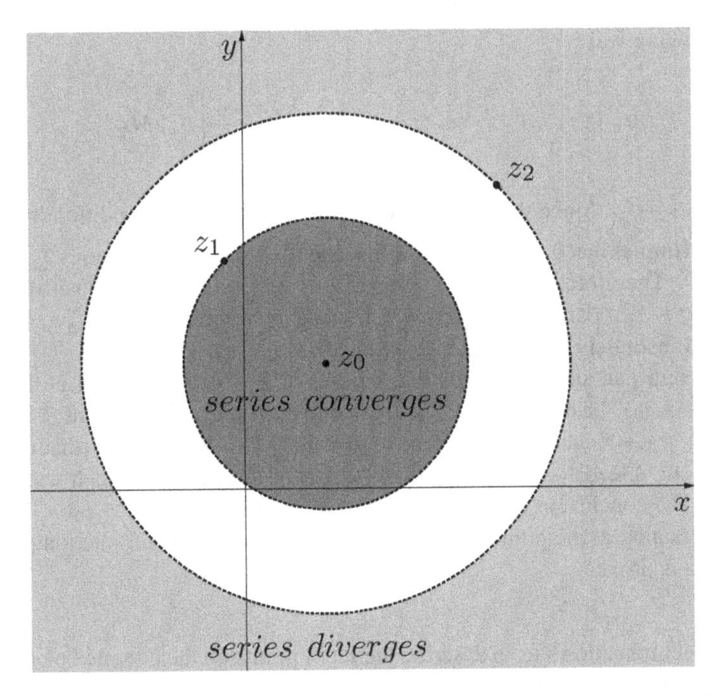

Fig. 1.12 The regions of the convergence and divergence regarding the convergent z_1 and divergent z_2 points

Proof Let us start with the first part of the Lemma. The convergence of the series $\sum_{n=0}^{\infty} c_n (z - z_0)^n$ at a point $z_1 \neq z_0$ means that the numerical series $\sum_{n=0}^{+\infty} c_n (z_1 - z_0)^n$ converges. Therefore, due to the necessity condition of the convergence, the general term of this numerical series approaches zero as $n \to \infty$: $c_n (z_1 - z_0)^n \underset{n\to\infty}{\to} 0$. This implies that for $\varepsilon = 1$ there exists N_1 such that $\forall n > N_1$ it follows that $\left|c_n (z_1 - z_0)^n\right| < 1$. The number of the elements of the numerical series that, possibly, do not satisfy this inequality is finite: $c_0, c_1 (z_1 - z_0), \ldots,$ $c_{N_1} (z_1 - z_0)^{N_1}$. One always can find the largest number among the absolute values of these elements and 1

$$M = \max\left\{ |c_0|,\ |c_1 (z_1 - z_0)|,\ \ldots\ ,\ \left|c_{N_1} (z_1 - z_0)^{N_1}\right|,\ 1\right\}.$$

Hence, for $\forall n \in \mathbb{N}$ the following inequality is true :

$$\left|c_n (z_1 - z_0)^n\right| \leq M.$$

Using the obtained inequality, we can show that the numerical series $\sum_{n=0}^{+\infty} \left|c_n (z - z_0)^n\right|$ is convergent at any fixed point z that belongs to the disk $|z - z_0| < |z_1 - z_0|$. Indeed, the general term of the numerical series can be evaluated in the following way:

$$\left|c_n (z - z_0)^n\right| = \left|c_n (z_1 - z_0)^n\right| \cdot \left|\frac{z - z_0}{z_1 - z_0}\right|^n \leq Mq^n,$$

where $q = \left|\frac{z-z_0}{z_1-z_0}\right|$. Since the chosen fixed point z satisfies the condition $|z - z_0| < |z_1 - z_0|$, it imples that $0 < q < 1$, and consequently, the geometric series $\sum_{n=0}^{\infty} Mq^n$ converges. Therefore, by the comparison test, the real positive series $\sum_{n=0}^{+\infty} \left|c_n (z - z_0)^n\right|$ also converges, which means that the series $\sum_{n=0}^{+\infty} c_n (z - z_0)^n$ converges (absolutely) at each point z of the disk $|z - z_0| < |z_1 - z_0|$.

The second part of the Lemma can be proved by contradiction. Let the series $\sum_{n=0}^{\infty} c_n (z - z_0)^n$ be divergent at z_2 and assume that there exists a point z^* such that $|z^* - z_0| > |z_2 - z_0|$ and the series converges at z^*. This assumption (that the series converges at z^*) leads (according to the first part of the Lemma, which was already proved) to the conclusion that this series converges in the disk $|z - z_0| < |z^* - z_0|$ and, in particular, at the point z_2. In this way, we arrive at a contradiction, and thus the Lemma is proved. □

Abel's Lemma allows us to describe the set of points on the complex plane where a power series is convergent.

Corollary *Let $E \subset \mathbb{C}$ be a set of points of the complex plane where a power series converges and denote $R = \sup\limits_{\forall z \in E} |z - z_0|$. There are three possibilities:*
(1) if $R = \infty$, then the series converges absolutely on the entire (finite) complex plane;
(2) if $R = 0$, then the series converges only at the point z_0 and diverges at all other points;
(3) if $0 < R < \infty$, then the series converges absolutely in the disk $|z - z_0| < R$ and diverges outside this disk.
The number R is referred to as the radius of convergence of a power series, the disk $|z - z_0| < R$ as the disk of convergence and the circle $|z - z_0| = R$ as the circle of convergence. (See illustration in Fig. 1.13.)

Proof (1) Let $R = \sup\limits_{\forall z \in E} |z - z_0| = \infty$. We will show that the power series $\sum_{n=0}^{+\infty} c_n (z - z_0)^n$ converges at any point $z \in \mathbb{C}$. Pick an arbitrary point of the complex plane and fix it: $z^* \in \mathbb{C}$. Since $\sup\limits_{\forall z \in E} |z - z_0| = \infty$, it means (according to the definition of an unbounded set) that for any number M (in particular, one can choose $M = |z^* - z_0|$) $\exists z_M \in E$ such that $|z_M - z_0| > M = |z^* - z_0|$. By the definition of E, the condition $z_M \in E$ means that the power series converges at this point, which implies (by Abel's Lemma) absolute convergence of this series at all points that satisfy the condition $|z - z_0| < |z_M - z_0|$. In particular, it includes the point z^*, since $|z^* - z_0| < |z_M - z_0|$. In this way, the first part is proved.

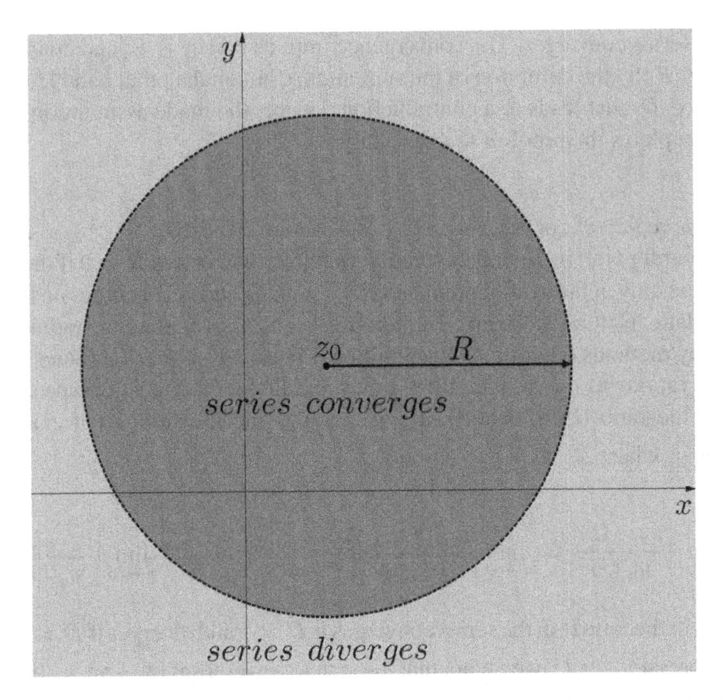

Fig. 1.13 The disk $|z - z_0| < R$ and the radius R of convergence of a power series (the case $R \neq \infty$, $R \neq 0$)

(2) Let $R = \sup\limits_{\forall z \in E} |z - z_0| = 0$. Any power series converges at its central point. Let us show that $\sum_{n=0}^{+\infty} c_n (z - z_0)^n$ diverges at any other point $z \neq z_0$. Suppose, by contradiction, that there exists a point $z_1 \neq z_0$ where the series converges. Then the point z_1 belongs to E (by the definition of E) and one has: $R = \sup\limits_{\forall z \in E} |z - z_0| \geq |z_1 - z_0| > 0$. The last inequality contradicts the given condition $R = 0$, which shows that the series diverges at any point z other than z_0.

(3) Let $0 < R < +\infty$ (general case). We divide an analysis of this situation into two subcases. First, we show that the series $\sum_{n=0}^{+\infty} c_n (z - z_0)^n$ converges in the disk $|z - z_0| < R$. To this end, take any point $\tilde{z}$ of this disk and fix it: $|\tilde{z} - z_0| < R$. By the definition of supremum, one has

$$R = \sup_{\forall z \in E} |z - z_0| : \begin{cases} 1)\ \forall z \in E \ \Rightarrow \ |z - z_0| \leq R\,; \\ 2)\ \forall \varepsilon > 0\,(\text{take } \varepsilon = R - |\tilde{z} - z_0|\,)\, \exists z_\varepsilon \in E \text{ such that } |z_\varepsilon - z_0| > |\tilde{z} - z_0|\,. \end{cases}$$

Since $z_\varepsilon \in E$, the series converges at this point and, due to the inequality $|\tilde{z} - z_0| < |z_\varepsilon - z_0|$, Abel's Lemma guarantees that it converges absolutely at the point $\tilde{z}$. Therefore, the series converges absolutely at any point of the disk $|z - z_0| < R$.

Second, we prove that the series diverges outside the disk, that is, on the set $D = \{|z - z_0| > R\}$. Assume, for contradiction, that there exists a point $z_1 \in D$

where the series converges. The convergence implies that $z_1 \in E$, and consequently $|z_1 - z_0| \le R$ (by the definition of the supremum), but on the other hand $|z_1 - z_0| > R$ since $z_1 \in D$, that leads to a contradiction. Hence, the made assumption is false.

This completes the proof of the Corollary. □

Thus, the region of convergence of a power series is the disk $|z - z_0| < R$, where z_0 is the central point, including as singular (pathological) cases $R = 0$ if the convergence occurs only at the central point and $R = \infty$ if the series converges on the entire complex plane. Naturally, we arrive now at the problem of finding the radius of convergence by methods other than the definition. Frequently, R can be found applying the root or ratio tests (also called the Cauchy and D'Alembert tests, respectively).

To use the ratio (D'Alembert) test we should find the limit (if it exists) $D = \lim_{n\to\infty}\left|\frac{u_{n+1}(z)}{u_n(z)}\right|$, where $u_n(z) = c_n (z - z_0)^n$:

$$D = \lim_{n\to\infty}\left|\frac{u_{n+1}(z)}{u_n(z)}\right| = \lim_{n\to\infty}\left|\frac{c_{n+1}(z - z_0)^{n+1}}{c_n (z - z_0)^n}\right| = |z - z_0| \cdot \lim_{n\to\infty}\left|\frac{c_{n+1}}{c_n}\right|. \quad (1.15)$$

According to the ratio test, the series converges if $D < 1$ and diverges if $D > 1$. Using the last expression for D, we obtain that the series converges if $|z - z_0| < \lim_{n\to\infty}\left|\frac{c_n}{c_{n+1}}\right|$ and it diverges if $|z - z_0| > \lim_{n\to\infty}\left|\frac{c_n}{c_{n+1}}\right|$. Hence, it follows from the last Corollary that the radius of convergence can be defined as $R = \lim_{n\to\infty}\left|\frac{c_n}{c_{n+1}}\right|$, if the last limit exists. This is a simple formula for the radius of convergence in the terms of the coefficients of a power series.

In a similar manner, we can apply the root (Cauchy) test. If there exists the limit

$$K = \lim_{n\to\infty}\sqrt[n]{|u_n(z)|} = \lim_{n\to\infty}\sqrt[n]{|c_n (z - z_0)^n|} = |z - z_0| \cdot \lim_{n\to\infty}\sqrt[n]{|c_n|}, \quad (1.16)$$

then for $K < 1$ the power series converges, while for $K > 1$ the series diverges. In other words, the series converges if $|z - z_0| < \frac{1}{\lim_{n\to\infty}\sqrt[n]{|c_n|}}$ and it diverges if $|z - z_0| > \frac{1}{\lim_{n\to\infty}\sqrt[n]{|c_n|}}$. Therefore, the radius of convergence can be determined by the formula $R = \frac{1}{\lim_{n\to\infty}\sqrt[n]{|c_n|}}$ if the last limit exists.

However, the two above limits may not exist. Nevertheless, there is a general form to find the radius of convergence of a power series presented in the next result.

Cauchy–Hadamard Theorem *Given a power series* $\sum_{n=0}^{+\infty} c_n (z - z_0)^n$, *consider the following upper limit* $l = \overline{\lim_{n\to\infty}}\sqrt[n]{|c_n|}$ *(recall that the upper limit of a sequence always exists). There are three possibilities:*
(1) if $l = 0$, *then the power series converges absolutely on the entire (finite) complex plane;*

(2) if $l = \infty$*, then the power series converges at* z_0 *and diverges at all other points;*
(3) if $0 < l < +\infty$ *then the power series converges absolutely in the disk* $|z - z_0| < \frac{1}{l}$ *and diverges outside this disk (i.e., on the set* $|z - z_0| > \frac{1}{l}$*).*
In this way, the radius of convergence R *is equal to* $\frac{1}{l}$

$$R = \frac{1}{l} = \frac{1}{\overline{\lim\limits_{n\to\infty}} \sqrt[n]{|c_n|}}. \tag{1.17}$$

This formula is referred to as the Cauchy–Hadamard formula.

Proof (1) Let $l = \overline{\lim\limits_{n\to\infty}} \sqrt[n]{|c_n|} = 0$. Since the sequence $\sqrt[n]{|c_n|}$ is non-negative, all its partial limits are also non-negative (according to the limit properties). Further, since the upper limit (which is greater than or equal to any other partial limit) is 0, it implies that all the partial limits are 0, and consequently, in this case there exists a general limit equal to 0 : $\lim\limits_{n\to\infty} \sqrt[n]{|c_n|} = 0$. Therefore, taking any point z of the complex plane, we obtain the following result:

$$\lim_{n\to\infty} \sqrt[n]{\left|c_n (z - z_0)^n\right|} = |z - z_0| \cdot \lim_{n\to\infty} \sqrt[n]{|c_n|} = 0 < 1 .$$

It means that, according to the root test, the series is absolutely convergent at any point z, that is, $R = \infty$.

(2) Let $l = \overline{\lim\limits_{n\to\infty}} \sqrt[n]{|c_n|} = +\infty$. By the definition of the upper limit, there exists a subsequence c_{n_k} such that $\lim\limits_{k\to\infty} \sqrt[n_k]{\left|c_{n_k}\right|} = \overline{\lim\limits_{n\to\infty}} \sqrt[n]{|c_n|} = +\infty$. Then, for any fixed point $z \neq z_0$, we have:

$$\lim_{k\to\infty} \sqrt[n_k]{\left|c_{n_k} (z - z_0)^{n_k}\right|} = |z - z_0| \cdot \lim_{k\to\infty} \sqrt[n_k]{\left|c_{n_k}\right|} = +\infty ,$$

that is, $\lim\limits_{k\to\infty} \sqrt[n_k]{\left|c_{n_k}(z-z_0)^{n_k}\right|} > 1$. Therefore, $\exists K_0$ such that $\forall k \geq K_0$, $\sqrt[n_k]{\left|c_{n_k}(z-z_0)^{n_k}\right|} > 1$ or $\left|c_{n_k} (z - z_0)^{n_k}\right| > 1$. It means that the general term of the series does not converge to 0, that is, the necessary condition of the convergence does not hold, and consequently, the series is divergent. Since this reasoning is valid for any $z \neq z_0$, it follows that $R = 0$.

(3) Let $0 < l = \overline{\lim\limits_{n\to\infty}} \sqrt[n]{|c_n|} < +\infty$. We start by showing that the power series converges in the disk $|z - z_0| < \frac{1}{l}$. Indeed, take any point z_1 of this disk and fix it: $|z_1 - z_0| < \frac{1}{l}$, or equivalently, $\frac{1}{|z_1 - z_0|} > l$. Since $\frac{1}{|z_1 - z_0|}$ and l are two different real numbers, there exists another real number l_1 such that $\frac{1}{|z_1 - z_0|} > l_1 > l$ (according to the density of $\mathbb{R}$ between any two real numbers there are infinitely many other real numbers). Denote $\varepsilon_1 = l_1 - l > 0$ and write down the property of the upper limit $l = \overline{\lim\limits_{n\to\infty}} \sqrt[n]{|c_n|}$ using the parameter ε_1

$$\forall \varepsilon > 0 \text{ (take } \varepsilon = \frac{\varepsilon_1}{2}) \; \exists N_\varepsilon, \; \forall n \geq N_\varepsilon \; \Rightarrow \; \sqrt[n]{|c_n|} < l + \frac{\varepsilon_1}{2}.$$

Then for all the indices $n \geq N_\varepsilon$ one obtains

$$\sqrt[n]{\left|c_n (z_1 - z_0)^n\right|} = \sqrt[n]{|c_n|} \cdot |z_1 - z_0| < \frac{l + \frac{\varepsilon_1}{2}}{l_1} = \frac{l + \frac{\varepsilon_1}{2}}{l + \varepsilon_1} = q_1, \text{ where } 0 < q_1 < 1.$$

Thus, the condition of convergence of the root test holds, and therefore, the series converges absolutely at z_1. Since z_1 is an arbitrary point of the disk $|z - z_0| < \frac{1}{l}$, it means that the series converges absolutely in the disk $|z - z_0| < \frac{1}{l}$.

Now we switch to the points outside the disk, that is, to the points z in the set $|z - z_0| > \frac{1}{l}$. Choose any fixed point z_2 such that $|z_2 - z_0| > \frac{1}{l}$, or equivalently, $\frac{1}{|z_2 - z_0|} < l$. By density of the real numbers, there exists a real number l_2 such that $\frac{1}{|z_2 - z_0|} < l_2 < l$. Since $\overline{\lim\limits_{n\to\infty}} \sqrt[n]{|c_n|} = l$, there exists a subsequence with the limit equal to l: $\lim\limits_{k\to\infty} \sqrt[n_k]{\left|c_{n_k}\right|} = \overline{\lim\limits_{n\to\infty}} \sqrt[n]{|c_n|} = l$. Writing the definition of the limit of the subsequence with $\varepsilon_2 = l - l_2 > 0$ instead of ε, we obtain

$$\forall \varepsilon > 0 \text{ (take } \varepsilon = \varepsilon_2) \; \exists K_\varepsilon, \; \forall k \geq K_\varepsilon \; \Rightarrow \; \left| \sqrt[n_k]{\left|c_{n_k}\right|} - l \right| < \varepsilon_2 .$$

The last inequality can be equivalently expressed in the form $l - \varepsilon_2 < \sqrt[n_k]{\left|c_{n_k}\right|} < l + \varepsilon_2$, where the left-hand side is of our interest. For the same indices $k \geq K_\varepsilon$, we have

$$\sqrt[n_k]{\left|c_{n_k} (z_2 - z_0)^{n_k}\right|} = \sqrt[n_k]{\left|c_{n_k}\right|} \cdot |z_2 - z_0| > \frac{l - \varepsilon_2}{l_2} = \frac{l_2}{l_2} = 1,$$

which shows that $\left|c_{n_k} (z_2 - z_0)^{n_k}\right| > 1$. Therefore, the general term of the series does not converge to 0 (which is a necessary condition for convergence), and consequently, the power series is divergent at z_2. Since z_2 is an arbitrary point of the set $|z - z_0| > \frac{1}{l}$, the series diverges on this entire region.

Thus, we have shown the desired result: the series is (absolutely) convergent in the disk $|z - z_0| < \frac{1}{l}$ and divergent outside this disk, which means that the radius of convergence is $R = \frac{1}{l}$.

This completes the proof of the Theorem. □

Remark Although the radius of convergence can always be calculated by the formula $R = \frac{1}{\overline{\lim\limits_{n\to\infty}} \sqrt[n]{|c_n|}}$, if the general limits $\lim\limits_{n\to\infty} \left|\frac{c_{n+1}}{c_n}\right|$ or $\lim\limits_{n\to\infty} \sqrt[n]{|c_n|}$ exist, then the radius of convergence is usually found employing one of these two limits.

The following example illustrates the application of the Cauchy–Hadamard formula (1.17).

Example Find the radius of convergence of the power series $\sum_{n=0}^{+\infty}\left(4+(-1)^n\right)^n z^n$.

Notice that the limits $\lim\limits_{n\to\infty}\left|\frac{c_{n+1}}{c_n}\right|$ and $\lim\limits_{n\to\infty}\sqrt[n]{|c_n|}$ do not exist: the ratio $\left|\frac{c_{n+1}}{c_n}\right|$ has partial limits ∞ and 0, and the expression $\sqrt[n]{|c_n|}$ has partial limits 3 and 5. Therefore, we should use formula (1.17). Since the coefficients of the series are $c_n=\begin{cases}5^n, & n=2k\\ 3^n, & n=2k+1\end{cases}$, it follows that

$$R=\frac{1}{\overline{\lim\limits_{n\to\infty}}\sqrt[n]{|c_n|}}=\frac{1}{\lim\limits_{k\to\infty}\sqrt[2k]{|c_{2k}|}}=\frac{1}{5},$$

that is, the series converges in the disk $|z|<1/5$.

Let us consider some other properties of power series.

Theorem 1.1 *A power series converges normally in the disk of convergence, that is, the series converges uniformly in any closed disk $|z-z_0|\le r$, where $r<R$ (R is the radius of convergence).*

Proof Take an arbitrary point of the circle $|z-z_0|=r$, $\forall r<R$. Since this point belongs to the disk of convergence, the power series converges absolutely at this point, that is, the series $\sum_{n=0}^{+\infty}\left|c_n(z-z_0)^n\right|=\sum_{n=0}^{+\infty}|c_n|\cdot r^n$ converges. For any point z of the disk $|z-z_0|\le r$ one has $\left|c_n(z-z_0)^n\right|\le|c_n|\,r^n$, and consequently, by the Weierstrass test, the series $\sum_{n=0}^{+\infty}c_n(z-z_0)^n$ converges uniformly in the disk $|z-z_0|\le r$, $\forall r<R$. □

The next property follows directly from the last Theorem and the Theorem on continuity of the limit function (formulated in the previous section).

Theorem 1.2 *If the radius of convergence of a power series $\sum_{n=0}^{+\infty}c_n(z-z_0)^n$ is different from 0, then the sum of this series is a continuous function in the disk of convergence.*

Proof Pick any point z_1 in the disk of convergence: $|z_1-z_0|=r_1<R$. The density of the real numbers implies that $\exists r\in\mathbb{R}$ such that $r_1<r<R$. Then the previous Theorem guarantees that the power series converges uniformly in the disk $|z-z_0|\le r$. Therefore, noting that the functions $u_n(z)=c_n(z-z_0)^n$ are continuous on the entire complex plane (and, in particular, on any disk) and using Corollary 1.2 of the previous section, we can conclude that the sum of the power series is a continuous function on the disk $|z-z_0|\le r$. In particular, the sum is continuous at the point z_1, because this point belongs to the considered disk: $|z_1-z_0|=r_1<r$. Since z_1 is an arbitrary point in the disk of convergence, the sum of the power series is continuous in this disk. □

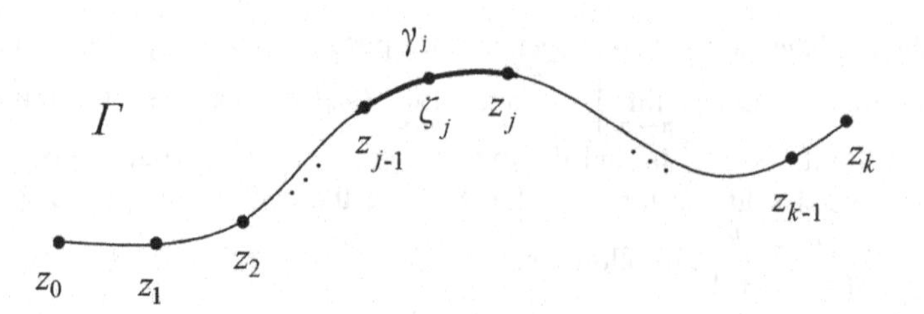

Fig. 1.14 Partition of a curve for the definition of the integral along the curve

1.8 Line Integral of a Complex Function

Let $\Gamma : z = z(t),\ a \le t \le b$ be a simple continuous rectifiable curve on the complex plane, and let $w = f(z)$ be a complex function defined on Γ. We introduce now the *integral of* $f(z)$ *along the curve* Γ (also called the *line integral*) denoted by $\int_\Gamma f(z)\,dz$. In general terms, the construction of this integral follows the traditional fashion: first, we divide the curve into small parts; then in each part we choose some point and calculate the function value at this point; next we multiply this value by some quantity corresponding to the chosen part of the curve and add up all these products; finally, we pass to the limit in the obtained sum as a partition measure approaches 0.

Now we will detail this procedure. To divide the curve Γ into parts, we choose a *partition* $\tau[a, b]$ of the interval $[a, b]$ with the parameter t: $\tau[a, b] = \{t_j\}_{j=0}^{k}$: $a = t_0 < t_1 < t_2 < \ldots < t_k = b$. The corresponding points $z_j = z(t_j)$ of the curve Γ divide this curve into arcs, which we denote by γ_j: each arc γ_j is the part of Γ lying between the points z_{j-1} and z_j (see Fig. 1.14). Next, choose an arbitrary point ζ_j in each arc, $\zeta_j \in \gamma_j$, and find the function value at this point – $f(\zeta_j)$ (the point $\zeta_j \in \gamma_j$ corresponds to the parameter value $\theta_j \in [t_{j-1}, t_j]$). Multiply each value $f(\zeta_j)$ by the complex number $\Delta z_j = z_j - z_{j-1}$ (the difference between the endpoints of the arc γ_j) and add up the products obtained in all the parts: $\sigma_\tau(f) = \sum_{j=1}^{k} f(\zeta_j) \cdot \Delta z_j$. The sum $\sigma_\tau(f)$ is called the *integral sum* of the function $f(z)$.

Since the curve Γ is rectifiable, all the arcs γ_j are also rectifiable. Denote by δ_τ the largest length among all the lengths of the arcs γ_j (since the number of arcs is finite, it is possible to find the largest length). The quantity δ_τ is called the *length of the partition* of Γ: $\delta_\tau = \max\limits_{\forall j=1,\ldots,k} \{l_{\gamma_j}\}$.

Definition If there exists a finite limit of integral sums as the length of the partition approaches 0, and this limit does not depend on the way the partition is chosen neither on the choice of the points ζ_j, then this limit is called the *integral of* $f(z)$ *along the curve* Γ and is denoted by $\int_\Gamma f(z)\,dz$:

$$\lim_{\delta_\tau \to 0} \sigma_\tau(f) = \lim_{\delta_\tau \to 0} \sum_{j=1}^{k} f(\zeta_j) \cdot \Delta z_j = \int_\Gamma f(z)\, dz. \tag{1.18}$$

In a particular case, when a simple continuous rectifiable curve Γ is closed and defined by the parameterization $z = z(t), t \in [a, b]$ (see (1.4)) with the positive (counterclockwise) orientation, the usual notation of the integral of $f(z)$ along Γ is $\oint_\Gamma f(z)\, dz$.

Consider the properties of the integral of $f(z)$ along the curve and its relation to the integrals of real and imaginary parts of $f(z)$.

Properties

Property 1.1 *If $f(z) = u(x, y) + iv(x, y)$ is integrable along a curve then*

$$\int_\Gamma f(z)\, dz = \int_\Gamma u(x, y)\, dx - v(x, y)\, dy + i \int_\Gamma u(x, y)\, dy + v(x, y)\, dx, \tag{1.19}$$

where the integrals on the right-hand side are the second kind line integrals of the real-valued functions.

Proof By the definition of the integral along the curve, one has

$$\int_\Gamma f(z)\, dz = \lim_{\delta_\tau \to 0} \sum_{j=1}^{k} f(\zeta_j) \cdot \Delta z_j.$$

Using notations

$$z = x + iy, \ z_j = x_j + iy_j, \ \zeta_j = \xi_j + i\eta_j; \ f(z) = u(x, y) + iv(x, y),$$

and subsequent relations

$$\Delta z_j = z_j - z_{j-1} = (x_j - x_{j-1}) + i(y_j - y_{j-1}) = \Delta x_j + i\Delta y_j; \ f(\zeta_j) = u(\xi_j, \eta_j) + iv(\xi_j, \eta_j),$$

we can rewrite the definition of the integral separating the real and imaginary parts:

$$\int_\Gamma f(z)\, dz = \lim_{\delta_\tau \to 0} \sum_{j=1}^{k} f(\zeta_j) \cdot \Delta z_j = \lim_{\delta_\tau \to 0} \sum_{j=1}^{k} (u(\xi_j, \eta_j) + iv(\xi_j, \eta_j)) \cdot (\Delta x_j + i\Delta y_j)$$

$$= \lim_{\delta_\tau \to 0} \left(\sum_{j=1}^{k} (u(\xi_j, \eta_j)\Delta x_j - v(\xi_j, \eta_j)\Delta y_j) + i \sum_{j=1}^{k} (u(\xi_j, \eta_j)\Delta y_j + v(\xi_j, \eta_j)\Delta x_j) \right)$$

$$= \lim_{\delta_\tau \to 0} \sum_{j=1}^{k} (u(\xi_j, \eta_j)\Delta x_j - v(\xi_j, \eta_j)\Delta y_j) + i \lim_{\delta_\tau \to 0} \sum_{j=1}^{k} (u(\xi_j, \eta_j)\Delta y_j + v(\xi_j, \eta_j)\Delta x_j).$$

In the last equality, we have also used the result that the existence of the limit of a complex function is equivalent to the existence of the limits of its real and imaginary parts. Noticing that the two sums on the right-hand side are exactly the integral sums for the line integrals along Γ of the second kind (i.e., the integrals along curve Γ with respect to x and y) for the functions $u(x, y)$ and $v(x, y)$, we arrive at formula (1.19). □

Thus, an *integration of* $f(z)$ *along* Γ *is equivalent to integration of* $u(x, y)$ *and* $v(x, y)$ *along* Γ (the latter is made with respect to x and y).

Since the line integrals of the second kind possess the additive property, the immediate consequence of formula (1.19) is the *additive property of the line integral of a complex function*

$$\int_{\Gamma} f(z)\,dz = \int_{\Gamma_1} f(z)dz + \int_{\Gamma_2} f(z)dz,$$

where $\Gamma_1 \cup \Gamma_2 = \Gamma$ and $\Gamma_1 \cap \Gamma_2 = P_0$, P_0 being a single point.

Another direct consequence of relation (1.19) is the change of the sign of a line integral when the orientation of a curve Γ is changed to opposite one

$$\int_{\Gamma^{-1}} f(z)\,dz = -\int_{\Gamma} f(z)dz,$$

where Γ^{-1} is the same curve Γ traversed in the opposite direction.

Property 1.2 *Let* Γ *be a simple piecewise smooth (and consequently rectifiable) curve and* $w = f(z)$ *be a continuous function on* Γ. *In this case, the integral* $\int_{\Gamma} f(z)\,dz$ *exists and can be calculated through a definite integral by the formula*

$$\int_{\Gamma} f(z)\,dz = \int_a^b f(z(t)) \cdot z'(t)\,dt\,. \tag{1.20}$$

Proof In fact, according to the previous property, the existence of the line integral of a complex function is equivalent to the existence of the integrals of the real $u(x, y)$ and imaginary $v(x, y)$ parts of $f(z)$ along the same curve (see formula (1.19)). As shown earlier (in Sect. 1.4), the continuity of $f(z)$ on Γ is equivalent to the continuity of $u(x, y)$ and $v(x, y)$ on Γ. It is known from Real Analysis that the integral of a continuous function along a piecewise smooth curve exists and can be reduced to a definite integral. Therefore, we can conclude that $\int_{\Gamma} f(z)\,dz$ exists. Besides, to reduce this integral to the definite one, we can use relation (1.19) and transform the two line integrals of real-valued functions in this formula to the definite integrals. Since $\Gamma : z = z(t)\,,\ t \in [a, b]$ is a piecewise smooth curve, it means that $z(t)$ in the parameterization (1.4) is differentiable on $[a, b]$. Therefore, we obtain

$$\int_{\Gamma} f(z)\,dz = \int_{\Gamma} u(x, y)\,dx - v(x, y)\,dy + i \int_{\Gamma} u(x, y)\,dy + v(x, y)\,dx$$

$$= \int_a^b \left(u\left(x(t), y(t)\right) \cdot x'(t)\, dt - v\left(x(t), y(t)\right) \cdot y'(t)\, dt\right)$$

$$+ i \int_a^b \left(u\left(x(t), y(t)\right) \cdot y'(t)\, dt + v\left(x(t), y(t)\right) \cdot x'(t)\, dt\right)$$

$$= \int_a^b \left(u\left(x(t), y(t)\right) \cdot \left(x'(t) + iy'(t)\right) + iv\left(x(t), y(t)\right) \cdot \left(x'(t) + iy'(t)\right)\right) dt$$

$$= \int_a^b \left(u\left(x(t), y(t)\right) \cdot + iv\left(x(t), y(t)\right)\right) \cdot z'(t)\, dt = \int_a^b f\left(z(t)\right) \cdot z'(t)\, dt.$$

Thus, under the imposed conditions, *the line integral of a complex function exists and can be calculated by formula* (1.20). □

Property 1.3 *Let Γ be a simple continuous and rectifiable curve and a function $w = f(z)$ be defined on Γ. Then the following inequality*

$$\left| \int_\Gamma f(z)\, dz \right| \leq \int_\Gamma |f(z)|\, |dz|$$

is true if the involved integrals exist. Notice that $z = x + iy$ means that $dz = dx + idy$, and consequently $|dz| = \sqrt{dx^2 + dy^2} = dl$, which is the differential of the arc length (known from Real Analysis).

Proof If both integrals exist, we have the following chain of the inequalities:

$$\left| \int_\Gamma f(z)\, dz \right| = \left| \lim_{\delta_T \to 0} \sum_{j=1}^k f(\zeta_j) \cdot \Delta z_j \right| = \lim_{\delta_T \to 0} \left| \sum_{j=1}^k f(\zeta_j) \cdot \Delta z_j \right| \leq \lim_{\delta_T \to 0} \sum_{j=1}^k |f(\zeta_j)| \cdot |\Delta z_j| = I.$$

Since $\left|\Delta z_j\right|$ is the length of the line segment joining the points z_{j-1} and z_j, it cannot be larger than the length of arc Γ_j between the same points: $\left|\Delta z_j\right| \leq l_{\gamma_j} = \Delta l_j$. Therefore,

$$\left| \int_\Gamma f(z)\, dz \right| \leq I \leq \lim_{\delta_\tau \to 0} \sum_{j=1}^k \left|f(\zeta_j)\right| \cdot \Delta l_j = \int_\Gamma |f(z)|\, dl = \int_\Gamma |f(z)|\, |dz|.$$

Notice that the integral on the right-hand side is the line integral of the first kind (i.e., the integral along the curve Γ with respect to arc length) of the real function $|f(z)|$. □

Property 1.4 *Let Γ be a simple continuous and rectifiable curve. If A is the initial point and B is the final point of Γ, then*

$$\int_\Gamma dz = B - A.$$

Proof In this integral, the function $f(z) \equiv 1$ is continuous, which guarantees the existence of the integral. Calculating the integral by the definition, one obtains the desired result

$$\int_{\Gamma} dz = \lim_{\delta_\tau \to 0} \sum_{j=1}^{k} f(\zeta_j) \cdot \Delta z_j = \lim_{\delta_\tau \to 0} \sum_{j=1}^{k} 1 \cdot (z_j - z_{j-1})$$

$$= \lim_{\delta_\tau \to 0} (z_1 - z_0 + z_2 - z_1 + z_3 - z_2 + \ldots + z_k - z_{k-1})$$

$$= \lim_{\delta_\tau \to 0} (z_k - z_0) = \lim_{\delta_\tau \to 0} (B - A) = B - A.$$

□

Property 1.5 *Let Γ be a simple continuous and rectifiable curve. Then*

$$\int_{\Gamma} |dz| = l_{\Gamma}.$$

Proof We can calculate this integral by the definition, using the property of additivity of the arc length

$$\int_{\Gamma} |dz| = \int_{\Gamma} dl = \lim_{\delta_\tau \to 0} \sum_{j=1}^{k} \Delta l_j = \lim_{\delta_\tau \to 0} \sum_{j=1}^{k} l_{\gamma_j} = \lim_{\delta_\tau \to 0} l_{\Gamma} = l_{\Gamma}.$$

□

Property 1.6 *Let Γ be a simple closed and rectifiable curve defined by the parameterization $z = z(t)$, $t \in [a, b]$ (see (1.4)) with the positive (counterclockwise) orientation. If a function $f(z)$ is continuous on Γ, then the integral along the closed curve Γ can be calculated by the formula*

$$\oint_{\Gamma} f(z)\, dz = \int_{AB} f(z)\, dz + \int_{BA} f(z)\, dz,$$

where A and B are arbitrary points on Γ, the first integral on the right-hand side is the integral along the part of Γ between A and B and the second integral is the integral along the part of Γ between B and A, both traversed in the positive direction.

Proof First, notice that the continuity of $f(z)$ on Γ assures the existence of the integral along Γ. Choose two arbitrary points A and B on Γ. The positive traversal along Γ corresponds to the variation of the parameter t from a to b and is indicated by the arrow in Fig. 1.15. The positive traversal on the part of Γ from A to B is marked by AmB and that on the part from B to A is marked by BpA (see Fig. 1.15).

Let us show that the integral along the closed curve Γ can be defined by the formula

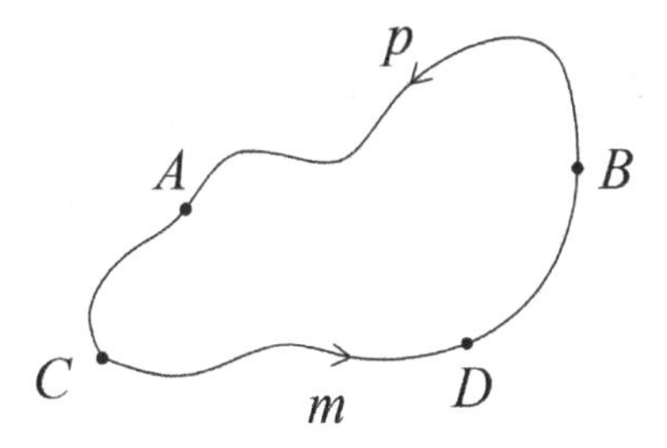

Fig. 1.15 Construction of the integral along a closed curve

$$\oint_{\Gamma} f(z)\,dz = \int_{AmB} f(z)\,dz + \int_{BpA} f(z)\,dz\,.$$

To show that the last formula is true, we should prove that the value of the integral along Γ does not depend on the choice of the points A and B. In fact, taking any other two points C and D on Γ (for example, as it is indicated in Fig. 1.15) and using the additive property of the line integral of a complex function, we obtain

$$\int_{AmB} f(z)\,dz + \int_{BpA} f(z)\,dz = \int_{AC} f(z)\,dz + \int_{CmD} f(z)\,dz + \int_{DB} f(z)\,dz + \int_{BpA} f(z)\,dz$$
$$= \int_{CmD} f(z)\,dz + \int_{DpC} f(z)\,dz\,.$$

Hence, the value of $\oint_{\Gamma} f(z)\,dz$ does not depend on the choice of the points on Γ. □

Property 1.7 ***Theorem on approximation of integral along a curve by integral along a polygonal line.*** *Let Γ be a continuous rectifiable curve contained in a domain D. If a function $f(z)$ is continuous in D, then the integral of $f(z)$ along Γ can be approximated with any desired accuracy by the integral of $f(z)$ along a polygonal line inscribed in Γ and contained in D. Using a formal expression: for $\forall\varepsilon > 0$ there exists a polygonal line $P_\varepsilon \subset D$, P_ε inscribed in Γ (all its vertices belong to Γ) such that*

$$\left|\int_{\Gamma} f(z)\,dz - \int_{P_\varepsilon} f(z)\,dz\right| < \varepsilon\,.$$

Proof First of all, let us determine the distance between two closed sets: the curve Γ and the boundary ∂D of the domain D. Since Γ is a rectifiable curve with the length l_Γ, it implies that Γ is a bounded and closed set. Besides, Γ and ∂D do not have common points ($\Gamma \cap \partial D = \varnothing$), because $\Gamma \subset D$ and D is an open set. Therefore, as it is established in Real Analysis, the distance between the two sets Γ and ∂D exists, positive and is defined by the formula

$$\rho = \rho(\Gamma, \partial D) = \inf_{\forall z' \in \Gamma, \forall z'' \in \partial D} \left|z' - z''\right| > 0\,.$$

Consider together with the curve Γ all the points of the complex plane whose distance to Γ is equal to or less than $\frac{\varrho}{2}$ (that is, place Γ inside a strip) and denote the obtained set by G (see Fig. 1.16). By construction, G is a bounded and closed set (compact), and $\Gamma \subset G \subset D$. Since $f(z)$ is continuous in D, it is also continuous on the compact set G, and consequently, by the Cantor Theorem, $f(z)$ is uniformly continuous on G. According to the definition of the uniform continuity, one has

$$\forall \varepsilon > 0 \left(\text{take } \varepsilon_1 = \frac{\varepsilon}{2l_\Gamma} \right) \exists \delta(\varepsilon) > 0 \,:\, \forall z_1, z_2 \in G\,,\ |z_1 - z_2| < \delta \Rightarrow |f(z_1) - f(z_2)| < \varepsilon_1\,.$$

Pick now any $z \in \Gamma$ and consider the open disk $|\zeta - z| < r$, where $r = \inf\left\{\frac{\delta}{4}, \frac{\varrho}{2}\right\}$. Varying point z, we construct a set of such disks, which contains the curve . Since Γ is a compact set, any open cover of Γ has a finite subcover. In our case, it means that from the original cover of Γ by the infinite set of open disks one can select a finite set of such disks $K_j = \left\{ |z - z_j| < r \right\}$, $j = 1, \ldots, n$, whose union still completely contains Γ: $\Gamma \subset \bigcup_{j=1}^{n} K_j \subset G$. Using the centers z_j of the disks K_j, $j = 1, \ldots, n$, and also the initial point z_0 of Γ (which belongs to the disk K_1) and the final point z_{n+1} of Γ (which belongs to K_n), we construct now the polygonal line joining first the points z_0 and z_1 by the line segment, then joining z_1 and z_2 by the second line segment, and so on, finally joining z_n and z_{n+1} by the last line segment (see Fig. 1.16).

In this manner, we have constructed the polygonal line P_ε consisting of $n + 1$ line segments. It follows from the construction that all the vertices of P_ε belong to Γ and that $P_\varepsilon \subset \bigcup_{j=1}^{n} K_j \subset G$. To show that this polygonal line is the desired one, we should evaluate the following difference:

$$I \equiv \left| \int_\Gamma f(z)\,dz - \int_{P_\varepsilon} f(z)\,dz \right|.$$

Notice that the points z_j divide all the curve Γ in arcs γ_j such that $\Gamma = \bigcup_{j=0}^{n} \gamma_j$, and also divide the polygonal line P_ε in line segments p_j such that $P_\varepsilon = \bigcup_{j=0}^{n} p_j$. Taking into account that

$$\int_{\gamma_j} f(z_j)\,dz = f(z_j) \int_{\gamma_j} dz = f(z_j)(z_{j+1} - z_j)\,,\ \int_{p_j} f(z_j)\,dz = f(z_j) \int_{p_j} dz = f(z_j)(z_{j+1} - z_j)\,,$$

and also using the additive property of the integral along a curve and the Properties 1.3 and 1.4, we obtain:

$$I = \left| \int_\Gamma f(z)\,dz - \int_{P_\varepsilon} f(z)\,dz \right| = \left| \sum_{j=0}^{n} \int_{\gamma_j} f(z)\,dz - \sum_{j=0}^{n} \int_{p_j} f(z)\,dz \right|$$

$$= \left| \sum_{j=0}^{n} \left(\int_{\gamma_j} f(z)\,dz - \int_{p_j} f(z)\,dz \right) \right| = \left| \sum_{j=0}^{n} \left(\int_{\gamma_j} (f(z) - f(z_j))dz - \int_{p_j} (f(z) - f(z_j))dz \right) \right|$$

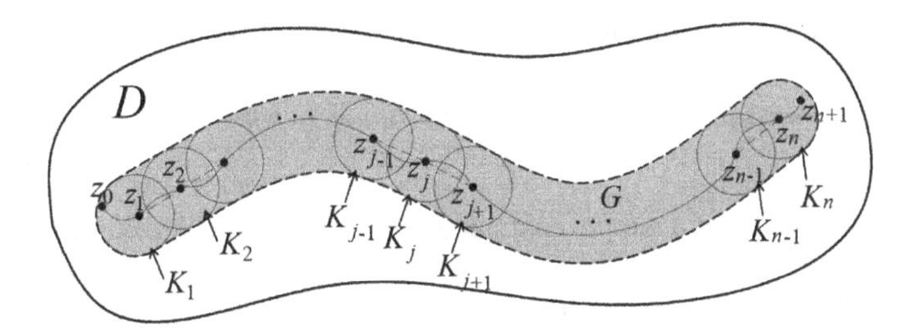

Fig. 1.16 Construction of a polygonal line for approximation of integral along a curve

$$\leq \sum_{j=0}^{n}\left(\int_{\gamma_j}|f(z)-f(z_j)|\,|dz|+\int_{p_j}|f(z)-f(z_j)|\,|dz|\right)\equiv I_1\,.$$

By construction, the point z_j is the center of the disk K_j, and γ_j is the arc of Γ between the points z_j and z_{j+1}. Therefore, for $\forall z \in \gamma_j$ and for $\forall z \in p_j$, we have $|z - z_j| < 4r \leq \delta$ and, by the definition of the uniform continuity, we obtain $|f(z) - f(z_j)| < \varepsilon_1$, $\forall z \in \gamma_j$ and $\forall z \in p_j$. Applying this result to integrals in I_1, taking into account that the constant ε_1 can be put out of the integral and sum, and also using the property 1.5 and the additive property, we can evaluate I_1 in the form

$$I_1 < \varepsilon_1\left(\sum_{j=0}^{n}\int_{\gamma_j}|dz|+\sum_{j=0}^{n}\int_{p_j}|dz|\right)=\varepsilon_1\left(\int_{\Gamma}|dz|+\int_{P_\varepsilon}|dz|\right)=\varepsilon_1\left(l_\Gamma+l_{P_\varepsilon}\right)$$

Recalling that $\varepsilon_1 = \frac{\varepsilon}{2l_\Gamma}$ and that the polygonal line P_ε is inscribed in Γ, which implies that $l_{P_\varepsilon} \leq l_\Gamma$, we arrive at the estimation

$$I_1 < \varepsilon_1\left(l_\Gamma+l_{P_\varepsilon}\right)\leq \varepsilon_1\left(l_\Gamma+l_\Gamma\right)=\frac{\varepsilon}{2l_\Gamma}2l_\Gamma=\varepsilon\,,$$

that finishes the proof. □

Generalized versions of this Theorem are presented in the following two corollaries.

Notice first that the change of the curve or polygonal line in one disk K_j or in two disks-neighbors does not modify the made evaluation, since the inequality $|z' - z''| < 2r < \delta$ (or $|z' - z''| < 4r \leq \delta$) continues to be true, and consequently $|f(z') - f(z'')| < \varepsilon_1$, that is, the proof of the Theorem remains the same. From this observation, we can deduce the following two results.

Corollary 1.1 *(1) If Γ is a simple continuous and rectifiable curve contained in D, then the polygonal line used in the Theorem need not be inscribed, except for the initial and terminal points, which should coincide with the corresponding points of Γ.*

(2) If Γ is a simple continuous closed and rectifiable curve, then there exists a polygonal line such that none of its vertices belongs to Γ, but still the integral along this polygonal line approximates the integral along Γ with any desired degree of accuracy.

Proof (1) In the case of a simple continuous and rectifiable curve contained in D, according to the Theorem, for any given $\varepsilon > 0$ we can first construct the polygonal line $P_\varepsilon^{(1)}$ inscribed in Γ such that $\left|\int_\Gamma f(z)\,dz - \int_{P_\varepsilon^{(1)}} f(z)\,dz\right| < \frac{\varepsilon}{2}$. In the next step, we take the polygonal line $P_\varepsilon^{(1)}$ as the original curve, and using the same value of ε, we construct another polygonal line $P_\varepsilon^{(2)}$ such that $\left|\int_{P_\varepsilon^{(1)}} f(z)\,dz - \int_{P_\varepsilon^{(2)}} f(z)\,dz\right| < \frac{\varepsilon}{2}$. Therefore,

$$\left|\int_\Gamma f(z)\,dz - \int_{P_\varepsilon^{(2)}} f(z)\,dz\right| \le \left|\int_\Gamma f(z)\,dz - \int_{P_\varepsilon^{(1)}} f(z)\,dz\right| + \left|\int_{P_\varepsilon^{(1)}} f(z)\,dz - \int_{P_\varepsilon^{(2)}} f(z)\,dz\right|$$
$$< \frac{\varepsilon}{2} + \frac{\varepsilon}{2} = \varepsilon\,.$$

Since $P_\varepsilon^{(2)}$ is inscribed in $P_\varepsilon^{(1)}$, its vertices belongs to $P_\varepsilon^{(1)}$, but not necessarily to Γ.

(2) If Γ is a simple continuous closed and rectifiable curve, then we start with the constructed above polygonal line $P_\varepsilon^{(2)}$, whose vertices do not belong to Γ, except for the initial z_0 and final z_{n+1} points. Then denote the initial disk (that contains z_0) by K_0 and take any two points of $P_\varepsilon^{(2)}$, which belong to K_0 and are different from z_0. Connect these two points by a line segment and keep the remaining part of $P_\varepsilon^{(2)}$ unchanged. This gives a new polygonal line $\tilde{P}_\varepsilon$. Since we have already shown that $\left|\int_{P_\varepsilon^{(2)}} f(z)\,dz - \int_{\tilde{P}_\varepsilon} f(z)\,dz\right| < \frac{\varepsilon}{2}$, it follows that

$$\left|\int_\Gamma f(z)\,dz - \int_{\tilde{P}_\varepsilon} f(z)\,dz\right| \le \left|\int_\Gamma f(z)\,dz - \int_{P_\varepsilon^{(2)}} f(z)\,dz\right| + \left|\int_{P_{\varepsilon(2)}} f(z)\,dz - \int_{\tilde{P}_\varepsilon} f(z)\,dz\right|$$
$$< \frac{\varepsilon}{2} + \frac{\varepsilon}{2} = \varepsilon\,.$$

Notice that none of the vertices of $\tilde{P}_\varepsilon$ belong to Γ. □

Corollary 1.2 *Let a domain D be bounded by a finite number of piecewise smooth curves with folds. If a function $w = f(z)$ is continuous on D up to its boundary, then the integral of $f(z)$ along the boundary ∂D can be approximated with any desired accuracy by the integral along a set of polygonal lines contained in D.*

Proof As was shown in Sect. 1.5, a continuous on D up to ∂D function $f(z)$ can be defined on ∂D in a continuous way. If the boundary ∂D consists of m components $\Gamma_j,\ j = 1, \ldots, m$, each of which is a piecewise smooth curve with folds, then applying the Theorem and Corollary 1.1 to each component separately, we can construct for each Γ_j and $\forall \varepsilon > 0$ a polygonal line $P_j \subset D$ such that $\left|\int_{\Gamma_j} f(z)\,dz - \int_{P_j} f(z)\,dz\right| < \frac{\varepsilon}{m}$. Then the union of the obtained polygonal lines

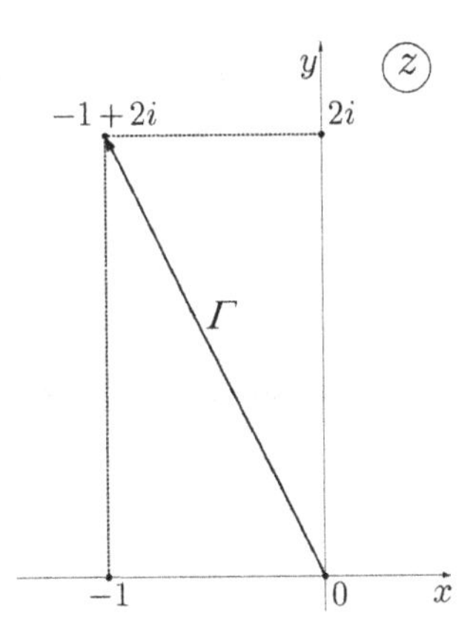

Fig. 1.17 The integration path Γ

$P_\varepsilon = \bigcup_{j=1}^{m} P_j$ is also contained in D and the following evaluation holds:

$$\left|\int_{\partial D} f(z)\,dz - \int_{P_\varepsilon} f(z)\,dz\right| = \left|\sum_{j=1}^{m}\left(\int_{\Gamma_j} f(z)\,dz - \int_{P_j} f(z)\,dz\right)\right|$$

$$\leq \sum_{j=1}^{m}\left|\int_{\Gamma_j} f(z)\,dz - \int_{P_j} f(z)\,dz\right| < \frac{\varepsilon}{m}\cdot m = \varepsilon.$$

This completes the proof of Corollary 1.2. □

Notice that the boundary ∂D in the integral $\int_{\partial D} f(z)\,dz$ is traversed in the positive sense, that is, the domain D always lies on the left with respect to the moving point.

We finalize this section with an example of the evaluation of a line integral.

Example Evaluate the integral $\int_\Gamma (\bar{z}-1)^2\,dz$ along the curve Γ defined as the straight-line segment from the origin to the point $-1+2i$ (see Fig. 1.17).

We start with the equation of the straight-line segment Γ, which can be given in the form $y=-2x$, with x varying from 0 to -1, which corresponds to the straight-line move from the point $(0,0)$ to the point $(-1,2)$. In the complex form, it means that $z = x - 2ix$ and consequently $\bar{z} = x + 2ix$, $dz = (1-2i)\,dx$. Substituting this expressions in the line integral, we get

$$\int_\Gamma (\bar{z}-1)^2\,dz = \int_0^{-1} ((x-1)+2ix)^2\,(1-2i)\,dx$$

$$= (1-2i)\int_0^{-1}\left[\left(x^2-2x+1-4x^2\right)+4i\left(x^2-x\right)\right]dx$$

$$= (1-2i)\left[\left(-x^3-x^2+x\right)+4i\left(\frac{x^3}{3}-\frac{x^2}{2}\right)\right]\Bigg|_0^{-1} = (1-2i)\left[(1-1-1)+4i\left(-\frac{1}{3}-\frac{1}{2}\right)\right]$$

$$= (1-2i)\left(-1-\frac{10i}{3}\right) = -1-\frac{20}{3}+i\left(2-\frac{10}{3}\right) = -\frac{23}{3}-\frac{4i}{3}.$$

1.9 Integrals Depending on a Parameter

In this section, we consider integrals depending on a parameter and their properties.

Definition Suppose $f(z, w)$ is a function of two complex variables defined on $\Gamma \times E$ $(z \in \Gamma,\ w \in E)$, where Γ is a continuous rectifiable curve in $\mathbb{C}$ and E is a subset of $\mathbb{C}$. If the integral $\int_\Gamma f(z, w)\, dz = F(w)$ exists for each $w \in E$, then it is called the *integral depending on a parameter.*

Theorem 1.1 *Let $f(z, w)$ be a function defined on $\Gamma \times E$ and w_0 be a limit point of E. If*
(1) $f(z, w)$ is continuous in z on Γ for any fixed $w \in E$;
(2) $f(z, w) \overset{\Gamma}{\underset{w\to w_0}{\rightrightarrows}} \varphi(z)$,
then

$$\lim_{w\to w_0} \int_\Gamma f(z, w)\, dz = \int_\Gamma \varphi(z)\, dz = \int_\Gamma \lim_{w\to w_0} f(z, w)\, dz. \tag{1.21}$$

Thus, under the given conditions, it is possible to pass to the limit inside the integral depending on a parameter, or, in other words, it is possible to interchange the limit and integration signs.

Proof First, we show that the two integrals considered in the Theorem exist. Indeed, since $f(z, w)$ is continuous in z on Γ (for each fixed $w \in E$) and Γ is a continuous rectifiable curve, the integral $\int_\Gamma f(z, w)\, dz$ exists for each fixed $w \in E$. Also, since $f(z, w) \overset{\Gamma}{\underset{w\to w_0}{\rightrightarrows}} \varphi(z)$, by the Theorem on the continuity of the limit function (see Sect. 1.6), the function $\varphi(z)$ is continuous on Γ, and consequently, the integral $\int_\Gamma \varphi(z)\, dz$ also exists.

Now let us prove the equality between the left-hand and right-hand sides in (1.21). For this, we consider the difference between the two integrals

$$I = \left| \int_\Gamma f(z, w)\, dz - \int_\Gamma \varphi(z)\, dz \right|.$$

Recalling the definition of the uniform convergence $f(z, w) \overset{\Gamma}{\underset{w\to w_0}{\rightrightarrows}} \varphi(z)$, we have that for $\forall \varepsilon > 0$ (take $\varepsilon_1 = \frac{\varepsilon}{l_\Gamma}$, where l_Γ is the length of Γ) there exists $\delta = \delta(\varepsilon) > 0$

such that for $\forall w \in E,\ 0 < |w - w_0| < \delta$ and simultaneously for $\forall z \in \Gamma$ it follows that

$$|f(z, w) - \varphi(z)| < \varepsilon_1. \tag{1.22}$$

Using the properties of the line integral and the last inequality, we obtain

$$I = \left| \int_\Gamma f(z, w)\, dz - \int_\Gamma \varphi(z)\, dz \right| = \left| \int_\Gamma (f(z, w) - \varphi(z))\, dz \right|$$

$$\leq \int_\Gamma |f(z, w) - \varphi(z)|\, |dz| < \varepsilon_1 \int_\Gamma |dz| = \varepsilon_1 \cdot l_\Gamma = \varepsilon.$$

Thus, the Theorem is proved. □

Corollary 1.1 *Let $\{f_n(z)\}_{n=1}^\infty$ be a sequence of functions defined on a continuous rectifiable curve Γ. If*
(1) the functions $f_n(z)$ are continuous on Γ (for $\forall n \in \mathbb{N}$);
(2) $f_n(z) \underset{n\to\infty}{\overset{\Gamma}{\rightrightarrows}} \varphi(z)$,
then

$$\lim_{n\to\infty} \int_\Gamma f_n(z)\, dz = \int_\Gamma \varphi(z)\, dz = \int_\Gamma \lim_{n\to\infty} f_n(z)\, dz,$$

that is, one can pass to the limit under the integral sign, or equivalently, one can interchange the limit and integration signs.

This result is a particular case of the last Theorem with $E = \mathbb{N}$.

Corollary 1.2 *A series of continuous functions can be integrated term-by-term along a continuous rectifiable curve, if the series converges uniformly on this curve.*

Proof This statement can be reduced to that of Corollary 1.1. If the functions $u_n(z)$ in the series $\sum_{n=1}^{+\infty} u_n(z)$ are continuous on Γ, then the partial sums $f_n(z) = \sum_{k=1}^{n} u_k(z)$ are also continuous on Γ. Besides, the series converges uniformly on Γ which implies that the sum of the series $\varphi(z) = \sum_{n=1}^{+\infty} u_n(z)$ is also continuous on Γ (see Corollary 2 in Sect. 1.6). Therefore, using Corollary 1.1 and the linear property of the integral applied to a finite sum, we have:

$$\int_\Gamma \sum_{k=1}^{+\infty} u_k(z)\, dz = \int_\Gamma \varphi(z)\, dz = \int_\Gamma \lim_{n\to\infty} f_n(z)\, dz = \lim_{n\to\infty} \int_\Gamma f_n(z)\, dz$$

$$= \lim_{n\to\infty} \int \sum_{k=1}^{n} u_k(z)\, dz = \lim_{n\to\infty} \sum_{k=1}^{n} \int_\Gamma u_k(z)\, dz = \sum_{k=1}^{+\infty} \int_\Gamma u_k(z)\, dz.$$

This completes the proof. □

Corollary 1.3 *A power series can be integrated term-by-term along any continuous rectifiable curve contained inside its circle of convergence.*

This Corollary is a particular case of Corollary 1.2, since a power series consists of continuous functions and converges uniformly inside the circle of convergence.

Theorem 1.2 *Let Γ be a continuous rectifiable curve and E be a compact set in $\mathbb{C}$. If a function $f(z, w)$ is continuous on $\Gamma \times E$, then the function $F(w) = \int_\Gamma f(z, w)\, dz$ is continuous on E.*

Proof First, notice that the set $\Gamma \times E$ is compact, since Γ is a compact set (by being a continuous rectifiable curve) and E is also compact (by the condition of the Theorem). Since $f(z, w)$ is a continuous function on the compact set $\Gamma \times E$, by the Cantor Theorem this function is uniformly continuous on $\Gamma \times E$, that is for $\forall \varepsilon > 0$ (choose $\varepsilon_1 = \frac{\varepsilon}{l_\Gamma}$, where l_Γ is the length of the curve Γ) $\exists \delta = \delta(\varepsilon) > 0$ such that

$$\text{for } \forall z, z' \in \Gamma,\ \left|z - z'\right| < \delta \text{ and for } \forall w, w' \in E,\ \left|w - w'\right| < \delta \Rightarrow \left|f(z, w) - f(z', w')\right| < \varepsilon_1 .$$

Let us pick any point $w_0 \in E$ and show that the function $F(w)$ is continuous at that point. To this end, we consider the points of E, which belong to the δ-neighborhood of w_0 (i.e., $|w - w_0| < \delta$), where δ is the value found in the definition of the uniform continuity of $f(z, w)$, and evaluate the difference in the definition of the continuity of $F(w)$

$$|F(w) - F(w_0)| = \left|\int_\Gamma f(z, w)\, dz - \int_\Gamma f(z, w_0)\, dz\right| = \left|\int_\Gamma (f(z, w) - f(z, w_0))\, dz\right|$$
$$\leq \int_\Gamma |f(z, w) - f(z, w_0)|\, |dz| \equiv I_1 .$$

According to the definition of the uniform continuity, $|f(z, w) - f(z, w_0)| < \varepsilon_1$ if $z' = z$, $\forall z \in \Gamma$ (in this case $\left|z - z'\right| = 0 < \delta$) and $w \in E$, $|w - w_0| < \delta$. Therefore,

$$|F(w) - F(w_0)| \leq I_1 < \varepsilon_1 \int_\Gamma |dz| = \varepsilon_1 \cdot l_\Gamma = \frac{\varepsilon}{l_\Gamma} l_\Gamma = \varepsilon,$$

which shows that $F(w)$ is continuous at w_0. Since w_0 is an arbitrary point of E, it means that $F(w)$ is continuous on E. □

Theorem 1.3 *Let Γ and Υ be continuous rectifiable curves. If $f(z, w)$ is continuous on $\Gamma \times \Upsilon$, then*

$$\int_\Upsilon \int_\Gamma f(z, w)\, dz dw = \int_\Gamma \int_\Upsilon f(z, w)\, dw dz, \tag{1.23}$$

that is, the order of the integration is interchangeable.

Proof Since $\Gamma \times \Upsilon$ is a compact set, by Theorem 1.2, the continuity of $f(z, w)$ on $\Gamma \times \Upsilon$ implies the continuity of $F(w) = \int_\Gamma f(z, w)\, dz$ on Υ and the continuity of $G(z) = \int_\Upsilon f(z, w)\, dw$ on Γ. Consequently, both integrals $\int_\Upsilon F(w)\, dw = \int_\Upsilon \int_\Gamma f(z, w)\, dzdw$ and $\int_\Gamma G(z)\, dz = \int_\Gamma \int_\Upsilon f(z, w)\, dwdz$ exist. It remains to show that the two integrals are equal.

By the Cantor Theorem, the continuity of $f(z, w)$ on the compact set $\Gamma \times \Upsilon$ guarantees its uniform continuity on the same set. Specifying the definition of the uniform continuity in this case, we have: for $\forall \varepsilon > 0$ (choose $\varepsilon_1 = \frac{\varepsilon}{l_\Gamma \cdot l_\Upsilon}$, where l_Γ and l_Υ are the lengths of the curves Γ and Υ, respectively) $\exists \delta = \delta(\varepsilon) > 0$ such that for $\forall z', z'' \in \Gamma$, $|z' - z''| < \delta$ and for $\forall w', w'' \in \Upsilon$, $|w' - w''| < \delta$ it follows that $|f(z', w') - f(z'', w'')| < \varepsilon_1$. Let us divide the curve Υ in small arcs Υ_k whose lengths are less than δ (where δ is from the definition of the uniform continuity). For instance, starting with the initial point w_0 of Υ, we take the arc Υ_1 of the length $\frac{\delta}{2}$ and mark its endpoint by w_1, then we choose the next arc Υ_2 of the same length starting at the point w_1 and so on. Since Υ is rectifiable, the number of such arcs Υ_k is finite: $\Upsilon = \bigcup_{k=1}^{n} \Upsilon_k$. By the definition of the line integral, one has

$$\int_\Upsilon \int_\Gamma f(z, w)\, dzdw = \int_\Upsilon F(w)\, dw = \lim_{\delta_\tau \to 0} \sum_{k=1}^{n} F(\theta_k) \cdot \Delta w_k, \tag{1.24}$$

where $\Delta w_k = w_k - w_{k-1}$ and θ_k is an arbitrary point of Υ_k: $\theta_k \in \Upsilon_k$ $(k = 1, \ldots, n)$. Then we can prove the validity of (1.23) by showing that the right-hand side in (1.24) is equal to $\int_\Gamma \int_\Upsilon f(z, w)\, dwdz$. This can be made by evaluating the difference between the corresponding sum and the integral (the same technique that we have used previously). Recalling that $F(w) = \int_\Gamma f(z, w)\, dz$ and that the points w_k divide the entire curve Υ in the arcs Υ_k, we obtain

$$I \equiv \left| \sum_{k=1}^{n} F(\theta_k) \cdot \Delta w_k - \int_\Gamma \int_\Upsilon f(z, w)\, dwdz \right|$$

$$= \left| \sum_{k=1}^{n} \Delta w_k \cdot \int_\Gamma f(z, \theta_k)\, dz - \sum_{k=1}^{n} \int_\Gamma \int_{\Upsilon_k} f(z, w)\, dwdz \right|. \tag{1.25}$$

Bringing $\Delta w_k = w_k - w_{k-1}$ inside the first integral in (1.25) and representing $w_k - w_{k-1}$ in the integral form $w_k - w_{k-1} = \int_{\Upsilon_k} dw$ (by Property 1.4 of the integral along a curve in Sect. 1.8), we can transform the general term of the first sum in (1.25) to the form $\Delta w_k \cdot \int_\Gamma f(z, \theta_k)\, dz = \int_\Gamma f(z, \theta_k) \cdot (w_k - w_{k-1})\, dz = \int_\Gamma f(z, \theta_k) \cdot \int_{\Upsilon_k} dwdz$. Using additionally the fact that $f(z, \theta_k)$ does not depends on w, we can bring it inside the integral in w and evaluate (1.25) in the following form:

$$I = \left| \sum_{k=1}^{n} \int_\Gamma \int_{\Upsilon_k} f(z, \theta_k)\, dwdz - \sum_{k=1}^{n} \int_\Gamma \int_{\Upsilon_k} f(z, w)\, dwdz \right|$$

$$= \left| \sum_{k=1}^{n} \int_{\Gamma} \int_{\Upsilon_k} (f(z, \theta_k) - f(z, w))\, dw dz \right|$$

$$\leq \sum_{k=1}^{n} \int_{\Gamma} \int_{\Upsilon_k} |f(z, \theta_k) - f(z, w)|\, |dw|\, |dz| \equiv I_1. \tag{1.26}$$

In the integral along Υ_k, one has $\theta_k,\ w \in \Upsilon_k$, and consequently $|\theta_k - w| \leq l_{\Upsilon_k} = \frac{\delta}{2} < \delta$. Using the definition of the uniform continuity with $z' = z'' = z, \forall z \in \Gamma, w' = \theta_k,\ w'' = w$ (where $\theta_k,\ w \in \Upsilon_k$, $k = 1, \ldots, n$), we obtain $|f(z, \theta_k) - f(z, w)| < \varepsilon_1$ for $\forall w \in \Upsilon_k, \forall z \in \Gamma$. Applying the last inequality and the properties of a line integral to the evaluation of I_1 in (1.26), we get

$$I \leq I_1 < \varepsilon_1 \cdot \sum_{k=1}^{n} \int_{\Gamma} \int_{\Upsilon_k} |dw|\, |dz| = \varepsilon_1 \cdot \int_{\Gamma} \sum_{k=1}^{n} \int_{\Upsilon_k} |dw|\, |dz| = \varepsilon_1 \cdot \int_{\Gamma} \int_{\Upsilon} |dw|\, |dz|$$

$$= \varepsilon_1 \cdot \int_{\Gamma} l_{\Upsilon}\, |dz| = \varepsilon_1 l_{\Upsilon} l_{\Gamma} = \varepsilon.$$

Thus, the Theorem is proved. □

In this chapter, we have considered some properties of complex functions, such as passage to the limit, continuity, and integrability, which are similar to those of real functions. In the remaining parts of this text, we will study the properties of complex functions, which have no analogous in Real Analysis.

Exercises

1. Perform the arithmetic operations and represent the answer in the form of a single complex number:

   (1) $\left(\frac{1+i}{1-i}\right)^5$;
   (2) $\frac{(2-i)^6}{(1+i)^4}$;
   (3) $\left(\frac{3+i^{18}}{2-i^{15}}\right)^2$;
   (4) $\left(\frac{1}{\sqrt{2}} - i\frac{1}{\sqrt{2}}\right)^6$.

2. Find all the roots:

   (1) $\sqrt[5]{\sqrt{3} - i}$;
   (2) $\sqrt[4]{\sqrt{3} + i}$;
   (3) $\sqrt[4]{1}$;
   (4) $\sqrt[6]{i}$;

(5) $\sqrt[6]{-1}$;
(6) $\sqrt{1-i\sqrt{3}}$;
(7) $\sqrt[3]{-1-i}$;
(8) $\sqrt[4]{-3+4i}$.

3. Show that the presented equalities are true for any complex numbers z_1 and z_2:

(1) $|z_1+z_2|^2+|z_1-z_2|^2=2|z_1|^2+2|z_2|^2$;
(2) $|z_1\bar{z}_2-1|^2-|z_1-z_2|^2=\left(|z_1|^2-1\right)\left(|z_2|^2-1\right)$.
Hint: use the identity $z\cdot\bar{z}=|z|^2$.

4*. Show that the following formula is true for all the complex numbers z:

$$\left|\sqrt{z^2-1}+z\right|+\left|\sqrt{z^2-1}-z\right|=|z-1|+|z+1|\ .$$

5. Find the set of points defined by the given equation or inequality and sketch this set on the complex plane:

(1) $|z-i|=2$;
(2) $|z-i|>2$;
(3) $|z-i|<2$;
(4) $|z+2|+|z-2|=5$;
(5) $|z+2|-|z-2|=3$;
(6) $\mathrm{Re}\,(z\,(1-i))<\sqrt{2}$;
(7*) $\mathrm{Re}\,z^4<\mathrm{Im}\,z^4$.

6. Sketch the following set of curves ($\forall c\in\mathbb{R}$):

(1) $\mathrm{Im}\,\frac{1}{z}=c$;
(2) $\mathrm{Re}\,z^2=c$;
(3) $\mathrm{Im}\,z^2=c$;
(4) $\mathrm{Re}\,\frac{1}{z-i}=c$;
(5) $\mathrm{Im}\,\frac{1}{z-i}=c$.

7. Find the images of the lines $x=c\,,\ y=c$ ($\forall c\in\mathbb{R}$), representing the rectangular net, for the following functions:

(1) $w=z^2$;
(2) $w=\bar{z}^2$;
(3) $w=\frac{1}{z}$.

8. Suppose the three complex numbers $z_1,\ z_2,\ z_3$ satisfy the equations $z_1+z_2+z_3=0$, and $|z_1|=|z_2|=|z_3|=1$. Show that these numbers are the vertices of an equilateral triangle inscribed in the unit circle.
9. Sketch the following curves defined in the parametric form on the complex plane:

(1) $z=t+it^2\,,\ t\in\mathbb{R}$;
(2) $z=t^2+it^4\,,\ t\in\mathbb{R}$;

(3) $z = \cos t + i \sin t\,,\ t \in [0, 2\pi]\,;$
(4) $z = \cos t + i \sin t\,,\ t \in [0,\ \pi/2]\,;$
(5) $z = \sqrt{1-t^2} + it\,,\ t \in [0, 1]\,;$
(6) $z = t + \frac{i}{t}\,,\ t \in (0, +\infty)\,;$
(7) $z = t + 2i\,,\ t \in [0, +\infty)\,;$
(8) $z = te^{i\frac{\pi}{4}} + \frac{1}{t}e^{-i\frac{\pi}{4}}\,,\ t \in (0, +\infty)\,.$

10. Analyze where the given function is continuous:
(1) $f(z) = \frac{\mathrm{Im}\,(z-2i)}{z-2i}\,;$
(2) $f(z) = \frac{z^2}{|z|^2}\,;$
(3) $f(z) = \frac{|z|^2}{z}\,;$
(4) $f(z) = \frac{z\mathrm{Re}\,z}{|z|}\,.$

11. Analyze whether the numerical series is convergent or divergent:
(1) $\sum_{n=1}^{+\infty} \frac{i^n}{n^2}\,;$
(2) $\sum_{n=1}^{+\infty} \frac{i^n}{n^n}\,;$
(3) $\sum_{n=1}^{+\infty} \frac{i^n}{\sqrt[n]{n}}\,;$
(4) $\sum_{n=1}^{+\infty} \frac{i^n}{n}\,;$
(5*) $\sum_{n=1}^{+\infty} \frac{e^{i\frac{n\pi}{4}}}{n}\,;$
(6) $\sum_{n=1}^{+\infty} \frac{\cos in}{e^n}\,.$

12. Let $\sum_{n=1}^{+\infty} c_n$ and $\sum_{n=1}^{+\infty} c_n^2$ be convergent series. Show that if $\mathrm{Re}\, c_n \geq 0\,,\ \forall n \in \mathbb{N}$, then the series $\sum_{n=1}^{+\infty} |c_n|^2$ is also convergent.

13. Find the radius of convergence of the power series:
(1) $\sum_{n=0}^{+\infty} (2 + i^n)\, z^n\,;$
(2) $\sum_{n=0}^{+\infty} (2 + i^n)^n\, z^n\,;$
(3) $\sum_{n=0}^{+\infty} 3^n z^{2n}\,;$
(4) $\sum_{n=0}^{+\infty} e^{in} z^{3n}\,;$
(5) $\sum_{n=1}^{+\infty} \frac{n!}{n^n} z^n\,;$
(6) $\sum_{n=0}^{+\infty} \left(1 + (-1)^n\right)^n z^{5n}\,;$
(7) $\sum_{n=0}^{+\infty} (\ln(n+2))^k z^n\,;$
(8) $\sum_{n=0}^{+\infty} 5^n z^{n!}\,;$
(9) $\sum_{n=1}^{+\infty} \frac{(2n)!}{(n!)^2} z^n\,.$

14*. Let $R_a\,,\ R_b,\ R_{a+b}$ and R_{ab} be the radii of convergence of the power series $\sum_{n=0}^{+\infty} a_n z^n,\ \sum_{n=0}^{+\infty} b_n z^n,\ \sum_{n=0}^{+\infty} (a_n + b_n)\, z^n$ and $\sum_{n=0}^{+\infty} a_n b_n z^n$, respectively. Show that

$$R_{a+b} \geq \min(R_a, R_b)\ ,\ R_{ab} \geq R_a \cdot R_b.$$

15. Evaluate the integrals along the given curves:
(1) $\int_\Gamma z\mathrm{Re}\, z\, dz$, where Γ is the radius-vector of the point $3 - i$;
(2) $\int_\Gamma z\mathrm{Re}\, z\, dz$, where Γ is the half-circle: $\begin{cases} |z| = 2 \\ -\pi/2 \leq \arg z \leq \pi/2 \end{cases};$

(3) $\int_\Gamma z\text{Im}\, z\, dz$, where Γ is the radius-vector of the point $-2+5i$;
(4) $\int_\Gamma z^2 dz$, where Γ is the radius-vector of the point $-1-5i$;
(5) $\int_\Gamma \text{Re}\, z^2 dz$, where Γ is the radius-vector of the point $-2+3i$;
(6) $\int_\Gamma \bar{z}^3 dz$, where Γ is the arc of the circle: $\begin{cases} |z| = 3 \\ -\pi/4 \le \arg z \le 0 \end{cases}$;
(7) $\int_\Gamma \text{Im}\, z^2 dz$, where Γ is the arc of the circle: $\begin{cases} |z| = 5 \\ \pi/4 \le \arg z \le \pi/2 \end{cases}$;
(8) $\int_\Gamma z^2 dz$, where Γ is a closed curve with fold, which consists of the unit circle $|z| = 1$ and the segment $[0, 1]$ of the real axis traversed twice in the opposite directions;
(9) $\int_\Gamma \sqrt{z} dz$, where Γ is a closed curve with fold, which consists of the unit circle $|z| = 1$ and the segment $[0, 1]$ of the real axis traversed twice in the opposite directions.

16. Prove that a convergent series $\sum_{n=1}^{+\infty} c_n$ with $|\arg c_n| \le \alpha < \frac{\pi}{2}$ converges absolutely.
17. Prove that the function $f(z) = e^{-\frac{1}{|z|}}$ is uniformly continuous on the set $0 < |z| \le R, \forall R > 0$.
18. Show that the function $f(z) = e^{-\frac{1}{z^2}}$ is:

(1) continuous, but not uniformly continuous on $D = \{0 < |z| \le R\}, \forall R > 0$;
(2) uniformly continuous on the sector $G = \{0 < |z| \le R, |\arg z| \le \frac{\pi}{6}\}, \forall R > 0$.
Hint for (2): use the Cantor theorem on uniform continuity.

19*. Let $f(z) = e^{\frac{i}{z}}$ and $D = \{0 < |z| \le 1, -\pi \le \arg z \le 0\}$. Prove that $f(z)$ is:

(1) bounded in $\overline{D}$, but not continuous in $\overline{D}$;
(2) continuous, but not uniformly continuous in D;
(3) uniformly continuous on the sector $G = \{0 < |z| \le 1, \alpha - \pi \le \arg z \le -\alpha\}$, $\alpha \in \left(0, \frac{\pi}{2}\right)$.
Hint for (3): use the Cantor theorem on uniform continuity.

20. Suppose $f(z)$ has a power series expansion $f(z) = \sum_{n=0}^{+\infty} c_n z^n$ in the disk $|z| < R$. Show that $\frac{1}{2\pi}\int_0^{2\pi} |f(re^{i\varphi})|^2 d\varphi = \sum_{n=0}^{+\infty} |c_n|^2 r^{2n}, \forall r \in (0, R)$.

Chapter 2
Analytic Functions and Their Properties

This chapter is devoted to the study of analytic functions and their properties. In Sect. 2.1, we introduce the first definition that radically separates the behavior of real and complex functions—the definition of complex-differentiability. In spite of the formal similarity between the definitions of real- and complex-differentiability, the presented elementary examples show that complex-differentiability is a much stricter requirement on complex function than just differentiability of its real and imaginary parts. The concepts of holomorphy and analyticity of a function are also considered in this section, and the first step in the investigation of the fundamental question of the relationship between differentiability and analyticity is made: it is shown that a function analytic at a point is holomorphic at this point.

In Sects. 2.2 and 2.3, we study Cauchy's and Goursat's theorems, and establish the Cauchy integral formula for holomorphic functions. The first two theorems are similar to the well-known result in Real Analysis: a line integral (along a piecewise smooth curve) of a continuous conservative vector field defined in a connected domain is independent of path. However, for a complex function an analogous property holds under the much weaker assumption of complex-differentiability in a finite simply connected domain. Further, the Cauchy integral formula shows that the values of a smooth complex function inside a domain are completely determined by its values on the boundary of this domain. This is one more striking feature of complex functions, which clearly differs from the behavior of real functions. Besides the fact that these results are important in their own right, they are also useful for evaluation of some types of line integrals and are essential part for the proof of the fundamental statement on the equivalence of differentiability and analyticity of a complex function in an arbitrary domain (the main criterion for analyticity) derived in Sect. 2.4. Then, in Sect. 2.5, we consider harmonic functions and prove one more criterion for analyticity: a function is analytic in a domain if and only if its real and imaginary parts are conjugate harmonic functions in a considered domain.

We continue this chapter by introducing an indefinite integral of an analytic function in Sect. 2.6 and demonstrating that in a finite simply connected domain, all the

A. Bourchtein and L. Bourchtein, *Complex Analysis*, Hindustan Publishing Corporation,
https://doi.org/10.1007/978-981-15-9219-5_2

primitives differ by a constant. Different properties of analytic functions are considered in Sect. 2.7, such as Morera's theorem (the converse to Cauchy's theorem), analyticity of a limit function of uniformly convergent family of analytic functions, and analyticity of integral depending on a parameter. The last three sections are concerned with analytic continuation. The theorem on uniqueness of an analytic function determined in a small part of the domain of analyticity and the concept of analytic continuation are used to extend the definition of the three elementary functions (e^x, $\cos x$ and $\sin x$) from the real axis to the entire complex plane. (An extensive investigation of these and other elementary functions will be performed in Chap. 4.) The procedure of a general construction of analytic continuation is described, including the definition of single-valued and multi-valued analytic functions on extended domains. A preliminary discussion of the Riemann surfaces is also provided.

2.1 Differentiable and Analytic Functions

Definition 2.1 Let $f(z)$ be defined in a neighborhood of a point z_0. The function $f(z)$ is *differentiable at the point* z_0 if there exists the finite limit

$$\lim_{\Delta z\to 0}\frac{f(z_0+\Delta z)-f(z_0)}{\Delta z}=f'(z_0), \tag{2.1}$$

which is called the *derivative* of $f(z)$ at z_0.

Remark As it was indicated in Sect. 1.4, the limit of a complex function possesses all the algebraic properties of the limits of real functions. For this reason, the existence of the limit in the definition of a differentiable function can also be expressed as follows:

$$\lim_{\Delta z\to 0}\frac{f(z_0+\Delta z)-f(z_0)}{\Delta z}=f'(z_0)\;\Leftrightarrow\;\frac{f(z_0+\Delta z)-f(z_0)}{\Delta z}=f'(z_0)+\alpha(\Delta z),$$

where $\alpha(\Delta z)\underset{\Delta z\to 0}{\to}0$, that is, $\alpha(\Delta z)$ is an infinitesimal function (or simply infinitesimal) as $\Delta z\to 0$. Therefore, we obtain one more *equivalent form of the definition of a differentiable function* at the point z_0

$$\Delta f(z_0)\equiv f(z_0+\Delta z)-f(z_0)=f'(z_0)\cdot\Delta z+\alpha(\Delta z)\cdot\Delta z. \tag{2.2}$$

The equality (2.2) can be written in a different form. Notice that $|\Delta z|=\sqrt{\Delta x^2+\Delta y^2}\equiv\rho$, and consequently, $\alpha(\Delta z)\cdot\Delta z=o(\rho)$, since $\alpha(\Delta z)$ is an infinitesimal as $\Delta z\to 0$. Then, the definition of a differentiable function takes the following form:

$$\Delta f(z_0)=f'(z_0)\cdot\Delta z+o(\rho), \tag{2.3}$$

where $\rho = \sqrt{\Delta x^2 + \Delta y^2} = |\Delta z|$. The formulations (2.1), (2.2) and (2.3) are equivalent, and each of them can be used as the definition of a differentiable function at z_0.

Example One of elementary examples of differentiable functions is the power function $f(z) = z^n$, $n \in \mathbb{N}$. Indeed, evaluating the limit in (2.1), we have

$$\lim_{\Delta z \to 0} \frac{f(z_0 + \Delta z) - f(z_0)}{\Delta z} = \lim_{\Delta z \to 0} \frac{(z_0 + \Delta z)^n - z_0^n}{\Delta z}$$

$$= \lim_{\Delta z \to 0} \left(n z_0^{n-1} + \frac{n(n-1)}{2} z_0^{n-2} \Delta z + \cdots + \Delta z^{n-1} \right) = n z_0^{n-1},$$

which means, according to Definition 2.1, that z^n is differentiable at every point $z_0 \in \mathbb{C}$ and its derivative is $(z^n)' = nz^{n-1}$, $\forall z \in \mathbb{C}$. It immediately implies that any complex polynomial $P_n(z) = c_n z^n + \cdots + c_1 z + c_0$ (c_j are complex constants, $c_n \neq 0$) is differentiable on the entire complex plane and $(P_n(z))' = nc_n z^{n-1} + \cdots + c_1$, $\forall z \in \mathbb{C}$. It is also easy to prove by definition that any rational function $R(z) = \frac{P_n(z)}{Q_m(z)}$ (where $P_n(z)$ and $Q_m(z)$ are polynomials of degree n and m, respectively) is differentiable at every point where it is defined, that is, at every point which is not a zero of $Q_m(z)$. Of course, the proof of the differentiability of rational functions (as well as of many other elementary functions) is much simpler by employing elementary arithmetic properties of differentiable functions presented later in this section.

In Chap. 1, we have studied the properties of complex functions (such as the existence of a limit, continuity, integrability), which, in a certain sense, are equivalent to the corresponding properties of real functions: a complex function satisfies one or several of these properties if and only if its real and imaginary parts possess the same properties.

Let us analyze whether the situation is the same with the differentiability property. At first glance, it seems that it might be so, since the definition of differentiability has apparently a usual form employed in Calculus and Real Analysis. Before to develop the relative theory, we consider simple examples involving the concepts of differentiability of complex and real functions. Naturally, the former is called *complex-differentiability* and the latter—*real-differentiability*. A complex function is called *real-differentiable* if its real and imaginary parts are real-differentiable functions (this definition amounts to that used for real vector-valued functions).

Example 2.1 Consider the function $f(z) = u(x, y) + iv(x, y) = \text{Re}\, z$, $z = x + iy$, that is, $u(x, y) = x$, $v(x, y) \equiv 0$. The real functions $u(x, y)$ and $v(x, y)$ are differentiable on the entire plane as the functions of two real variables, which implies the real-differentiability of $f(z)$. Moreover, both $u(x, y)$ and $v(x, y)$ are infinitely differentiable on the entire plane. Let us analyze whether the complex function $f(z) = \text{Re}\, z$ is complex-differentiable. According to the definition in the form (2.1), we evaluate $\lim\limits_{\Delta z \to 0} \frac{f(z_0+\Delta z) - f(z_0)}{\Delta z}$, where z_0 is an arbitrary point of the complex plane. Choosing the path $z = x_0 + iy$ (see Fig. 2.1), that is, $\Delta z = i\Delta y$ ($\Delta x = 0$), we obtain

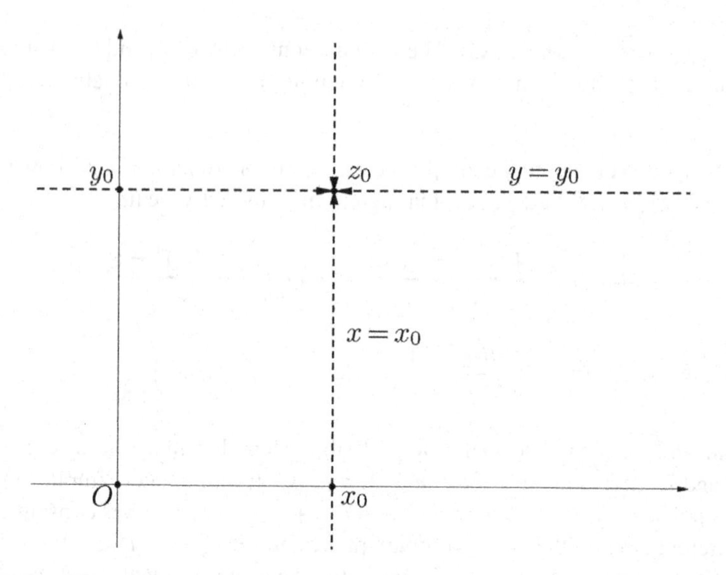

Fig. 2.1 Two different paths for approaching z_0

$$\lim_{\substack{\Delta z \to 0 \\ x = x_0}} \frac{f(z_0 + \Delta z) - f(z_0)}{\Delta z} = \lim_{\substack{\Delta z \to 0 \\ x = x_0}} \frac{(x_0 + \Delta x) - x_0}{\Delta z} = \lim_{\substack{\Delta y \to 0 \\ \Delta x = 0}} \frac{\Delta x}{\Delta x + i\Delta y} = \lim_{\Delta y \to 0} \frac{0}{i\Delta y} = 0.$$

However, taking another path $z = x + iy_0$ (see Fig. 2.1), that is, $\Delta z = \Delta x$ $(\Delta y = 0)$, we have a different result

$$\lim_{\substack{\Delta z \to 0 \\ z = x + iy_0}} \frac{f(z_0 + \Delta z) - f(z_0)}{\Delta z} = \lim_{\substack{\Delta z \to 0 \\ z = x + iy_0}} \frac{(x_0 + \Delta x) - x_0}{\Delta z} = \lim_{\substack{\Delta x \to 0 \\ \Delta y = 0}} \frac{\Delta x}{\Delta x + i\Delta y} = \lim_{\Delta x \to 0} \frac{\Delta x}{\Delta x} = 1.$$

Since the two partial limits as z approaches z_0 are different, it means that the limit $\lim\limits_{\Delta z \to 0} \frac{f(z_0+\Delta z)-f(z_0)}{\Delta z}$ does not exist. Therefore, the given complex function is not complex-differentiable at any point, whereas its real and imaginary parts are infinitely differentiable real functions on the entire plane.

Example 2.2 It is easy to show that the conjugate function $f(z) = \bar{z}$ is not differentiable at any point. In fact, in this case, the ratio in (2.1) takes the form

$$\frac{f(z_0 + \Delta z) - f(z_0)}{\Delta z} = \frac{\overline{\Delta z}}{\Delta z}$$

for any z_0. Therefore, for $\Delta z \in \mathbb{R}$ this ratio is 1 and the corresponding partial limit is also 1, while for $\Delta z \in i\mathbb{R}$, the ratio and the partial limit equal -1 (see Fig. 2.1). It shows that the general limit in (2.1) does not exist. However, the real $u(x, y) = x$ and imaginary $v(x, y) = -y$ parts of $f(z) = \bar{z}$ are infinitely differentiable and even analytic functions in $\mathbb{R}^2$.

Thus, these elementary examples show that *differentiability of a complex function $f(z) = u(x, y) + iv(x, y)$ and differentiability of its real $u(x, y)$ and imaginary $v(x, y)$ parts are different properties*. For this reason, if one needs to specify what kind of differentiability is considered, it is appropriate to use the terms *complex-differentiable* and *real-differentiable*. However, in this text we usually consider *complex-differentiable functions* and a sporadic use of *real-differentiability* will be clear from the context. Therefore, for the sake of brevity, in the most cases we will continue to use the term *differentiability*.

As we have seen, the differentiability of $u(x, y)$ and $v(x, y)$ is not sufficient to guarantee the differentiability of $f(z)$. Moreover, the infinite differentiability and even the analyticity of $u(x, y)$ and $v(x, y)$ do not imply the differentiability of $f(z)$. Nevertheless, the following simple theorem provides the necessary and sufficient conditions of differentiability of a complex function at a point.

Theorem 2.1 *A function $w = f(z) = u(x, y) + iv(x, y)$ is differentiable at a point $z_0 = x_0 + iy_0$ if and only if the following two conditions hold:*

(1) the functions $u(x, y)$ and $v(x, y)$ are differentiable at (x_0, y_0) as the functions of two real variables;
(2) the Cauchy–Riemann conditions

$$\begin{cases} \dfrac{\partial u}{\partial x}(x_0, y_0) = \dfrac{\partial v}{\partial y}(x_0, y_0) , \\ \dfrac{\partial u}{\partial y}(x_0, y_0) = -\dfrac{\partial v}{\partial x}(x_0, y_0) , \end{cases} \tag{2.4}$$

are satisfied at (x_0, y_0).

Proof *Necessity*. If a function $w = f(z)$ is differentiable at a point z_0, then, by the definition of differentiability (2.2), one has: $\Delta f(z_0) = f'(z_0) \cdot \Delta z + \alpha(\Delta z) \cdot \Delta z$. Since $f(z) = u(x, y) + iv(x, y)$ and $z_0 = x_0 + iy_0$, it follows that $\Delta z = \Delta x + i\Delta y$ and

$$\begin{aligned} \Delta f(z_0) &= f(z_0 + \Delta z) - f(z_0) \\ &= u(x_0 + \Delta x, y_0 + \Delta y) + iv(x_0 + \Delta x, y_0 + \Delta y) - u(x_0, y_0) - iv(x_0, y_0) \\ &= \Delta u(x_0, y_0) + i\Delta v(x_0, y_0). \end{aligned}$$

Denoting $f'(z_0) = A + iB, \alpha(\Delta z) = \alpha_1(\Delta x, \Delta y) + i\alpha_2(\Delta x, \Delta y)$, the equality (2.2) can be rewritten in the form

$$\begin{aligned} \Delta u(x_0, y_0) + i\Delta v(x_0, y_0) = (A + iB)(\Delta x + i\Delta y) \\ + (\alpha_1(\Delta x, \Delta y) + i\alpha_2(\Delta x, \Delta y)) \cdot (\Delta x + i\Delta y). \end{aligned}$$

Since the complex numbers are equal, it means that their real and imaginary parts are equal

$$\begin{aligned}\Delta u(x_0, y_0) &= A\Delta x - B\Delta y + \alpha_1 \Delta x - \alpha_2 \Delta y\,, \\ \Delta v(x_0, y_0) &= B\Delta x + A\Delta y + \alpha_2 \Delta x + \alpha_1 \Delta y\,.\end{aligned} \tag{2.5}$$

Observing that A and B are constant values, and the functions α_1 and α_2 converge to 0 as $(\Delta x, \Delta y)$ approaches $(0, 0)$, we conclude that the last two relations express the definitions of differentiability of $u(x, y)$ and $v(x, y)$ at the point (x_0, y_0). Furthermore, the definition of differentiability of the functions of two real variables implies that $A = \frac{\partial u}{\partial x}(x_0, y_0)$, $-B = \frac{\partial u}{\partial y}(x_0, y_0)$ due to the first formula in (2.5) and $B = \frac{\partial v}{\partial x}(x_0, y_0)$, $A = \frac{\partial v}{\partial y}(x_0, y_0)$ due to the second one. Therefore, $\frac{\partial u}{\partial x} = A = \frac{\partial v}{\partial y}$, $\frac{\partial u}{\partial y} = -B = -\frac{\partial v}{\partial x}$. This completes the proof of the first part.

Sufficiency. Suppose the functions $u(x, y)$ and $v(x, y)$ are differentiable at a point (x_0, y_0) as the functions of two real variables, and the Cauchy–Riemann conditions (2.4) are satisfied at (x_0, y_0). By the definition of differentiability of $u(x, y)$ and $v(x, y)$, we have

$$\begin{aligned}\Delta u(x_0, y_0) &= \frac{\partial u}{\partial x}(x_0, y_0) \cdot \Delta x + \frac{\partial u}{\partial y}(x_0, y_0) \cdot \Delta y + o(\rho), \\ \Delta v(x_0, y_0) &= \frac{\partial v}{\partial x}(x_0, y_0) \cdot \Delta x + \frac{\partial v}{\partial y}(x_0, y_0) \cdot \Delta y + o(\rho),\end{aligned}$$

where $\rho = \sqrt{\Delta x^2 + \Delta y^2} = |\Delta z|$. Multiplying the second equality by i and adding to the first, and also applying the Cauchy–Riemann conditions to the obtained expression, we get

$$\Delta f(z_0) = \Delta u(x_0, y_0) + i\Delta v(x_0, y_0) = \Delta x \cdot \left(\frac{\partial u}{\partial x} + i\frac{\partial v}{\partial x}\right) + i\Delta y \cdot \left(\frac{\partial v}{\partial y} - i\frac{\partial u}{\partial y}\right) + o(\rho)$$

$$= \Delta x \cdot \left(\frac{\partial u}{\partial x} + i\frac{\partial v}{\partial x}\right) + i\Delta y \cdot \left(\frac{\partial u}{\partial x} + i\frac{\partial v}{\partial x}\right) + o(\rho) = \left(\frac{\partial u}{\partial x} + i\frac{\partial v}{\partial x}\right)(\Delta x + i\Delta y) + o(\rho)$$

$$= f'(z_0) \cdot \Delta z + o(\rho).$$

The complex number $\frac{\partial u}{\partial x}(x_0, y_0) + i\frac{\partial v}{\partial x}(x_0, y_0)$ denoted by $f'(z_0)$ and $o(\rho)$ are the same quantities that appear in the definition (2.3). Therefore, the function $f(z)$ is differentiable at z_0. Thus, the Theorem is proved. □

Remark 2.1 In the course of the proof we derive the formula for the derivative of a complex function through the partial derivatives of $u(x, y)$ and $v(x, y)$

$$f'(z_0) = \frac{\partial u}{\partial x}(x_0, y_0) + i\frac{\partial v}{\partial x}(x_0, y_0)\,.$$

Applying the Cauchy–Riemann conditions to the last result, we obtain other representations for the derivative

$$f'(z_0) = \frac{\partial u}{\partial x} - i\frac{\partial u}{\partial y} = \frac{\partial v}{\partial y} + i\frac{\partial v}{\partial x} = \frac{\partial v}{\partial y} - i\frac{\partial u}{\partial y}\,. \tag{2.6}$$

Hence, the *derivative of a complex function can be found knowing only the real or imaginary part of a function*. Alternatively, this derivative can be obtained using the partial derivatives of $u(x, y)$ and $v(x, y)$ only with respect to one of the independent variables (x or y).

Remark 2.2 In the above Examples 2.1 and 2.2, the functions $u(x, y)$ and $v(x, y)$ were analytic in $\mathbb{R}^2$, but the Cauchy–Riemann conditions were not satisfied at any point: for the first function $f(z) = \text{Re}z = x$, one has

$$\frac{\partial u}{\partial x} = 1 \neq 0 = \frac{\partial v}{\partial y}\,,$$

and for second function $f(z) = \bar{z} = x - iy$, one gets

$$\frac{\partial u}{\partial x} = 1 \neq -1 = \frac{\partial v}{\partial y}\,.$$

Consider a simple example that uses the results of Theorem 2.1 and Remark 2.1.

Example 2.1 Verify if the function $f\,(z) = z^2 + |z|^2$ is differentiable, and if so, find its derivative.

To solve this problem, we use the conditions of differentiability formulated in Theorem 2.1. Using the algebraic form of the variable and function $z = x + iy,\ \ w = f\,(z) = u\,(x, y) + iv\,(x, y)$, we rewrite the definition of the function in the coordinate form: $w = f\,(z) = z^2 + |z|^2 = x^2 - y^2 + 2ixy + x^2 + y^2 = 2x^2 + 2ixy$, that is, $u\,(x, y) = 2x^2$ and $v\,(x, y) = 2xy$. Both functions $u\,(x, y)$ and $v\,(x, y)$ are differentiable in two real variables x, y in the entire plane. Calculating the partial derivatives

$$\frac{\partial u}{\partial x} = 4x\,, \quad \frac{\partial u}{\partial y} = 0\,, \quad \frac{\partial v}{\partial x} = 2y, \quad \frac{\partial v}{\partial y} = 2x$$

and substituting them in the Cauchy–Riemann conditions, we obtain the following system:

$$\begin{cases} \dfrac{\partial u}{\partial x} = 4x = 2x = \dfrac{\partial v}{\partial y}\,, \\ \dfrac{\partial u}{\partial y} = 0 = -2y = -\dfrac{\partial v}{\partial x}\,. \end{cases}$$

Evidently, the last system has the unique solution $(x, y) = (0\,, 0)$. Therefore, Theorem 2.1 implies that the function $f(z)$ is differentiable only at the point $z = 0$. The derivative at this point is $f'\,(0) = \left(\frac{\partial u}{\partial x} + i\frac{\partial v}{\partial x}\right)|_{z=0} = (4x + 2iy)\,|_{(0\,,\,0)} = 0$.

Remark 2.3 Notice that the Cauchy–Riemann conditions (2.4) by themselves do not guarantee differentiability of a function. The first condition of smoothness of real and imaginary parts in Theorem 2.1 is also important as is shown in the following example.

Example 2.2 Show that the Cauchy–Riemann conditions are satisfied for the function $f(z) = \sqrt{|z^2 - \bar{z}^2|}$ at $z_0 = 0$, but $f(z)$ is not differentiable at z_0.

First rewrite the given function in the form $f(z) = u(x, y) + iv(x, y) = \sqrt{|z^2 - \bar{z}^2|} = 2\sqrt{|xy|}$, which makes clear that $u(x, y) = 2\sqrt{|xy|}$ and $v(x, y) \equiv 0$. Then, we immediately conclude that $v(x, y)$ is a smooth function with $\frac{\partial v}{\partial x} \equiv \frac{\partial v}{\partial y} \equiv 0$ in the entire plane $\mathbb{R}^2$. The function $u(x, y)$ has both partial derivatives at the origin, which can be found by the definition

$$\frac{\partial u}{\partial x}(0,0) = \lim_{x \to 0} \frac{u(x,0) - u(0,0)}{x - 0} = \lim_{x \to 0} \frac{0 - 0}{x} = 0 ,$$

$$\frac{\partial u}{\partial y}(0,0) = \lim_{y \to 0} \frac{u(0,y) - u(0,0)}{y - 0} = \lim_{y \to 0} \frac{0 - 0}{y} = 0 .$$

Therefore, the Cauchy–Riemann conditions hold at (0, 0).

However, $f(z)$ is not differentiable at the origin, which can be seen from the following evaluation with the use of the polar coordinates $x = r\cos\varphi$, $y = r\sin\varphi$:

$$R(z) \equiv \frac{f(z) - f(0)}{z - 0} = \frac{2\sqrt{|xy|}}{x + iy} = \frac{2\sqrt{|\cos\varphi \sin\varphi|}}{\cos\varphi + i\sin\varphi} = 2\sqrt{|\cos\varphi \sin\varphi|}\, e^{-i\varphi} .$$

The expression on the right-hand side is independent of r and takes different values for different angles φ. For instance, approaching 0 along the path $\varphi = 0, r \to 0$ one has $R(z) = 0$ and the corresponding partial limit equal to 0, while along the path $\varphi = \frac{\pi}{4}, r \to 0$ one gets $R(z) = 1 - i$ and the partial limit $1 - i$. Therefore, the limit of $R(z)$ as z approaches 0 does not exist, which means that $f(z)$ is not differentiable at the origin.

The cause of this behavior is a non-smoothness of $u(x, y)$ at the origin. Indeed, $u(x, y)$ is not real-differentiable at (0, 0), that can be seen from the evaluation of the remainder in the definition of differentiability

$$\beta(x,y) = u(x,y) - \left(u(0,0) + \frac{\partial u}{\partial x}(0,0) \cdot x + \frac{\partial u}{\partial y}(0,0) \cdot y \right) = 2\sqrt{|xy|} .$$

Along the path $y = x$, this reminder does not approach zero faster than $\rho = \sqrt{x^2 + y^2}$

$$\lim_{\rho \to 0} \frac{\beta}{\rho} = \lim_{x \to 0} \frac{2|x|}{\sqrt{2}|x|} = \sqrt{2} \neq 0 ,$$

which implies that $u(x, y)$ is not real-differentiable at (0, 0).

Remark 2.4 Recall from Real Analysis that the existence of partial derivatives does not assure differentiability of a function. This weakness of partial derivatives was used in Example 2.2. However, the continuity of partial derivatives at some point guarantees differentiability at the same point. Therefore, the *sufficient condition of* *complex-differentiability* can be formulated as follows: if partial derivatives of $u(x, y)$ and $v(x, y)$ are continuous and satisfy the Cauchy–Riemann conditions at (x_0, y_0), then $f(z) = u + iv$ is differentiable at $z_0 = x_0 + iy_0$.

Remark 2.5 Due to much more restrictive conditions of complex-differentiability as compared to real-differentiability, a gap between complex-continuous and complex-differentiable functions is much larger than in the real case. In contrast to real-valued functions, which demand intricate constructions of everywhere continuous and nowhere differentiable functions (such as Weierstrass function or Takagi function), analogous examples for complex functions are quite simple: for instance, it is easy to show that the function $f(z) = 2x + iy$ is everywhere complex-continuous and nowhere complex-differentiable. The same is true for the functions $f(z) = \text{Re}z = x$ and $f(z) = \bar{z} = x - iy$ considered in the previous examples.

Elementary properties of differentiable functions. *Algebraic properties and the chain rule.*

Since the derivative of a complex function is defined through a limit and all the algebraic properties of the limits of real functions can be applied to complex functions (as it was noted before), we can translate *the properties of the derivatives of real functions into the statements about the derivatives of complex functions*. In particular, if functions $f_1(z)$ and $f_2(z)$ are differentiable at z_0, then the following functions are differentiable at the same point and the following formulas hold:

(1) $f_1(z) \pm f_2(z) \, : \, (f_1(z) \pm f_2(z))' = f_1'(z) \pm f_2'(z);$

(2) $f_1(z) \cdot f_2(z) \, : \, (f_1(z) \cdot f_2(z))' = f_1'(z) \cdot f_2(z) + f_1(z) \cdot f_2'(z);$

(3) $\dfrac{f_1(z)}{f_2(z)}, \; f_2(z_0) \neq 0 \, : \, \left(\dfrac{f_1(z)}{f_2(z)}\right)' = \dfrac{f_1'(z) \cdot f_2(z) - f_2'(z) \cdot f_1(z)}{f_2^2(z)}.$

Also, the chain rule has a familiar form

(4) if $f(z)$ is differentiable at z_0 and $h(w)$ is differentiable at $w_0 = f(z_0)$, then the composite function $g(z) = h(f(z))$ is differentiable at z_0 and $g'(z_0) = h'(w_0) \cdot f'(z_0)$.

Definition 2.2 A function $w = f(z)$ is *differentiable in a domain D* if it is differentiable at every point of D. Another frequently used term for a *complex-differentiable in a domain function* is a *holomorphic function*.

Definition 2.3 A function $w = f(z)$ is *holomorphic at a point* z_0 if there exists a neighborhood of z_0 in which $f(z)$ is complex-differentiable.

Notice that the concept of *holomorphic function* is referred to a domain. Consequently, the *complex differentiability* and *holomorphy* at a point are different concepts: the latter implies the former but the converse is not true. For instance, the function $f(z) = z^2 + |z|^2$ of Example 2.1 is complex-differentiable at the origin, but is not holomorphic.

To exemplify one more time the difference between complex-differentiability and holomorphy at a point and show that it may happen not only for pointwise sets, let us consider the following example.

Example 2.3 Show that the function $f(z) = x^3y^2 + ix^2y^3$ is complex-differentiable on the coordinate axes, but is holomorphic nowhere.

In fact, the functions $u(x, y) = x^3y^2$ and $v(x, y) = x^2y^3$ are real-differentiable on the entire plane $\mathbb{R}^2$, and the Cauchy–Riemann conditions

$$\frac{\partial u}{\partial x} = 3x^2y^2 = 3x^2y^2 = \frac{\partial v}{\partial y},$$

$$\frac{\partial u}{\partial y} = 2x^3y = -2xy^3 = -\frac{\partial v}{\partial x}$$

are reduced to the only relation $xy(x^2 + y^2) = 0$, whose solution is $x = 0$ and $y = 0$. Therefore, all the points at which $f(z)$ is complex-differentiable are the points of x- and y-axis. Since none of these points possesses a neighborhood in which $f(z)$ is complex-differentiable, there is no point where $f(z)$ is holomorphic.

Thus, the *complex differentiability* and *holomorphy* at a point are different concepts. However, when it concerns domains, the terms *complex-differentiable* and *holomorphic* are equivalent and we will use them interchangeably with the same meaning in the text.

This slightly complicated terminology is originated by historical evolution of the complex analysis terms, and the use of the given above definition of holomorphy (in a domain and at a point) seems to be justified by two reasons: first, the word holomorphic is a shorter specification of complex-differentiability in a domain as distinguished from real-differentiability; and second, in the further development of the theory (in Sect. 2.4) we will see that holomorphy and another important property—analyticity—are equivalent concepts for complex functions.

Definition 2.4 A function is called *entire* if it is holomorphic on the entire complex plane.

Evidently, every polynomial is an entire function. Among entire transcendental functions there are e^z, $\cos z$, $\sin z$, whose properties we will study in detail in Chap. 4 devoted to elementary functions. At this point, we can show that e^z is an entire function.

Example 2.4 Show that $f(z) = e^z$ is an entire function.

We can use one of possible definitions of the complex-exponential function in the form

$$f(z) = e^z = e^{x+iy} = e^x(\cos y + i \sin y).$$

Both the real part $u(x, y) = e^x \cos y$ and imaginary part $v(x, y) = e^x \sin y$ are real-differentiable on the entire plane $\mathbb{R}^2$. The Cauchy–Riemann conditions

$$\frac{\partial u}{\partial x} = e^x \cos y = \frac{\partial v}{\partial y},$$

$$\frac{\partial u}{\partial y} = -e^x \cos y = -\frac{\partial v}{\partial x}$$

are satisfied at every point $(x, y) \in \mathbb{R}^2$. Therefore, e^z is complex-differentiable on the entire complex plane $\mathbb{C}$, that is, e^z is an entire function.

Definition 2.5 Let a function $f(z)$ be defined in a neighborhood of z_0. The function is called *analytic at the point* z_0 if it can be expanded in a power series $f(z) = \sum_{n=0}^{+\infty} c_n(z - z_0)^n$ that converges in a neighborhood of z_0, that is, there exists $r > 0$ such that the series converges in the open disk $|z - z_0| < r$. In this series expansion, the constants c_n are called the coefficients and z_0—the c entral point.

Definition 2.5′. A function $f(z)$ is called *analytic at* ∞ if it can be expressed as a series $f(z) = \sum_{n=0}^{+\infty} \frac{a_n}{z^n}$ that converges in a neighborhood of the point ∞, that is, there exists R >0 such that the series converges in the set $|z| > R$. The constants a_n are called the coefficients of the series.

Definition 2.6 A function is called *analytic in a domain D* if it is analytic at every point of D.

A function analytic at a point (in a domain) is sometimes referred to as *regular* at this point (in this domain).

One of the important goals of this chapter is to reveal the *relationship between differentiability and analyticity* of a function. We begin with the following result.

Theorem 2.2 *If a function* $w = f(z)$ *is analytic at a point* $z_0 \neq \infty$*, then it is differentiable at this point.*

Proof Since $f(z)$ is analytic at z_0, it can be expressed as the power series $f(z) = \sum_{n=0}^{+\infty} c_n(z - z_0)^n$ that converges in a neighborhood of z_0: $|z - z_0| < r$. At the point z_0 one has $f(z_0) = \sum_{n=0}^{+\infty} c_n(z_0 - z_0)^n = c_0$. Now we find the derivative of $f(z)$ at z_0 using the definition (2.1). In the required limit z approaches z_0, which allows us to consider only the points z in a small disk $|z - z_0| < r$. Since a power series converges normally in its disk of convergence (see Sect. 1.7), we can calculate the limit in (2.1) term-by-term, which gives the following result:

$$f'(z_0) = \lim_{z \to z_0} \frac{f(z) - f(z_0)}{z - z_0} = \lim_{z \to z_0} \frac{\sum_{n=0}^{+\infty} c_n(z - z_0)^n - c_0}{z - z_0} = \lim_{z \to z_0} \frac{\sum_{n=1}^{+\infty} c_n(z - z_0)^n}{z - z_0}$$

$$= \lim_{z \to z_0} \sum_{n=1}^{+\infty} c_n(z - z_0)^{n-1} = \sum_{n=1}^{+\infty} \lim_{z \to z_0} c_n(z - z_0)^{n-1} = c_1.$$

Thus, the limit in the definition (2.1) exists and is finite, which means that the function is differentiable at z_0 and $f'(z_0) = c_1$. This completes the proof of the Theorem. □

The last Theorem provides only the pointwise information about the differentiability and the formula of derivative of an analytic at z_0 function. However, the analyticity of a function is not a pointwise property: $f(z)$ is analytic at z_0 if it can be expanded in a power series in some neighborhood of z_0. Besides, the well known result of Real Analysis states that a real-analytic function can be infinitely differentiated and each of its derivatives can be also expressed in the power series. All this suggests that similar non-pointwise statement may be true for complex functions. The following results show that this is so.

Theorem 2.3 (On differentiability of a power series) *If a power series*

$$f(z) = \sum_{n=0}^{+\infty} c_n (z - z_0)^n \tag{2.7}$$

converges in a disk $K = \{|z - z_0| < R\}$, then $f(z)$ is differentiable (holomorphic) in this disk and its derivative can be found by means of term-by-term differentiation, that is

$$f'(z) = \sum_{n=1}^{+\infty} n c_n (z - z_0)^{n-1}, \ \forall z \in K. \tag{2.8}$$

Proof First, we show that the power series

$$g(z) = \sum_{n=1}^{+\infty} n c_n (z - z_0)^{n-1} \tag{2.9}$$

converges in K. Indeed, take an arbitrary point $z \in K$ such that $|z - z_0| = r < R$ and fix this point. Since K is an open disk we can always choose another point $z_1 \in K$ such that $r < |z_1 - z_0| = r_1 < R$. The absolute value of the general term of (2.9) can be evaluated as follows:

$$|n c_n (z - z_0)^{n-1}| = |c_n (z_1 - z_0)^{n-1}| \cdot n \left| \frac{z - z_0}{z_1 - z_0} \right|^{n-1} \le |c_n (z_1 - z_0)^{n-1}| \cdot M = M |c_n| r_1^{n-1},$$

where M is a constant (such a constant M exists, because $\left| \frac{z - z_0}{z_1 - z_0} \right| = \frac{r}{r_1} = q < 1$ and consequently $n q^{n-1} \underset{n \to \infty}{\to} 0$). Recalling that the power series $\sum_{n=0}^{+\infty} c_n (z - z_0)^n$ converges absolutely in K, which implies that the series $\sum_{n=0}^{+\infty} \left| c_n (z_1 - z_0)^n \right| = \sum_{n=0}^{+\infty} |c_n| \cdot r_1^n$ converges, we conclude that the series $\sum_{n=1}^{+\infty} |n c_n (z - z_0)^{n-1}|$ also converges, which means that (2.9) converges absolutely at any $z \in K$.

Now we proceed with the considerations about the derivative. Again we fix an arbitrary point $z_1 \in K$ such that $|z_1 - z_0| = r < R$, and consider the closed disk $\overline{K}_1 = \{|z - z_0| \le r_1\}$, where $r < r_1 < R$. Let Δz be such small increments that $r + |\Delta z| < r_1$, that is, all the points $z = z_1 + \Delta z$ lie in K_1. We employ the definition (2.2) of the differentiable function rewritten for the point z_1 in the form:

$$f(z) - f(z_1) - f'(z_1) \cdot \Delta z = \alpha(\Delta z) \cdot \Delta z, \tag{2.10}$$

where $\alpha(\Delta z) \to 0$ as $\Delta z \to 0$. To show that $f(z)$ is differentiable at z_1 and its derivative can be written in the form (2.8), we need to represent the expression $f(z) - f(z_1) - g(z_1) \cdot \Delta z$ in the form compatible with the right-hand side of (2.10). This expression can be written in the terms of the power series and evaluated as follows:

$$f(z) - f(z_1) - g(z_1) \cdot \Delta z = \sum_{n=0}^{+\infty} c_n (z - z_0)^n - \sum_{n=0}^{+\infty} c_n (z_1 - z_0)^n - \sum_{n=1}^{+\infty} n c_n (z_1 - z_0)^{n-1} \cdot \Delta z$$

$$= \sum_{n=0}^{+\infty} c_n ((z - z_0)^n - (z_1 - z_0)^n) - \sum_{n=1}^{+\infty} n c_n (z_1 - z_0)^{n-1} \cdot \Delta z$$

$$= \sum_{n=1}^{+\infty} c_n (z - z_0 - (z_1 - z_0)) \left((z - z_0)^{n-1} + (z - z_0)^{n-2} (z_1 - z_0) + \cdots + (z_1 - z_0)^{n-1} \right)$$

$$- \sum_{n=1}^{+\infty} n c_n (z_1 - z_0)^{n-1} \cdot \Delta z$$

$$= \sum_{n=1}^{+\infty} c_n \left((z - z_0)^{n-1} + (z - z_0)^{n-2} (z_1 - z_0) + \cdots + (z_1 - z_0)^{n-1} - n (z_1 - z_0)^{n-1} \right) \cdot \Delta z$$

$$\equiv S(z) \cdot \Delta z \,. \tag{2.11}$$

Since the power series $\sum_{n=0}^{+\infty} c_n (z - z_0)^n$ converges in K, according to Theorem 1.1 of Sect. 1.7 this series converges uniformly on $\overline{K}_1$. The remaining two series for $f(z_1)$ and $g(z_1)$ are convergent numerical series since z_1 is a fixed point in K. Therefore, by the algebraic properties of convergent series, the series $S(z)$ on the right-hand side of (2.11) is uniformly convergent on $\overline{K}_1$. Noting that a general term of $S(z)$ is a continuous function on $\overline{K}_1$, we can conclude that, according to Corollary 1.2 in Sect. 1.6, the series $S(z)$ represents a continuous function on $\overline{K}_1$. In particular, $S(z)$ is continuous at the chosen point z_1, which means that $S(z_1 + \Delta z) \underset{\Delta z \to 0}{\to} S(z_1)$. Since $S(z_1) = \sum_{n=1}^{+\infty} c_n (n(z_1 - z_0)^{n-1} - n(z_1 - z_0)^{n-1}) = 0$, we can conclude that $S(z_1 + \Delta z)$ is the function $\alpha(\Delta z)$ and $g(z) = \sum_{n=1}^{+\infty} n c_n (z_1 - z_0)^{n-1}$ is $f'(z_1)$ in the definition of differentiability (2.10).

Thus, the Theorem is proved. □

Corollary 2.1 (On the infinite differentiability of a power series) *Since the derivative of a power series is again a power series, which converges in the same disk, we can, in turn, differentiate the latter power series to obtain the second derivative of the original series. Then we can apply the same reasoning to the power series of the second derivative and so on. Therefore, a power series has a derivative of an arbitrary order, that is, infinitely differentiable, in its disk of convergence and all these derivatives are represented through the power series convergent in the same* *disk as the original series. The expression for the kth derivative has the form*

$$f^{(k)}(z) = \sum_{n=k}^{+\infty} n \cdot (n-1) \cdot \ldots \cdot (n-k+1) c_n (z-z_0)^{n-k}, \ \forall z \in K . \tag{2.12}$$

Corollary 2.2 (On the coefficients of the power series of $f(z)$) *The coefficient c_0 is found by setting $z = z_0$ in the original power series: $f(z_0) = c_0$. The next coefficient c_1 is determined by using $z = z_0$ in the series of the first derivative: $f'(z_0) = 1 \cdot c_1$ (see also the result of Theorem 2.2). Generally, the coefficient c_n is determined by setting $z = z_0$ in the series of the nth derivative: $f^{(n)}(z_0) = n! \cdot c_n$. Thus, the power series* (2.7) *can be written in the form*

$$f(z) = \sum_{n=0}^{+\infty} \frac{f^{(n)}(z_0)}{n!}(z-z_0)^n , \tag{2.13}$$

which is called the Taylor series of $f(z)$ and the coefficients $c_n = \frac{f^{(n)}(z_0)}{n!}$ are the Taylor coefficients of $f(z)$.

Corollary 2.3 (On differentiability of an analytic function in its disk of convergence) *An analytic function $w = f(z)$ is holomorphic in its disk of convergence and its derivative can be found by differentiating term-by-term the power series of $f(z)$. More generally, an analytic function $w = f(z)$ is infinitely differentiable in its disk of convergence and its nth derivative can be found by n-time successive term-by-term differentiation of the power series of $f(z)$.*

Corollary 2.4 (On differentiability of an analytic function in a neighborhood) *If $w = f(z)$ is analytic at a point z_0 of a domain D, then it is holomorphic at z_0 and its derivative can be found by differentiating term-by-term the power series of $f(z)$ in a small neighborhood inside D. More generally, if $w = f(z)$ is analytic at a point z_0 of a domain D, then it is infinitely differentiable in a small neighborhood inside D and its nth derivative can be found by n-time successive term-by-term differentiation of the power series of $f(z)$.*

Corollary 2.5 (On differentiability of an analytic function on a domain) *If $w = f(z)$ is analytic in a domain D, it is analytic at every point of D. Therefore, according to the previous Corollary, $f(z)$ is (infinitely) differentiable at every point of D, which means that $f(z)$ is (infinitely) differentiable on D.*

Remark In this section, we have solved the problem of the relationship between differentiability and analyticity in one direction. The converse statement will be considered in Sect. 2.4.

2.2 Cauchy's Theorem. Goursat's Theorem

Lemma *An integral of a linear function along a closed polygonal chain, which consists of a finite number of line segments, is equal to zero.*

Proof First notice that a polygonal chain P with a finite number of segments is a piecewise smooth curve. A linear function is continuous, and therefore, the integral along the closed curve $\oint_P (az+b)dz$ exists. By definition, the integral is the limit of integral sums: $\oint_P (az+b)dz = \lim\limits_{\delta_\tau \to 0} \sum_{j=1}^n (a\zeta_j + b) \cdot \Delta z_j$. Since the integral exists, the limit keeps the same value under any type of partitions of P and any choices of the points ζ_j in which the function is evaluated. For this reason, we can choose such partitions of P that all the vertices of P belong to the set of point of these partitions. In this case, each arc γ_j of the partition τ is a line segment connecting the points z_{j-1} and z_j. The points ζ_j we choose as the midpoints of the segments γ_j, that is, $\zeta_j = \frac{z_{j-1}+z_j}{2}$. Recalling that the polygonal chain is closed, that is $z_0 = z_n$, we obtain

$$\oint_P (az+b)dz = \lim_{\delta_\tau \to 0} \sum_{j=1}^n (a\zeta_j + b) \cdot \Delta z_j$$

$$= \lim_{\delta_\tau \to 0} \sum_{j=1}^n \left(a\frac{z_j + z_{j-1}}{2} + b \right) \cdot (z_j - z_{j-1}) = \lim_{\delta_\tau \to 0} \left(\frac{a}{2} \sum_{j=1}^n \left(z_j^2 - z_{j-1}^2 \right) + b \sum_{j=1}^n (z_j - z_{j-1}) \right)$$

$$= \lim_{\delta_\tau \to 0} \left(\frac{a}{2} \left(z_1^2 - z_0^2 + z_2^2 - z_1^2 + \cdots + z_n^2 - z_{n-1}^2 \right) + b\,(z_1 - z_0 + z_2 - z_1 + \cdots + z_n - z_{n-1}) \right)$$

$$= \lim_{\delta_\tau \to 0} \left(\frac{a}{2} \left(z_n^2 - z_0^2 \right) + b\,(z_n - z_0) \right) = 0\,.$$

Hence, the Lemma is proved. □

Now we are ready to present two important results on the integrals along a closed curve.

Cauchy's Theorem *Let D be a finite simply connected domain. If a function $w = f(z)$ is analytic in D, then the integral of $f(z)$ along any closed continuous and rectifiable curve contained in D is equal to zero.*

Goursat's Theorem *Let D be a finite simply connected domain. If a function $w = f(z)$ is holomorphic on D, then the integral of $f(z)$ along any closed continuous and rectifiable curve contained in D is equal to zero.*

Remark 2.1 Recall that a domain is finite if it does not contain the point ∞, in other words, it is contained in the finite complex plane $\mathbb{C}$, but still can be both bounded and unbounded.

Remark 2.2 In the previous section it was shown that an analytic function at a point z_0 is holomorphic at this point. Consequently, if a function is analytic in a domain

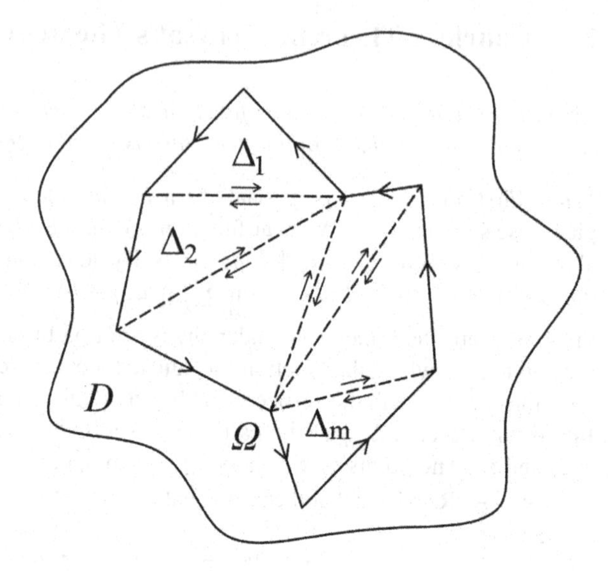

Fig. 2.2 Division of the polygon Ω into the triangles Δ_k and the directions of positive traversal along their boundaries

D, it is holomorphic on this domain. Therefore, by proving Goursat's Theorem we will demonstrate the statement of Cauchy's Theorem. Thus, to show the validity of the two theorems it is sufficient to prove Goursat's Theorem.

Proof of Goursat's Theorem We divide the proof into two parts.

The first step is to simplify the problem. A holomorphic on a domain D function $f(z)$ is continuous on this domain. According to Property 1.7 in Sect. 1.9, the integral of $f(z)$ along a continuous rectifiable curve Γ contained in D can be approximated with any desired accuracy by the integral of $f(z)$ along a polygonal line inscribed in Γ and contained in D. More precisely, for any $\varepsilon > 0$ there exists a polygonal line P_ε inscribed in Γ and $P_\varepsilon \in D$, such that $\left|\int_\Gamma f(z)dz - \int_{P_\varepsilon} f(z)dz\right| < \varepsilon$. Hence, if we manage to prove that $\int_{P_\varepsilon} f(z)dz = 0$ for any polygonal chain contained in D, then it implies that $\int_\Gamma f(z)dz = 0$, since the last integral is a fixed number and ε is an arbitrary positive number.

In turn, a closed polygonal chain with a finite number of segments can be divided into a finite number of simple closed polygonal chains. Therefore, it suffices to show that the integral of $f(z)$ along any simple closed polygonal line in D is equal to zero. A simple closed polygonal line represents the boundary of a polygon Ω, which lies entirely in D, because D is a simply-connected domain. Let us divide Ω into a a finite set of triangles Δ_k (see Fig. 2.2).

The integral along the boundary of Ω is equal to the sum of the integrals along the boundaries of all the triangles: $\int_{\partial\Omega} f(z)dz = \sum_k \int_{\partial\Delta_k} f(z)dz$, since the integration along each of the auxiliary segments, which divide Ω into triangles, is performed twice in opposite directions, and therefore, each pair of the integrals along the auxiliary boundaries sums to zero (see Fig. 2.2). Hence, we finally reduce the proof of

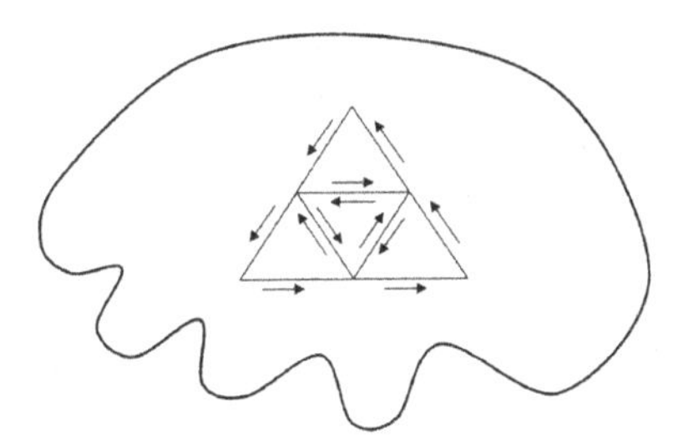

Fig. 2.3 The positive directions in the integration along the boundaries of the triangle and its subtriangles

the Theorem to the verification that an integral of $f(z)$ along the boundary of any triangle contained in D is equal to zero.

In the second part of the proof, we analyze the case of an arbitrary triangle lying in D. Assume, by contradiction, that there exists a triangle Δ_0 contained in D together with its boundary such that $\int_{\partial\Delta_0} f(z)dz \neq 0$. Then $\left|\int_{\partial\Delta_0} f(z)dz\right| = C,\ C > 0 (C = constant)$. Using the positive number $B = \frac{C}{A(\Delta_0)}$, where $A(\Delta_0)$ is the area of the triangle Δ_0, we can rewrite the last formula in the form

$$\left|\int_{\partial\Delta_0} f(z)dz\right| = B \cdot A(\Delta_0),\ B > 0\,. \tag{2.14}$$

The three midsegments of Δ_0 (i.e., the three line segments joining the midpoints of the sides of Δ_0) divide this triangle into four congruent subtriangles $\Delta^{(k)}$, $k = 1, 2, 3, 4$ (see Fig. 2.3). Among these four triangles $\Delta^{(k)}$, there exists at least one such that

$$\left|\int_{\partial\Delta^{(k)}} f(z)dz\right| \geq B \cdot A(\Delta^{(k)}) = B \cdot \frac{A(\Delta_0)}{4}\,, \tag{2.15}$$

because otherwise we would arrive at a contradiction. Indeed, if the last inequality is not true for $\forall k = 1, 2, 3, 4$, then

$$\left|\int_{\partial\Delta^{(k)}} f(z)dz\right| < B \cdot \frac{A(\Delta_0)}{4},\ \forall k = 1, 2, 3, 4\,. \tag{2.16}$$

The integral along $\partial\Delta_0$ is equal to the sum of integrals along all the boundaries $\Delta^{(k)}$, $k = 1, 2, 3, 4$, since the integrals along the three midsegments are traversed in opposite directions when considered as the boundaries of two different subtriangles $\Delta^{(k)}$, and consequently, they sum to zero (see Fig. 2.3). Then, using the properties of integrals and the supposition (2.16), we obtain the following contradictory result:

$$C = \left|\int_{\partial\Delta_0} f(z)dz\right| = \left|\sum_{k=1}^{4}\int_{\partial\Delta^{(k)}} f(z)dz\right| \leq \sum_{k=1}^{4}\left|\int_{\partial\Delta^{(k)}} f(z)dz\right| < B \cdot \frac{A(\Delta_0)}{4} \cdot 4 = C\,,$$

that is, $C < C$. It means that the supposition (2.16) is false, and consequently, the inequality (2.15) is true.

For convenience, denote the triangle (or one of the triangles) that satisfies (2.15) by Δ_1. In the chosen triangle perform the same procedure as it was applied to Δ_0: trace the midsegments and among obtained subtriangles denote by Δ_2 such that $\left|\int_{\partial\Delta_2} f(z)dz\right| \geq B \cdot \frac{A(\Delta_0)}{4^2}$ (the existence of such a subtriangle can be shown in the same way as it was made for Δ_1). Repeat this process successively to obtain a sequence of nested triangles (each current triangle is contained in the previous one), all of them contained in D together with their boundaries: $D \supset \bar{\Delta}_0 \supset \bar{\Delta}_1 \supset \bar{\Delta}_2 \supset \cdots \supset \bar{\Delta}_n \supset \cdots$. By construction, the following property holds for these triangles:

$$\left|\int_{\partial\Delta_n} f(z)\, dz\right| \geq B \cdot \frac{A(\Delta_0)}{4^n}, \quad \forall n \in \mathbb{N}. \tag{2.17}$$

Another property of this sequence of nested triangles is its convergence to the only point. In fact, in any triangle the length of each midsegment is a half of the length of the corresponding side of the triangle. Then, at each division, the perimeter of the obtained subtriangle is twice as small as the divided triangle: the perimeter of Δ_1 is twice as small as that of Δ_0, the perimeter of Δ_2 is twice as small as that of Δ_1 and so on. Therefore, the sequence of perimeters converges to zero: $p(\Delta_n) = \frac{p(\Delta_0)}{2^n} \underset{n\to\infty}{\to} 0$. For a triangle, its diameter d is the length of its longest side and consequently $d(\Delta_n) < p(\Delta_n)$, which implies that $d(\Delta_n) \underset{n\to\infty}{\to} 0$. (Recall that the diameter $d(E)$ of any plane set $E \in \mathbb{C}$ is defined as $d(E) = \sup\limits_{z_1,z_2\in E} |z_1 - z_2|$.) Therefore, according to the theorem of Real Analysis on a nested sequence of closed plane sets whose diameters converge to zero, the constructed sequence of the nested closed triangles with perimeters convergent to zero possesses the only point z_0, which belongs to all the triangles: $z_0 \in \bar{\Delta}_n, \forall n \in \mathbb{N}$. By construction, all the triangles are contained in the domain D, and consequently, the point z_0 also belongs to D: $z_0 \in D$. Since $f(z)$ is holomorphic on D, it is holomorphic at any point of D and, in particular, at z_0. Therefore, the following representation is valid in a neighborhood of z_0:

$$f(z) - f(z_0) = \Delta f(z_0) = f'(z_0) \cdot \Delta z + o(|\Delta z|) = f'(z_0) \cdot (z - z_0) + o(|z - z_0|). \tag{2.18}$$

Since the diameters $d(\Delta_n)$ converge to zero ($d(\Delta_n) \underset{n\to\infty}{\to} 0$), starting from some number N^* the triangles Δ_n are contained in the mentioned neighborhood of z_0, and consequently, for $\forall n > N^*$ the representation (2.18) holds for all the points $z \in \bar{\Delta}_n$.

We return now to the integrals along the boundary $\partial\Delta_n$ (for $\forall n > N^*$). Notice first that $|z - z_0| \leq d(\Delta_n)$ if $z \in \partial\Delta_n$ and that the quantity $d^2(\Delta_n)$ is of the same order as the area of Δ_n. Now using relations (2.17) and (2.18), and also the properties of the integral along a curve and the Lemma of this section, we obtain the following evaluation:

$$B \cdot \frac{A(\Delta_0)}{4^n} = B \cdot A(\Delta_n) \leq \left|\int_{\partial\Delta_n} f(z)dz\right| = \left|\int_{\partial\Delta_n} \left(f(z_0) + f'(z_0)(z - z_0) + o(|z - z_0|)\right) dz\right|$$

$$= \left| \int_{\partial \Delta_n} (f(z_0) + f'(z_0)(z - z_0))\, dz + \int_{\partial \Delta_n} o(|z - z_0|)\, dz \right|$$

$$= \left| \int_{\partial \Delta_n} o(|z - z_0|)\, dz \right| \le \int_{\partial \Delta_n} |o(|z - z_0|)|\, |dz|$$

$$\le o(d(\Delta_n)) \cdot \int_{\partial \Delta_n} |dz| = o(d(\Delta_n)) \cdot p(\Delta_n) \le o(d(\Delta_n)) \cdot 3d(\Delta_n) = o(A(\Delta_n)) = A(\Delta_n) \cdot \beta_n,$$

where $\beta_n \underset{n\to\infty}{\to} 0$. Thus, we get $B \cdot A(\Delta_n) \le A(\Delta_n) \cdot \beta_n$, which implies that $0 < B \le \beta_n$. Since $\beta_n \underset{n\to\infty}{\to} 0$, we conclude from the last inequality that $B = 0$, but this contradicts our assumption that $B > 0$ in (2.14). Hence, the Theorem is proved. □

The next goal is a generalization of Goursat's Theorem onto a wider class of domains. Notice that throughout the proof, we do not actually utilize the property of simple connectedness of D: in considering simple closed polygonal chains it was only important that the enclosed polygon was contained in D; in considering the triangles $\Delta_0, \Delta_1, \ldots$ again the important fact was that these triangles were lying in D. Therefore, the degree of connectedness of D was not important to perform the presented proof, whereas it was important that each constructed simple closed curve was contained in D together with the domain enclosed by this curve. Thus, we can propose the following generalization, dropping the requirement of simple connectedness of D, but keeping it for the domain enclosed by the considered curve.

Corollary 2.1 (The first generalization of Goursat's Theorem) *Suppose D is a finite domain and G is a simply connected domain enclosed by a continuous rectifiable curve and contained in D together with its boundary: $\overline{G} \subset D$. If $f(z)$ is holomorphic on D, then $\int_{\partial G} f(z)dz = 0$.*

The demonstration used for Goursat's Theorem is applicable with no change to this Corollary.

Next, we consider and prove even stronger result, that contains Corollary 2.1 as a particular case.

Corollary 2.2 (The second generalization of Goursat's Theorem) *Suppose D is a finite domain and G is a finitely connected domain bounded by a finite number of continuous rectifiable curves and contained in D together with its boundary: $\overline{G} \subset D$. If $f(z)$ is holomorphic on D, then $\int_{\partial G} f(z)dz = 0$.*

Proof In the n-connected domain G, we introduce $n - 1$ appropriately chosen cross-cuts to join successively n boundary components of ∂G (see Fig. 2.4). In this way, the domain G is transformed into a simply-connected domain G^*, $\overline{G}^* \subset D$ and, by Corollary 2.1, $\int_{\partial G^*} f(z)dz = 0$.

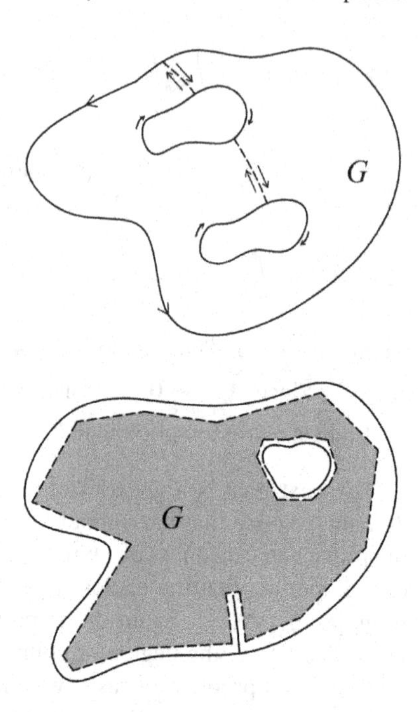

Fig. 2.4 Construction of auxiliary cuts in the domain G and the positive directions of the integration along the boundary of G^*

Fig. 2.5 The boundary of D—two piecewise smooth curves with a fold and their approximation by closed polygonal chains

Notice that the integration along the auxiliary cuts is performed twice in opposite directions (see Fig. 2.4), and consequently, the integrals along these crosscuts are canceled. Therefore

$$\int_{\partial G} f(z)dz = \int_{\partial G^*} f(z)dz = 0.$$

□

Corollary 2.3 (The third generalization of Goursat's Theorem) *Suppose D is a bounded finitely connected domain, whose boundary consists of a finite number of piecewise smooth curves with folds. If a function $f(z)$ is holomorphic on D and continuous on D up to its boundary, then $\int_{\partial D} f(z)dz = 0$.*

Proof In this case, according to Property 1.7 in Sect. 1.8, the integral along the boundary of D can be approximated with any desired accuracy by integrals along a finite number of closed polygonal lines contained in D (see Fig. 2.5). Each of these polygonal lines is a part of the boundary of a domain G contained in D together with its boundary ($\overline{G} \subset D$). Therefore, by Corollary 2.2, for the integral along ∂G we have $\int_{\partial G} f(z)dz = 0$, and since the last integral approximates the integral along ∂D with any degree of accuracy, we obtain $\int_{\partial D} f(z)dz = 0$. □

2.3 Cauchy Integral Formula

Let us start with a preliminary discussion about the meaning of the Cauchy integral formula. We have already seen that the differentiability of real and complex functions are different local properties. As a consequence of this, we will also find the differences in global behavior of real- and complex-differentiable functions. Suppose that a real function $y = f(x)$ is differentiable (and even analytic) on an interval $[a, b]$ and its values at the endpoints a and b are known. Then what can we say about the behavior of $f(x)$ inside the interval $[a, b]$? The only thing we can state is that the graph of $f(x)$ is a bounded smooth curve between the points $(a, f(a))$ and $(b, f(b))$, but we cannot determine specific values of $f(x)$ in the points of (a, b). The same is true for a function of two real variables $f(x, y)$ differentiable (and even analytic) on the closure of a two-dimensional domain, whose values are determined on the domain boundary. However, the situation is completely different for complex functions: *if a function $f(z)$ is differentiable on the closure of a domain $\overline{D}$ and the values of $f(z)$ are specified on the boundary ∂D, then the values of $f(z)$ are uniquely determined at each point of the domain D.*

Theorem (Cauchy integral formula) *Let D be a finite domain. If a function $w = f(z)$ is holomorphic in D, then at each point $z_0 \in D$ the following Cauchy integral formula is true*

$$f(z_0) = \frac{1}{2\pi i} \int_{\partial G} \frac{f(z)}{z - z_0} dz, \tag{2.19}$$

where G is an arbitrary finitely connected domain bounded by piecewise smooth curves and such that $z_0 \in G$, $\overline{G} \subset D$.

Proof Pick an arbitrary point z_0 of D and consider any finitely connected domain G bounded by piecewise smooth curves, which contains z_0 and lies together with its boundary in D. The integrand $\varphi(z) = \frac{f(z)}{z-z_0}$ in (2.19) is holomorphic on D, except for the point z_0, as the ratio of two holomorphic functions. If we consider a new domain $D_{z_0} = D \backslash \{z_0\}$ by removing z_0 from D, then $\varphi(z)$ is holomorphic in this domain. Since z_0 belongs to G and G is a domain, it implies that z_0 is an interior point of G, that is, z_0 is contained in G together with some neighborhood: $\exists r > 0$ such that $\{|z - z_0| < r\} \subset G$. Evidently, any disk centered at z_0 with radius $\rho \leq r$ also lies in G: $\{|z - z_0| < \rho\} \subset G, \forall \rho \leq r$. Let us introduce the auxiliary domain G_ρ by removing from G the disk $|z - z_0| \leq \rho$ for some $\rho < r$ (see Fig. 2.6). Since G_ρ is a finitely connected domain bounded by piecewise smooth curves, whose closure is contained in D_{z_0} ($\overline{G}_\rho \subset D_{z_0}$) and the function $\varphi(z) = \frac{f(z)}{z-z_0}$ is holomorphic on D_{z_0}, according to Corollary 2.2 to Goursat's Theorem, we have $\int_{\partial G_\rho} \frac{f(z)}{z-z_0} dz = 0$.

The boundary of G_ρ consists of ∂G and the circle $|z - z_0| = \rho$. Notice that, with respect to the domain G_ρ, the positive traversal along the part of its boundary that coincides with ∂G is the same as the positive traversal of ∂G with respect to G, whereas the positive traversal of the circle $|z - z_0| = \rho$ is clockwise, that is, for the circle itself this orientation is negative (see Fig. 2.6). If we change the traversal along

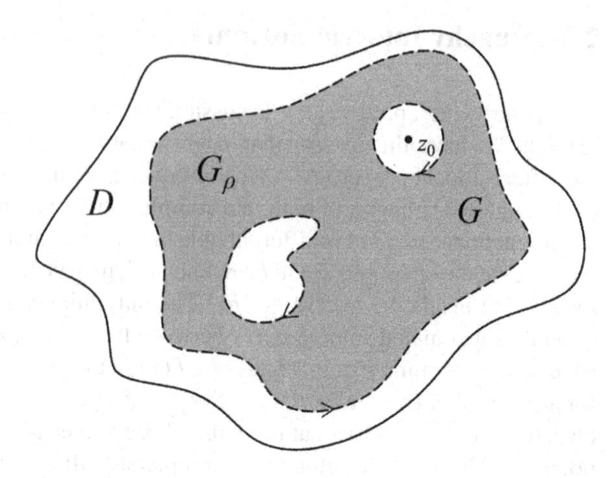

Fig. 2.6 The domains D, G and G_ρ, and the positive directions along the boundary of G_ρ

the circle to the opposite (counterclockwise) direction, then the sign of the integral along this part of the boundary also changes, and we obtain

$$0=\int_{\partial G_\rho}\frac{f(z)}{z-z_0}dz=\int_{\partial G}\frac{f(z)}{z-z_0}dz+\int_{(|z-z_0|=\rho)^{-1}}\frac{f(z)}{z-z_0}dz=\int_{\partial G}\frac{f(z)}{z-z_0}dz-\int_{|z-z_0|=\rho}\frac{f(z)}{z-z_0}dz.$$

Therefore

$$\int_{\partial G}\frac{f(z)}{z-z_0}dz=\int_{|z-z_0|=\rho}\frac{f(z)}{z-z_0}dz\,.$$

The last relation we can also rewrite in the form

$$\int_{\partial G}\frac{f(z)}{z-z_0}dz=\int_{|z-z_0|=\rho}\frac{f(z)-f(z_0)}{z-z_0}dz+\int_{|z-z_0|=\rho}\frac{f(z_0)}{z-z_0}dz\,. \tag{2.20}$$

To calculate the second integral on the right-hand side, we use the parametric form of the circle $z-z_0=\rho e^{it},\ 0\le t\le 2\pi,\ dz=\rho e^{it}idt$ and obtain

$$\begin{aligned}\int_{|z-z_0|=\rho}\frac{f(z_0)}{z-z_0}dz&=f(z_0)\int_{|z-z_0|=\rho}\frac{1}{z-z_0}dz\\&=f(z_0)\int_0^{2\pi}\frac{\rho e^{it}idt}{\rho e^{it}}=f(z_0)\cdot i\int_0^{2\pi}dt=2\pi i f(z_0)\,.\end{aligned} \tag{2.21}$$

To find the first integral on the right-hand side, we observe that in formula (2.20) the two integrals—the integral on the left-hand side and the second integral on the right-hand side (which is equal to $2\pi i f(z_0)$)—do not depend on ρ. Therefore, the first integral on the right-hand side also does not depend on ρ, that is, the value of this integral is the same for any choice of ρ $(0<\rho<r)$, and consequently, it can be evaluated passing to the limit as $\rho\to 0$. By the condition of the Theorem, the function $f(z)$ is holomorphic in D, in particular, at the point z_0, that is, $\frac{f(z)-f(z_0)}{z-z_0}=$

$f'(z_0) + \alpha(z - z_0)$, where $\alpha(z - z_0) \underset{z \to z_0}{\to} 0$. For this reason, $|\alpha(z - z_0)| < 1$ when $\rho = |z - z_0| \to 0$. Hence, we can make the following evaluations:

$$\left| \int_{|z-z_0|=\rho} \frac{f(z) - f(z_0)}{z - z_0} dz \right| = \left| \int_{|z-z_0|=\rho} \left(f'(z_0) + \alpha(z - z_0)\right) dz \right|$$

$$\leq \int_{|z-z_0|=\rho} \left(\left|f'(z_0)\right| + |\alpha(z - z_0)|\right) |dz| < \int_{|z-z_0|=\rho} \left(\left|f'(z_0)\right| + 1\right) |dz|$$

$$= \left(\left|f'(z_0)\right| + 1\right) \int_{|z-z_0|=\rho} |dz| = \left(\left|f'(z_0)\right| + 1\right) \cdot 2\pi\rho \underset{\rho \to 0}{\to} 0 \tag{2.22}$$

(in the last equality we have used the property that $\int_\Gamma |dz| = l_\Gamma$, where l_Γ is the length of the curve Γ, which is the circle in the considered case, and consequently $\int_{|z-z_0|=\rho} |dz| = 2\pi\rho$).

Finally, taking the limit in the equality (2.20) as $\rho \to 0$ and using the evaluations (2.21) and (2.22), we obtain:

$$\int_{\partial G} \frac{f(z)}{z - z_0} dz = 2\pi i f(z_0),$$

or equivalently,

$$f(z_0) = \frac{1}{2\pi i} \int_{\partial G} \frac{f(z)}{z - z_0} dz.$$

Thus, the Theorem is proved. □

Remark The Cauchy integral formula is frequently applied to the integrals along the boundary of a domain D. For this case, we can use the following statement.

Corollary *Let D be a bounded finitely connected domain enclosed by piecewise smooth curves with folds. If a function $w = f(z)$ is holomorphic in D and continuous on D up to its boundary, then at every point $z_0 \in D$ the following Cauchy integral formula is true:*

$$f(z_0) = \frac{1}{2\pi i} \int_{\partial D} \frac{f(z)}{z - z_0} dz.$$

The demonstration of this statement follows exactly the proof of the Theorem, except that the third generalization (Corollary 2.3) of Goursat's Theorem should be used instead of the second one.

2.4 The Main Criterion for Analyticity

Recall that the differentiability of real or complex function means the existence of its first derivative, and the analyticity means the power series expansion of a function. It is well known from Real Analysis that the expansion in a power series is a very strong requirement to a function: if a real function is analytic (that is, can be expanded in a power series), then this function is not only differentiable, it is infinitely differentiable. The same is true for an analytic complex function (see Corollary 2.1 to Theorem 2.3 in Sect. 2.1). On the other hand, even the existence of all the derivatives of a real function does not guarantee that this function can be expanded in a power series. The following theorem shows that the situation is completely different for complex functions.

Theorem (The main criterion for analyticity) *Let D be a finite domain. A function $w = f(z)$ is analytic in a domain D if and only if $f(z)$ is holomorphic in D.*

Proof *Necessity*. In the Corollary 2.5 to the Theorem 2.3 of Sect. 2.1, it was already shown that an analytic in a domain D function $f(z)$ is (infinitely) differentiable in D.

Sufficiency. Let us show that a holomorphic in a domain D function $w = f(z)$ is analytic at every point $z_0 \in D$. Since D is a domain, z_0 belongs to D together with some neighborhood: $\exists r > 0$ such that $K = \{|z - z_0| < r\} \subset D$ (we always can choose r such that $\overline{K} \subset D$). Now we pick an arbitrary point $z \in K$ (and consequently $z \in D$) and evaluate $f(z)$ using the Cauchy integral formula (2.19) with $G = K$ and variable of integration ζ

$$f(z) = \frac{1}{2\pi i}\int_{\partial K}\frac{f(\zeta)}{\zeta - z}d\zeta = \frac{1}{2\pi i}\int_{|\zeta - z_0| = r}\frac{f(\zeta)}{\zeta - z}d\zeta\,. \tag{2.23}$$

The ratio $\frac{1}{\zeta - z}$ where $z \in K, \zeta \in \partial K$ can be expressed as follows:

$$\frac{1}{\zeta - z} = \frac{1}{(\zeta - z_0) - (z - z_0)} = \frac{1}{\zeta - z_0}\cdot\frac{1}{1 - \frac{z - z_0}{\zeta - z_0}}\,.$$

Notice that the point z of the disk K is fixed (once it has been chosen) and the point ζ moves along the circle $|\zeta - z_0| = r$. Then $\left|\frac{z - z_0}{\zeta - z_0}\right| = q$, $q < 1$, because $|z - z_0| < r$ and $|\zeta - z_0| = r, \forall\zeta \in \partial K$. Therefore, the expression $\frac{1}{1 - \frac{z - z_0}{\zeta - z_0}}$ can be considered as the sum of the convergent geometric series with the ratio $\frac{z - z_0}{\zeta - z_0}$ whose absolute value is less than 1

$$\frac{1}{1 - \frac{z - z_0}{\zeta - z_0}} = \sum_{n=0}^{+\infty}\left(\frac{z - z_0}{\zeta - z_0}\right)^n. \tag{2.24}$$

Hence, we have a series of functions with respect to the variable ζ. Since $\left|\frac{z-z_0}{\zeta-z_0}\right| = q$, $q < 1$, $\forall\zeta \in \partial K$, this series has a majorant convergent numerical series $\sum_{n=0}^{+\infty} q^n$, and therefore, by the Weierstrass test, it converges uniformly on the boundary of the disk K.

Using the last formula, we can rewrite the integrand in (2.23) in the form

$$\frac{f(\zeta)}{\zeta - z} = \frac{f(\zeta)}{\zeta - z_0} \cdot \frac{1}{1 - \frac{z-z_0}{\zeta-z_0}} = \frac{f(\zeta)}{\zeta - z_0} \sum_{n=0}^{+\infty} \frac{(z-z_0)^n}{(\zeta-z_0)^n} = \sum_{n=0}^{+\infty} \frac{f(\zeta)}{(\zeta-z_0)^{n+1}} (z-z_0)^n . \tag{2.25}$$

Notice that all the terms of the series (2.24) are multiplied by the same function $\frac{f(\zeta)}{\zeta-z_0}$ in the last formula. The function $f(\zeta)$ is holomorphic in D (by the Theorem conditions), and consequently is continuous in D. In particular, it is continuous on the circle $|\zeta - z_0| = r$, which is a bounded and closed set, that is, a compact set. Then, by the Weierstrass Theorem, $f(\zeta)$ is bounded on this circle. The function $\frac{1}{\zeta-z_0}$ is also continuous and bounded (its absolute value is constant) on $|\zeta - z_0| = r$, since $\frac{1}{|\zeta-z_0|} = \frac{1}{r}$, $\forall\zeta \in \partial K$. A multiplication of all the terms of a uniformly convergent series by a continuous bounded function results in another uniformly convergent series, that is, the series on the right-hand side of (2.25) converges uniformly on the circle $|\zeta - z_0| = r$. Substituting the series (2.25) inside the integral in (2.23), we obtain

$$f(z) = \frac{1}{2\pi i} \int_{|\zeta-z_0|=r} \frac{f(\zeta)}{\zeta - z} d\zeta = \frac{1}{2\pi i} \int_{|\zeta-z_0|=r} \sum_{n=0}^{+\infty} \frac{f(\zeta)}{(\zeta-z_0)^{n+1}} (z-z_0)^n \, d\zeta . \tag{2.26}$$

The uniform convergence of the series inside the integral in (2.26) on the circle $|\zeta - z_0| = r$ allows the term-by-term integration along this circle. Then, interchanging the integral and sum signs and noting that the factor $(z - z_0)^n$ does not depend on the variable of integration, we can transform (2.26) to

$$f(z) = \frac{1}{2\pi i} \sum_{n=0}^{+\infty} \int_{|\zeta-z_0|=r} \frac{f(\zeta)}{(\zeta-z_0)^{n+1}} (z-z_0)^n \, d\zeta$$

$$= \sum_{n=0}^{+\infty} (z-z_0)^n \frac{1}{2\pi i} \int_{|\zeta-z_0|=r} \frac{f(\zeta)}{(\zeta-z_0)^{n+1}} d\zeta = \sum_{n=0}^{+\infty} c_n (z-z_0)^n .$$

Here, the coefficients c_n are introduced by the integral formula

$$c_n = \frac{1}{2\pi i} \int_{|\zeta-z_0|=r} \frac{f(\zeta)}{(\zeta-z_0)^{n+1}} d\zeta , \tag{2.27}$$

which does not depend on z. Thus, we construct the expansion of $f(z)$ in the power series $f(z) = \sum_{n=0}^{+\infty} c_n (z - z_0)^n$, which is valid for $\forall z \in K$, that is, this series con-

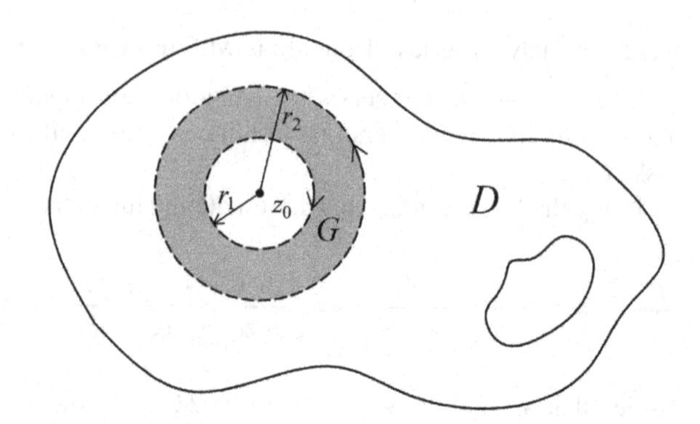

Fig. 2.7 The ring G and the positive direction of integration along its boundary

verges in the disk $|z - z_0| < r$. In other words, we show that $f(z)$ is analytic at the point z_0. Since z_0 is an arbitrary point in D, the function $f(z)$ is analytic in the entire domain D. This completes the proof of the Theorem. □

Remark 2.1 Consider again the coefficients c_n introduced in the course of the proof by formula (2.27). At first glance, it may seem that these coefficients depend on the radius r. Let us show that it is not so. Take two closed disks $K_1 = \{|\zeta - z_0| \le r_1\} \subset D$ and $K_2 = \{|\zeta - z_0| \le r_2\} \subset D$ with $r_1 < r_2$, and consider the open ring between their boundaries $G = \{r_1 < |\zeta - z_0| < r_2\}$ (see Fig. 2.7). The function $\frac{f(\zeta)}{(\zeta - z_0)^{n+1}}$ is defined on the closed ring $\overline{G}$ for any $n \in \mathbb{N}$ (since the denominator is different from 0), differentiable on $\overline{G}$ (as the ratio of two differentiable functions), and consequently, it is continuous on $\overline{G}$. Then, by Goursat's Theorem

$$\int_{\partial G} \frac{f(\zeta)}{(\zeta - z_0)^{n+1}} d\zeta = 0 . \tag{2.28}$$

The boundary of the ring G contains the two components (circles) which are traversed in opposite directions (see Fig. 2.7): the circle $|\zeta - z_0| = r_2$ is traveled counterclockwise (which is also the positive traversal of this circle itself), whereas $|\zeta - z_0| = r_1$ is traveled clockwise (which is the negative direction for the second circle itself). Separating the traversal along ∂G into two circles and changing the direction of the traversal of the second circle (which implies the change of the sign of the corresponding integral), we obtain

$$\int_{\partial G} \frac{f(\zeta)}{(\zeta - z_0)^{n+1}} d\zeta = \int_{|\zeta - z_0| = r_2} \frac{f(\zeta)}{(\zeta - z_0)^{n+1}} d\zeta + \int_{(|\zeta - z_0| = r_1)^{-1}} \frac{f(\zeta)}{(\zeta - z_0)^{n+1}} d\zeta$$

$$= \int_{|\zeta - z_0| = r_2} \frac{f(\zeta)}{(\zeta - z_0)^{n+1}} d\zeta - \int_{|\zeta - z_0| = r_1} \frac{f(\zeta)}{(\zeta - z_0)^{n+1}} d\zeta .$$

Hence, from the last formula and equality (2.28) we arrive at the claimed result on the independence of c_n from the radius of the disk K:

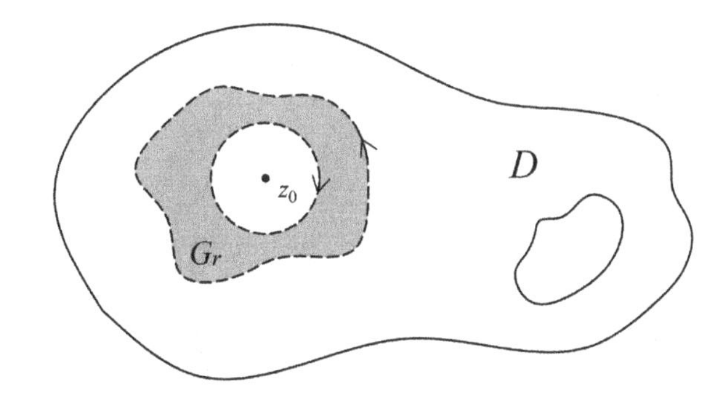

Fig. 2.8 The auxiliary domain G_r and the positive direction of integration along its boundary

$$\int_{|\zeta-z_0|=r_1} \frac{f(\zeta)}{(\zeta-z_0)^{n+1}} d\zeta = \int_{|\zeta-z_0|=r_2} \frac{f(\zeta)}{(\zeta-z_0)^{n+1}} d\zeta .$$

Remark 2.2 The integration path in (2.27) can be more general: not only any circle, that bounds a closed disk lying in the domain D, is admissible, but any sufficiently smooth closed curve, which is contained in D together with the enclosed domain, can also be used. Even more general case is described in the following result: if a domain G is bounded by a finite number of piecewise smooth curves, its closure $\overline{G}$ is contained in a finite domain D and the point z_0 belongs to G, then the coefficients c_n defined by (2.27) for a holomorphic in D function $f(z)$ can be calculated by the formula

$$c_n = \frac{1}{2\pi i} \int_{\partial G} \frac{f(\zeta)}{(\zeta-z_0)^{n+1}} d\zeta . \tag{2.29}$$

Indeed, consider the auxiliary domain $G_r = G \backslash \{|\zeta - z_0| \le r\}$ (see Fig. 2.8). Since $\overline{G}_r \subset D$ and the function $\frac{f(\zeta)}{(\zeta-z_0)^{n+1}}$ is differentiable in $\overline{G}_r$ for any $n \in \mathbb{N}$, we can apply Goursat's Theorem to this function in $\overline{G}_r$ and obtain

$$0 = \int_{\partial G_r} \frac{f(\zeta)}{(\zeta-z_0)^{n+1}} d\zeta = \int_{\partial G} \frac{f(\zeta)}{(\zeta-z_0)^{n+1}} d\zeta + \int_{(|\zeta-z_0|=r)^{-1}} \frac{f(\zeta)}{(\zeta-z_0)^{n+1}} d\zeta$$

$$= \int_{\partial G} \frac{f(\zeta)}{(\zeta-z_0)^{n+1}} d\zeta - \int_{|\zeta-z_0|=r} \frac{f(\zeta)}{(\zeta-z_0)^{n+1}} d\zeta .$$

It follows immediately from the last formula that

$$\int_{\partial G} \frac{f(\zeta)}{(\zeta-z_0)^{n+1}} d\zeta = \int_{|\zeta-z_0|=r} \frac{f(\zeta)}{(\zeta-z_0)^{n+1}} d\zeta ,$$

and consequently, formula (2.29) is true.

Corollary 2.1 *In Corollary 2.2 to Theorem 2.3 in Sect. 2.1, we have found that the coefficients c_n in the power series expansion of an analytic function are the Taylor* *coefficients, that is,* $f^{(n)}(z_0) = n! \cdot c_n$. *Comparing this expression with formula* (2.27) *for* c_n *derived in the last Theorem, we obtain the integral formula for the nth derivative of* $f(z)$

$$f^{(n)}(z_0) = \frac{n!}{2\pi i} \int_{|z-z_0|=r} \frac{f(z)}{(z-z_0)^{n+1}} dz, \tag{2.30}$$

where $|z - z_0| \le r$ *is a closed disk lying in the domain* D *of the analyticity of* $f(z)$.

Corollary 2.2 *More general integral formula for* $f^{(n)}$ *is obtained if we compare the Taylor coefficients with more general integral representation of* c_n (2.29)*:*

$$f^{(n)}(z_0) = \frac{n!}{2\pi i} \int_{\partial G} \frac{f(z)}{(z-z_0)^{n+1}} dz, \tag{2.31}$$

where ∂G *consists of a finite number of piecewise smooth curves and a domain* G *contains* z_0 *and its closure is a subset of the domain* D *where* $f(z)$ *is analytic.*

Corollary 2.3 *The last Theorem states that the differentiability of* $f(z)$ *in a domain* D *implies the analyticity of* $f(z)$ *in* D. *Since the analyticity of* $f(z)$ *implies the infinite differentiability of* $f(z)$ *and analyticity of any derivative of* $f^{(n)}(z)$ *(as was shown in Theorem 2.3 of Sect. 2.1), we conclude that the differentiability of* $f(z)$ *in* D *implies its infinite differentiability and the analyticity of all the derivatives* $f^{(n)}(z)$ *in* D.

Corollary 2.4 *The criterion for analyticity states that a holomorphic on a domain* D *function* $f(z)$ *is analytic at every point of* D, *that is,* $f(z)$ *can be expanded in a power series at each* $z_0 \in D$. *In the course of the proof of this criterion, we show that such a power series converges in any disk contained in* D. *Then, the radius of convergence* R *of this power series is not less than the distance from* z_0 *to the boundary of* D*:* $R \ge \rho(z_0, \partial D)$, *since the series certainly converges in any disk contained in* D, *but it may possibly converge in a larger disk.*

The following example illustrates the use of Goursat's Theorem, the Cauchy integral formula and formula for derivatives of analytic functions.

Example Evaluate the line integral along Γ, where Γ is an arbitrary simple closed curve, using Goursat's Theorem, the Cauchy integral formula or the Theorem on infinite differentiability of analytic function

$$\oint_\Gamma \frac{e^z dz}{z(2-z)^3}; \quad 0, 2 \notin \Gamma.$$

Notice first that there are three possible options of the location of Γ with respect to the points 0 and 2

(a) the points 0 and 2 are lying out of the region D enclosed by Γ (Fig. 2.9);
(b) one of the points $z = 0$ or $z = 2$ is inside the region D enclosed by Γ while another is outside ($D = D_2$ or $D = D_3$ in Fig. 2.10);

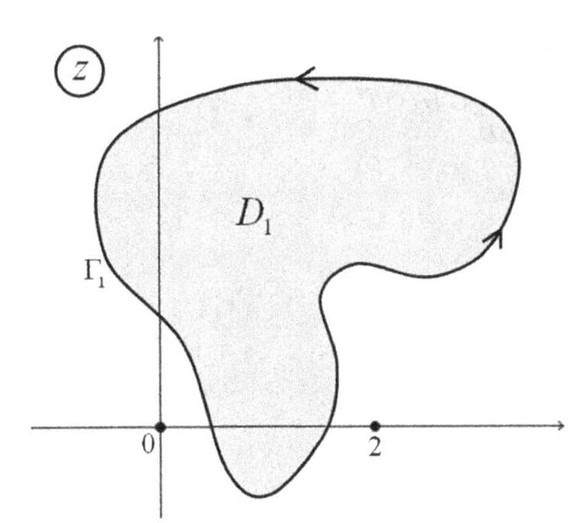

Fig. 2.9 The case when both points $0,\ 2 \notin D_1$, $\Gamma_1 = \partial D_1$

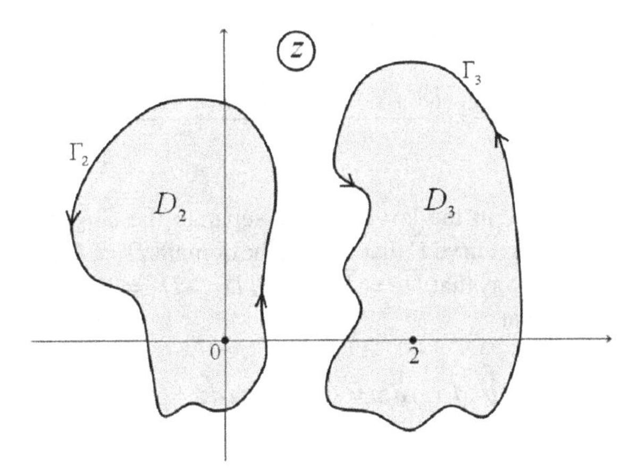

Fig. 2.10 The case when one of the points is inside while another is outside D, $\Gamma = \partial D$: $0 \in D_2$, $2 \notin D_2$; $2 \in D_3$, $0 \notin D_3$

(c) both points 0 and 2 are inside the region D enclosed by Γ (Fig. 2.11).

Denote by D_k, $k = 1, 2, 3, 4$ the domain bounded by the curve Γ_k, $k = 1, 2, 3, 4$. In the first case (Fig. 2.9), the function $f(z) = \frac{e^z}{z(2-z)^3}$ is analytic in $\overline{D}_1$, and therefore, by Goursat's Theorem, $\oint_{\Gamma_1} f(z)\,dz = 0$.

In the case when 0 is in and 2 out of D (see the domain D_2 in Fig. 2.10), the function $f(z)$ can be written in the form $f(z) = \frac{g(z)}{z}$, where $g(z) = \frac{e^z}{(2-z)^3}$ is analytic in $\overline{D}_2$. Then, using the Cauchy integral formula (see formula (2.19) in Sect. 2.3), we obtain

$$\oint_{\Gamma_2} f(z)\,dz = \oint_{\partial D_2} \frac{g(z)}{z-0}dz = 2\pi i \cdot g(0) = 2\pi i \frac{e^0}{(2-0)^3} = \frac{\pi i}{4}.$$

In the opposite situation, when 2 is in and 0 out of D (domain D_3 in Fig. 2.10), we represent $f(z)$ in the form $f(z) = \frac{h(z)}{(z-2)^3}$, where $h(z) = -\frac{e^z}{z}$ is an analytic function in $\overline{D}_3$. In this case, using the Theorem on infinite differentiability of an analytic function (formula (2.31) in Sect. 2.4 with $n = 2$), we get

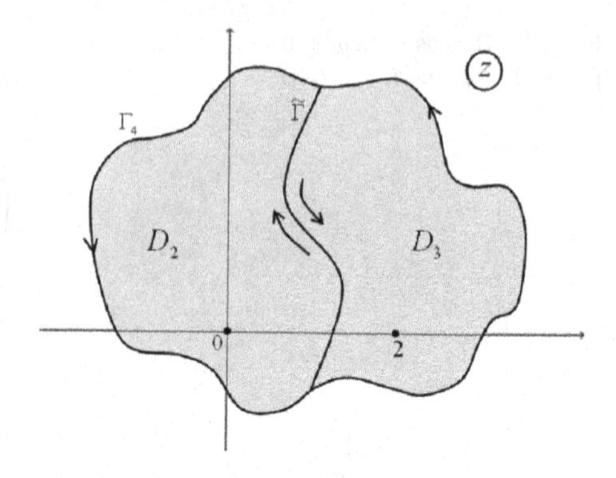

Fig. 2.11 The case when both points 0 and 2 are inside $D_4 = D_2 \cup D_3 \cup \tilde{\Gamma}$, $\Gamma_4 = \partial D_4$

$$\oint_{\Gamma_3} f(z)\,dz = \oint_{\partial D_3} \frac{h(z)}{(z-2)^3} dz = \frac{2\pi i}{2!} h''(z)|_{z=2} = -\pi i \left(\frac{e^z}{z}\right)'' |_{z=2} = -\pi i \left(\frac{ze^z - e^z}{z^2}\right)' |_{z=2}$$

$$= -\pi i \frac{(e^z + ze^z - e^z)\,z^2 - 2z\,(z-1)\,e^z}{z^4} |_{z=2} = -ie^2\pi \frac{4-2}{8} = -\frac{i\pi}{4} e^2 .$$

Finally, in the last case, we separate the singular points 0 and 2 by tracing a continuous curve $\tilde{\Gamma}$ that divides the domain $D = D_4$ into two subdomains D_2 and D_3 in such a way that $\overline{D}_4 = \overline{D}_2 \cup \overline{D}_3$, $D_2 \cap D_3 = \varnothing$ and $0 \in D_2, 2 \in D_3$ (see Fig. 2.11). Notice that

$$\oint_{\Gamma_4} f(z)\,dz = \oint_{\partial D_4} f(z)\,dz = \oint_{\partial D_2} f(z)\,dz + \oint_{\partial D_3} f(z)\,dz ,$$

since the integration along $\tilde{\Gamma}$ is performed twice in opposite directions. Applying the results obtained in the second case, we obtain

$$\oint_{\Gamma_4} f(z)\,dz = \oint_{\partial D_2} f(z)\,dz + \oint_{\partial D_3} f(z)\,dz = \frac{\pi i}{4} - \frac{\pi i}{4} e^2 = \frac{\pi i}{4} (1 - e^2) .$$

Thus, we arrive at the following formula:

$$\oint_{\Gamma} f(z)\,dz = \oint_{\partial D} f(z)\,dz = \begin{cases} 0 , & 0, 2 \notin \overline{D} ; \\ i\pi/4 , & 0 \in D,\ 2 \notin \overline{D} ; \\ -ie^2\pi/4 , & 2 \in D,\ 0 \notin \overline{D} ; \\ i\pi\left(1 - e^2\right)/4 , & 0, 2 \in D . \end{cases}$$

Remark 2.3 The criterion for analyticity shows that *the differentiability and the analyticity of a complex function considered in a domain are equivalent concepts.* However, *at a single point these two concepts are different.* Recall that in Sect. 2.1, we prove that the analyticity at a point implies the differentiability at the same point.

It happens that the converse is false, which is what we expect due to the criterion for analyticity. In fact, this criterion states that analyticity and holomorphy are equivalent concepts in any finite domain, in particular, in a neighborhood of a point, that is, at a point z_0. However, in Example 2.1 to Theorem 2.1 of Sect. 2.1, we have already shown that the function $f(z) = z^2 + |z|^2$ is differentiable but not holomorphic at the origin, which implies that this function is not analytic at the origin. Let us consider a similar example, using a prove of non-analyticity by the definition.

Example Consider the function $f(z) = |z|^2 = z \cdot \bar{z} = x^2 + y^2$. To find a set where this function is differentiable, we can use the Theorem on necessary and sufficient condition of differentiability at a point (Theorem 2.1 in Sect. 2.1). The real and imaginary parts of $f(z)$ are the functions $u(x, y) = x^2 + y^2$ and $v(x, y) = 0$. Both functions are infinitely differentiable at each point of the plane, because they are polynomial functions ($u(x, y)$ is a polynomial of degree two and $v(x, y)$ is a constant). Calculating the partial derivatives

$$\frac{\partial u}{\partial x} = 2x\,,\ \frac{\partial u}{\partial y} = 2y\,,\ \frac{\partial v}{\partial x} = 0\,,\ \frac{\partial v}{\partial y} = 0$$

and substituting in the Cauchy–Riemann conditions, we get

$$2x = \frac{\partial u}{\partial x} = \frac{\partial v}{\partial y} = 0\,,\ 2y = \frac{\partial u}{\partial y} = -\frac{\partial v}{\partial x} = 0\,.$$

Hence, the Cauchy–Riemann conditions hold only at the origin $(x, y) = (0, 0)$, that is, at $z = 0$, which implies that $f(z) = |z|^2$ is differentiable only at the origin and is not differentiable at any other point.

Let us see if $f(z)$ is analytic at $z = 0$ (at other points of the plane it is not analytic because is not differentiable), or, equivalently, if it is possible to represent $f(z)$ in a power series $f(z) = z \cdot \bar{z} = \sum_{n=0}^{+\infty} c_n z^n$ convergent in a neighborhood of the origin. Suppose, by contradiction, that the power series $f(z) = z \cdot \bar{z} = \sum_{n=0}^{+\infty} c_n z^n$ converges. Then we can take the limit of this series term by term as $z \to 0$, which gives $0 = c_0$. Consequently, the series can be rewritten as $z \cdot \bar{z} = \sum_{n=1}^{+\infty} c_n z^n$ and for any $z \neq 0$ we can divide both sides by z: $\bar{z} = \sum_{n=1}^{+\infty} c_n z^{n-1}$. Again taking the limit of the last series term by term as $z \to 0$, we obtain $0 = c_1$. Then $z \cdot \bar{z} = \sum_{n=2}^{+\infty} c_n z^n$ and, dividing by z^2 (for any $z \neq 0$), we arrive at the equality $\frac{\bar{z}}{z} = \sum_{n=2}^{+\infty} c_n z^{n-2}$. Representing z in the exponential form $z = |z|\, e^{i\varphi}$, $\varphi = \arg z$, we can rewrite the last equality in the form $\frac{|z|e^{-i\varphi}}{|z|e^{i\varphi}} = e^{-2i\varphi} = \sum_{n=2}^{+\infty} c_n z^{n-2}$. One more time, on the right-hand side we have the convergent power series, and consequently, we can calculate the term-by-term limit as $z \to 0$, which results in $e^{-2i\varphi} = c_2$. If now we approach $z = 0$ along the positive part of the real axis (i.e., take the direction $\varphi = 0$), then $c_2 = 1$; but if we approach $z = 0$ along the positive part of the imaginary axis (i.e., take the direction $\varphi = \frac{\pi}{2}$), then $c_2 = -1$. Hence, we arrive at the contradiction, since the coefficient c_2 is a constant and cannot depend on the way of approximation to the origin. Therefore, our supposition on a possibility to expand $f(z) = z \cdot \overline{z}$ in a power series in a neighborhood of the origin is false. Consequently, the function $f(z) = z \cdot \overline{z}$ is differentiable at $z = 0$, but is not analytic at this point.

2.5 Harmonic and Conjugate Harmonic Functions

Definition 2.1 Consider a real function of two real variables $\omega = \omega(x, y)$ twice continuously differentiable on a finite domain D. The function $\omega(x, y)$ is called *harmonic in a domain* D if it satisfies the Laplace equation in D

$$\Delta\omega \equiv \frac{\partial^2 \omega}{\partial x^2} + \frac{\partial^2 \omega}{\partial y^2} = 0\,, \ \forall\,(x, y) \in D\,.$$

The function $\omega(x, y)$ is called *harmonic at a point* (x_0, y_0) if it satisfies the Laplace equation at this point. The operator $\Delta \equiv \frac{\partial^2}{\partial x^2} + \frac{\partial^2}{\partial y^2}$ is called the *Laplace operator or Laplacian*.

Consider the following example.

Example Verify if there exists a harmonic function different of constant in the form $u = g\,(xy)$.

To analyze restrictions imposed on a given function by the condition of harmonicity, we find its partial derivatives

$$\frac{\partial u}{\partial x} = g'\,(xy) \cdot y\,, \ \frac{\partial u}{\partial y} = g'\,(xy) \cdot x\,; \ \frac{\partial^2 u}{\partial x^2} = g''\,(xy) \cdot y^2\,, \ \frac{\partial^2 u}{\partial y^2} = g''\,(xy) \cdot x^2$$

and bring them into the Laplace equation

$$\frac{\partial^2 u}{\partial x^2} + \frac{\partial^2 u}{\partial y^2} = g''\,(xy) \cdot y^2 + g''\,(xy) \cdot x^2 = \left(x^2 + y^2\right) g''\,(xy) = 0\,.$$

If $z \neq 0$, then the last relation implies that $g''\,(xy) = 0$, which means that $g'\,(t) = c_1$ and consequently $g\,(t) = c_1 t + c_2$, where $t = xy$ and c_1, c_2 are arbitrary real constants, that is, $g\,(xy) = c_1 xy + c_2$, $\forall c_1, c_2 \in \mathbb{R}$. Notice that this function satisfies the Laplace equation at (0, 0), and consequently, it is also harmonic at the origin. Thus, the function $u = g\,(xy) = c_1 xy + c_2$, $\forall c_1, c_2 \in \mathbb{R}$ is harmonic in the entire plane.

Let $w = f(z) = u\,(x, y) + iv\,(x, y)$ be a holomorphic function in a domain D. It was already shown that such a function is infinitely differentiable in D, and consequently, the real functions $u(x, y)$ and $v(x, y)$ are infinitely differentiable in D. Besides, at each point of D the functions $u(x, y)$ and $v(x, y)$ satisfy the Cauchy–Riemann conditions

$$\begin{cases} \dfrac{\partial u}{\partial x} = \dfrac{\partial v}{\partial y}, \ \dfrac{\partial u}{\partial y} = -\dfrac{\partial v}{\partial x} \end{cases}.$$

Differentiating the first equality in x, the second in y, summing up the results and recalling that the mixed derivatives coincide, we obtain

$$\frac{\partial^2 u}{\partial x^2} + \frac{\partial^2 u}{\partial y^2} = \frac{\partial^2 v}{\partial y \partial x} - \frac{\partial^2 v}{\partial x \partial y} = 0 .$$

In the same way, differentiating the second equation in x, the first in y and subtracting the first result from the second, we get

$$\frac{\partial^2 v}{\partial y^2} + \frac{\partial^2 v}{\partial x^2} = \frac{\partial^2 u}{\partial x \partial y} - \frac{\partial^2 u}{\partial y \partial x} = 0 .$$

Hence, the functions $u(x, y)$ and $v(x, y)$ are solutions of the Laplace equation in the domain D. Summarizing, if $f(z) = u(x, y) + iv(x, y)$ is holomorphic in D, then its real and imaginary parts are harmonic functions in D.

Let us consider the converse problem. Let $u(x, y)$ and $v(x, y)$ be two harmonic functions in D. If we compose the complex function $f(z) = u(x, y) + iv(x, y)$, would this function be holomorphic in D or would not? It happens that such $f(z)$ is not necessarily holomorphic and the following example shows this.

Example Let $u(x, y) = x$, $v(x, y) \equiv 0$. Both functions are infinitely differentiable and harmonic on the entire plane

$$\frac{\partial u}{\partial x} = 1 , \ \frac{\partial u}{\partial y} = 0 , \ \frac{\partial v}{\partial x} = 0 , \ \frac{\partial v}{\partial y} = 0 , \ \frac{\partial^2 u}{\partial x^2} + \frac{\partial^2 u}{\partial y^2} = 0 , \ \frac{\partial^2 v}{\partial x^2} + \frac{\partial^2 v}{\partial y^2} = 0 .$$

However, the complex function $f(z) = u + iv = x$ is not differentiable at any point of the complex plane as was shown in Sect. 2.1. The reason of this "bad" behavior of $f(z)$ is that the Cauchy–Riemann conditions are not satisfied for this function.

Definition 2.2 Functions $u(x, y)$ and $v(x, y)$ are called *conjugate harmonic in a domain* D if they are harmonic in D and satisfy the Cauchy–Riemann conditions in D.

We have already seen that a differentiable in D function $f(z) = u(x, y) + iv(x, y)$ determines the pair of conjugate harmonic in D functions $u(x, y)$ and $v(x, y)$. The converse is also true: if functions $u(x, y)$ and $v(x, y)$ are conjugate harmonic in D, then the complex function $f(z) = u(x, y) + iv(x, y)$ is differentiable in D. Indeed, any harmonic function possesses continuous second-order partial derivatives and consequently is differentiable in D. Besides, $u(x, y)$ and $v(x, y)$ satisfy the Cauchy–Riemann conditions in D. Therefore, by the Theorem 2.1 of Sect. 2.1, the function $f(z)$ is differentiable in D.

Joining this result with that on the relationship between differentiability and analyticity, we arrive at the following statement.

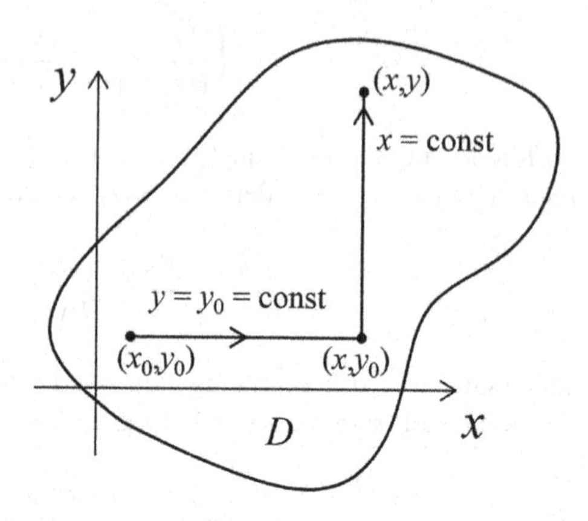

Fig. 2.12 The integration path composed of the two straight-line segments parallel to the coordinate axes

Theorem *The following conditions are equivalent:*

$$\begin{array}{ccc} f(z) \text{ is holomorphic in } D & \Leftrightarrow & f(z) \text{ is analytic in } D \\ \Updownarrow & & \Updownarrow \\ \multicolumn{3}{c}{u(x,y) \text{ and } v(x,y) \text{ are conjugate harmonic in } D} \end{array}$$

Consider now the problem of finding a harmonic function $v(x, y)$, which is conjugate to a given harmonic function $u(x, y)$ in a simply connected domain D.

To solve this task, we choose two points in the domain D: the first one (x_0, y_0) will be fixed and the second (x, y)—variable, and we will try to find the function $v(x, y)$ in the form of the integral of dv along a curve lying in D that connects the points (x_0, y_0) and (x, y).

The function $v(x, y)$ is twice continuously differentiable and dv is its first differential. Therefore, the integral $\int_{(x_0,y_0)}^{(x,y)} dv = \int_{(x_0,y_0)}^{(x,y)} \frac{\partial v}{\partial x} dx + \frac{\partial v}{\partial y} dy$ does not depend on the form of a curve. For convenience, let us compose the integration path of two (or more) straight-line segments parallel to the coordinate axes. Along such segments, we have either $dx = 0$ when $x = constant$ or $dy = 0$ when $y = constant$ (see Fig. 2.12). Taking this into account and noting that $u(x, y)$ and $v(x, y)$ should satisfy the Cauchy–Riemann conditions, we can express the function $v(x, y)$ as follows:

$$v(x, y) = \int_{(x_0,y_0)}^{(x,y)} dv = \int_{(x_0,y_0)}^{(x,y)} \frac{\partial v}{\partial x} dx + \frac{\partial v}{\partial y} dy = \int_{(x_0,y_0)}^{(x,y)} -\frac{\partial u}{\partial y} dx + \frac{\partial u}{\partial x} dy$$

$$= \int_{(x_0,y_0)}^{(x,y_0)} -\frac{\partial u}{\partial y} dx + \frac{\partial u}{\partial x} dy + \int_{(x,y_0)}^{(x,y)} -\frac{\partial u}{\partial y} dx + \frac{\partial u}{\partial x} dy = \int_{x_0}^{x} -\frac{\partial u(x,y_0)}{\partial y} dx + \int_{y_0}^{y} \frac{\partial u(x,y)}{\partial x} dy.$$

The obtained integral determines the function $v(x, y)$ up to an arbitrary constant, since the point (x_0, y_0) is chosen arbitrarily. Hence, for a given harmonic function $u(x, y)$, its conjugate harmonic $v(x, y)$ is defined up to an arbitrary constant. In a similar way, one can find $u(x, y)$ if $v(x, y)$ is given. In practice, the integration along a curve is frequently substituted by calculation of the indefinite integral.

The next two examples illustrate the technique of reconstruction of a conjugate harmonic function.

Example 2.1 Verify that the function $u(x, y) = 2xy$ is harmonic, find the corresponding conjugate harmonic function and recover the complex function $f(z) = u(x, y) + iv(x, y)$.

For any point of the plane one has

$$\frac{\partial u}{\partial x} = 2y\,, \ \frac{\partial u}{\partial y} = 2x\,, \ \frac{\partial^2 u}{\partial x^2} = 0\,, \ \frac{\partial^2 u}{\partial y^2} = 0\,, \ \frac{\partial^2 u}{\partial x^2} + \frac{\partial^2 u}{\partial y^2} = 0\,,$$

which means that $u(x, y)$ is harmonic on the entire plane. Let us find the conjugate harmonic function $v(x, y)$ by integrating one of the Cauchy–Riemann conditions

$$\frac{\partial v}{\partial x} = -\frac{\partial u}{\partial y} = -2x \Rightarrow v = \int \frac{\partial v}{\partial x} dx + \varphi(y) = -2\int x dx + \varphi(y) = -x^2 + \varphi(y)\,.$$

An arbitrary function $\varphi(y)$ appeared, because $v(x, y)$ is a function of two variables, whereas the integration was performed with respect to x. To specify $\varphi(y)$, we should use the second Cauchy–Riemann condition

$$\frac{\partial v}{\partial y} = \varphi'(y) = 2y = \frac{\partial u}{\partial x}\,.$$

Then

$$\varphi(y) = \int 2y dy = y^2 + C\,, \ C = constant\,.$$

Hence, $v(x, y) = -x^2 + y^2 + C$, and the complex function $f(z)$ takes the form

$$f(z) = u(x, y) + iv(x, y) = 2xy - ix^2 + iy^2 + iC = -i\left(x^2 - y^2 + 2ixy\right) + iC = -iz^2 + iC\,.$$

Example 2.2 Verify if the function $u(x, y) = \frac{5y}{x^2+y^2} - 2x^3 + 6xy^2 - 5x$ is harmonic. If so, find the corresponding conjugate harmonic function and recover the complex function $f(z) = u(x, y) + iv(x, y)$.

First, we check if the function $u(x, y)$ is harmonic. Calculating its partial derivatives we have

$$\frac{\partial u}{\partial x} = -\frac{10xy}{\left(x^2 + y^2\right)^2} - 6x^2 + 6y^2 - 5\,,$$

$$\frac{\partial u}{\partial y} = 5\frac{x^2 + y^2 - 2y^2}{\left(x^2 + y^2\right)^2} + 12xy = 5\frac{x^2 - y^2}{\left(x^2 + y^2\right)^2} + 12xy\,,$$

and

$$\frac{\partial^2 u}{\partial x^2} = -10y\frac{\left(x^2 + y^2\right)^2 - 4x^2\left(x^2 + y^2\right)}{\left(x^2 + y^2\right)^4} - 12x = 10\frac{3x^2y - y^3}{\left(x^2 + y^2\right)^3} - 12x\,,$$

$$\frac{\partial^2 u}{\partial y^2} = 5\frac{-2y\left(x^2+y^2\right)^2 - 4y\left(x^2+y^2\right)\left(x^2-y^2\right)}{\left(x^2+y^2\right)^4} + 12x = 10\frac{-3x^2y+y^3}{\left(x^2+y^2\right)^3} + 12x\,.$$

Substituting these results in the Laplace equation $\frac{\partial^2 u}{\partial x^2} + \frac{\partial^2 u}{\partial y^2} = 0$, we obtain

$$\frac{\partial^2 u}{\partial x^2} + \frac{\partial^2 u}{\partial y^2} = 10\frac{3x^2y-y^3}{\left(x^2+y^2\right)^3} - 12x + 10\frac{-3x^2y+y^3}{\left(x^2+y^2\right)^3} + 12x = 0\,,\ \forall (x,y) \neq (0,0)$$

(notice that the function $u\,(x, y)$ itself is also not defined at the point $z = 0$). Hence, the function $u(x, y)$ is harmonic in the entire finite complex plane except at the point $z = 0$. Therefore, we can find the corresponding conjugate harmonic function $v(x, y)$ in $\mathbb{C}_z \setminus \{0\}$ using the Cauchy–Riemann conditions

$$\begin{cases} \dfrac{\partial v}{\partial y} = \dfrac{\partial u}{\partial x} = -\dfrac{10xy}{\left(x^2+y^2\right)^2} - 6x^2 + 6y^2 - 5\,, \\ \dfrac{\partial v}{\partial x} = -\dfrac{\partial u}{\partial y} = -5\dfrac{x^2-y^2}{\left(x^2+y^2\right)^2} - 12xy\,. \end{cases} \tag{2.32}$$

Integrating the first equation in the last system, we obtain the following expression for $v(x, y)$:

$$v = \int \frac{\partial v}{\partial y} dy + \psi\,(x) = \int \left(-\frac{10xy}{\left(x^2+y^2\right)^2} - 6x^2 + 6y^2 - 5 \right) dy + \psi\,(x)$$

$$= \frac{5x}{x^2+y^2} - 6x^2y + 2y^3 - 5y + \psi\,(x)\,.$$

Differentiating this expression in x and using the second condition of the system (2.32), we obtain

$$5\frac{x^2+y^2-2x^2}{\left(x^2+y^2\right)^2} - 12xy + \psi'\,(x) = -5\frac{x^2-y^2}{\left(x^2+y^2\right)^2} - 12xy\,,$$

which implies that $\psi'\,(x) = 0$, that is, $\psi\,(x) = C = constant$. Thus

$$v\,(x,y) = \frac{5x}{x^2+y^2} - 6x^2y + 2y^3 - 5y + C\,,\ \forall z \in \mathbb{C}_z \setminus \{0\}\,.$$

Finally, we find the complex function $f\,(z) = u\,(x, y) + iv\,(x, y)$, applying the relations $z = x + iy$ and $\bar{z} = x - iy$ in $u(x, y)$ and $v(x, y)$

$$f(z) = u(x, y) + iv(x, y) = \frac{5y}{x^2 + y^2} - 2x^3 + 6xy^2 - 5x + \left(\frac{5x}{x^2 + y^2} - 6x^2 y + 2y^3 - 5y + C \right) i$$

$$= \frac{5i(x - iy)}{x^2 + y^2} - 2\left(x^3 + 3x^2 yi + 3x(yi)^2 + (yi)^3\right) - 5(x + iy) + Ci$$

$$= \frac{5i\,\bar{z}}{z \cdot \bar{z}} - 2(x + iy)^3 - 5(x + iy) + Ci = \frac{5i}{z} - 2z^3 - 5z + Ci\,.$$

The function $f(z)$ is analytic in the entire (finite) complex plane except at the point $z = 0$.

We can conclude that *starting just from the real or imaginary part of a differentiable complex function, it is possible to completely reconstruct this function up to an arbitrary constant.*

Remark The Theorem formulated in this section shows that *differentiability of $f(z)$, analyticity of $f(z)$ and conjugate harmonicity of $u(x, y)$ and $v(x, y)$ are equivalent conditions in a domain*. However, *these conditions are not equivalent at a single point*, which is illustrated in the two following examples.

Example 2.1 Let $f(z) = z \cdot \bar{z} = x^2 + y^2$. Then the functions $u(x, y) = x^2 + y^2$ and $v(x, y) \equiv 0$ are real-differentiable on the entire plane

$$\frac{\partial u}{\partial x} = 2x\,, \quad \frac{\partial u}{\partial y} = 2y\,, \quad \frac{\partial v}{\partial x} = 0\,, \quad \frac{\partial v}{\partial y} = 0\,,$$

but the Cauchy–Riemann conditions are satisfied only at the origin

$$\frac{\partial u}{\partial x} = \frac{\partial v}{\partial y} \Rightarrow x = 0\,, \quad \frac{\partial u}{\partial y} = -\frac{\partial v}{\partial x} \Rightarrow y = 0\,.$$

Therefore, the function is differentiable at the point $z = 0$. However, calculating the Laplacian of $u(x, y)$, we have

$$\Delta u = \frac{\partial^2 u}{\partial x^2} + \frac{\partial^2 u}{\partial y^2} = 2 + 2 = 4\,,$$

that is, the function $u(x, y)$ is not harmonic at any point of the plane. Previously it was shown that this function is not analytic at any point (see Sect. 2.4).

Example 2.2 Consider the function $f(z) = \text{Re}z^2 = x^2 - y^2$. In this case, the functions $u(x, y) = x^2 - y^2$ and $v(x, y) \equiv 0$ are real-differentiable on the entire plane

$$\frac{\partial u}{\partial x} = 2x\,, \quad \frac{\partial u}{\partial y} = -2y\,, \quad \frac{\partial v}{\partial x} = 0\,, \quad \frac{\partial v}{\partial y} = 0\,,$$

but the Cauchy–Riemann conditions are satisfied only at the origin

$$\frac{\partial u}{\partial x} = \frac{\partial v}{\partial y} \Rightarrow x = 0\,, \quad \frac{\partial u}{\partial y} = -\frac{\partial v}{\partial x} \Rightarrow y = 0\,.$$

Therefore, the function $f(z)$ is differentiable only at the point $z = 0$. Calculating the Laplacian of $u(x, y)$ and $v(x, y)$, we have

$$\frac{\partial^2 u}{\partial x^2} + \frac{\partial^2 u}{\partial y^2} = 2 - 2 = 0\,, \quad \frac{\partial^2 v}{\partial x^2} + \frac{\partial^2 v}{\partial y^2} = 0 + 0 = 0\,.$$

Hence, the functions $u(x, y)$ and $v(x, y)$ are harmonic in the entire plane, but the Cauchy–Riemann conditions hold only at the origin, and consequently, these two functions are conjugate harmonic only at the origin. Moreover, the function $f(z)$ is not analytic at any point of the plane.

2.6 Indefinite Integral of Analytic Function

Theorem *Let D be a finite simply connected domain. If a function $f(z)$ is analytic in D, then in this domain there exists an analytic primitive (or antiderivative) of $f(z)$, that is, there exists an analytic in D function $F(z)$ such that $F'(z) = f(z)$, $\forall z \in D$.*

Proof We pick two different points in D—one of them, say a, will be fixed, while another, say z, will be variable. We connect these points by a continuous rectifiable curve Γ contained in D, and consider the integral along this curve $\int_\Gamma f(\zeta)\,d\zeta$. First, we demonstrate that the integral value does not depend on a form of the curve Γ. Consider two different curves joining a and z, which lie in D. Their union $\Gamma_1 \cup \Gamma_2^{-1}$ is a continuous rectifiable closed curve contained in D (see Fig. 2.13). Since D is a finite simply connected domain and $f(z)$ is analytic in D, by Goursat's Theorem $\int_{\Gamma_1 \cup \Gamma_2^{-1}} f(\zeta)\,d\zeta = 0$. Using the additive property and changing the direction of the traversal along Γ_2, we obtain

$$0 = \int_{\Gamma_1 \cup \Gamma_2^{-1}} f(\zeta)\,d\zeta = \int_{\Gamma_1} f(\zeta)\,d\zeta + \int_{\Gamma_2^{-1}} f(\zeta)\,d\zeta = \int_{\Gamma_1} f(\zeta)\,d\zeta - \int_{\Gamma_2} f(\zeta)\,d\zeta\,,$$

or

$$\int_{\Gamma_1} f(\zeta)\,d\zeta = \int_{\Gamma_2} f(\zeta)\,d\zeta\,,$$

which means that the integral value does not depend on the form of a curve, but is only determined by the choice of the starting and ending points a and z. Since a is fixed, the integral value depends only on z.

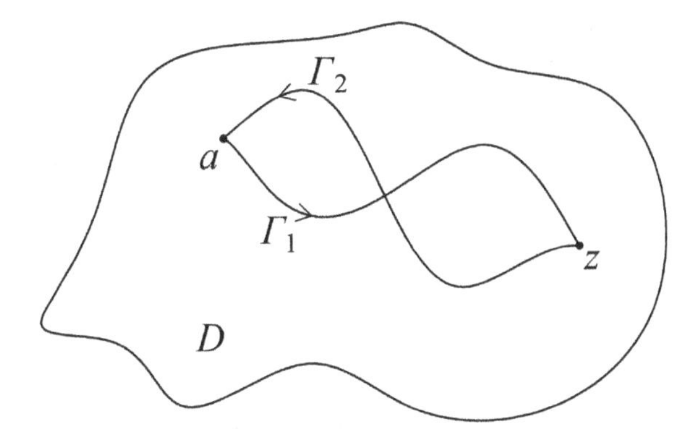

Fig. 2.13 The curves Γ_1 and Γ_2 joining the points a and z and the closed curve $\Gamma_1 \cup \Gamma_2^{-1}$

Now we introduce the function $F(z) = \int_a^z f(\zeta)\, d\zeta$ and show that this is the desired antiderivative, that is, $F'(z) = f(z),\ \forall z \in D$. Let us take an arbitrary point $z_0 \in D$ and evaluate the following difference:

$$R \equiv \left| \frac{F(z_0 + \Delta z) - F(z_0)}{\Delta z} - f(z_0) \right| = \frac{1}{|\Delta z|} \cdot \left| \int_a^{z_0+\Delta z} f(\zeta)\, d\zeta - \int_a^{z_0} f(\zeta)\, d\zeta - f(z_0)\, \Delta z \right|. \tag{2.33}$$

The point z_0 is contained in D together with some its neighborhood: there exists $r > 0$ such that $K = \{|z - z_0| < r\} \subset D$. In (2.33) we are interested only in small values of Δz and, for this reason, we can impose a restriction $|\Delta z| < r$ to guarantee that the point $z_0 + \Delta z$ belongs to K. In this case, the straight-line segment connecting z_0 and $z_0 + \Delta z$ lies in the same disk K, and consequently, it is contained in D. Since we have already demonstrated that the integral value does not depend on the form of a curve contained in D, we will choose a specific path for the first integral in (2.33), while keeping an arbitrary continuous rectifiable curve Γ contained in D for the second integral. We compose the curve for the first integral from the two parts: the first going from a to z_0 along the curve Γ, and the second going from z_0 to $z_0 + \Delta z$ along the straight-line segment γ contained in D (see Fig. 2.14). Using these curves in the integrals in (2.33) and recalling that $\int_{z_0}^{z_0+\Delta z} d\zeta = (z_0 + \Delta z) - z_0 = \Delta z$, we can evaluate (2.33) as follows:

$$R = \frac{1}{|\Delta z|} \left| \int_\Gamma f(\zeta)\, d\zeta + \int_{z_0}^{z_0+\Delta z} f(\zeta)\, d\zeta - \int_\Gamma f(\zeta)\, d\zeta - \int_{z_0}^{z_0+\Delta z} f(z_0)\, d\zeta \right|$$

$$= \frac{1}{|\Delta z|} \left| \int_{z_0}^{z_0+\Delta z} (f(\zeta) - f(z_0))\, d\zeta \right| \le \frac{1}{|\Delta z|} \int_{z_0}^{z_0+\Delta z} |f(\zeta) - f(z_0)|\, |d\zeta| \equiv R_1.$$

The function $f(z)$ is continuous in D (since it is analytic in D) and, in particular, it is continuous at z_0, that is

$$\forall \varepsilon > 0\ \exists \delta(\varepsilon) > 0 \text{ such that } \forall z \in D,\ |z - z_0| < \delta \text{ it follows that } |f(z) - f(z_0)| < \varepsilon.$$

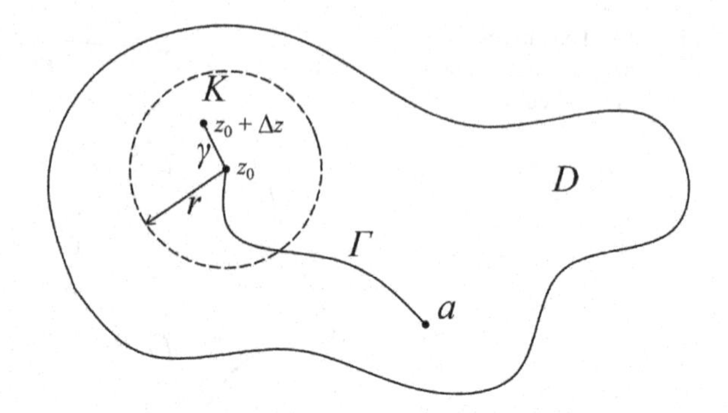

Fig. 2.14 The integration path composed of the curve Γ and the line segment γ

Choosing $|\Delta z| < \delta$ and recalling that γ is the straight-line segment, for $\forall \zeta \in \gamma$, we have

$$|\zeta - z_0| \le |(z_0 + \Delta z) - z_0| = |\Delta z| < \delta .$$

Therefore, by the definition of continuity, $|f(\zeta) - f(z_0)| < \varepsilon, \ \forall \zeta \in \gamma$. Then, returning to the evaluation of R_1, we obtain

$$R_1 < \frac{1}{|\Delta z|} \int_\gamma \varepsilon \, |d\zeta| = \frac{\varepsilon}{|\Delta z|} \int_\gamma |d\zeta| = \frac{\varepsilon}{|\Delta z|} l_\gamma = \frac{\varepsilon}{|\Delta z|} |\Delta z| = \varepsilon$$

(by the property of the integral along a curve, $\int_\gamma |d\zeta| = l_\gamma = |\Delta z|$, since γ is a straight-line segment between the points z_0 and $z_0 + \Delta z$). Hence, we arrive at the definition of the limit

$$\forall \varepsilon > 0 \, \exists \delta(\varepsilon) > 0 \text{ such that } \forall \Delta z, \ 0 < |\Delta z| < \delta \Rightarrow \left| \frac{F(z_0 + \Delta z) - F(z_0)}{\Delta z} - f(z_0) \right| < \varepsilon .$$

Therefore

$$\lim_{\Delta z \to 0} \frac{F(z_0 + \Delta z) - F(z_0)}{\Delta z} = f(z_0) .$$

By the definition of derivative, the last limit is the derivative of $F(z)$ at z_0, which is equal to $f(z_0)$

$$F'(z_0) = \lim_{\Delta z \to 0} \frac{F(z_0 + \Delta z) - F(z_0)}{\Delta z} = f(z_0) .$$

Since z_0 is an arbitrary point in D, it follows that $F'(z) = f(z), \ \forall z \in D$, which means that $F(z)$ is differentiable in D, is a primitive of $f(z)$ in D, and by the criterion for analyticity, is analytic in D.

Thus, the Theorem is proved. □

Corollary 2.1 *If a function $f(z)$ is analytic in a finite simply connected domain D, then its primitive is determined up to a constant. More precisely, all the primitives of $f(z)$ differ by a constant.*

Proof Let $F_1(z)$ and $F_2(z)$ be two primitives of $f(z)$ in D, that is, $F_1'(z) = f(z)$ and $F_2'(z) = f(z)$. Consider the function $G(z) = F_1(z) - F_2(z)$. This function is analytic in D (as the difference of two analytic functions) and also $G'(z) = F_1'(z) - F_2'(z) = 0$ for $\forall z \in D$. Then, representing $G(z)$ in the form $G(z) = u(x, y) + iv(x, y)$ and using the formula of derivative $G'(z) = \frac{\partial u}{\partial x} + i\frac{\partial v}{\partial x}$, we see that $\frac{\partial u}{\partial x} \equiv 0$ and $\frac{\partial v}{\partial x} \equiv 0$ in D. From the Cauchy–Riemann conditions, it follows that $\frac{\partial u}{\partial y} \equiv 0$ and $\frac{\partial v}{\partial y} \equiv 0$. Therefore, the real function $u(x, y)$ satisfies the two equations in D: $\frac{\partial u}{\partial x} \equiv 0$ and $\frac{\partial u}{\partial y} \equiv 0$. As is well known from Real Analysis, the only function that satisfies such relations is a constant: $u(x, y) \equiv C_1$. In the same manner, $v(x, y) \equiv C_2$, and consequently $G(z) \equiv C_1 + iC_2 = C$ and $F_1(z) = F_2(z) + C$, that is, the difference between two primitives is a constant. □

Corollary 2.2 *If a function $f(z)$ is analytic in a finite simply connected domain D, then the following Newton–Leibniz formula is true:*

$$\int_a^b f(\zeta)\, d\zeta = \Phi(b) - \Phi(a),$$

where $\Phi(z)$ is a primitive of $f(z)$ in D. One can see that formally this result is the same as for real functions.

Proof Let $\Phi(z)$ be a primitive of $f(z)$ in D. By the Theorem, the function $F(z) = \int_a^z f(\zeta)\, d\zeta$ is also the primitive of $f(z)$ in D. Then, by Corollary 2.1, we have $\Phi(z) = F(z) + C = \int_a^z f(\zeta)\, d\zeta + C$, where C is a constant. Setting $z = a$ in this equality, we get

$$\Phi(a) = \int_a^a f(\zeta)\, d\zeta + C = C$$

(the integral $\int_a^a f(\zeta)\, d\zeta = 0$ by definition). Therefore, $C = \Phi(a)$ and $\Phi(z) = \int_a^z f(\zeta)\, d\zeta + \Phi(a)$. Setting in the last equality $z = b$, we obtain $\Phi(b) = \int_a^b f(\zeta)\, d\zeta + \Phi(a)$, and finally,

$$\int_a^b f(\zeta)\, d\zeta = \Phi(b) - \Phi(a).$$

□

Remark Notice that the conditions of finiteness and simple connectedness of a domain D are relevant in the Theorem on primitive. The following example illustrates what may happen if one of these two conditions is not satisfied.

Example Consider the function $f(z) = \frac{1}{z}$. This function is analytic in the entire plane except for the origin. We will show that in the simply connected (but not finite)

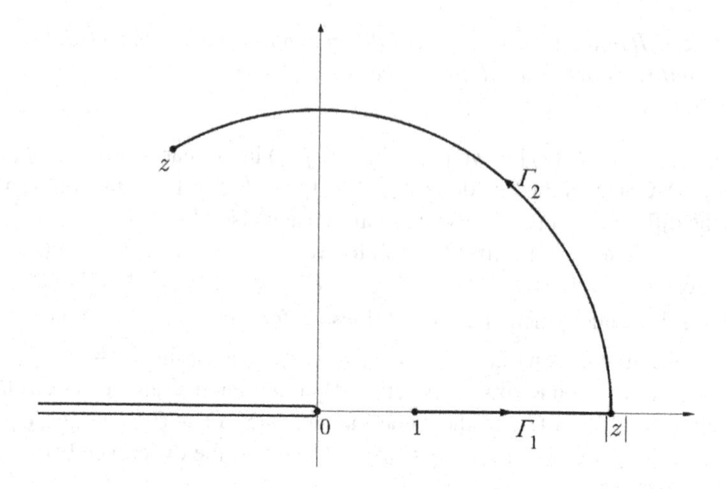

Fig. 2.15 The integration path in the domain $D = \mathbb{C}\setminus\{(-\infty, 0]\}$

domain $\overline{\mathbb{C}}\setminus\{0\}$ and in the finite (but not simply connected) domain $\overline{\mathbb{C}}\setminus(\{0\}\cup\{\infty\}) = \mathbb{C}\setminus\{0\}$, the function $f(z)$ does not possess an analytic primitive.

Since the primitives, in the case when they exist, differ only by a constant, we can restrict our search by such a primitive that is represented in the form of the integral along a curve given in the Theorem of this section. Take $a = 1$ and consider $\int_\Gamma \frac{d\zeta}{\zeta}$ where Γ is a continuous rectifiable curve connecting the points 1 and z, which does not pass through 0. Suppose, initially, that the curve Γ does not complete a circuit around the origin. Then we can make a crosscut joining the points 0 and ∞ (for instance, along the negative part of the real axis), and obtain the auxiliary domain D, which is finite and simply connected (the plane with a crosscut). The function $f(z) = \frac{1}{z}$ is analytic in this domain, and consequently (as it was shown in the Theorem), $\int_\Gamma \frac{d\zeta}{\zeta}$ does not depend on the form of the curve Γ ($\Gamma \in D$). We will choose the path from 1 to z consisting of two parts: the first part Γ_1 joins the initial point 1 with the point $|z|$ going along the positive part of the x-axis, and the second part Γ_2 starts at $|z|$ and terminates at $z = |z|\, e^{i \arg z}$ sliding along the arc of the circle of the radius $|z|$ centered at the origin (see Fig. 2.15). Then we have

$$\int_\Gamma \frac{d\zeta}{\zeta} = \int_{\Gamma_1} \frac{d\zeta}{\zeta} + \int_{\Gamma_2} \frac{d\zeta}{\zeta}.$$

We calculate the two integrals using parametric representation of the curves Γ_1 and Γ_2. For Γ_1, we have $\zeta = x \quad (y = 0)\,,\ 1 \le x \le |z|\,,\ d\zeta = dx$, and then

$$\int_{\Gamma_1} \frac{d\zeta}{\zeta} = \int_1^{|z|} \frac{dx}{x} = \ln|x|\Big|_1^{|z|} = \ln|z| - \ln 1 = \ln|z|\,.$$

For Γ_2 we use the parameterization $\zeta = |z|\, e^{i\varphi},\ 0 \le \varphi \le \arg z,\ d\zeta = |z|\, e^{i\varphi} i d\varphi$ that gives

Fig. 2.16 The integration path for the curve $\Gamma^{(1)}$ making one circuit around the origin in the counterclockwise direction

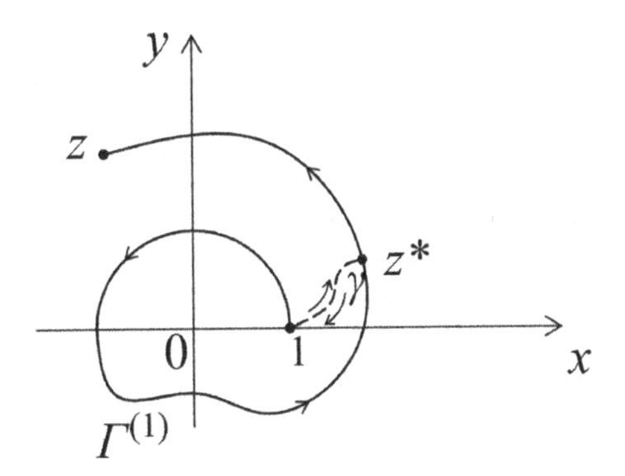

$$\int_{\Gamma_2} \frac{d\zeta}{\zeta} = \int_0^{\arg z} \frac{|z|\, e^{i\varphi} i d\varphi}{|z|\, e^{i\varphi}} = i\varphi \Big|_0^{\arg z} = i \arg z\,.$$

Therefore, we obtain

$$\int_{\Gamma} \frac{d\zeta}{\zeta} = \int_1^z \frac{d\zeta}{\zeta} = \ln|z| + i \arg z \;\; (-\pi < \arg z \leq \pi)\,,$$

for any curve Γ which does not complete a circuit around the origin.

Suppose now that curve, that connects 1 and z, makes one complete circuit around the origin in the counterclockwise direction and denote this curve by $\Gamma^{(1)}$. Pick some point z^* on $\Gamma^{(1)}$ after one circuit and join it with the point 1 by a continuous rectifiable curve γ (see Fig. 2.16). A part of the curve $\Gamma^{(1)}$ from 1 to z^* together with γ form the closed curve C, and the curve γ^{-1} together with the remaining part of $\Gamma^{(1)}$ from the point z^* to z compose the curve Γ of the same type as we have already considered.

Since the integration along γ is performed twice in opposite directions, the integral along $\Gamma^{(1)}$ is represented as

$$\int_{\Gamma^{(1)}} \frac{d\zeta}{\zeta} = \oint_C \frac{d\zeta}{\zeta} + \int_{\Gamma} \frac{d\zeta}{\zeta}\,.$$

The last integral on the right-hand side has already been calculated

$$\int_{\Gamma} \frac{d\zeta}{\zeta} = \ln|z| + i \arg z\,.$$

The integral along C satisfies the conditions of the Cauchy integral formula if we choose $g(\zeta) \equiv 1$ and $z_0 = 0$, in which case

$$\oint_C \frac{d\zeta}{\zeta} = \oint_C \frac{g(\zeta)}{\zeta - 0} d\zeta = 2\pi i \cdot g(0) = 2\pi i\,.$$

Therefore

$$\int_{\Gamma^{(1)}} \frac{d\zeta}{\zeta} = \ln|z| + i \arg z + 2\pi i .$$

If the curve completes k circuits around the origin in the counterclockwise direction, then the Cauchy integral formula should be applied k times, which results in the additional term $2k\pi i$ (if the k circuits are made in the clockwise direction, then the additional term is $-2k\pi i$). Hence, in general situation

$$\int_1^z \frac{d\zeta}{\zeta} = \ln|z| + i \arg z + 2k\pi i \quad (-\pi < \arg z \le \pi),$$

where k is the number of the circuits around the origin that takes into account the used direction—if the direction is counterclockwise, then k is positive; if the direction is clockwise, then k is negative. Thus, in this case, the integral $\int_1^z \frac{d\zeta}{\zeta}$ is a discontinuous function, since the function values are different when the point z is approached using different curves. Consequently, this function cannot be analytic and differentiable. Since the integral $\int_1^z \frac{d\zeta}{\zeta}$ takes different values at the same points of the complex plane, it is said that this integral determines a *multi-valued function*. Notice that for real positive values of z ($z = x > 0$), this integral represents the well-known real function $\int_1^x \frac{d\zeta}{\zeta} = \ln x$. For this reason, the constructed multi-valued function is called the complex logarithm and denoted by $\ln z$

$$\ln z = \int_1^z \frac{d\zeta}{\zeta} = \ln|z| + i \arg z + 2k\pi i .$$

2.7 Analyticity Conditions

Property 2.1 *Algebraic properties and composite function.*

We have already established that differentiability and analyticity of a function in a domain are equivalent conditions. For this reason, *the properties of differentiable functions can be transferred to analytic functions*. In particular, the following properties are true: if $f(z)$ and $g(z)$ are analytic functions in a domain D, then the following functions are also analytic in this domain:

(1) $f(z) \pm g(z)$,
(2) $f(z) \cdot g(z)$,
(3) $\dfrac{f(z)}{g(z)}$, if $g(z) \ne 0$ in D.

For a composite function, we have: if $w = f(z)$ is analytic at a point z_0 and $\zeta = g(w)$ is analytic at the corresponding point $w_0 = f(z_0)$, then the composite function $h(z) = g(f(z))$ is analytic at z_0 and its derivative can be calculated by the formula $h'(z_0) = g'(w_0) \cdot f'(z_0)$. In the case of domains, the result on analyticity of a composite function can be formulated as follows: if $w = f(z)$ is analytic in a domain D and $\zeta = g(w)$ is analytic in the corresponding domain $G = f(D)$, then the composite function $h(z) = g(f(z))$ is analytic in D.

Remark At the moment, the condition that the image $f(D)$ is a domain in the last statement should be considered as hypothesis. However, later (in Sect. 5.2) we will show that any analytic non-constant function maps a domain onto a domain.

Property 2.2 (Theorem on analyticity of the inverse function) *Let function $w = f(z)$ be analytic in a domain D and $f'(z_0) \neq 0$ at some point $z_0 \in D$. Then, in some neighborhood of $w_0 = f(z_0)$ there exists the inverse function $z = g(w) \equiv f^{-1}(w)$, which is analytic and whose derivative can be found by the formula*

$$g'(w) = \frac{1}{f'(z)} .$$

Proof Since $w = f(z) = u(x, y) + iv(x, y)$ is analytic in D, it is infinitely differentiable in D, and consequently, the functions $u = u(x, y)$ and $v = v(x, y)$ are also infinitely differentiable in D. Consider the two auxiliary functions of four variables

$$\begin{cases} F_1(x, y, u, v) = u(x, y) - u \\ F_2(x, y, u, v) = v(x, y) - v \end{cases}$$

and show that these functions satisfy the conditions of the Theorem on the system of implicit functions in a neighborhood of the point $M_0(x_0, y_0, u_0, v_0)$, where $u_0 = u(x_0, y_0)$ and $v_0 = v(x_0, y_0)$.

In fact:

(1) first, the functions F_1 and F_2 are continuous and have continuous partial derivatives in a neighborhood of M_0, because the functions $u = u(x, y)$ and $v = v(x, y)$ are infinitely differentiable and the functions F_1 and F_2 are linear with respect to u and v, which implies that F_1 and F_2 have the continuous partial derivatives of any order with respect to u and v.

(2) second, the functions F_1 and F_2 vanish at the point M_0

$$F_1(M_0) = F_1(x_0, y_0, u_0, v_0) = u(x_0, y_0) - u_0 = 0 ,$$

$$F_2(M_0) = F_2(x_0, y_0, u_0, v_0) = v(x_0, y_0) - v_0 = 0 .$$

(3) third, the Jacobian of F_1, F_2 with respect to x, y is different from 0

$$\Delta = \frac{D(F_1, F_2)}{D(x, y)}(M_0) = \left| \begin{matrix} \frac{\partial F_1}{\partial x} & \frac{\partial F_1}{\partial y} \\ \frac{\partial F_2}{\partial x} & \frac{\partial F_2}{\partial y} \end{matrix} \right|_{M_0} = \left| \begin{matrix} \frac{\partial u}{\partial x} & \frac{\partial u}{\partial y} \\ \frac{\partial v}{\partial x} & \frac{\partial v}{\partial y} \end{matrix} \right|_{M_0}$$

$$= \left(\frac{\partial u}{\partial x} \cdot \frac{\partial v}{\partial y} - \frac{\partial u}{\partial y} \cdot \frac{\partial v}{\partial x} \right) \Big|_{z_0} = \left(\left(\frac{\partial u}{\partial x} \right)^2 + \left(\frac{\partial v}{\partial x} \right)^2 \right) \Big|_{z_0} = \left| f'(z_0) \right|^2 \neq 0 .$$

Here, we have used the Cauchy–Riemann conditions and the following expression for the derivative:

$$f'(z) = \frac{\partial u}{\partial x} + i \frac{\partial v}{\partial x} .$$

Hence, all the conditions of the Theorem on the system of implicit functions are satisfied, which implies that

(1) in some neighborhood of the point $M_0 (x_0, y_0, u_0, v_0)$ the system of the equations

$$\begin{cases} F_1 (x, y, u, v) = 0 \\ F_2 (x, y, u, v) = 0 \end{cases}$$

determines single-valued functions of the variables u and v: $x = x(u, v)$, $y = y(u, v)$. Consequently, the function $z = g(w) = x(u, v) + iy(u, v)$ is defined and, by construction, it is the inverse function to $w = f(z)$. Therefore, the original $w = f(z)$ and inverse $z = g(w)$ functions establish a one-to-one correspondence between a neighborhood of z_0 and the corresponding neighborhood of w_0. We denote these neighborhoods by $U (z_0)$ and $U (w_0)$, respectively;

(2) the functions $x = x(u, v)$ and $y = y(u, v)$ are infinitely differentiable in the neighborhood $U (w_0)$. Notice that the condition $f' (z_0) \neq 0$ guarantees that in the entire neighborhood $U (z_0)$, the derivative $f'(z)$ is different from 0 due to its continuity.

Let us derive now the formula for the derivative $g'(w_0)$. Pick an arbitrary point $w \in U (w_0)$ and add an increment $\Delta w \neq 0$ such that $(w + \Delta w) \in U (w_0)$. Then the corresponding increment of the function $g(w)$ is non-zero due to one-to-one correspondence: $\Delta z = \Delta g(w) = g (w + \Delta w) - g(w) \neq 0$. Moreover, continuity of $x = x(u, v)$ and $y = y(u, v)$ in $U (w_0)$ guarantees continuity of $g(w)$ in $U (w_0)$, that is, $\Delta z = \Delta g(w) \to 0$ as $\Delta w \to 0$. By the definition, the derivative $g'(w)$, if it exists, is the limit $g'(w) = \lim\limits_{\Delta w \to 0} \frac{\Delta g(w)}{\Delta w}$. As it was noted, $\Delta g(w) = \Delta z \neq 0$ if $\Delta w \neq 0$, and consequently, the last limit can be expressed as follows:

$$g'(w) = \lim_{\Delta w \to 0} \frac{\Delta z}{\Delta w} = \lim_{\Delta z \to 0} \frac{1}{\Delta w / \Delta z} = \frac{1}{\lim\limits_{\Delta z \to 0} \frac{\Delta w}{\Delta z}} = \frac{1}{f'(z)} .$$

The hypotheses of the Theorem assure that $f(z)$ is differentiable in D and $f' (z) \neq 0$ in $U (z_0)$. Hence, the last formula shows that the function $g(w)$ is differentiable in the neighborhood $U (w_0)$ and $g'(w) = \frac{1}{f'(z)}$ at every point of this neighborhood. In particular, at the point w_0 we have $g'(w_0) = \frac{1}{f'(z_0)}$.

This completes the proof of the Theorem. □

Property 2.3 (Morera's Theorem) *Let $w = f(z)$ be a continuous function in a finite domain D. If an integral along the boundary of any polygon Ω, contained in D together with its boundary ($\overline{\Omega} \subset D$), is equal to zero, then the function $f(z)$ is analytic in D. (This is the converse of Cauchy's Theorem, or, equivalently, of Goursat's Theorem).*

Proof We show that $f(z)$ is analytic in D by constructing its primitive and showing that the last is a holomorphic function at any point $z_0 \in D$. Since D is a domain, z_0 is contained in D together with its neighborhood, that is, there exists $r > 0$ such that $K = \{|z - z_0| < r\} \subset D$. We take two arbitrary points a and z in the disk K and connect them by a continuous rectifiable curve Γ contained in K. Let us show that $\int_\Gamma f(\zeta)\, d\zeta$ does not depend on the form of the curve Γ. Indeed, consider two continuous rectifiable curves Γ_1 and Γ_2 contained in K, which connect the points a and z. These two curves together form the closed curve $\Gamma_1 \cup \Gamma_2^{-1}$. It was demonstrated in Sect. 1.8 that the integral of a continuous function along a continuous rectifiable curve can be approximated with any desired accuracy by the integral along a polygonal chain contained in K with a finite number of segments. In ϵ-notation, $\forall \varepsilon > 0$ there exists a closed polygonal chain P_ε with a finite number of segments, which is contained in K and such that

$$\left| \int_{\Gamma_1 \cup \Gamma_2^{-1}} f(\zeta)\, d\zeta - \int_{P_\varepsilon} f(\zeta)\, d\zeta \right| < \varepsilon . \tag{2.34}$$

The polygonal chain P_ε can be divided in a finite number of simple closed polygonal lines P_k ($k = 1, \ldots, m$), each of which represents the boundary of a polygon contained in K and consequently in D: $P_k = \partial\Omega_k$, $\overline{\Omega}_k \subset K \subset D$ (see Fig. 2.17). Using the Theorem conditions, we obtain

$$\int_{P_\varepsilon} f(\zeta)\, d\zeta = \sum_{k=1}^{m} \int_{P_k} f(\zeta)\, d\zeta = \sum_{k=1}^{m} \int_{\partial\Omega_k} f(\zeta)\, d\zeta = 0 .$$

Since ε in (2.34) takes an arbitrary positive value, it follows that $\int_{\Gamma_1 \cup \Gamma_2^{-1}} f(\zeta)\, d\zeta = 0$, and then

$$0 = \int_{\Gamma_1 \cup \Gamma_2^{-1}} f(\zeta)\, d\zeta = \int_{\Gamma_1} f(\zeta)\, d\zeta + \int_{\Gamma_2^{-1}} f(\zeta)\, d\zeta = \int_{\Gamma_1} f(\zeta)\, d\zeta - \int_{\Gamma_2} f(\zeta)\, d\zeta .$$

Hence

$$\int_{\Gamma_1} f(\zeta)\, d\zeta = \int_{\Gamma_2} f(\zeta)\, d\zeta ,$$

which means that the integral $\int_\Gamma f(\zeta)\, d\zeta$ does not depend on the form of the curve Γ, it is completely determined by the choice of the initial and final points of the curve.

If we fix the point a and let z be variable, we obtain the function of $z \in K$

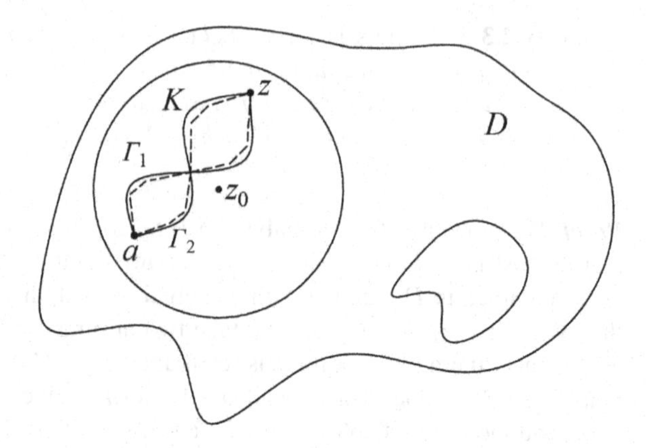

Fig. 2.17 The curves Γ_1 and Γ_2 joining the points a and z, the closed curve $\Gamma_1 \cup \Gamma_2^{-1}$ and the closed polygonal chain P_ε (dashed lines)

$$g(z) = \int_a^z f(\zeta)\, d\zeta = \int_\Gamma f(\zeta)\, d\zeta\,.$$

Using a similar procedure as in the Theorem about a primitive in Sect. 2.6, we will show now that the function $g(z)$ is holomorphic in K and $g'(z) = f(z),\ \forall z \in K$. Notice that z is contained in K together with some neighborhood (since z is an interior point): there exists $\rho > 0$ such that $\{|\zeta - z| < \rho\} \subset K$. Add an increment Δz to z in such a way that $(z + \Delta z) \in \{|\zeta - z| < \rho\}$, and consequently $(z + \Delta z) \in K$. We can always impose this restriction on Δz, because we are interested in the behavior of $g(z)$ when $\Delta z \to 0$. Let us evaluate the following difference:

$$\left|\frac{\Delta g(z)}{\Delta z} - f(z)\right| = \left|\frac{g(z + \Delta z) - g(z)}{\Delta z} - f(z)\right|$$

$$= \frac{1}{|\Delta z|}\left|\int_a^{z+\Delta z} f(\zeta)\, d\zeta - \int_a^z f(\zeta)\, d\zeta - f(z)\Delta z\right| \equiv R\,.$$

Since the integral does not depend on the form of the curve, we choose the following integration path in the first integral: initially, we move from a to z along the same curve Γ, which is used in the second integral, and then we add the line segment γ from z to $z + \Delta z$. Noting that $\Delta z = (z + \Delta z) - z = \int_z^{z+\Delta z} d\zeta$ and that $f(z)$ can be placed inside the integral, since it does not depend on the variable of integration, we continue with the evaluation of R in the form

$$R = \frac{1}{|\Delta z|}\left|\int_\Gamma f(\zeta)\, d\zeta + \int_\gamma f(\zeta)\, d\zeta - \int_\Gamma f(\zeta)\, d\zeta - \int_z^{z+\Delta z} f(z)\, d\zeta\right|$$

$$= \frac{1}{|\Delta z|}\left|\int_\gamma (f(\zeta) - f(z))\, d\zeta\right| \le \frac{1}{|\Delta z|}\int_\gamma |f(\zeta) - f(z)|\, |d\zeta| \equiv R_1\,.$$

The continuity of $f(z)$ at any point $z \in D$ means that

$$\forall \varepsilon > 0 \, \exists \delta(\varepsilon) > 0 \text{ such that } \forall \omega \in D, \; |\omega - z| < \delta \Rightarrow |f(\omega) - f(z)| < \varepsilon .$$

Let us choose Δz such that $|\Delta z| < \delta$. Since γ is the line segment, for any point $\zeta \in \gamma$ we have $|\zeta - z| \le |(z + \Delta z) - z| = |\Delta z| < \delta$. Therefore, $|f(\zeta) - f(z)| < \varepsilon, \forall \zeta \in \gamma$. Using the last inequality in R_1, we obtain

$$R_1 < \frac{1}{|\Delta z|} \cdot \varepsilon \int_\gamma |d\zeta| = \frac{1}{|\Delta z|} \cdot \varepsilon \cdot l_\gamma = \frac{1}{|\Delta z|} \cdot \varepsilon \cdot |\Delta z| = \varepsilon .$$

Hence, we arrive at the following result:

$$\forall \varepsilon > 0 \;\; \exists \delta(\varepsilon) > 0 \text{ such that } \forall \Delta z, \; |\Delta z| < \delta \Rightarrow \left| \frac{\Delta g(z)}{\Delta z} - f(z) \right| < \varepsilon .$$

This means that there exists the limit $\lim\limits_{\Delta z \to 0} \frac{\Delta g(z)}{\Delta z} = f(z)$, that is, there exists the derivative $g(z)$: $g'(z) = f(z), \; \forall z \in K$. Thus, $g(z)$ is holomorphic and consequently analytic in K. Since a derivative of any order of an analytic function is again an analytic function, $f(z) = g'(z)$ is analytic in K and, in particular, at z_0. The proof is completed by recalling that z_0 is an arbitrary point in D, which means that $f(z)$ is analytic in D. □

Property 2.4 (Theorem on analyticity of the limit function) *Suppose $f(z, w)$ is defined in $D \times E$, where D is a domain and E is a set in $\mathbb{C}$ with a limit point w_0. If*

(1) $f(z, w)$ is analytic with respect to z in D for any fixed $w \in E$;
(2) $f(z, w) \underset{w \to w_0}{\overset{normal}{\rightrightarrows}} g(z)$,

then

(1) the limit function $g(z)$ is analytic in D;
(2) $\dfrac{\partial^n f(z, w)}{\partial z^n} \underset{w \to w_0}{\overset{normal}{\rightrightarrows}} \dfrac{d^n g(z)}{dz^n}, \; \forall n \in \mathbb{N}$.

Proof Recall first that the normal convergence on a domain D means the uniform convergence on every compact set B contained in D.

We start the proof with the verification of the conditions of Morera's Theorem for the function $g(z)$ in the domain D. Since $f(z, w)$ is analytic with respect to z in D, it is continuous with respect to z in D. The continuity of $f(z, w)$ together with the normal convergence of $f(z, w)$ to $g(z)$ on D guarantee the continuity of the limit function $g(z)$ in D (see the corresponding Theorem on continuity in Sect. 1.6). Now take any polygon Ω contained in D together with its boundary ($\overline{\Omega} \subset D$) and calculate the integral

$$I \equiv \int_{\partial\Omega} g(z)dz = \int_{\partial\Omega} \lim_{w\to w_0} f(z,w)dz\,.$$

The boundary $\partial\Omega$ is a bounded closed set contained in D, and therefore, $f(z,w)$ converges uniformly to $g(z)$ on $\partial\Omega$: $f(z,w) \underset{w\to w_0}{\overset{\partial\Omega}{\rightrightarrows}} g(z)$. Therefore, we can interchange the limit and integration signs (see Theorem 1.1 in Sect. 1.9) in I and then apply Goursat's Theorem (taking into account that $f(z,w)$ is analytic with respect to z in D and that $\partial\Omega$ is the boundary of a polygon contained in D)

$$I = \lim_{w\to w_0} \int_{\partial\Omega} f(z,w)dz = 0\,.$$

Hence, both conditions of Morera's Theorem are satisfied for $g(z)$, and consequently, $g(z)$ is an analytic function in D.

Now we analyze the behavior of the derivatives. The functions $f(z,w)$ and $g(z)$ have the derivatives of any order in D, since both functions are analytic with respect to z in D. Let us consider a compact set B, which is contained in D: $B \subset D$. According to the hypothesis of the normal convergence of $f(z,w)$ on D, the function $f(z,w)$ converges uniformly on B. We will show that the derivatives of $f(z,w)$ also converges uniformly on B to the corresponding derivatives of $g(z)$ $\frac{\partial^n f(z,w)}{\partial z^n} \underset{w\to w_0}{\overset{B}{\rightrightarrows}} \frac{d^n g(z)}{dz^n}$. Denote by $d = \rho(B,\partial D) > 0$ the distance between the compact set B and the boundary ∂D. This distance is positive, because B and ∂D are closed sets without common points and at least one of them (B) is bounded. Construct a new domain G by adding to the set B all the points whose distance to B is less than $\frac{d}{2}$. The new domain satisfies the following relations: $B \subset G, \overline{G} \subset D$ and $\rho(B,\partial G) = \frac{d}{2}$. Using the expressions (2.31) for nth derivatives of the analytic functions $f(z,w)$ and $g(z)$, we evaluate the difference between the two derivatives

$$\left|\frac{\partial^n f(z,w)}{\partial z^n} - \frac{d^n g(z)}{dz^n}\right| = \left|\frac{n!}{2\pi i}\int_{\partial G} \frac{f(\zeta,w)}{(\zeta-z)^{n+1}}d\zeta - \frac{n!}{2\pi i}\int_{\partial G} \frac{g(\zeta)}{(\zeta-z)^{n+1}}d\zeta\right|$$

$$= \frac{n!}{2\pi}\left|\int_{\partial G} \frac{f(\zeta,w)-g(\zeta)}{(\zeta-z)^{n+1}}d\zeta\right| \le \frac{n!}{2\pi}\int_{\partial G} \frac{|f(\zeta,w)-g(\zeta)|}{|\zeta-z|^{n+1}}|d\zeta| \equiv R\,.$$

According to the construction, the domain G is bounded by a finite number of piecewise smooth curves, that is, there exists a finite length $l_{\partial G}$ of the boundary ∂G. Moreover, the distance from ∂G to any point z of the set B is not less than $\frac{d}{2}$: $|\zeta - z| \ge \frac{d}{2}$, $\forall z \in B$ and $\forall \zeta \in \partial G$. The boundary ∂G is a compact set, and consequently $f(z,w) \underset{w\to w_0}{\overset{\partial G}{\rightrightarrows}} g(z)$, that is, $\forall \varepsilon > 0$ (choose $\varepsilon_1 = \frac{\varepsilon\cdot 2\pi\cdot(d/2)^{n+1}}{n!\cdot l_{\partial G}}$) $\exists\delta = \delta(\varepsilon)$ such that for all $w \in E$, $0 < |w - w_0| < \delta$ and simultaneously for all $\zeta \in \partial G$ it follows $|f(\zeta,w) - g(\zeta)| < \varepsilon_1$. Notice that in this definition δ depends only on ε and does not depend on the choice of the points $\zeta \in \partial G$. Employing the last inequality for the evaluation of R, we obtain

$$R < \frac{n!}{2\pi} \cdot \varepsilon_1 \cdot \frac{1}{(d/2)^{n+1}} \int_{\partial G} |d\zeta| = \frac{n!}{2\pi} \cdot \varepsilon_1 \cdot \frac{l_{\partial G}}{(d/2)^{n+1}} = \varepsilon ,$$

that is,

$$\left| \frac{\partial^n f(z,w)}{\partial z^n} - \frac{d^n g(z)}{dz^n} \right| < \varepsilon .$$

Since this evaluation is true simultaneously for all points $z \in B$, it means the uniform convergence on B

$$\frac{\partial^n f(z,w)}{\partial z^n} \underset{w \to w_0}{\overset{B}{\rightrightarrows}} \frac{d^n g(z)}{dz^n} .$$

Recalling that B is an arbitrary compact set in D, we obtain the normal convergence on D.

This terminates the proof of the Theorem. □

Corollary 2.1 *Let a sequence $\{f_n(z)\}_{n=1}^{+\infty}$ be defined on a domain D. If*

(1) the functions $f_n(z)$ are analytic in D, $\forall n \in \mathbb{N}$;
(2) $f_n(z) \underset{n\to\infty}{\overset{normal}{\rightrightarrows}} g(z)$,

then

(1) the limit function $g(z)$ is analytic in D;
(2) $\dfrac{d^k f_n(z)}{dz^k} \underset{n\to\infty}{\overset{normal}{\rightrightarrows}} \dfrac{d^k g(z)}{dz^k}$, $\forall k \in \mathbb{N}$.

This Corollary is a specific case of the last Theorem when $E = \mathbb{N}$.

Corollary 2.2 *Let a series $\sum_{n=1}^{+\infty} u_n(z)$ be defined on a domain D. If*

(1) the functions $u_n(z)$ are analytic in D, $\forall n \in \mathbb{N}$;
(2) the series converges normally on D,

then

(1) the sum of series is an analytic function in D;
(2) the nth derivative of the series exists in D and can be found by term-by-term differentiation.

Proof Denote by $g(z)$ the sum of the series—$g(z) = \sum_{n=1}^{+\infty} u_n(z)$, and by $f_n(z)$ the partial sums—$f_n(z) = \sum_{k=1}^{n} u_k(z)$. The functions $f_n(z)$ are analytic in D as the sums of a finite number of analytic functions, and by the definition of uniform convergence of a series, the sequence $f_n(z)$ converges normally on D: $f_n(z) \underset{n\to\infty}{\overset{normal}{\rightrightarrows}} g(z)$. Therefore, by Corollary 2.1, the function $g(z)$ is analytic in D and

$$\frac{d^m f_n(z)}{dz^m} \overset{normal}{\underset{n\to\infty}{\rightrightarrows}} \frac{d^m g(z)}{dz^m}, \forall m \in \mathbb{N}. \tag{2.35}$$

At the same time, $\frac{d^m f_n(z)}{dz^m} = \sum_{k=1}^{n} \frac{d^m u_k(z)}{dz^m}$. Passing to the limit as $n \to \infty$, we obtain the series $\sum_{k=1}^{+\infty} \frac{d^m u_k(z)}{dz^m}$ and the condition (2.35) means that this series converges uniformly to the function $\frac{d^m g(z)}{dz^m}$, that is,

$$\sum_{k=1}^{+\infty} \frac{d^m u_k(z)}{dz^m} = \frac{d^m g(z)}{dz^m} = \frac{d^m}{dz^m}\left(\sum_{k=1}^{+\infty} u_k(z)\right).$$

□

Corollary 2.3 *The sum of a power series is an analytic function inside the circle of convergence and the power series is infinitely differentiable in the disk of convergence.*

This is just a partial statement of Corollary 2.3 in Sect. 2.1 reiterated for the completeness of the results.

Property 2.5 (Theorem on analyticity of an integral depending on a parameter) *Suppose $f(z,w)$ is defined on $D \times \Gamma$, where D is a domain and Γ is a continuous rectifiable curve. If*

(1) $f(z,w)$ is continuous in $D \times \Gamma$;
(2) $f(z,w)$ is analytic with respect to z in D for each fixed $w \in \Gamma$,

then

(1) the function $F(z) = \int_\Gamma f(z,w)dw$ is analytic in D;
(2) $F^{(k)}(z) = \left(\int_\Gamma f(z,w)dw\right)^{(k)} = \int_\Gamma f^{(k)}_{z^k}(z,w)dw$, where $F^{(k)}(z) \equiv \frac{d^k F(z)}{dz^k}$ and $f^{(k)}_{z^k}(z,w) \equiv \frac{\partial^k f(z,w)}{\partial z^k}$, that is, the integral $\int_\Gamma f(z,w)dw$ depending on the parameter z can be differentiated k times inside the integral sign, $\forall k \in \mathbb{N}$.

Proof To prove the analyticity of $F(z)$ in D, we show that it satisfies the conditions of Morera's Theorem. Indeed, continuity of $f(z,w)$ on $D \times \Gamma$ implies continuity of $F(z) = \int_\Gamma f(z,w)dw$ in D (see Theorem 1.2 in Sect. 1.9 on continuity of an integral depending on a parameter). Consider now any polygon Ω contained in D together with its boundary and calculate $\int_{\partial\Omega} F(z)dz = \int_{\partial\Omega}\int_\Gamma f(z,w)dwdz$. Since $\partial\Omega \times \Gamma$ is a compact set and the function $f(z,w)$ is continuous on this set, the order of integration can be interchanged (according to Theorem 1.3 in Sect. 1.9) and, using analyticity of $f(z,w)$ with respect to z, we can apply Goursat's Theorem with respect to z to obtain

$$\int_{\partial\Omega} F(z)dz = \int_{\partial\Omega}\int_{\Gamma} f(z,w)dwdz = \int_{\Gamma}\int_{\partial\Omega} f(z,w)dzdw = 0\,.$$

Therefore, $F(z)$ satisfies both conditions of Morera's Theorem, and consequently, it is analytic in D.

Now we will find the kth derivative ($\forall k \in \mathbb{N}$) of $F(z)$ employing the Theorem on infinite differentiability of an analytic function (see Corollary 2.1 to Theorem 2.3 in Sect. 2.1). For any point z in D there exists its neighborhood contained in D: $\exists r > 0$ such that $\{|\zeta - z| \le r\} \subset D$. Applying formula (2.30) (Sect. 2.4) for the analytic function $F(z)$, we can write

$$F^{(k)}(z) = \frac{k!}{2\pi i}\int_{|\zeta - z|=r} \frac{F(\zeta)}{(\zeta - z)^{k+1}}d\zeta$$

$$= \frac{k!}{2\pi i}\int_{|\zeta - z|=r} \frac{1}{(\zeta - z)^{k+1}}\int_{\Gamma} f(\zeta, w)\,dwd\zeta = \frac{k!}{2\pi i}\int_{|\zeta - z|=r}\int_{\Gamma} \frac{f(\zeta, w)}{(\zeta - z)^{k+1}}dwd\zeta \equiv R\,.$$

Consider the function $g(\zeta, w) = \frac{f(\zeta, w)}{(\zeta - z)^{k+1}}$ where z is a fixed point. Since $|\zeta - z| = r$, the denominator is different from zero, and consequently, the function $g(\zeta, w)$ is continuous on the compact set $\{|\zeta - z| = r\} \times \Gamma$ as the ratio of two continuous functions. Therefore, we can interchange the order of integration in R and then use formula (2.30) for the kth derivative of the analytic function $f(z, w)$ with respect to z

$$R = \frac{k!}{2\pi i}\int_{\Gamma}\int_{|\zeta - z|=r} \frac{f(\zeta, w)}{(\zeta - z)^{k+1}}d\zeta dw = \int_{\Gamma} \frac{k!}{2\pi i}\int_{|\zeta - z|=r} \frac{f(\zeta, w)}{(\zeta - z)^{k+1}}d\zeta dw = \int_{\Gamma} f_{z^k}^{(k)}(z, w)dw\,.$$

Hence,

$$F^{(k)}(z) = \int_{\Gamma} f_{z^k}^{(k)}(z, w)dw\,.$$

The Theorem is proved. □

Corollary *Consider the function* $g(z, w) = \frac{f(w)}{w - z}$ *defined on* $D \times \Gamma$*, where* D *is a domain and* Γ *is a continuous rectifiable curve having no common points with* D*:* $\Gamma \cap D = \varnothing$*. If* $f(w)$ *is continuous on* Γ*, then* $g(z, w)$ *is continuous on* $D \times \Gamma$*, because* $g(z, w)$ *is the ratio of two continuous functions and the denominator is different of zero for any* z *in* D *and* w *in* Γ*. Moreover,* $g(z, w)$ *is differentiable in* z *and consequently analytic in* D*:* $g_z'(z, w) = \frac{f(w)}{(w - z)^2}$*. Therefore, the function* $g(z, w)$ *satisfies the conditions of the last Theorem and then*

(1) the function

$$F(z) = \frac{1}{2\pi i}\int_{\Gamma} \frac{f(w)}{w - z}dw \tag{2.36}$$

is analytic in D*;*

(2) the derivatives of $F(z)$ *can be found by the formula*

$$F^{(k)}(z) = \frac{1}{2\pi i}\int_{\Gamma}\left(\frac{f(w)}{w-z}\right)^{(k)}_{z^k} dw = \frac{k!}{2\pi i}\int_{\Gamma}\frac{f(w)}{(w-z)^{k+1}}dw\,. \tag{2.37}$$

Notice that the expression in (2.36) is similar to the Cauchy integral formula (2.19), and (2.37) resembles formula (2.31) for the kth derivative of an analytic function. Due to this reason, the integral in (2.36) is called the *integral of the Cauchy type*. However, in formulas (2.36) and (2.37), the restrictions on $f(z)$ are weaker than in (2.19) and (2.31), since $f(z)$ needs only be continuous on Γ. If $\Gamma = \partial D$ and $f(z)$ is analytic in D and continuous on $\overline{D}$, then the integral of the Cauchy type (2.36) turns into the Cauchy integral formula and in this case $F(z) = f(z), \forall z \in D$.

2.8 Uniqueness of an Analytic Function

Theorem (Uniqueness of an analytic function) *Assume a function $w = f(z)$ is analytic in a domain D and E is a subset of D possessing a limit point that belongs to D. If $f(z) = 0$ on E, then $f(z) \equiv 0$ on D.*

Proof Let a be a limit point of E that lies in D. Since $a \in D$, there exists its neighborhood contained in D: $\exists r > 0$ such that $K_a = \{|z-a| < r\} \subset D$. According to Corollary 2.4 to the Theorem in Sect. 2.4, the function $f(z)$ can be expanded in the power series convergent in the disk K_a

$$f(z) = \sum_{n=0}^{+\infty} c_n (z-a)^n\,. \tag{2.38}$$

Since a is a limit point of E, there exists a sequence $\{z_k\}_{k=1}^{\infty}$, contained in E, such that $z_k \neq a$, $z_k \underset{k\to\infty}{\to} a$. Consequently, starting from some index N_0, all the points z_k, $\forall k \geq N_0$, belong to the disk K_a and the series expansion (2.38) is valid at each of these points. Therefore, we obtain the equations $f(z_k) = \sum_{n=0}^{+\infty} c_n (z_k - a)^n = 0$. Since the power series converges uniformly, in the last equality we can calculate the limit term by term as $k \to \infty$ and get $c_0 = 0$. Hence, the series (2.38) can be rewritten in the form $f(z) = \sum_{n=1}^{+\infty} c_n (z-a)^n$. We can divide the last series by $z_k - a$ (recall that $z_k - a \neq 0$) and evaluate it at z_k

$$\frac{f(z_k)}{z_k - a} = \sum_{n=1}^{+\infty} c_n (z_k - a)^{n-1} = 0\,.$$

Again taking the limit term by term in the last equation as $k \to \infty$, we have $c_1 = 0$ and the series (2.38) takes the form $f(z) = \sum_{n=2}^{+\infty} c_n (z-a)^n$. Similarly, dividing the last series by $(z_k - a)^2$, evaluating at z_k and taking term-by-term limit as $k \to \infty$, we get $c_2 = 0$; and so on. Applying this procedure successively for all indices n, we

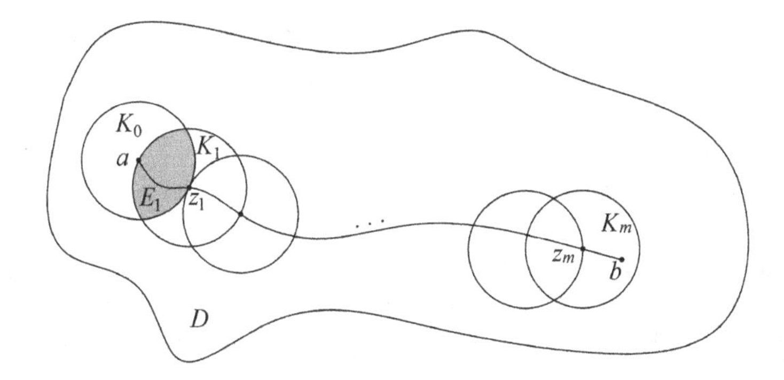

Fig. 2.18 The disks K_j, $j = 1, \ldots, m$ such that $\Gamma \subset \cup_{j=1}^{m} K_j \subset D$

conclude that $c_n = 0$, $\forall n$, and consequently $f(z) \equiv 0$ in the disk K_a. Notice that K_a is an arbitrary disk centered at a and contained in D.

We will show now that $f(b) = 0$ for an arbitrary point $b \in D$. Since both a and b are in the domain D, they can be connected by a continuous rectifiable curve Γ contained in D. The distance $\rho = \rho(\Gamma, \partial D)$ from Γ to the boundary of D is positive, because Γ and ∂D are closed sets with no common points and the curve Γ is a bounded set. Construct the disk K_0 centered at a with the radius ρ: $K_0 = \{|z - a| \leq \rho\}$ (see Fig. 2.18). Let us start the traversal of Γ from a to b and denote by z_1 the first point of the intersection between Γ and the circle $|z - a| = \rho$. Construct the next disk K_1 centered at z_1 with the radius ρ. Continue the traversal of Γ from z_1 to b and denote by z_2 the first point of the intersection between Γ and the circle $|z - z_1| = \rho$; and so on. Since Γ is a rectifiable curve and each disk K_j contains a part of the curve with the length equal or greater than 2ρ, then in a finite number of these steps, say in m steps, we reach the final point b of the curve Γ. Hence, $\Gamma \subset \cup_{j=1}^{m} K_j \subset D$, as shown in Fig. 2.18.

Denote $E_1 = K_0 \cap K_1$ and observe that $E_1 \subset D$, z_1 is a limit point of E_1, $z_1 \in D$ and $f(z) = 0$, $\forall z \in E_1$ since $f(z) \equiv 0$ in K_0 (see Fig. 2.18). Then, according to the demonstrated part of the Theorem, $f(z) \equiv 0$ in any disk centered at z_1 and contained in D, in particular, $f(z) \equiv 0$ in K_1. Next, denote $E_2 = K_1 \cap K_2$ and note that $E_2 \subset D$, z_2 is a limit point of E_2, $z_2 \in D$ and $f(z) = 0$, $\forall z \in E_2$. Therefore, $f(z) \equiv 0$ in K_2; and so on. In a finite number of similar steps we reach the disk K_m and show that $f(z) \equiv 0$ in K_m, and in particular, $f(b) = 0$. Hence, $f(z) = 0$, $\forall z \in D$, that is, $f(z) \equiv 0$ in D. □

Corollary *Suppose $f(z)$ and $g(z)$ are analytic functions in a domain D, a set E is contained in D and a is a limit point of E, $a \in D$. If $f(z) = g(z)$, $\forall z \in E$, then $f(z) \equiv g(z)$ in D.*

Proof Considering the function $F(z) = f(z) - g(z)$, we can see that it satisfies all the conditions of the last Theorem: $F(z)$ is analytic in D and $F(z) = 0, \forall z \in E$, where E is a subset of D, which has a limit point belonging to D. Therefore, $F(z) \equiv 0$ in D, that is, $f(z) \equiv g(z)$ in D. □

This Corollary states that *an analytic in a domain D function is defined uniquely by its values in a set $E \subset D$ that has a limit point contained in D.*

2.9 Analytic Continuation

Definition Let D be a domain and E be a subset of D, $E \subset D$. Suppose the function $f(z)$ is defined in E and the function $F(z)$ is defined and analytic in D. If $f(z) = F(z), \forall z \in E$, then $F(z)$ is said to be an *analytic continuation (or extension)* of $f(z)$ from E onto D.

Theorem (Uniqueness of an analytic continuation) *Let $f(z)$ be defined on a set E, $E \subset D$. If the set E has a limit point in D and if an analytic continuation of $f(z)$ from E to D can be performed, then it is unique.*

This result is an immediate consequence of Corollary to the Theorem in Sect. 2.8.

Example 2.1 Consider the exponential function e^x defined on the real axis, which can be represented in the power series $e^x = \sum_{n=0}^{+\infty} \frac{x^n}{n!}$, $\forall x \in \mathbb{R}$. Consider also the complex series $f(z) = \sum_{n=0}^{+\infty} \frac{z^n}{n!}, z \in \mathbb{C}$, which converges on the entire complex plane $\mathbb{C}$ because its radius of the convergence is infinite:

$$R = \lim_{n\to\infty} \left| \frac{c_n}{c_{n+1}} \right| = \lim_{n\to\infty} \frac{(n+1)!}{n!} = \lim_{n\to\infty} (n+1) = +\infty .$$

Consequently, the sum $f(z)$ of this series is an analytic function in the entire complex plane $\mathbb{C}$. Notice that $\mathbb{R} \subset \mathbb{C}$ and $f(x) = \sum_{n=0}^{+\infty} \frac{x^n}{n!} = e^x$ for $\forall z \in \mathbb{R}$ (that is, for $z = x$). This means, according to the definition, that $f(z)$ is the analytic continuation of e^x from the real axis onto the entire complex plane (in this case, $E = \mathbb{R}$ and $D = \mathbb{C}$). Since $f(x) = e^x$ when $z = x$, it is natural to denote $f(z) = e^z$. Notice that the set $E = \mathbb{R}$ has limit points in $D = \mathbb{C}$ (actually, every $x \in \mathbb{R} \subset \mathbb{C}$ is a limit point of E) and for this reason the analytic continuation $f(z) = e^z$ is unique according to the Uniqueness Theorem.

Consider some properties of the function e^z. First, we show that $e^{z_1+z_2} = e^{z_1} \cdot e^{z_2}$ for $\forall z_1, z_2 \in \mathbb{C}$. Indeed, this equality is true for real values

$$e^{x_1+x_2} = e^{x_1} \cdot e^{x_2}, \quad \forall x_1, x_2 \in \mathbb{R} . \tag{2.39}$$

Let us fix for a moment any value of $x_2 \in \mathbb{R}$ and perform an analytic extension of the functions $e^{x_1+x_2}$ and $e^{x_1} \cdot e^{x_2}$ from the real axis ($x_1 \in \mathbb{R}$) onto the complex plane ($z_1 \in \mathbb{C}$). As was just shown, such analytic continuation exists and is unique for each of the functions. Since for the real values $z_1 = x_1$ these two functions coincide (see the equality (2.39)), by the Uniqueness Theorem, they have also the same values for any z_1

$$e^{z_1+x_2} = e^{z_1} \cdot e^{x_2}, \quad \forall z_1 \in \mathbb{C}. \tag{2.40}$$

Now, in relation (2.40), let us fix an arbitrary value of $z_1 \in \mathbb{C}$ and extend analytically the functions $e^{z_1+x_2}$ and $e^{z_1} \cdot e^{x_2}$ from the real axis ($x_2 \in \mathbb{R}$) onto the complex plane ($z_2 \in \mathbb{C}$). Such analytic continuation exists and is unique for both functions. Since the functions are equal on the real axis (see formula (2.40) where $z_2 = x_2$), by the Uniqueness Theorem, they coincide for all the values of $z_2 \in \mathbb{C}$. Hence, the equality $e^{z_1+z_2} = e^{z_1} \cdot e^{z_2}$ is true for $\forall z_1, z_2 \in \mathbb{C}$.

Some immediate consequences of the property $e^{z_1+z_2} = e^{z_1} \cdot e^{z_2}$ are the following. Using the shown equality and the Euler formula $e^{iy} = \cos y + i \sin y$ for real argument y, we can write $w = e^z = e^{x+iy} = e^x \cdot e^{iy} = e^x (\cos y + i \sin y)$. Then $|w| = |e^z| = e^x$ and $\arg w = \arg e^z = y$. Since $e^x \neq 0, \ \forall x \in \mathbb{R}$, it follows that $|e^z| \neq 0$ and consequently $e^z \neq 0, \ \forall z \in \mathbb{C}$.

Example 2.2 It is well known that the real functions $\sin x$ and $\cos x$ can be expanded in the power series for $\forall x \in \mathbb{R}$

$$\cos x = \sum_{n=0}^{+\infty} (-1)^n \frac{x^{2n}}{(2n)!}, \quad \sin x = \sum_{n=0}^{+\infty} (-1)^n \frac{x^{2n+1}}{(2n+1)!}.$$

We carry out the analytic continuation of these functions onto the complex plane $\mathbb{C}$ by introducing the complex power series

$$f_1(z) = \sum_{n=0}^{+\infty} (-1)^n \frac{z^{2n}}{(2n)!}, \tag{2.41}$$

$$f_2(z) = \sum_{n=0}^{+\infty} (-1)^n \frac{z^{2n+1}}{(2n+1)!}. \tag{2.42}$$

Both series (2.41) and (2.42) converge on the entire complex plane $\mathbb{C}$ since their radii of convergence are infinite. Therefore, the sums of these series $f_1(z)$ and $f_2(z)$ are analytic functions in $\mathbb{C}$ and, by the definition, $f_1(z)$ and $f_2(z)$ are the analytic continuations of $\cos x$ and $\sin x$, respectively, from $\mathbb{R}$ onto $\mathbb{C}$. For $z = x$, $f_1(x) = \cos x$, $f_2(x) = \sin x$ that justifies the notations $f_1(z) = \cos z$ and $f_2(z) = \sin z$. Since $E = \mathbb{R}$ has limit points in $D = \mathbb{C}$, according to the Uniqueness Theorem both continuations are unique.

Let us show that the Euler formula is true for $\forall z \in \mathbb{C}$

$$e^{iz} = \cos z + i \sin z. \tag{2.43}$$

In fact, the Euler formula is valid for any real number ($z = x$)

$$e^{ix} = \cos x + i \sin x. \tag{2.44}$$

Considering the analytic continuations of the functions e^{ix}, $\cos x$ and $\sin x$ from the real axis onto the complex plane, we obtain the functions e^{iz}, $\cos z$ and $\sin z$. Since the equality (2.44) is true for any real number, according to the uniqueness of the analytic continuation, the same relation is also valid for any complex number, and consequently, formula (2.43) is satisfied. It follows immediately from (2.43) that both formulas $e^{iz} = \cos z + i \sin z$ and $e^{-iz} = \cos z - i \sin z$ are valid, as well as their linear combinations

$$\cos z = \frac{e^{iz} + e^{-iz}}{2}, \quad \sin z = \frac{e^{iz} - e^{-iz}}{2i}.$$

Later (in Chap. 4) we will discuss the properties of the functions e^z, $\cos z$ and $\sin z$ in detail.

2.10 Multi-valued Analytic Functions and Riemann Surfaces

Let a function $f(z)$ be analytic in a domain D. Take an arbitrary $z_0 \in D$ and expand $f(z)$ in the power series at the point z_0

$$f(z) = \sum_{n=0}^{+\infty} c_n (z - z_0)^n . \tag{2.45}$$

It was shown in Sect. 2.4 that the radius of convergence R_0 of this series is not less than the distance from z_0 to the boundary of D: $R_0 \geq \rho(z_0, \partial D)$. Hence, we have two possibilities:

(1) $R_0 = \rho(z_0, \partial D)$;
   or
(2) $R_0 > \rho(z_0, \partial D)$.

In the first case, the series (2.45) converges in the disk $K_0 = \{|z - z_0| < \rho(z_0, \partial D)\}$, but does not converge in any disk centered at z_0 with a greater radius. Therefore, an analytic continuation from D in the direction of the point $z^* \in \partial D \cap \partial K_0$ is impossible. Hence, there is no analytic element at z^* and this point is called a *singular point* (see Fig. 2.19 where z_0 is denoted by $\widetilde{z}_0$ and K_0 by $\widetilde{K}_0$ in the case of a singular point).

Let us consider the second option: $R_0 > \rho(z_0, \partial D)$. In this case, the series (2.45) converges in the entire disk $K_0 = \{|z - z_0| < R_0\}$ extended beyond the domain D, that is, the sum of the series is an analytic function in the disk K_0 and, in particular, at the points of K_0 lying outside D where the function $f(z)$ was not originally defined (see Fig. 2.19). Denote by $f_0(z) = \sum_{n=0}^{+\infty} c_n^{(0)} (z - z_0)^n$ the sum of this series in K_0 and introduce the function

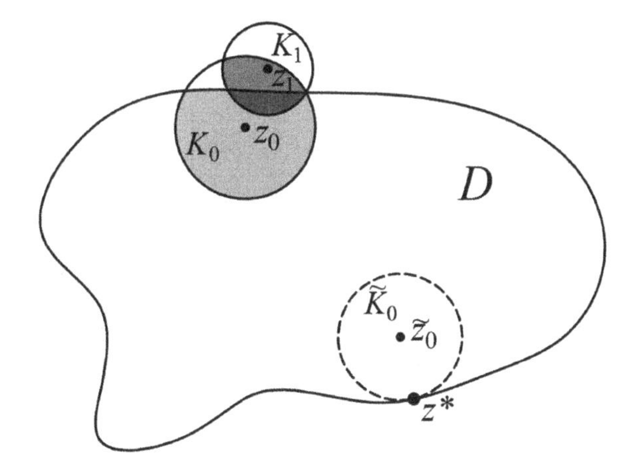

Fig. 2.19 Analytic continuation from original domain D onto the disks K_0 and K_1, and a singular point $z^* \in \partial D \cap \partial \widetilde{K}_0$ without an analytic element

$$F_0(z) = \begin{cases} f(z), z \in D \\ f_0(z), z \in K_0 \end{cases}$$

on the domain $D_0 = D \cup K_0$. Suppose for a moment that $D \cap K_0$ is a connected set. Then, $f(z) = f_0(z)$ for any $z \in D \cap K_0$, and consequently, by the Theorem in Sect. 2.8, the function $F_0(z)$ is uniquely defined and analytic in $D_0 = D \cup K_0$. Therefore, $F_0(z)$ is the *analytic continuation (or extension)* of $f(z)$ from D onto D_0, which is also called *direct analytic continuation* to distinguish from the case of multi-stage continuation. The function $f(z)$ is called the *analytic (original) element of the function* $F_0(z)$ in D, and $f_0(z)$ is the *analytic element* of $F_0(z)$ in K_0 (or the analytic element at the point z_0). Since for $\forall z \in D \cap K_0$ the elements $f(z)$ and $f_0(z)$ coincide, we glue together D and K_0 along their common part.

Take now an arbitrary point $z_1 \in D_0$ and expand again the constructed function $F_0(z)$ (by definition, it is equal to $f(z)$ if $z_1 \in D$ or $f_0(z)$ if $z_1 \in K_0$) in the power series: $F_0(z) = \sum_{n=0}^{+\infty} c_n^{(1)} (z - z_1)^n$. The radius of convergence of the last series satisfies the inequality $R_1 \geq \rho(z_1, \partial D_0)$. Suppose again that $R_1 > \rho(z_1, \partial D_0)$. Then we have the analytic function $f_1(z) = \sum_{n=0}^{+\infty} c_n^{(1)} (z - z_1)^n$ in the disk $K_1 = \{|z - z_1| < R_1\}$, which represents the analytic element in K_1 of the following function extended on the set $D \cup K_0 \cup K_1$

$$F_1(z) = \begin{cases} f(z), z \in D \\ f_0(z), z \in K_0 \\ f_1(z), z \in K_1 \end{cases}.$$

Again assuming (for a moment) that $D_0 \cap K_1$ is a connected set, we define the only analytic function $F_1(z)$ on the entire set $D_1 = D \cup K_0 \cup K_1$. Geometrically, in the z-plane, we connect D_0 and K_1 over their common part, that is, we consider each point of $D_0 \cap K_1$ only one time (see Fig. 2.19). And so on: we extend the original

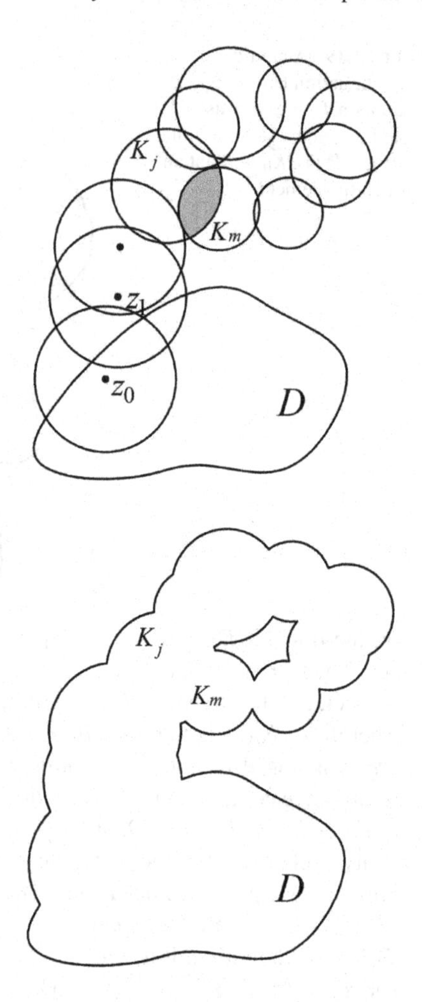

Fig. 2.20 The disks K_j and K_m with the corresponding analytic elements f_j and f_m, which have common points

Fig. 2.21 The domain obtained by gluing the disks K_j and K_m over their common part in the case $f_m(z) = f_j(z)$ for $\forall z \in K_m \cap K_j$

function analytically in all possible directions and construct new analytic elements, by gluing together the common parts of the successive disks in the z-plane.

Let us see what happens if two non-consecutive disks overlap in some part (this is the case of non-direct analytic continuation). Suppose that at the mth step of this procedure we obtain an analytic element $f_m(z)$ in the disk K_m, which has common points with one of the previously constructed disks K_j, $m \neq j+1$ ($K_m \cap K_j \neq \varnothing$) or with the domain D ($K_m \cap D \neq \varnothing$). This situation is illustrated in Fig. 2.20.

Recall that in the disk K_j there exists its own analytic element. If $f_m(z) = f_j(z)$ for $\forall z \in K_m \cap K_j$, then the extended function is uniquely defined at the common points of the disks K_m and K_j, and in this case we glue together the disks K_m and K_j over their common part (see Fig. 2.21). Otherwise, if there exists a point $z \in K_m \cap K_j$ such that $f_m(z) \neq f_j(z)$, then the analytically continued function (defined as the set of the analytic elements) is not single-valued. If this happens, we do not paste together

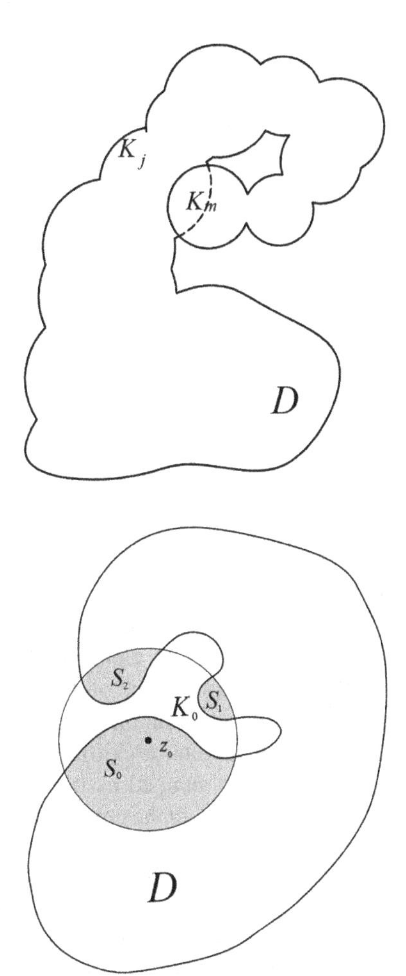

Fig. 2.22 A part of the Riemann surface in the case $f_m(z) \neq f_j(z)$, $z \in K_m \cap K_j$: the disk K_m lies above the disk K_j

Fig. 2.23 The case when $D \cap K_0$ is not a connected set

the disks K_m and K_j, that is, each point belonging to their intersection we consider geometrically as two different points—one in K_j and another in K_m (see Fig. 2.22).

If now we drop the assumption of connectedness of the intersection at each step of this procedure, then we notice that a similar situation may occur even on the first direct analytic continuation. In fact, let us return to the definition of the function $F_0(z)$ on the domain $D_0 = D \cup K_0$ and suppose that the intersection $D \cap K_0$ consists of a few separated sets S_i, $i = 0, \dots, p$, each of which is a connected set, and the point z_0 lies in S_0 (see Fig. 2.23). According to the construction of $F_0(z)$, we have $f(z) = f_0(z)$ for each $z \in D \cap S_0$. However, $f(z)$ and $f_0(z)$ can take different values in other parts S_i, $i = 1, \dots, p$. In all the sets S_i (including, at least, S_0) where $f(z)$ and $f_0(z)$ coincide, we glue together common point of S_i and D, while for the remaining sets S_i where $f(z) \neq f_0(z)$, we don't paste together the points of the intersection of S_i

Fig. 2.24 An example of the Riemann surface: the parts where the function extension has the same values are joined together and the parts with different values at the same points are placed one above another

with D, instead we consider these points twice—one exemplar in D and another in S_i.

In this way, we proceed with the analytic continuation of the original element $f(z)$ and all the analytic elements already obtained in all possible directions, pasting together the common parts of different disks if the corresponding analytic elements coincide on these common parts, and not gluing together the overlapping disks if their analytic elements are different (see Fig. 2.24).

The function obtained as the result of all possible analytic continuations and defined as the set of all the constructed analytic elements is called a *complete analytic function*, which may be a *multi-valued analytic function*. The surface obtained as the result of connecting the corresponding disks with common parts is called the *Riemann surface* of this function. In the case of a multi-valued analytic function, it is more convenient to consider the Riemann surface than a set on the complex plane as the function domain. Indeed, at each point of the Riemann surface the function is extended and defined uniquely, since such a point is determined not only by its location on the plane, but also by the indication to what sheet or disk K_j of the Riemann surface this point belongs and, in this manner, the extended function is defined by the analytic single-valued element $f_j(z)$ in K_j.

Notice that an analytic continuation need not to be performed in the form of convergent series. A similar procedure can be performed independently of a form used for definition of function elements. The only thing that important is a smoothness of the elements: each of them should be holomorphic on the corresponding domain. In this way, one can think about an arbitrary holomorphic (analytic) function $f(z)$ defined in a domain D in an arbitrary form and try to extend it to a domain K_0 that does not lie in D and overlaps D over a set E_0, which has limit points both in D and K_0. If it is possible, we define a (direct) analytic continuation $F_0(z)$ of $f(z)$ from D onto $D_0 = D \cup K_0$ by the formula $F_0(z) = \begin{cases} f(z), z \in D \\ f_0(z), z \in K_0 \end{cases}$. If $D \cap K_0$ is connected, by the Theorem in Sect. 2.8, this continuation is unique and analytic in D_0. The obtained function $F_0(z)$ may have a (direct) analytic continuation from D_0 to a larger domain $D_1 = D_0 \cup K_1$, and so on. It may happen that multi-stage analytic continuation of the original element leads eventually to different values in some overlapping domains, in particular, we could return to the original region with different values, and in this case we obtain a multi-valued function, which is usually represented using the Riemann surface.

Another generalization of the above described process is directly related to the formulation of Theorem on Uniqueness of an analytic continuation (Sect. 2.9), which allows us to use an arbitrary set E as an original set of definition of $f(z)$. If domain D contains E and an analytic extension $F(z)$ of $f(z)$ onto D is possible (in other words, if there exists an analytic function $F(z)$ defined in D such that $f(z) = F(z)$ on E), then $F(z)$ is called analytic continuation. If, additionally, E has a limit point in D, then this analytic continuation is unique (according to Theorem of Sect. 2.9). This type of analytic continuation was used in Sect. 2.9 to extend the well-known definitions of exponential and trigonometric functions from the real axis onto the entire complex plane.

We end this section with one more introductory example of the analytic continuation using the function $f(z) = \sqrt{z}$.

Example As was shown in Sect. 1.1, the square root $\sqrt{z}$ takes two different values for each complex point $z \neq 0$, that is, the function $f(z) = \sqrt{z}$ is two-valued. In what follows, we will define this function by an analytic continuation starting with the real (single-valued) roots defined on the positive real axis: $f(x) = \sqrt{x} > 0, \forall x > 0$. Notice that for any point $x_0 > 0$ the function $f(x) = \sqrt{x}$ has the following expansion in the power series:

$$f(x) = \sqrt{x} = \sqrt{x_0}\sqrt{1 + \frac{x - x_0}{x_0}} = \sqrt{x_0}\left(1 + \frac{x - x_0}{2x_0} + \sum_{n=2}^{+\infty}(-1)^{n-1}\frac{(2n-3)!!}{2^n n!} \cdot \frac{(x - x_0)^n}{x_0^n}\right).$$

Consider the function $f(z)$, $z \in \mathbb{C}$ defined by the corresponding series

$$f(z) = \sqrt{x_0}\left(1 + \frac{z - x_0}{2x_0} + \sum_{n=2}^{+\infty}(-1)^{n-1}\frac{(2n-3)!!}{2^n n!} \cdot \frac{(z - x_0)^n}{x_0^n}\right),$$

where $x_0 > 0$, $\sqrt{x_0} > 0$. The radius of convergence of the last series is

$$R = \lim_{n\to\infty}\left|\frac{c_n}{c_{n+1}}\right| = \lim_{n\to\infty}\frac{(2n-3)!! \cdot 2^{n+1}(n+1)!}{(2n-1)!! \cdot 2^n n!} = \lim_{n\to\infty}\frac{2(n+1)}{2n-1} = 1\,.$$

Therefore, the series converges in the disk $|z - x_0| < x_0$ and its sum is an analytic function in this disk. Since this result holds for any $x_0 > 0$, approaching x_0 to $+\infty$ we obtain the analytic continuation of $f(x) = \sqrt{x}$ on the entire right half-plane $\text{Re}z > 0$. It is natural to denote the extended function by $f(z) = \sqrt{z}$.

In a similar manner, taking $\forall z_0 \neq 0$ and fixing one of the two values of $\sqrt{z_0}$, we expand it in the corresponding power series

$$f(z) = \sqrt{z} = \sqrt{z_0}\left(1 + \frac{z - z_0}{2z_0} + \sum_{n=2}^{+\infty}(-1)^{n-1}\frac{(2n-3)!!}{2^n n!}\cdot\frac{(z-z_0)^n}{z_0^n}\right)$$

convergent in the disk $|z - z_0| < |z_0|$. Therefore, the sum of this series is an analytic function in this disk, which represents an analytic element of the function $f(z) = \sqrt{z}$ in the disk $|z - z_0| < |z_0|$. Constructing analytic continuations centered at different points z_0, we extend $f(x) = \sqrt{x}$ on the entire complex plane, except at the origin. At the point $z_0 = 0$ there is no analytic element, because $\sqrt{z}$ has two different values at every point $z \neq 0$ in any neighborhood of 0, which means that $z_0 = 0$ is a singular point of $f(z) = \sqrt{z}$. This type of singularity is called a *branch point.*

Let us analyze how the process of analytic continuation leads to the two-valued function. If we take any $z_0 = r_0 e^{i\varphi_0} \neq 0$ and choose one of the values of $\sqrt{z_0}$, for instance, $w_1 = \sqrt{z_0} = \sqrt{r_0}e^{i\varphi_0/2}$, then in the disk $|z - z_0| < |z_0|$ there exists an analytic element of $f(z) = \sqrt{z}$. Now we move the disk $|z - z_0| < |z_0|$ counterclockwise around the origin in such a way that the centers of all the obtained disks lie on the circle $|z| = |z_0|$ and the argument of the centerpoints changes continuously. By previous considerations, the corresponding analytic element is defined in each of these disks. However, completing one circuit around the origin, we return to the point z_0 with the argument $\arg z_0 = \varphi_0 + 2\pi$, and consequently, we arrive at another value of the square root $w_2 = \sqrt{z_0} = -\sqrt{r_0}e^{i\varphi_0/2} = -w_1$, which corresponds to another analytic element of $\sqrt{z}$. Hence, the values of two analytic elements defined on the same points are different and we obtain the multi-valued (two-valued) function $f(z) = \sqrt{z}$.

To avoid different roots, that is, to obtain a single-valued analytic function as the result of analytic continuation, we should exclude a possibility to go all the way around the branch point 0. It can be made by introducing a cut in the complex plane (called a *branch cut*) along a curve starting at the origin and going to infinity. The shape and direction of the curve are not fixed, they are chosen for convenience in the problem at hand. One of the standard forms of the branch cut for $f(z) = \sqrt{z}$ is the negative real axis. Then we can define the single-valued analytic function (continuation) on the entire complex plane with this cut. For instance, we can choose the analytic element to be $f_1(z) = \sqrt{z} = \sqrt{r}e^{i\varphi/2}$, $\forall\varphi \in [-\pi, \pi)$, which contains the value w_1. For a point on the upper edge of the cut, this function has the value $\sqrt{r}e^{i\pi/2} = i\sqrt{r}$, while on the lower edge of the cut it takes the value $\sqrt{r}e^{-i\pi/2} = -i\sqrt{r}$. Therefore, there is a discontinuity across the cut and to keep the function analytic we define that it doesn't cross the cut.

Finally, we give a brief description of the construction of the Riemann surface of the multi-valued function $f(z) = \sqrt{z}$. Notice that after two counterclockwise circuits around the origin the argument of the point $z_0 = r_0 e^{i\varphi}$ receives the increment 4π and we return to the first value w_1 of $\sqrt{z}$: $\sqrt{r_0 e^{i(\varphi_0+4\pi)}} = \sqrt{r_0}e^{i(\varphi_0/2+2\pi)} = \sqrt{r_0}e^{i\varphi_0/2} = w_1$. The same happens if the point moves in the clockwise direction. Therefore, after two loops around the origin we return to the original analytic element $f_1(z) =$

$\sqrt{r}e^{i\varphi/2}$ of the function $\sqrt{z}$, which means that this function has exactly two analytic elements at each point $z_0 \neq 0$ and its Riemann surface contains two sheets. Each sheet represents the complex plane with a cut connecting the points 0 and ∞, which allows us to pick out a single-valued analytic element (branch) of $f(z) = \sqrt{z}$. The Riemann surface then obtained by connecting these two sheets along their cuts: the upper edge of the first sheet, which is the domain of the analytic element $f_1(z)$ is glued with the lower edge of the second sheet, which is the domain for $f_2(z)$, and then the upper edge of the second sheet is glued with the lower edge of the first sheet. More detailed study and construction of the Riemann surface for the functions $f(z) = \sqrt[n]{z}, \forall n \in \mathbb{N}, n \geq 2$ (including $f(z) = \sqrt{z}$) will be provided in Sect. 4.6.

Other specific examples of multi-valued functions and construction of the corresponding Riemann surfaces will be given in Sects. 4.6–4.9 of Chap. 4 devoted to the study of elementary functions. An application of the direct analytic continuation in the form of the Riemann–Schwarz symmetry principle will be described in Sect. 5.4.

Exercises

1. Verify if the function is differentiable, and if so, find the derivative:

   (1) $f(z) = z + 2\bar{z}^2$;
   (2) $f(z) = z^3 \cdot \operatorname{Re} z$;
   (3) $f(z) = \operatorname{Re} z^4$;
   (4) $f(z) = \operatorname{Im} z^4$;
   (5) $f(z) = z^3 + \bar{z}$;
   (6) $f(z) = z^2 + 5\operatorname{Re} z$;
   (7) $f(z) = z^2 + \operatorname{Re}\,(z-2)^2$;
   (8) $f(z) = \cos z$;
   (9) $f(z) = e^z = e^x\,(\cos y + i \sin y)$.

2*. Show that the function $f(z)$ satisfies the Cauchy–Riemann conditions at the point $z = 0$, but it is not differentiable at this point:

   (1) $f(z) = \sqrt[3]{x^2 y}$;
   (2) $f(z) = \begin{cases} \frac{z^2 - \bar{z}^2}{|z|}, & z \neq 0 \\ 0, & z = 0 \end{cases}$.

   Explain this result.

3. Write the Cauchy–Riemann conditions in the polar coordinates $z = re^{i\varphi}$.
4. Prove the following version of the mean value theorem for complex functions: if $f(z)$ is a holomorphic function on a convex domain D, then for any two points $a, b \in D$ there exist a pair of points z_1, z_2 belonging to line segment (a, b) such that $\operatorname{Re} f'(z_1) = \operatorname{Re} \frac{f(b)-f(a)}{b-a}$ and $\operatorname{Im} f'(z_2) = \operatorname{Im} \frac{f(b)-f(a)}{b-a}$.

5. Verify whether the given function is harmonic. If it is, find the corresponding conjugate harmonic function and reconstruct the complex function $f(z) = u(x, y) + iv(x, y)$:

   (1) $u(x, y) = x^3 + 6x^2y - 3xy^2 - 2y^3$;
   (2) $u(x, y) = \sin x \cdot \cosh y - \cosh x \cdot \cos y$;
   (3) $v(x, y) = y \cos y \cdot \sinh x + x \sin y \cdot \cosh x$;
   (4) $u(x, y) = e^x (x \cos y - y \sin y)$;
   (5) $u(x, y) = x \cos x \cdot \cosh y + y \sin x \cdot \sinh y$;
   (6) $u(x, y) = x^5 - 10x^3y^2 + 5xy^4 + xy - 3y$;
   (7) $v(x, y) = \cos x \cdot \sinh y - \sinh x \cdot \sin y$;
   (8) $v(x, y) = \frac{2y}{x^2+y^2} - 3x + 2y$;
   (9) $v(x, y) = 9x^2y - 3y^3 - 7x + 5y$;
   (10) $v(x, y) = e^x (y \cos y + x \sin y)$;
   (11) $u(x, y) = \frac{x}{x^2+y^2} + 2x^2 - 2y^2 - 3xy + 5x$.

6. Analyze whether there exists a non-constant harmonic function of the indicated type. If it exists, find this function:

   (1) $u = g(x)$;
   (2) $u = g(x^2 + y^2)$;
   (3) $u = g(x^2 + y)$;
   (4) $u = g\left(x + \sqrt{x^2 + y^2}\right)$;
   (5) $u = g(y/x)$;
   (6) $u = g(x^2 - y^2)$.

7. Evaluate the integrals along a closed simple curve Γ employing Goursat's Theorem, the Cauchy integral formula or the Theorem on infinite differentiability of an analytic function:

   (1) $\oint_\Gamma (z - a)^n \, dz$, $a \notin \Gamma$, $\forall n \in \mathbb{Z}$;
   (2) $\oint_\Gamma \frac{dz}{z^2+25}$, $\pm 5i \notin \Gamma$;
   (3) $\oint_\Gamma \frac{dz}{z(z^2-4)}$, $0, \pm 2 \notin \Gamma$;
   (4) $\oint_{|z-a|=a} \frac{2z^2}{z^4-1} dz$, $a > 1$;
   (5) $\oint_\Gamma \frac{e^z}{z^2+a^2} dz$, $\pm ia \notin \Gamma$;
   (6) $\oint_\Gamma \frac{ze^z}{(z-a)^3} dz$, $a \notin \Gamma$;
   (7) $\oint_\Gamma \frac{z \sin z}{(z-1)(z-2)^2} dz$, $1, 2 \notin \Gamma$;
   (8) $\oint_\Gamma \frac{\cos z \, dz}{z^2(z^2+2z+2)}$, $0, -1 \pm i \notin \Gamma$.

8. Suppose a function $f(z)$ is analytic in the disk $|z| \le 1$ and takes real values on its boundary. Prove that $f(z) \equiv constant$.
9. Let $u(x, y)$ and $v(x, y)$ be a pair of conjugate harmonic functions in a domain D. Prove that the following pair of functions $g(x, y)$ and $h(x, y)$ are also conjugate harmonic in D:

(1) $g = au - bv$, $h = bu + av$, $\forall a, b \in \mathbb{C}$, $a, b = constants$;
(2) $g = e^u \cos v$, $h = e^u \sin v$;
(3) $g = e^{u^2-v^2} \cos 2uv$, $h = e^{u^2-v^2} \sin 2uv$;
(4) $g = e^{uv} \cos \frac{u^2-v^2}{2}$, $h = e^{uv} \sin \frac{v^2-u^2}{2}$.

10. Derive Laplace's equation in polar coordinates.
Hint: $\Delta u(r, \varphi) = \frac{\partial^2 u}{\partial r^2} + \frac{1}{r}\frac{\partial u}{\partial r} + \frac{1}{r^2}\frac{\partial^2 u}{\partial \varphi^2} = 0$.

11. Let $R > 0$ and θ be given numbers. Show that:

(1) the function $f(z) = \frac{Re^{i\theta}+(z-a)}{Re^{i\theta}-(z-a)}$, $z = re^{i\varphi} + a$, $r < R$ is analytic in the disk $|z - a| < R$;
(2) the function $u(r, \varphi) = \mathrm{Re} f(z)$ can be expressed in the form $u(r, \varphi) = \frac{R^2-r^2}{R^2+r^2-2Rr\cos(\varphi-\theta)}$ and is harmonic in the disk $|z - a| < R$ (verify by the definition);
(3) $\frac{1}{2\pi}\int_0^{2\pi} u(r, \varphi)d\varphi = 1$.
Hint for (3): use the Cauchy integral formula.

12. Suppose $f(z)$ is an analytic function in a finite convex domain D, and it satisfies the condition $|f(z)| \le M$, $\forall z \in D$. Prove that the inequality $\left|\int_{z_1}^{z_2} f(z)dz\right| \le M \cdot |z_2 - z_1|$ holds for any $z_1, z_2 \in D$.
Hint: use the fact that the line segment joining z_1 and z_2 is contained in D.

13*. Suppose $f(z)$ is analytic in a finite convex domain D and satisfies the condition $\mathrm{Re} f(z) \ge M > 0$ (or $\mathrm{Im} f(z) \ge M > 0$), $\forall z \in D$. Prove that the inequality $\left|\int_{z_1}^{z_2} f(z)dz\right| \ge M \cdot |z_2 - z_1|$ holds for any $z_1, z_2 \in D$. Verify whether the inequality is true if the condition $\mathrm{Re} f(z) \ge M > 0$ (or $\mathrm{Im} f(z) \ge M > 0$) is substituted by $|f(z)| \ge M > 0$, $\forall z \in D$.
Hint: use the fact that the line segment joining z_1 and z_2 is contained in D.

14. Let $f(z)$ be analytic in $|z| < R$ and continuous in $|z| \le R$. Calculate the integral $\iint_{r \le |z| \le R} f(z)dxdy$, $r < R$.

15. Let $f(z)$ be analytic in $|z - a| < R$ and continuous in $|z - a| \le R$. Demonstrate that $\frac{1}{2\pi}\int_0^{2\pi} f(a + Re^{i\varphi})d\varphi = f(a)$. This formula is called the mean value theorem.

16*. Let $f(z)$ be an analytic function in a simply connected domain D containing the point ∞. Demonstrate that an analytic primitive of $f(z)$ in D exists if and only if $\lim\limits_{z\to\infty} zf(z) = 0$.
Hint: show that $\int_\Gamma f(\zeta)d\zeta = \int_a^z f(\zeta)d\zeta$ does not depend on the choice of curve $\Gamma \subset D$ that joins the points a and z.

17. Suppose $f(z)$ has a power series expansion $f(z) = \sum_{n=0}^{+\infty} c_n z^n$ in the disk $|z| < R$. Show that:

(1) if $\max\limits_{|z|=r} |f(z)| = M(r)$, then $|c_n| \le \frac{M(r)}{r^n}$ $(r < R)$, $\forall n$;
(2) if there exists k such that $|c_k| = \frac{M(r)}{r^k}$, then $f(z) = c_k z^k$.
Hint for (2): make use of the result of Exercise 20 in Chap. 1.

Chapter 3
Singular Points, Laurent Series, and Residues

This chapter is devoted to the systematic study of singular points, Laurent series expansion, and application of the theory of residues by the evaluation of integrals. We start with the result that reveals an important difference between the behavior of real and complex series—the theorem on the existence of at least one singular point on the circle of convergence. Then, in Sect. 3.2, we formulate a sufficient condition to identify a singular boundary point and show that there is no relationship between the behavior of a complex power series on the circle of convergence and the number of singular boundary points. The latter characteristic is crucial to determine the possibility of analytic continuation of the sum of a power series outside the disk of convergence. In Sect. 3.3 we consider a generalization of the power series—the expansion of functions in Laurent series and analyze uniform convergence of this series.

The following five sections treat with different level of depth singular points of various kinds. Section 3.4 presents the classification of singular points, while the next four sections provide a detailed investigation of single-valued isolated singular points, which traditionally is the focus of study in undergraduate-level courses. We consider the three types of single-valued isolated singularities—removable singularities, poles, and essential singularities, and study their connection with the form of the corresponding Laurent series.

In Sect. 3.9 we introduce the notion of residues and prove the fundamental theorem on residues, which sets the stage for developing the methods of evaluation of different types of line integrals. To this end, we establish different formulas of evaluation of residues for all kinds of single-valued isolated singularities (removable, poles, and essential), both finite and infinite. Then these formulas are applied to the calculation of line integrals of various families of functions—rational, trigonometric, exponential, etc. We also provide an extension of these results for the evaluation of improper integrals.

In the last two sections, we consider the argument principle and Rouché's theorem. The former expresses the relationship between the numbers of zeros and poles in a

A. Bourchtein and L. Bourchtein, *Complex Analysis*, Hindustan Publishing Corporation,
https://doi.org/10.1007/978-981-15-9219-5_3

domain on the one hand and a line integral along the boundary of this domain on the other hand. The latter is an important tool to determine the number of zeros of a holomorphic function inside a chosen curve. The discussed theoretical applications of Rouché's Theorem include a simple proof of the fundamental theorem of algebra and Hurwitz's theorem on the univalence of the limit function of a sequence of univalent analytic functions.

3.1 Singular Points on the Boundary of the Disk of Convergence

In Sect. 2.10, we have defined a singular point as a point $z_0 \in \partial D$ at which the analytic continuation of an analytic in a domain D function $f(z)$ does not exist, that is, there is no analytic element of $f(z)$ at z_0. In this section, we will study singular points in detail.

Definition Let $f(z)$ be an analytic function in a domain D. A point $z_0 \in \partial D$ is called *singular boundary point* of $f(z)$ if the function $f(z)$ has no analytic continuation at this point or $f(z)$ is not defined at z_0.

We consider again the expansion of a function $f(z)$ in power series. Recall that for real functions the radius of convergence of a power series is not generally related to the properties of the functions for which this series is constructed. For example, the function $f(x) = \frac{1}{1+x^2}$ is represented by the power series $f(x) = \frac{1}{1+x^2} = \sum_{n=0}^{\infty} (-1)^n x^{2n}$ that converges only on the interval $(-1, 1)$ and consequently has the radius of convergence $R = 1$, whereas the original function $f(x) = \frac{1}{1+x^2}$ is defined and infinitely differentiable on $\mathbb{R}$. However, if we pass from the real axis to the complex plane, then we see that the radius of convergence of a power series is intimately tied to the behavior of the function expanded in this series.

Theorem *On the circle of convergence of a power series there exists at least one singular point of the sum of this series.*

Proof Suppose the function $f(z)$ is represented in the power series in a neighborhood of some point a: $f(z) = \sum_{n=0}^{\infty} c_n (z-a)^n$ and the disk of convergence of this series is $K_0 = \{|z-a| < R_0\}$ with the radius of convergence $R_0 \neq 0$, $R_0 \neq \infty$. Let us show that there exists at least one singular point of $f(z)$ on the circle $|z-a| = R_0$. Assume, for contradiction, that the function $f(z)$ has no singular points on the circle $|z-a| = R_0$, which means that we can extend analytically the function $f(z)$ to any point of this circle. In other words, for any point ζ of the circle $|\zeta - a| = R_0$ there exists an analytic function (analytic element) $f_\zeta(z)$ in a disk $K_\zeta = \{|z-\zeta| < r_\zeta\}$ such that the equality $f_\zeta(z) = f(z)$ holds for $\forall z \in K_0 \cap K_\zeta$. Then, the entire circle $|\zeta - a| = R_0$ is covered by the disks K_ζ. Since a circle is a compact set, its open cover by these disks has a finite subcover, that is, there exists a finite number of the

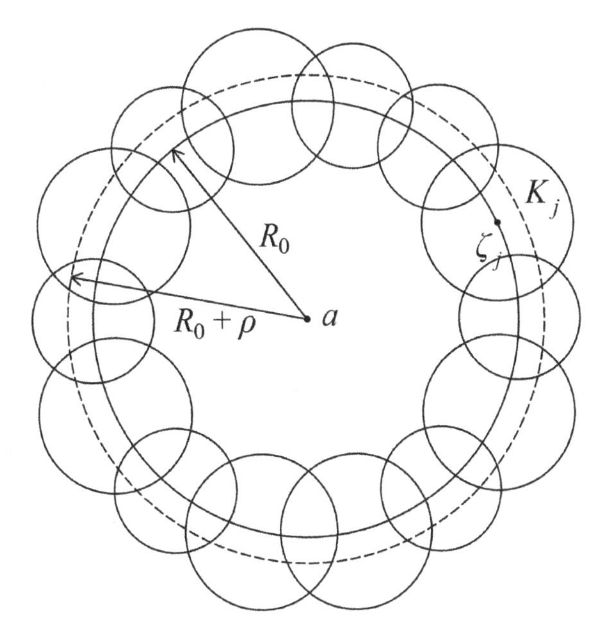

Fig. 3.1 Cover of the circle $|\zeta - a| = R_0$ by the disks $K_j = \{|z - \zeta_j| < r_j\}$, $j = 1, \ldots, m$

disks K_ζ, say $K_1, K_2, \ldots, K_m$, centered at the points $\zeta_1, \zeta_2, \ldots, \zeta_m$, respectively, which cover the entire circle $|\zeta - a| = R_0$: $\{|\zeta - a| = R_0\} \subset \bigcup\limits_{j=1}^{m} K_j$ (see Fig. 3.1).

Introduce the function

$$F(z) = \begin{cases} f(z), \ z \in K_0, \\ f_j(z), z \in K_j, \ j = 1, 2, \ldots, m, \end{cases}$$

which is, by the definition, the analytic continuation of $f(z)$ from the disk K_0 onto the region $D = K_0 \cup \left(\bigcup\limits_{j=1}^{m} K_j\right)$. Let us show that $F(z)$ is single-valued (and consequently analytic) in a domain D. A problem with the uniqueness of the function definition could arise only in the common parts of the different disks K_j. Let us denote $G_j = K_{j-1} \cap K_j$ (see Fig. 3.2) and show that $f_{j-1}(z) = f_j(z)$ for $\forall z \in G_j$. Consider one more set $E_j = G_j \cap K_0 = K_{j-1} \cap K_j \cap K_0$. Since each point of E_j belongs to the intersection $K_{j-1} \cap K_0$, by construction of the analytic element $f_{j-1}(z)$, we have $f_{j-1}(z) = f(z)$, $\forall z \in E_j$. At the same time, any point of E_j belongs to another intersection $K_j \cap K_0$ (see Fig. 3.2), and consequently $f_j(z) = f(z)$, $\forall z \in E_j$. Therefore, $f_{j-1}(z) = f(z) = f_j(z)$, $\forall z \in E_j$. The functions $f_{j-1}(z)$ and $f_j(z)$ are analytic in the domain G_j, the set E_j is contained in G_j and has limit points which belong to G_j. Therefore, by the Theorem on uniqueness of an analytic function, the equality $f_{j-1}(z) = f_j(z)$, $\forall z \in E_j$ implies that the same equality holds for all the points $z \in G_j = K_{j-1} \cap K_j$, that is, the function $F(z)$ is defined as a single-valued function in the domain D. Since $F(z)$ is analytic at every

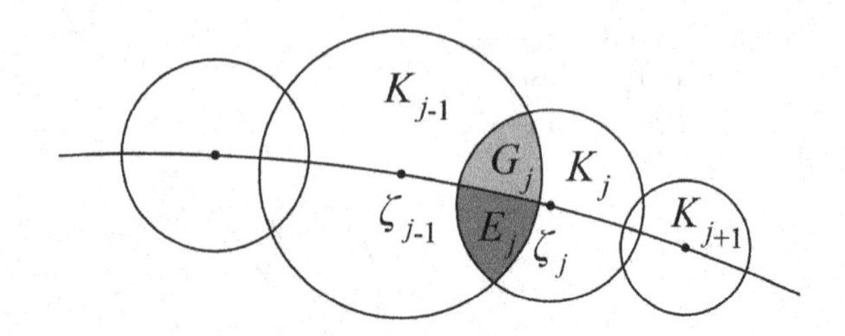

Fig. 3.2 Construction of the sets $G_j = K_{j-1} \cap K_j$ and $E_j = G_j \cap K_0$

point of D (there exists an analytic element at every point of D), this function is analytic in the domain D.

We will show now that D contains a disk centered at a with a radius greater than R_0. Denote by ρ the distance between the two boundaries ∂K_0 and ∂D. This distance is positive ($\rho = \rho(\partial K_0, \partial D) > 0$) since ∂K_0 and ∂D are two compact sets with no common points. Consider the disk $K^* = \{|z - a| < R^*\}$ with the radius $R^* = R_0 + \rho$ (see Fig. 3.1). By construction, this disk is contained in the domain D: $K^* \subset D$. Therefore, $F(z)$ is analytic in K^* (since it is analytic in D) and can be expanded in the power series in this disk: $F(z) = \sum_{n=0}^{\infty} b_n (z - a)^n$. Since the equality $F(z) = f(z)$ is true at any point $z \in K_0$, we have $F(z) = f(z) = \sum_{n=0}^{\infty} c_n (z - a)^n, \forall z \in K_0$, that is, $b_n = c_n$, $\forall n$. Since the series $F(z) = \sum_{n=0}^{\infty} c_n (z - a)^n$ converges in the disk $|z - a| < R^*$ with the radius $R^* = R_0 + \rho > R_0$ and it coincides with the power series of $f(z)$ in the disk K_0, we conclude that the original series of $f(z)$ converges in the disk larger than the disk of convergence. Hence, we arrive at a contradiction with the definition of the radius of convergence of a power series, which means that our assumption that there is no singular point on the boundary $|z - a| = R_0$ of the disk K_0 is false. Hence, there exists at least one singular point of $f(z)$ on the circle $|z - a| = R_0$.

Thus, the Theorem is proved. □

A straightforward consequence of the proved Theorem is the following simple rule for finding the radius of convergence of a power series.

Corollary *If a function $f(z)$ is analytic at a point z_0, that is, $f(z)$ is expended in a power series in a neighborhood of z_0, then the radius of convergence of this power series is equal to the distance between z_0 and the nearest singular point of $f(z)$.*

Remark Let us return to the function $f(x) = \frac{1}{1+x^2}$ and consider the corresponding complex function $f(z) = \frac{1}{1+z^2}$. The last function has singular points when the denominator is zero: $1 + z^2 = 0$, that is, $z = \pm i$. These two singular points of $f(z)$ belong to the circle $|z| = 1$ and it makes clear why the radius of convergence of the series $f(z) = \frac{1}{1+z^2} = \sum_{n=0}^{\infty} (-1)^n z^{2n}$ is equal to 1.

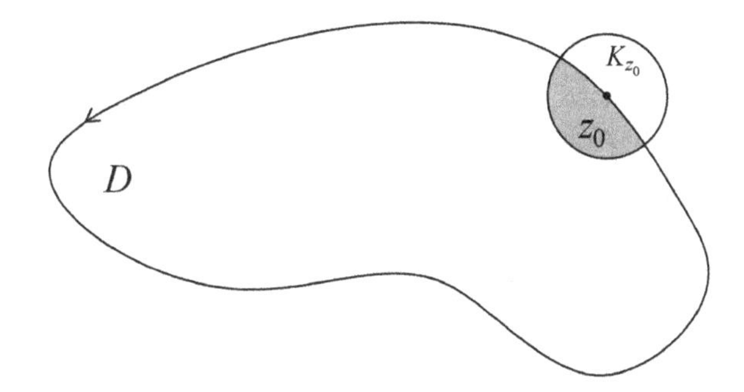

Fig. 3.3 The analyticity domain D of the function $f(z)$ and the analyticity disk K_{z_0} of the element $f_{z_0}(z)$

3.2 Sufficient Condition for a Singular Boundary Point

Theorem *Let $f(z)$ be an analytic function in a domain D and z_0 be a boundary point of D ($z_0 \in \partial D$). If there exists a number $k, k = 0, 1, 2, \ldots$, such that $\overline{\lim\limits_{z \to z_0,\, z \in D}} \left| f^{(k)}(z) \right| = +\infty$, then z_0 is a singular boundary point of $f(z)$.*

Proof We argue by contradiction and suppose that z_0 is not a singular point of $f(z)$. Then we can extend analytically the function $f(z)$ at the point z_0, that is, there exists an analytic function (analytic element) $f_{z_0}(z)$ in a disk $K_{z_0} = \{|z - z_0| < r_0\}$ such that for any point $z \in D \cap K_{z_0}$ we have $f_{z_0}(z) = f(z)$. Take any number k ($k = 0, 1, 2, \ldots$) and find $\lim\limits_{z \to z_0, z \in D} f^{(k)}(z)$. Since z approaches z_0, it is sufficient to consider only the points of the domain D close to the point z_0, that is, the points contained in K_{z_0}. However, if $z \in D \cap K_{z_0}$, then $f(z) = f_{z_0}(z)$, and consequently $f^{(k)}(z) = f_{z_0}^{(k)}(z)$, $\forall k = 0, 1, 2, \ldots$ (see Fig. 3.3). The function $f_{z_0}(z)$ is analytic in the entire disk K_{z_0}, in particular, at the point z_0 and, by the Theorem on infinite differentiability of an analytic function (Theorem 2.3 in Sect. 2.1), $f_{z_0}(z)$ has analytic derivative of any order, which means that for any k, $k = 0, 1, 2, \ldots$ it follows

$$\lim_{z \to z_0,\, z \in D} f^{(k)}(z) = \lim_{z \to z_0,\, z \in D \cap K_{z_0}} f^{(k)}(z) = \lim_{z \to z_0,\, z \in D \cap K_{z_0}} f_{z_0}^{(k)}(z) = f_{z_0}^{(k)}(z_0).$$

Hence, the finite limit exists for any k, $k = 0, 1, 2, \ldots$, but this contradicts the hypothesis of the Theorem. Therefore, our assumption is false, and consequently, z_0 is a singular point of $f(z)$. □

The following example shows that there exist power series, whose circle of convergence consists only of singular points.

Example 3.1 Consider the function defined by the following series:

$$f(z) = \sum_{n=0}^{\infty} z^{2^n} = z + z^2 + z^4 + z^8 + z^{16} + \cdots + z^{2^n} + \ldots.$$

First, let us find the radius of convergence of this series. Notice that $c_k = \begin{cases} 1, k = 2^n \\ 0, k \neq 2^n \end{cases}$, which shows that a general limit of c_k does not exist. Then, we can use the Cauchy–Hadamard formula (see formula (1.17) in Sect. 1.7): $R = \frac{1}{\overline{\lim\limits_{k\to\infty}} \sqrt[k]{|c_k|}} = \frac{1}{\lim\limits_{n\to\infty} \sqrt[2^n]{1}} = 1$, which means that the series converges in the disk $|z| < 1$, and consequently, the function $f(z)$ is analytic in this disk. Notice that on the boundary points $|z| = 1$ one has $z = e^{i\varphi}$, $\varphi \in [0, 2\pi]$ and the corresponding series

$$f\left(e^{i\varphi}\right) = \sum_{n=0}^{\infty} e^{i2^n\varphi}$$

diverges for any $\varphi \in [0, 2\pi]$ since the general term $e^{i2^n\varphi}$ does not approach 0.

By the Theorem of the previous section, there exists at least one singular point of $f(z)$ on the circle $|z| = 1$. Applying the sufficient condition of a singular boundary point (proved in this section) to the point $z = 1$ and using in this condition $k = 0$, we obtain

$$\overline{\lim_{z\to 1,\,|z|<1}} |f(z)| \geq \lim_{z=x\to 1,\,0<x<1} |f(z)| = \lim_{x\to 1^-} \left(x + x^2 + x^4 + \ldots\right)$$

$$\geq \lim_{x\to 1^-} \sum_{n=0}^{N} x^{2^n} = N + 1 \underset{N\to\infty}{\to} +\infty\,.$$

Therefore, $z = 1$ is a singular boundary point of $f(z)$.

Let us rewrite the power series of $f(z)$ in the following form:

$$f(z) = z + z^2 + z^4 + \cdots + z^{2^{n-1}} + z^{2^n} + z^{2^{n+1}} + \cdots + z^{2^{n+k}} + \ldots$$

$$= z + z^2 + z^4 + \cdots + z^{2^{n-1}} + \sum_{k=0}^{+\infty} \left(z^{2^n}\right)^{2^k} = z + z^2 + z^4 + \cdots + z^{2^{n-1}} + f\left(z^{2^n}\right).$$

Notice that the first n terms—the functions $z, z^2, \ldots, z^{2^{n-1}}$—are analytic in the entire (finite) complex plane $\mathbb{C}$, and consequently, the function $f(z)$ has the same singular points as the function $f\left(z^{2^n}\right)$. As was shown, the last function has singular points when $\zeta = z^{2^n} = 1$, that is, all the solutions of the equation $z^{2^n} = 1$: $z_p = e^{\frac{2p\pi i}{2^n}}$, $p = 0, 1, \ldots, 2^n - 1$ are singular points. In this manner, we obtain 2^n singular points z_p of the function $f(z)$. All these points z_p lie on the unit circle and divide it into 2^n equal arcs. Notice that this reasoning is true for $\forall n \in \mathbb{N}$.

Finally, we will show that each point of the circle $|z| = 1$ is a singular point of the function $f(z)$. Take any point z^*, $|z^*| = 1$. Whatever neighborhood of this point is chosen ($\forall r^* > 0$), there exists a natural number N such that for all $n > N$ there are points z_p belonging to this neighborhood. Therefore, there exists no analytic element of $f(z)$ in the neighborhood $|z - z^*| < r^*$, which means that $f(z)$ has no analytic continuation at the point z^*, that is, z^* is a singular boundary point of $f(z)$. Thus, the interesting feature of the function $f(z)$ is that all the points of the unit circle are singular boundary points, which means that the function $f(z)$ has no analytic continuation in any direction outside the open unit disk.

Usually, on the circle of convergence of a power series, there are both singular and regular points, the latter indicating directions of possible analytic continuation of the original function (determined by the given power series). The above Example 3.1 may suggest that there exists a relationship between a possibility of analytic continuation beyond the disk of convergence and the convergence of the corresponding power series at the points of the circle of convergence. The next examples show that this impression is false.

Example 3.2 We start with the case when a power series diverges on the entire circle of convergence, but the function defined by this series can be analytically continued in any direction but one. Consider the following power series:

$$f(z) = \frac{1}{1-z} = \sum_{n=0}^{\infty} z^n . \tag{3.1}$$

Since all the coefficients $c_n = 1$, the radius of convergence can be readily found using the D'Alembert formula $R = \lim\limits_{n\to\infty} \left|\frac{c_n}{c_{n+1}}\right| = 1$. The points of the unit circle $|z| = 1$ can be defined parametrically $z = e^{i\varphi}$, $\varphi \in [0, 2\pi]$ and the series (3.1) can be rewritten at these points in the form $\sum_{n=0}^{\infty} e^{in\varphi}$. The last series diverges for any $\varphi \in [0, 2\pi]$, because $\left|e^{in\varphi}\right| = 1 \underset{n\to\infty}{\not\to} 0$. However, the corresponding function $f(z) = \frac{1}{1-z}$ has the only singular point $z = 1$ in the entire plane $\mathbb{C}$, and consequently, it can be analytically extended beyond the unit disk in any direction but in the direction of $z = 1$. Notice that the point $z = 1$ satisfies the sufficient condition of a singular boundary point with $k = 0$

$$\overline{\lim_{z\to z_0,\, z\in D}} |f(z)| = \lim_{z\to 1,\, |z|<1} \frac{1}{|1-z|} = +\infty .$$

Example 3.3 This example reveals that a power series may be convergent at every point of the circle of convergence, although the corresponding function possesses singularities at some of these points. In Example of Sect. 2.10, we have studied some properties of the two-valued function $\sqrt{z}$. Here we consider the intimately related function $\sqrt{1-z}$ which has a similar behavior. First, we choose the analytic branch (element) of the two-valued function $f(z) = \sqrt{1-z}$ in a neighborhood of $z = 0$ which takes the value $f(0) = 1$. This function is holomorphic in a neighborhood of $z = 0$ and its derivatives can be found by the formula $f^{(n)}(z) = -\frac{(2n-3)!!}{2^n}(1-z)^{-(2n-1)/2}$. In particular, at the point $z = 0$ we get $f^{(n)}(0) = -\frac{(2n-3)!!}{2^n}$. Then we obtain the series expansion of $f(z)$ in the following form:

$$f(z) = \sqrt{1-z} = 1 - \frac{z}{2} - \sum_{n=2}^{\infty} \frac{(2n-3)!!}{2^n \cdot n!} z^n . \tag{3.2}$$

The radius of convergence can be found using the D'Alembert formula

$$R = \lim_{n\to\infty} \left| \frac{c_n}{c_{n+1}} \right| = \lim_{n\to\infty} \frac{(2n-3)!! \cdot 2^{n+1} \cdot (n+1)!}{(2n-1)!! \cdot 2^n \cdot n!} = \lim_{n\to\infty} \frac{2(n+1)}{2n-1} = 1 ,$$

that is, the series converges in the unit disk $|z| < 1$.

On the boundary of this disk ($z = e^{i\varphi}$, $\varphi \in [0, 2\pi]$) the series in (3.2) takes the form

$$\sum_{n=2}^{\infty} \frac{(2n-3)!!}{2^n \cdot n!} e^{in\varphi} . \tag{3.3}$$

Let us show that the auxiliary positive numerical series

$$\sum_{n=2}^{\infty} \frac{(2n-3)!!}{2^n \cdot n!} \tag{3.4}$$

converges. Indeed, using mathematical induction, we can arrive at the following evaluation of the general term of the last series:

$$a_n = \frac{(2n-3)!!}{2^n \cdot n!} < \frac{1}{n^{3/2}} , \ \forall n \geq 2 . \tag{3.5}$$

For $n = 2$ this is evident: $a_2 = \frac{1}{4 \cdot 2} < \frac{1}{2\sqrt{2}}$. Assuming that (3.5) is true for $n = m$, we obtain for $n = m + 1$

$$a_{m+1} = \frac{(2m-1)!!}{2^{m+1} \cdot (m+1)!} = \frac{(2m-3)!!}{2^m \cdot m!} \cdot \frac{2m-1}{2(m+1)} < \frac{1}{m^{3/2}} \frac{2m-1}{2(m+1)}$$

$$= \frac{1}{(m+1)^{3/2}} \frac{(m+1)^{3/2} \cdot (2m-1)}{m^{3/2} \cdot 2(m+1)} = \frac{1}{(m+1)^{3/2}} \frac{(m+1)^{1/2} \cdot (2m-1)}{m^{1/2} \cdot 2m} .$$

To show that $\frac{(m+1)^{1/2} \cdot (2m-1)}{m^{1/2} \cdot 2m} < 1$ we can rewrite it in the following equivalent form:

$$(2m-1)(m+1)^{1/2} < 2m \cdot m^{1/2}$$

and squaring the last relation

$$(2m-1)^2(m+1) = 4m^3 - 3m + 1 < 4m^3, \ \forall m \geq 1 .$$

Since the last inequality is true, it follows that $\frac{(m+1)^{1/2}\cdot(2m-1)}{m^{1/2}\cdot 2m} < 1$. Therefore, $a_{m+1} < \frac{1}{(m+1)^{3/2}}$ and (3.5) holds. Then, the convergence of the series $\sum_{n=2}^{\infty}\frac{1}{n^{3/2}}$ implies the convergence of (3.4), which means that the series (3.3) converges absolutely for any $\varphi \in [0, 2\pi]$. Hence, the series (3.2) converges absolutely and uniformly (due to the Weierstrass test) on the closed unit disk $|z| \le 1$.

At the same time, the sum of this series $f(z) = \sqrt{1-z}$ has a singular point on the unit circle $|z| = 1$. We can see that this singular point is $z = 1$ by applying the Theorem of this section with $k = 1$

$$\overline{\lim_{z\to z_0,\, z\in D}} \left|f'(z)\right| = \lim_{z\to 1,\, |z|<1} \frac{1}{2|\sqrt{1-z}|} = +\infty\,.$$

Notice that $f(z)$ has no other singularities in $\mathbb{C}$ and can be analytically continued in any direction except the direction passing through the point $z = 1$. If we introduce a cut emanating from the point $z = 1$, for instance, the ray $[1, +\infty)$ of the real axis, then the analytic continuation of $f(z)$ onto the entire plane $\mathbb{C}$ with such a cut will be an analytic (single-valued) function.

Example 3.4 This example is similar to Example 3.3, but the function defined by the power series has four boundary singularities instead of only one. (It is possible to construct a similar example with any finite number of boundary singularities.) Consider the function $f(z) = (1 - z^4)\ln(1 - z^4)$ (recall that the logarithmic function $\ln z$ was defined in Sect. 2.6). This function is multi-valued because of $\ln(1 - z^4)$, and we choose such analytic branch of the last term that takes the value $\ln(1 - z^4)|_{z=0} = 0$. Then, this branch of the function $f(z)$ is holomorphic in a neighborhood of $z = 0$ and can be expanded in the power series at $z = 0$. To find out the form of this series we first consider the auxiliary function $g(z) = (1 - z)\ln(1 - z)$, which has similar properties for the branch with the value $\ln(1 - z)|_{z=0} = 0$. The derivatives of $g(z)$ have the form $g'(z) = -\ln(1 - z) - 1$, $g''(z) = (1 - z)^{-1}$, $g'''(z) = (1 - z)^{-2}$, ..., $g^{(n)}(z) = (n-2)!(1 - z)^{-(n-1)}$, and at $z = 0$ one has $g'(0) = -1$, $g''(0) = 1$, $g'''(0) = 1!$, ..., $g^{(n)}(0) = (n-2)!$. Therefore, $g(z)$ can be expanded in the power series

$$g(z) = -z + \sum_{n=2}^{\infty} \frac{(n-2)!}{n!} z^n = -z + \sum_{n=2}^{\infty} \frac{z^n}{n(n-1)}\,.$$

Accordingly, the power series for the chosen branch of $f(z)$ can be written as follows

$$f(z) = -z^4 + \sum_{n=2}^{\infty} \frac{z^{4n}}{n(n-1)}\,. \tag{3.6}$$

The coefficients of this series are $c_k = \begin{cases} \frac{1}{n(n-1)},\ k = 4n \\ 0,\ k \ne 4n \end{cases}$ and the radius of convergence can be found by the Cauchy–Hadamard formula: $R = \frac{1}{\overline{\lim\limits_{k\to\infty}} \sqrt[k]{|c_k|}} =$

$\frac{1}{\lim\limits_{n\to\infty} \sqrt[4n]{1/n(n-1)}} = 1$. Hence, the series converges in the disk $|z| < 1$. Notice that for any $|z| \leq 1$ one has $\frac{|z|^{4n}}{n(n-1)} \leq \frac{1}{n(n-1)}$ and the numerical series $\sum_{n=2}^{\infty} \frac{1}{n(n-1)}$ converges. Therefore, by the Weierstrass test, the series (3.6) converges absolutely and uniformly on the closed disk $|z| \leq 1$, in particular, it converges on the boundary of this disk.

At the same time, there are singular points of $f(z)$ on the circle $|z| = 1$, which are the branch points of $\ln(1 - z^4)$: $1 - z^4 = 0$, that is, $z_j = e^{j\pi i/2}$, $j = 0, 1, 2, 3$. The singularity of these points can also be verified by the Theorem of this section used with $k = 1$

$$\overline{\lim_{z\to z_j,\, z\in D}} \left|f'(z)\right| = \lim_{z\to z_j,\, |z|<1} 4|z^3| \cdot |\ln(1 - z^4) + 1| = +\infty\,.$$

Therefore, the function $f(z)$ can be analytically continued beyond the unit disk in any direction except those passing through the points $\pm 1, \pm i$. If we introduce the cuts in $\mathbb{C}$ along the rays $[1, +\infty)$, $(-\infty, -1]$ of the real axis and $[i, +i\infty)$, $(-i\infty, -i]$ of the imaginary axis, then the analytic continuation of $f(z)$ on the complex plane with these cuts will be an analytic (single-valued) function.

The above examples show that convergence/divergence properties of a power series on the boundary of the disk of convergence themselves usually do not determine the number of singular boundary points and their distribution. On the one hand, a power series may be divergent on the entire boundary, while the sum of this series has only one singular boundary point. On the other hand, a power series may be convergent (absolutely and uniformly) on the entire boundary, but the corresponding function has different singular boundary points.

3.3 Laurent Series Expansion

Laurent Theorem *If $f(z)$ is an analytic function in the ring $D = \{r < |z - a| < R\}$, then $f(z)$ can be represented in this ring by the series*

$$f(z) = \sum_{n=-\infty}^{+\infty} c_n (z - a)^n\,, \tag{3.7}$$

which is called the Laurent series. Moreover, the Laurent series converges uniformly in any closed ring contained in the ring D and all the coefficients c_n can be found by the formula

$$c_n = \frac{1}{2\pi i} \int_{|\zeta-a|=\rho} \frac{f(\zeta)}{(\zeta - a)^{n+1}} d\zeta\,, \tag{3.8}$$

where ρ is any real number such that $r < \rho < R$.

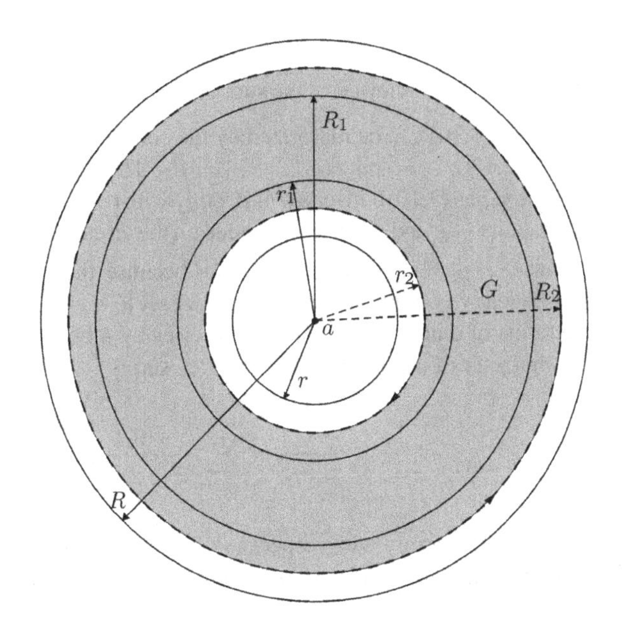

Fig. 3.4 The rings $D = \{r < |z - a| < R\}$, $B = \{r_1 < |z - a| < R_1\}$ and $G = \{r_2 < |z - a| < R_2\}$ such that $\bar{B} \subset G$, $\bar{G} \subset D$

Proof Consider any open ring $B = \{r_1 < |z - a| < R_1\}$ whose closure is contained in the open ring D: $\bar{B} \subset D$, that is, $r < r_1 < R_1 < R$. Due to the density of the real numbers there exist numbers r_2 and R_2 such that $r < r_2 < r_1 < R_1 < R_2 < R$. Denote by G the auxiliary ring $G = \{r_2 < |z - a| < R_2\}$, which contains the closure of B and whose closure is contained in D: $\bar{B} \subset G$, $\bar{G} \subset D$ (see Fig. 3.4). Pick now an arbitrary point $\forall z \in \bar{B}$ and evaluate $f(z)$ at this point using the Cauchy integral formula

$$f(z) = \frac{1}{2\pi i} \int_{\partial G} \frac{f(\zeta)}{\zeta - z} d\zeta .$$

In this formula the two circles, which compose the boundary of G, are traversed in opposite directions: the circle $|\zeta - a| = R_2$ is traversed counterclockwise, while $|\zeta - a| = r_2$ clockwise (see Fig. 3.4). To calculate the integral, we change the direction of the traversal along the circle $|\zeta - a| = r_2$ and obtain

$$f(z) = \frac{1}{2\pi i} \int_{\partial G} \frac{f(\zeta)}{\zeta - z} d\zeta = \frac{1}{2\pi i} \int_{|\zeta - a| = R_2} \frac{f(\zeta)}{\zeta - z} d\zeta - \frac{1}{2\pi i} \int_{|\zeta - a| = r_2} \frac{f(\zeta)}{\zeta - z} d\zeta . \tag{3.9}$$

The ratio $\frac{1}{\zeta - z}$ in the first integral on the right-hand side of (3.9) can be represented in the form

$$\frac{1}{\zeta - z} = \frac{1}{(\zeta - a) - (z - a)} = \frac{1}{\zeta - a} \cdot \frac{1}{1 - \frac{z-a}{\zeta-a}} = \frac{1}{\zeta - a} \cdot \sum_{n=0}^{+\infty} \left(\frac{z - a}{\zeta - a} \right)^n . \tag{3.10}$$

Since $z \in \bar{B}$ and $\zeta \in \{|\zeta - a| = R_2\}$, one has $\left|\frac{z-a}{\zeta-a}\right| \leq \frac{R_1}{R_2} = q_1 < 1$. Therefore, the ratio $\frac{1}{1-\frac{z-a}{\zeta-a}}$ can be viewed as the sum of the convergent geometric series. The series of functions (3.10) can be majorized by the convergent numerical series $\sum_{n=0}^{+\infty} q_1^n$, $0 < q_1 < 1$ for $\forall z \in \bar{B}$ and $\forall \zeta,\ |\zeta - a| = R_2$. Therefore, according to the Weierstrass test, the series (3.10) converges uniformly with respect to $z \in \bar{B}$ and $\zeta : |\zeta - a| = R_2$. The factors $\frac{1}{\zeta-a}$ and $f(\zeta)$ are bounded on the circle $|\zeta - a| = R_2$: the first is bounded because $\frac{1}{|\zeta-a|} = \frac{1}{R_2}$, and the second—because the function $f(\zeta)$ is analytic and consequently continuous on the compact set $|\zeta - a| = R_2$. The multiplication of all the terms of a uniformly convergent series by a bounded function does not change the character of convergence. Hence, the series

$$\frac{f(\zeta)}{\zeta - z} = \sum_{n=0}^{+\infty} \frac{f(\zeta)}{(\zeta - a)^{n+1}} (z - a)^n \tag{3.11}$$

converges uniformly with respect to ζ and z such that $|\zeta - a| = R_2$ and $z \in \bar{B}$.

In a similar manner, we can evaluate the function inside the second integral on the right-hand side of (3.9). First, we represent the ratio $\frac{1}{\zeta-z}$ in the form

$$\frac{1}{\zeta - z} = \frac{1}{(\zeta - a) - (z - a)} = -\frac{1}{z - a} \cdot \frac{1}{1 - \frac{\zeta-a}{z-a}} = -\frac{1}{z - a} \cdot \sum_{k=0}^{+\infty} \left(\frac{\zeta - a}{z - a}\right)^k. \tag{3.12}$$

Taking into account that $z \in \bar{B}$ and $\zeta \in \{|\zeta - a| = r_2\}$, we have $\left|\frac{\zeta-a}{z-a}\right| \leq \frac{r_2}{r_1} = q_2 < 1$. Therefore, the ratio $\frac{1}{1-\frac{\zeta-a}{z-a}}$ is the sum of the convergent geometric series. The series of functions (3.12) can be majorized by the convergent numerical series $\sum_{k=0}^{+\infty} q_2^k$, $0 < q_2 < 1$, for $\forall z \in \bar{B}$ and $\forall \zeta,\ |\zeta - a| = r_2$, which implies, by the Weierstrass test, the uniform convergence of the series (3.12) with respect to $z \in \bar{B}$ and $\zeta : |\zeta - a| = r_2$. Since $\frac{1}{|z-a|} \leq \frac{1}{r_1}$, $\forall z \in \bar{B}$, and the function $f(\zeta)$ is bounded due to its analyticity on the compact set $|\zeta - a| = r_2$, the series

$$\frac{f(\zeta)}{\zeta - z} = -\sum_{k=0}^{+\infty} \frac{f(\zeta) \cdot (\zeta - a)^k}{(z - a)^{k+1}} \tag{3.13}$$

converges uniformly with respect to the variables ζ and z such that $|\zeta - a| = r_2$ and $z \in \bar{B}$.

Now we substitute the series (3.11) and (3.13) in the integrals on the right-hand side of (3.9). Due to the uniform convergence of these series with respect to ζ, we can integrate them term-by-term along the circles $|\zeta - a| = R_2$ and $|\zeta - a| = r_2$, respectively (notice that the uniform convergence with respect to $z \in \bar{B}$ is preserved under these transformations)

$$f(z) = \frac{1}{2\pi i}\int_{|\zeta-a|=R_2}\frac{f(\zeta)}{\zeta-z}d\zeta - \frac{1}{2\pi i}\int_{|\zeta-a|=r_2}\frac{f(\zeta)}{\zeta-z}d\zeta = \frac{1}{2\pi i}\int_{|\zeta-a|=R_2}\sum_{n=0}^{+\infty}\frac{f(\zeta)(z-a)^n}{(\zeta-a)^{n+1}}d\zeta$$

$$+\frac{1}{2\pi i}\int_{|\zeta-a|=r_2}\sum_{k=0}^{+\infty}\frac{f(\zeta)(\zeta-a)^k}{(z-a)^{k+1}}d\zeta = \frac{1}{2\pi i}\sum_{n=0}^{+\infty}(z-a)^n\int_{|\zeta-a|=R_2}\frac{f(\zeta)}{(\zeta-a)^{n+1}}d\zeta$$

$$+\frac{1}{2\pi i}\sum_{k=0}^{+\infty}\frac{1}{(z-a)^{k+1}}\int_{|\zeta-a|=r_2}f(\zeta)(\zeta-a)^k\,d\zeta = I\,.$$

Denoting the integrals inside the series by

$$c_n = \frac{1}{2\pi i}\int_{|\zeta-a|=R_2}\frac{f(\zeta)}{(\zeta-a)^{n+1}}d\zeta\,,\quad b_k = \frac{1}{2\pi i}\int_{|\zeta-a|=r_2}f(\zeta)(\zeta-a)^k\,d\zeta\,,$$

we obtain

$$f(z) = I = \sum_{n=0}^{+\infty}c_n(z-a)^n + \sum_{k=0}^{+\infty}\frac{b_k}{(z-a)^{k+1}}$$

$$= \sum_{n=0}^{+\infty}c_n(z-a)^n + \sum_{n=-1}^{-\infty}b_{-n-1}(z-a)^n = I_1.$$

Changing the index in the second series $k+1=-n$ and denoting again $b_{-n-1}=c_n$ (without any confusion regarding the coefficients c_n in the first series, since in the second series n takes only negative values), we obtain the Laurent series in the ring $r<|z-a|<R$

$$f(z) = I_1 = \sum_{n=0}^{+\infty}c_n(z-a)^n + \sum_{n=-1}^{-\infty}c_n(z-a)^n = \sum_{n=-\infty}^{+\infty}c_n(z-a)^n\,.$$

As is shown, the last series converges uniformly on the set $\bar{B}$, that is, in any closed ring contained in the original ring D.

Notice that the formulas of c_n derived so far are different for $n\geq 0$ and $n<0$. We will now prove that these coefficients can be calculated by the same formula. Consider first $n\geq 0$, that is, the formula

$$c_n = \frac{1}{2\pi i}\int_{|\zeta-a|=R_2}\frac{f(\zeta)}{(\zeta-a)^{n+1}}d\zeta\,.$$

Take any ρ such that $r<\rho<R$ and consider the ring $\Omega=\{\rho<|\zeta-a|<R_2\}$ if $\rho<R_2$ or the ring $\Omega=\{R_2<|\zeta-a|<\rho\}$ if $\rho>R_2$. The function $\frac{f(\zeta)}{(\zeta-a)^{n+1}}$ is analytic on the closed ring $\bar{\Omega}$ for any integer n, and consequently, by Goursat's

Theorem, we have

$$0 = \int_{\partial\Omega} \frac{f(\zeta)}{(\zeta - a)^{n+1}} d\zeta = \int_{|\zeta - a| = R_2} \frac{f(\zeta)}{(\zeta - a)^{n+1}} d\zeta - \int_{|\zeta - a| = \rho} \frac{f(\zeta)}{(\zeta - a)^{n+1}} d\zeta .$$

Hence

$$c_n = \frac{1}{2\pi i} \int_{|\zeta - a| = R_2} \frac{f(\zeta)}{(\zeta - a)^{n+1}} d\zeta = \frac{1}{2\pi i} \int_{|\zeta - a| = \rho} \frac{f(\zeta)}{(\zeta - a)^{n+1}} d\zeta , \; n{=}0, 1, 2, \ldots .$$

In a similar way, if $n = -1, -2, \ldots$, then, choosing the ring $\Omega = \{r_2 < |\zeta - a| < \rho\}$ if $r_2 < \rho$ or $\Omega = \{\rho < |\zeta - a| < r_2\}$ if $r_2 > \rho$, we obtain

$$c_n = b_{-n-1} = \frac{1}{2\pi i} \int_{|\zeta - a| = r_2} f(\zeta) (\zeta - a)^{-n-1} d\zeta = \frac{1}{2\pi i} \int_{|\zeta - a| = \rho} \frac{f(\zeta)}{(\zeta - a)^{n+1}} d\zeta .$$

Thus, all the coefficients of the Laurent series are found by the same formula for any integer n.

The Theorem is proved. □

Remark 3.1 If $f(z)$ is an analytic function in the ring $r < |z - a| < R$, then, by the Laurent Theorem, $f(z)$ can be expanded in the Laurent series in this ring: $f(z) = \sum_{n=-\infty}^{+\infty} c_n (z - a)^n$. Let us divide the last expression into two series

$$f(z) = \sum_{n=-\infty}^{+\infty} c_n (z - a)^n = \sum_{n=0}^{+\infty} c_n (z - a)^n + \sum_{n=-1}^{-\infty} c_n (z - a)^n . \tag{3.14}$$

The first series on the right-hand side in (3.14) is the power series. Since this power series converges in the ring $r < |z - a| < R$, it also converges on the entire disk $|z - a| < R$ and its sum is an analytic function in the disk of convergence. Hence, the function $f_1(z) = \sum_{n=0}^{+\infty} c_n (z - a)^n$ is analytic in the entire disk $|z - a| < R$.

Consider now the second series on the right-hand side in (3.14). Using the substitution $t = \frac{1}{z-a}$, we rewrite this series as the following power series with respect to t:

$$\sum_{n=-1}^{-\infty} c_n (z - a)^n = \sum_{n=-1}^{-\infty} \frac{c_n}{t^n} = \sum_{k=1}^{\infty} c_{-k} t^k ,$$

where n was changed by $-k$. Since the series $\sum_{n=-1}^{-\infty} c_n (z - a)^n$ converges on the ring $r < |z - a| < R$, the corresponding series $\sum_{k=1}^{\infty} c_{-k} t^k$ converges on the ring $\frac{1}{R} < |t| < \frac{1}{r}$. However, the last series is a power series, and consequently, it converges in the entire disk $|t| < \frac{1}{r}$ and its sum $g(t) = \sum_{k=1}^{+\infty} c_{-k} t^k$ is an analytic function in this disk. Returning to the series $\sum_{n=-1}^{-\infty} c_n (z - a)^n$, we conclude that it converges on the region $|z - a| > r$ and the function $f_2(z) = g\left(\frac{1}{z-a}\right) = \sum_{n=-1}^{-\infty} c_n (z - a)^n$ is analytic in this region. Thus, we have

$$f(z)=\sum_{n=0}^{+\infty}c_n(z-a)^n+\sum_{n=-1}^{-\infty}c_n(z-a)^n=f_1(z)+f_2(z)\,.$$

Since $f_1(z)$ is analytic in the disk $|z-a|<R$ and $f_2(z)$ is analytic in the region $|z-a|>r$, the sum $f(z)$ is an analytic function on the intersection of these two sets, that is, $f(z)$ is analytic in the ring $r<|z-a|<R$.

The following example illustrates the analysis of convergence of the Laurent series:

Example Find the set of convergence of the Laurent series

$$\sum_{n=-\infty}^{+\infty}\frac{z^n}{3^n+1}\,.$$

Let us divide the given Laurent series into the two parts: the *regular (analytic) part* (containing non-negative indices n) and the *principal part* (containing only negative n)

$$\sum_{n=-\infty}^{+\infty}\frac{z^n}{3^n+1}=\sum_{n=0}^{+\infty}\frac{z^n}{3^n+1}+\sum_{n=-1}^{-\infty}\frac{z^n}{3^n+1}\,. \tag{3.15}$$

The first part is a power series and its radius of convergence can be found by the Cauchy formula (see formula (1.16) in Sect. 1.7)

$$R_1=\frac{1}{\lim\limits_{n\to\infty}\sqrt[n]{|c_n|}}=\frac{1}{\lim\limits_{n\to\infty}\sqrt[n]{(3^n+1)^{-1}}}=\lim_{n\to\infty}\sqrt[n]{3^n+1}=3\,.$$

Therefore, the regular (first) part in (3.15) converges in the disk $|z|<3$. If $|z|=3$, then $z=3e^{i\varphi}$, $\varphi\in[0\,,\,2\pi]$ and the series $\sum_{n=0}^{+\infty}\frac{3^ne^{in\varphi}}{3^n+1}$ diverges, since its general term does not approach zero

$$|a_n|=\left|\frac{3^ne^{in\varphi}}{3^n+1}\right|=\frac{3^n}{3^n+1}\underset{n\to+\infty}{\to}1\neq 0\,.$$

In the second series (principal part) in (3.15), we change the summation index ($k=-n$) and the independent variable $t=\frac{1}{z}$ in order to transform this part into the power series

$$\sum_{n=-1}^{-\infty}\frac{z^n}{3^n+1}=\sum_{k=1}^{+\infty}\frac{z^{-k}}{3^{-k}+1}=\sum_{k=1}^{+\infty}\frac{t^k}{3^{-k}+1}\,. \tag{3.16}$$

Using again the Cauchy formula to find the radius of convergence of the series (3.16)

$$R_2 = \frac{1}{\lim\limits_{k\to+\infty} \sqrt[k]{\left(3^{-k}+1\right)^{-1}}} = \lim_{k\to+\infty} \sqrt[k]{3^{-k}+1} = 1\,,$$

we see that the series $\sum_{k=1}^{+\infty} \frac{t^k}{3^{-k}+1}$ converges in the disk $|t| < 1$. If $|t| = 1$, that is, $t = e^{i\theta}$, $\theta \in [0\,,\, 2\pi]$, then the series

$$\sum_{k=1}^{+\infty} \frac{t^k}{3^{-k}+1}\Bigg|_{|t|=1} = \sum_{k=1}^{+\infty} \frac{e^{ik\theta}}{3^{-k}+1}$$

diverges due to the divergence test:

$$\left|\frac{e^{ik\theta}}{3^{-k}+1}\right| = \frac{1}{3^{-k}+1} \underset{k\to+\infty}{\to} 1 \neq 0\,.$$

Returning to the original variable, we see that the last series in (3.15) converges in the set $|z| > 1$. Joining the two analyzed parts, we conclude that the given Laurent series converges in the ring $1 < |z| < 3$.

Remark 3.2 Let $f(z)$ be an analytic function in the ring $0 < |z-a| < R$ (here $r = 0$). The Laurent series expansion of $f(z)$ in this ring is as follows:

$$f(z) = \sum_{n=-\infty}^{+\infty} c_n (z-a)^n = \sum_{n=0}^{+\infty} c_n (z-a)^n + \sum_{n=-1}^{-\infty} c_n (z-a)^n = f_1(z) + f_2(z)\,.$$

This series is referred to as *the Laurent series of* $f(z)$ *at the point* a (since the ring $0 < |z-a| < R$ is a deleted neighborhood of a). According to Remark 3.1, the function $f_1(z)$ is analytic in the entire disk $|z-a| < R$, and for this reason, it is named the *regular or analytic part of the Laurent series of* $f(z)$ *at the point* a. The function $f_2(z)$ is analytic in the region $|z-a| > 0$. Since $f_1(z)$ is analytic at a, it means that $f(z)$ has at the point a the same singularity as $f_2(z)$. Therefore, the function $f_2(z)$ (that is, the sum of all the negative powers) is named the *principal part of the Laurent series at the point* a.

Remark 3.3 Let $f(z)$ be an analytic function in the ring $r < |z| < +\infty$ (here $R = +\infty$), that is, in the deleted neighborhood of the point ∞. The Laurent series expansion of $f(z)$ in this ring goes as follows:

$$f(z) = \sum_{n=-\infty}^{+\infty} c_n z^n = \sum_{n=1}^{+\infty} c_n z^n + \sum_{n=0}^{-\infty} c_n z^n = f_1(z) + f_2(z)\,.$$

Notice, that we have transferred, for convenience, the term c_0 (corresponding to $n = 0$) to the second sum, but this will not change the reasoning. This series is called the *Laurent series of* $f(z)$ *at infinity*. By Remark 3.1, the function $f_2(z)$ is analytic in the region $|z| > r$, in particular, it is analytic at the point ∞. Therefore, in this case, $f_2(z)$ (that is, the sum of all the non-positive powers of the Laurent series) is called the *regular part of the Laurent series at the point* ∞. The function $f(z)$ has the same singularity at infinity as $f_1(z)$, and correspondingly, the function $f_1(z)$ (that is, the sum of all the positive powers) is called the *principal part of the Laurent series at* ∞.

3.4 Classification of Singular Points

In this section, we specify different classes of singular points.

Definitions A point z_0 is called a *singular point* or *singularity* of a function $f(z)$, if this function cannot be analytically extended to the point z_0 or $f(z)$ is not defined at this point.

There are two kinds of singular points—*isolated* and *not isolated.*

A point z_0 is called an *isolated singular point* of a function $f(z)$, if there exists a neighborhood of this point that does not contain other singular points of this function, different from z_0, that is, there exists $r > 0$ such that the function $f(z)$ can be analytically extended to the ring $0 < |z - z_0| < r$.

A point z_0 is called a *non-isolated singular point* of a function $f(z)$, if in any neighborhood of z_0 there exist other singular points of $f(z)$, different from z_0.

Now we present a classification of isolated singularities.

A point z_0 is said to be *a single-valued isolated singular point* of a function $f(z)$, if there exists $r > 0$ such that $f(z)$ is analytic in deleted r-neighborhood of z_0, that is, as a result of the analytic continuation of $f(z)$ onto the ring $0 < |z - z_0| < r$ we obtain a single-valued analytic function.

Single-valued isolated singular points can be of three types: *removable singularities, poles*, and *essential singularities.*

A *single-valued isolated singular point* z_0 of a function $f(z)$ is said to be

(1) a *removable singular point (removable singularity)* if there exists a finite limit of $f(z)$ as z approaches z_0: $\lim\limits_{z\to z_0} f(z) = A$;

(2) a *pole* if $\lim\limits_{z\to z_0} f(z) = \infty$;

(3) an *essential singular point (essential singularity)* if there is no limit (either finite or infinite) of $f(z)$ at the point z_0.

Consider now *multi-valued isolated singular points* also referred to as *branch points.*

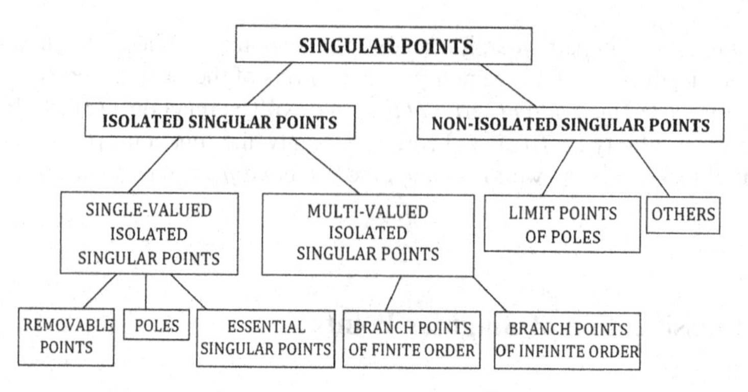

Fig. 3.5 Classification of singular points

A point z_0 is said to be a *multi-valued isolated singular point* or a *branch point* of a function $f(z)$, if $f(z)$ has analytic continuation onto the ring $0 < |z - z_0| < r$, but for any $r > 0$ the result of this continuation is a multi-valued analytic function.

There are two types of *branch points*: *branch points of a finite order* and *branch points of an infinite order*.

A point z_0 is a *branch point of a finite order* or an *algebraic branch point* of a function $f(z)$, if there exists a finite number of different analytic elements of $f(z)$ at every point of the ring $0 < |z - z_0| < r$, that is, if $f(z)$ has a finite number of different values at every point of this ring.

A point z_0 is a *branch point of an infinite order* or a *logarithmic branch point* of a function $f(z)$, if for any $r > 0$ there are points in the ring $0 < |z - z_0| < r$ at which $f(z)$ has an infinite number of different analytic elements, that is, the function $f(z)$ takes an infinite number of different values at points of the ring $0 < |z - z_0| < r, \forall r > 0$.

Finally, we briefly mention *non-isolated singular points*. There are many types of these points, but we consider only *singularities representing limit points of poles*. For a point z_0 of this type, there exists $R > 0$ such that in any ring $0 < |z - z_0| < r$, $r < R$ the function $f(z)$ has an infinite number of poles and no singular points of another type.

The scheme in Fig. 3.5 illustrates the *classification of singular points*.

Let us consider an example of searching for and classification of singular points.

Example Find singular points of the function $f(z) = \frac{2+e^z}{1-e^z}$ and classify them.

First, we find the zeros of the denominator solving the equation $1 - e^z = 0$. Representing the number 1 in the exponential form, we have $e^z = 1 = e^{2k\pi i}$, $\forall k \in \mathbb{Z}$, from which it follows that $z_k = 2k\pi i$, $\forall k \in \mathbb{Z}$. All these points are single-valued isolated singularities of $f(z)$, because both the numerator and denominator of $f(z)$ are analytic functions in $\mathbb{C}$. Since the numerator of $f(z)$ is different from 0 at these points: $2 + e^{z_k} = 2 + e^{2k\pi i} = 3 \neq 0$, we obtain $\lim\limits_{z \to z_k} f(z) = \lim\limits_{z \to z_k} \frac{2+e^z}{1-e^z} = \infty$,

$\forall k \in \mathbb{Z}$, which means that z_k are the poles of $f(z)$. Notice that $z_k \to \infty$ as $k \to \pm\infty$, meaning that $z = \infty$ is a singularity representing the limit point of the poles of $f(z)$.

We finalize this section with two examples of multi-valued isolated singular points and in the next sections, we will study single-valued isolated singular points in detail.

Example 3.1 Consider $f(z) = \sqrt[n]{z}$. The singular points of this function are 0 and ∞, and both of them are isolated. In Sect. 1.1, it was shown that there exist n different roots $\sqrt[n]{z}$ of an arbitrary complex number $z \neq 0$ and the formulas for these roots were found. The different roots are obtained making circuits around the point $z = 0$. Notice that a loop around the origin means at the same time a circuit around infinity. Hence, the singular points $z = 0$ and $z = \infty$ are branch points of a finite (n-th) order of $\sqrt[n]{z}$.

Example 3.2 The function $f(z) = \ln z = \int_1^z \frac{d\zeta}{\zeta} = \ln|z| + i\arg z + 2k\pi i$, $\forall k \in \mathbb{Z}$ was introduced in Sect. 2.6 in the example on a primitive of an analytic function. By construction, $f(z)$ is extendable analytically onto the entire complex plane, except for the points 0 and ∞ (that is, the analytic continuation exists in the ring $0 < |z| < +\infty$). In this way, the points 0 and ∞ are isolated singular points of $\ln z$. Since the function $f(z) = \ln z$ has an infinite number of different values at any point z, except for $z = 0$ and $z = \infty$, the last two points are branch points of an infinite order of $\ln z$.

3.5 Removable Singular Points

Theorem (The first criterion for a removable singular point) *Let z_0 be a single-valued isolated singular point of a function $f(z)$. The point z_0 is removable if and only if the principal part of the Laurent series of $f(z)$ is equal to zero.*

Proof *Necessity.* Let us start with the case when the removable singular point is not infinity: $z_0 \neq \infty$. Since z_0 is a single-valued isolated singular point, the function $f(z)$ is analytic in a deleted neighborhood of z_0, that is, $f(z)$ can be expanded in the Laurent series in the ring $0 < |z - z_0| < r$:

$$f(z) = \sum_{n=-\infty}^{+\infty} c_n (z - z_0)^n, \ \forall z : 0 < |z - z_0| < r.$$

According to the Laurent Theorem, the coefficients c_n are found by formula (3.8)

$$c_n = \frac{1}{2\pi i} \int_{|\zeta - z_0| = \rho} \frac{f(\zeta)}{(\zeta - z_0)^{n+1}} d\zeta, \tag{3.17}$$

where ρ is an arbitrary number such that $0 < \rho < r$.

By the definition of a removable singular point, there exists a finite limit $\lim_{z\to z_0} f(z) = A$, which means that for $\forall\varepsilon > 0$ (take $\varepsilon = 1$) there exists $\delta(\varepsilon) > 0$ (one can always choose $\delta < r$) such that $|f(z) - A| < 1$ whenever $z \in \overset{0}{U}_\delta(z_0)$. Therefore, $|f(z)| = |f(z) - A + A| \le |f(z) - A| + |A| < |A| + 1$, that is, the function $f(z)$ is limited in some deleted neighborhood of z_0. Let us evaluate the coefficients c_n in the Laurent series. Since ρ in (3.17) is an arbitrary number such that $0 < \rho < r$, we can choose $0 < \rho < \delta(\varepsilon)$. Then we obtain

$$|c_n| = \left|\frac{1}{2\pi i}\int_{|\zeta - z_0| = \rho} \frac{f(\zeta)}{(\zeta - z_0)^{n+1}} d\zeta\right| \le \frac{1}{2\pi}\int_{|\zeta - z_0| = \rho} \frac{|f(\zeta)|}{|\zeta - z_0|^{n+1}} |d\zeta|$$

$$< \frac{|A| + 1}{2\pi \cdot \rho^{n+1}} 2\pi\rho = (|A| + 1)\,\rho^{-n}.$$

If n is negative, then $(|A| + 1) \cdot \rho^{-n} \underset{\rho\to 0}{\to} 0$, that is, $|c_n| \underset{\rho\to 0}{\to} 0$. On the other hand, the coefficients c_n are constants that do not depend on ρ, and consequently $c_n = 0$ for all $n = -1, -2, -3, \ldots$. This means that all the negative powers do not appear in the Laurent series, that is, the principal part of the Laurent series of $f(z)$ is equal to zero.

If a single-valued isolated singular point is $z_0 = \infty$, then a deleted neighborhood of z_0 where $f(z)$ is analytic is a ring $R < |z| < +\infty$, that is, $f(z)$ can be expanded in the following Laurent series:

$$f(z) = \sum_{n=-\infty}^{+\infty} c_n z^n, \ \forall z : R < |z| < +\infty.$$

By the definition of a removable singular point, there exists a finite limit $\lim_{z\to\infty} f(z) = A$, which implies that $f(z)$ is bounded—$|f(z)| < |A| + 1$—in a deleted neighborhood of the infinity point (in the same way as for $z_0 \ne \infty$). Using formula (3.17) with $\rho > R$ for the evaluation of the coefficients c_n in the Laurent series, we obtain for any $n \in \mathbb{N}$

$$|c_n| = \left|\frac{1}{2\pi i}\int_{|\zeta| = \rho} \frac{f(\zeta)}{\zeta^{n+1}} d\zeta\right| \le \frac{1}{2\pi}\int_{|\zeta| = \rho} \frac{|f(\zeta)|}{|\zeta|^{n+1}} |d\zeta| < \frac{1}{2\pi} \cdot \frac{|A| + 1}{\rho^{n+1}} \cdot 2\pi\rho = \frac{|A| + 1}{\rho^n} \underset{\rho\to\infty}{\to} 0.$$

Since c_n are constants, it means that $c_n = 0$, $\forall n \in \mathbb{N}$. Therefore, all the positive powers are not present in the Laurent series, that is, the principal part of this series is equal to zero.

Sufficiency. Let z_0 be a single-valued isolated singular point and the principal part of the Laurent series of $f(z)$ at z_0 be equal to zero. If $z_0 \ne \infty$, then the Laurent series takes the form

$$f(z) = \sum_{n=0}^{+\infty} c_n (z - z_0)^n \tag{3.18}$$

in the ring $0 < |z - z_0| < r$. Therefore, the Laurent series of $f(z)$ is a power series. However, any power series converges in its disk of convergence, and consequently, the last series converges in the disk $|z - z_0| < r$. Besides, the limit of a power series can be calculated term-by-term. Applying this rule to the last series as z approaches z_0, we obtain $\lim_{z \to z_0} f(z) = c_0$, that is, $f(z)$ has a finite limit as z approaches z_0, and consequently, z_0 is a removable singular point. If we define $f(z)$ at z_0 using this limit: $f(z_0) = \lim_{z \to z_0} f(z) = c_0$, then the function $f(z)$ is analytic in the entire disk $|z - z_0| < r$, in particular, at the point z_0. For this reason, such a point z_0 is called a removable singularity.

If $z_0 = \infty$, then the Laurent series with the zero principal part takes the form

$$f(z) = \sum_{n=-\infty}^{0} c_n z^n$$

in the ring $R < |z| < +\infty$. Changing the variable $t = \frac{1}{z}$, we obtain the function $g(t) = f\left(\frac{1}{t}\right)$, which is analytic in the ring $0 < |t| < \frac{1}{R}$ and can be expanded in the series

$$g(t) = f\left(\frac{1}{t}\right) = \sum_{n=0}^{+\infty} c_{-n} t^n$$

in this ring. The last series is a power series, which converges in the entire disk $|t| < \frac{1}{R}$. Passing to the term-by-term limit in the last series as t approaches 0, we obtain $\lim_{t \to 0} g(t) = c_0$, that is, the limit exists and is finite. Then $\lim_{z \to \infty} f(z) = \lim_{z \to \infty} g\left(\frac{1}{z}\right) = \lim_{t \to 0} g(t) = c_0$, and therefore, by the definition, the point ∞ is a removable singular point of $f(z)$.

This completes the proof of the Theorem. □

Corollary (The second criterion for a removable singular point) *Let z_0 be a single-valued isolated singular point of a function $f(z)$. The point z_0 is removable if and only if $f(z)$ is a bounded function in a deleted neighborhood of this point.*

Proof In fact, if z_0 (finite or infinite) is a removable singular point of $f(z)$, then it was shown in the proof of necessity part of the last Theorem that $f(z)$ is bounded in a deleted neighborhood of z_0.

Conversely, if $f(z)$ is bounded in a deleted neighborhood of z_0 (where z_0 is a single-valued isolated singularity), then the proof of necessity part of the last Theorem shows that the principal part of the Laurent series of $f(z)$ at the point z_0 vanishes, and consequently, by the sufficient condition of the Theorem, z_0 is a removable singular point. □

3.6 Liouville's Theorem

Liouville's Theorem *Let $f(z)$ be an entire function (that is, function $f(z)$ be analytic on the entire finite complex plane). Denote $M_f(R) = \sup_{|z|=R} |f(z)|$. If $\varliminf_{R\to\infty} R^{-n} \cdot M_f(R) = 0$, then $f(z)$ is a polynomial of degree less than or equal to $(n-1)$.*

Proof Since $\varliminf_{R\to\infty} R^{-n} M_f(R) = 0$, by the definition of the lower limit it means that there exists a sequence of real numbers R_p, $R_p \to \infty$ as $p \to \infty$ such that

$$\lim_{p\to\infty} R_p^{-n} M_f(R_p) = \varliminf_{R\to\infty} R^{-n} M_f(R) = 0\,. \tag{3.19}$$

The property of being an entire function means that $f(z)$ can be represented in the power series $f(z) = \sum_{n=0}^{+\infty} c_n z^n$ convergent on the entire (finite) complex plane (that is, the radius of convergence of this series is ∞). Let us evaluate the coefficients of this series using the integral formula:

$$c_k = \frac{1}{2\pi i} \int_{|z|=R} \frac{f(z)}{z^{k+1}}\, dz\,. \tag{3.20}$$

Since c_k does not depend on the radius of the circle R (this was shown in Sect. 3.3), we can calculate the integral along the circle $|z| = R_p$, where $\{R_p\}_{p=1}^{+\infty}$ is the sequence that satisfies the condition (3.19). For any $k \geq n$, we have

$$|c_k| = \left| \frac{1}{2\pi i} \int_{|z|=R_p} \frac{f(z)}{z^{k+1}} dz \right| \leq \frac{1}{2\pi} \int_{|z|=R_p} \frac{|f(z)|}{|z|^{k+1}} |dz| \leq \frac{M_f(R_p)}{2\pi \cdot R_p^{k+1}} \cdot 2\pi R_p$$

$$= \frac{M_f(R_p)}{R_p^k} = \frac{R_p^{-n} M_f(R_p)}{R_p^{k-n}} \underset{p\to\infty}{\to} 0\,,$$

since $R_p^{-n} M_f(R_p) \underset{p\to\infty}{\to} 0$ by the condition (3.19) and $k - n \geq 0$. Therefore, $c_k = 0$ for $\forall k \geq n$. Hence, the function $f(z)$ has the form $f(z) = \sum_{k=0}^{n-1} c_k z^k$, which is a polynomial of degree less than or equal to $(n-1)$. □

Corollary 3.1 *If a function $f(z)$ is analytic and bounded on the entire (finite) complex plane, then $f(z)$ is a constant.*

Proof Since $f(z)$ is bounded on the entire complex plane, there exists $M > 0$ such that $|f(z)| \leq M, \forall z \in \mathbb{C}$. Then $M_f(R) = \sup_{|z|=R} |f(z)| \leq M$ and we have that $\lim_{R\to\infty} R^{-1} M_f(R) = 0$. Applying Liouville's Theorem with $n = 1$, we conclude that $f(z)$ is a polynomial of degree zero, that is, $f(z)$ is a constant. □

Corollary 3.2 *If a function $f(z)$ is analytic on the entire extended complex plane $\overline{\mathbb{C}}$, then $f(z)$ is a constant.*

Proof Since $f(z)$ is analytic on $\overline{\mathbb{C}}$, then it has a representation in the power series at the centerpoint $z = 0$ with the infinite radius of convergence

$$f(z) = \sum_{n=0}^{+\infty} c_n z^n = c_0 + \sum_{n=1}^{+\infty} c_n z^n .$$

At the same time, this series is the Laurent series of $f(z)$ at ∞, and since $f(z)$ is analytic at ∞ (that is, ∞ is a removable singularity), the Theorem of Sect. 3.5 guarantees that the principal part of this Laurent series is equal to zero: $\sum_{n=1}^{+\infty} c_n z^n = 0$. Therefore, $f(z) = c_0$, $\forall z \in \overline{\mathbb{C}}$. □

Frequently both Corollaries are also called Liouville's Theorem.

3.7 Poles

Theorem 3.1 *If a (finite or infinite) point z_0 is a pole of the function $f(z)$, then the function $g(z) = \frac{1}{f(z)}$ is analytic at z_0 and $g(z_0) = 0$.*

Proof Let us first consider a finite point z_0. Since z_0 is a pole of $f(z)$, this function is analytic in the ring $0 < |z - z_0| < r$ and $\lim\limits_{z \to z_0} f(z) = \infty$. Recall the definition of an infinite limit: for $\forall \varepsilon > 0$ (take $\varepsilon = 1$) there exists $\delta > 0$ (choose $\delta \le r$) such that $|f(z)| > 1$ whenever $0 < |z - z_0| < \delta$. Then, the function $g(z) = \frac{1}{f(z)}$ is analytic in the ring $0 < |z - z_0| < \delta$ as the ratio of two analytic functions with denominator $f(z) \ne 0$ (since $|f(z)| > 1$) in this ring. Therefore, z_0 is a single-valued isolated singular point of $g(z)$. To determine what kind of singularity the function has at this point, we calculate the limit: $\lim\limits_{z \to z_0} g(z) = \lim\limits_{z \to z_0} \frac{1}{f(z)} = 0$. Hence, z_0 is a removable singular point of the function $g(z)$, and consequently, we can eliminate the singularity at the point z_0 by defining $g(z_0)$ through the limit: $g(z_0) = \lim\limits_{z \to z_0} g(z) = 0$. The function $g(z)$ defined in this manner is analytic in the entire disk $|z - z_0| < \delta$, that is, $g(z)$ is analytic at z_0 and $g(z_0) = 0$.

For an infinite z_0 the proof is practically the same and left to the reader. □

Definition 3.1 Let a function $g(z)$ be analytic at a point z_0, $z_0 \ne \infty$. We say that *$g(z)$ has a zero of order k ($k \in \mathbb{N}$) (or multiplicity k) at the point z_0*, if the function $g(z)$ can be represented in the form $g(z) = (z - z_0)^k \, g_1(z)$, where $g_1(z)$ is an analytic function at the point z_0 and $g_1(z_0) \ne 0$. If $k = 1$, then a zero is said to be simple.

Definition 3.2 Let $g(z)$ be an analytic function at the point ∞. We say that $g(z)$ *has a zero of order (multiplicity) k* $(k \in \mathbb{N})$ at this point, if $g(z)$ can be represented in the form $g(z) = z^{-k} g_1(z)$, where $g_1(z)$ is an analytic function at ∞ and $g_1(\infty) \neq 0$.

Theorem 3.2 (Criterion for multiplicity of a zero) *Let $g(z)$ be an analytic function at a point z_0, $z_0 \neq \infty$. The function $g(z)$ has a zero of order k at z_0 if and only if* $g(z_0) = g'(z_0) = \cdots = g^{(k-1)}(z_0) = 0$, $g^{(k)}(z_0) \neq 0$.

Proof *Necessity*. Since $g(z)$ has a zero of order k at the point z_0, then, by the definition, it can be represented in the form $g(z) = (z - z_0)^k g_1(z)$, where $g_1(z)$ is an analytic function at z_0 and $g_1(z_0) \neq 0$. For the analytic at z_0 function $g_1(z)$, we have the power series expansion $g_1(z) = \sum_{m=0}^{+\infty} a_m (z - z_0)^m$, where the coefficients a_m are the Taylor coefficients and, in particular, $a_0 = g_1(z_0) \neq 0$. Then

$$g(z) = (z - z_0)^k g_1(z) = (z - z_0)^k \cdot \sum_{m=0}^{+\infty} a_m (z - z_0)^m = \sum_{m=0}^{+\infty} a_m (z - z_0)^{m+k}$$

$$= \sum_{n=k}^{+\infty} a_{n-k} (z - z_0)^n = \sum_{n=k}^{+\infty} c_n (z - z_0)^n$$

(here we denote $m + k = n$ and $a_{n-k} = c_n$). Therefore, we obtain a power series with the coefficients $c_0 = c_1 = \cdots = c_{k-1} = 0$ and $c_k = a_0 = g_1(z_0) \neq 0$. Since c_j are the Taylor coefficients, that is, $c_j = \frac{g^{(j)}(z_0)}{j!}$, we have $g(z_0) = g'(z_0) = \cdots = g^{(k-1)}(z_0) = 0$ and $g^{(k)}(z_0) \neq 0$.

Sufficiency. An analytic at z_0 function $g(z)$ can be expanded in a power series in a neighborhood of the point z_0: $g(z) = \sum_{n=0}^{+\infty} c_n (z - z_0)^n$. Since c_n in this series are the Taylor coefficients, that is, $c_n = \frac{g^{(n)}(z_0)}{n!}$ and, by the hypothesis of the Theorem, $g(z_0) = g'(z_0) = \cdots = g^{(k-1)}(z_0) = 0$, $g^{(k)}(z_0) \neq 0$, we conclude that $c_0 = c_1 = \cdots = c_{k-1} = 0$ and $c_k \neq 0$. Therefore, the power series takes the form

$$g(z) = \sum_{n=k}^{+\infty} c_n (z - z_0)^n = (z - z_0)^k \sum_{n=k}^{+\infty} c_n (z - z_0)^{n-k} = (z - z_0)^k \sum_{m=0}^{+\infty} c_{m+k} (z - z_0)^m$$

$$= (z - z_0)^k \sum_{m=0}^{+\infty} a_m (z - z_0)^m = (z - z_0)^k \cdot g_1(z)$$

(here we denote $n - k = m$ and $c_{m+k} = a_m$). The function $g_1(z)$ is analytic in a neighborhood of z_0 as the sum of the convergent power series, and additionally $g_1(z_0) = a_0 = c_k \neq 0$. Thus, we arrive at the definition of a zero of order k of the function $g(z)$ at the point z_0. This completes the proof of the Theorem. □

In Theorem 3.1, we have shown that if a function $f(z)$ has a pole at a point z_0, then the function $g(z) = \frac{1}{f(z)}$ is analytic at z_0 and has a zero at this point.

Definition 3.3 Let a (finite or infinite) point z_0 be a pole of a function $f(z)$. We say that z_0 is *a pole of order (multiplicity) k* of the function $f(z)$ if the function $g(z) = \frac{1}{f(z)}$ has a zero of order k at this point. If $k = 1$ then a pole is said to be simple.

Theorem 3.3 (The first criterion for multiplicity of a pole) *A function $f(z)$ has a pole of multiplicity k at a finite point z_0 ($z_0 \neq \infty$) if and only if there exists a deleted neighborhood of z_0 in which $f(z)$ can be represented in the form $f(z) = (z - z_0)^{-k} f_1(z)$, where $f_1(z)$ is an analytic function at z_0 and $f_1(z_0) \neq 0$. A function $f(z)$ has a pole of multiplicity k at the point ∞ if and only if there exists a deleted neighborhood of ∞ in which $f(z)$ is expressed in the form $f(z) = z^k f_1(z)$, where $f_1(z)$ is an analytic function at ∞ and $f_1(\infty) \neq 0$.*

Proof We consider only the case of a finite point z_0, since for ∞ the proof is very similar.

Necessity. Let a finite point z_0 be a pole of order k of a function $f(z)$. Then, by the definition of a pole of order k, the function $g(z) = \frac{1}{f(z)}$ is analytic at z_0 and has a zero of order k at this point. By the definition of a zero of order k, we have $g(z) = (z - z_0)^k g_1(z)$, where the function $g_1(z)$ is analytic at z_0 and $g_1(z_0) \neq 0$. Therefore, $f(z) = \frac{1}{g(z)} = (z - z_0)^{-k} f_1(z)$, where $f_1(z) = \frac{1}{g_1(z)}$ is an analytic function at z_0 as the ratio of two analytic functions with the denominator $g_1(z)$ different from zero at the point z_0, and additionally $f_1(z_0) = \frac{1}{g_1(z_0)} \neq 0$.

Sufficiency. Suppose that in some neighborhood of z_0 a function $f(z)$ is represented in the form $f(z) = (z - z_0)^{-k} f_1(z)$, where $f_1(z)$ is an analytic function at z_0 and $f_1(z_0) \neq 0$. Then we consider the function

$$g(z) = \frac{1}{f(z)} = (z - z_0)^k \cdot g_1(z). \tag{3.21}$$

In the last formula, the function $g_1(z) = \frac{1}{f_1(z)}$ is analytic at z_0 as the ratio of two analytic functions with the denominator different from zero at the point z_0: $g_1(z_0) = \frac{1}{f_1(z_0)} \neq 0$. Therefore, the function $g(z)$ is analytic as the product of two analytic functions in the form (3.21). Then, by the definition, $g(z)$ has a zero of order k at z_0, and consequently, the function $f(z)$ has a pole of order k at the point z_0.

A similar proof can be made for the case $z_0 = \infty$ and it is left to the reader.

The Theorem is proved. □

Theorem 3.4 (The second criterion for multiplicity of a pole) *Let z_0 be a single-valued isolated singular point of a function $f(z)$. The function $f(z)$ has a pole of order k at z_0 if and only if the principal part of the Laurent series of $f(z)$ at z_0 does not vanish and contains a finite number of terms. More precisely: if $z_0 \neq \infty$, then the necessary and sufficient condition for z_0 to be a pole of order k is $c_n = 0, \forall n < -k$ and $c_{-k} \neq 0$ in the Laurent series of $f(z)$; if $z_0 = \infty$, then the necessary and sufficient condition for z_0 to be a pole of order k is $c_n = 0, \forall n > k$ and $c_k \neq 0$ in the Laurent series of $f(z)$.*

Proof The proofs for a finite point and infinity are practically the same. For this reason, we consider only the case of a finite point.

Necessity. Let $z_0 \neq \infty$ be a pole of order k of the function $f(z)$. According to the Theorem 3.3, we can express $f(z)$ in the form $f(z) = (z - z_0)^{-k} f_1(z)$, where the function $f_1(z)$ is analytic at z_0 and $f_1(z_0) \neq 0$. Since $f_1(z)$ is analytic at z_0, it can be expanded in the power series $f_1(z) = \sum_{m=0}^{+\infty} a_m (z - z_0)^m$ convergent in the disk $|z - z_0| < r$. Therefore, in the ring $0 < |z - z_0| < r$, we have

$$f(z) = (z - z_0)^{-k} f_1(z) = (z - z_0)^{-k} \sum_{m=0}^{+\infty} a_m (z - z_0)^m$$

$$= \sum_{m=0}^{+\infty} a_m (z - z_0)^{m-k} = \sum_{n=-k}^{+\infty} c_n (z - z_0)^n ,$$

where we denote $m - k = n$ and $a_{n+k} = c_n$. The last series is the Laurent series of $f(z)$ in a deleted neighborhood of z_0 in which all the terms with the indices less than $-k$ are absent, that is, $c_n = 0, \forall n < -k$. Calculating c_{-k}, we have $c_{-k} = a_0 = f_1(z_0) \neq 0$.

Sufficiency. Since z_0 is a single-valued isolated singular point of $f(z)$, we can use the Laurent series expansion of $f(z)$ in a deleted neighborhood of z_0, which has the form $f(z) = \sum_{n=-k}^{+\infty} c_n (z - z_0)^n$. The coefficients $c_n = 0$ for $\forall n < -k$ and $c_{-k} \neq 0$ by the hypothesis of the Theorem. We can transform this series as follows:

$$f(z) = \sum_{n=-k}^{+\infty} c_n (z-z_0)^n = (z-z_0)^{-k} \sum_{n=-k}^{+\infty} c_n (z-z_0)^{n+k} = (z-z_0)^{-k} \sum_{m=0}^{+\infty} c_{m-k} (z-z_0)^m$$

$$= (z - z_0)^{-k} \sum_{m=0}^{+\infty} a_m (z - z_0)^m = (z - z_0)^{-k} \cdot f_1(z) ,$$

where we denote $n + k = m$ and $c_{m-k} = a_m$. The function $f_1(z)$ is analytic at z_0 as the sum of the convergent power series and $f_1(z_0) = a_0 = c_{-k} \neq 0$. Then, by Theorem 3.3, we conclude that z_0 is a pole of order k of the function $f(z)$.

The case $z_0 = \infty$ can be demonstrated in a similar manner and is left to the reader. □

Theorem 3.5 *If $g(z)$ and $h(z)$ are analytic functions at a point z_0, then the function $f(z) = \frac{g(z)}{h(z)}$ is either analytic or has a pole at z_0. More precisely, if $g(z)$ has a zero of order $k \geq 0$ at z_0 and $h(z)$ has a zero of order $m \geq 0$ at z_0, then*
(1) in the case $k - m \geq 0$ the function $f(z) = \frac{g(z)}{h(z)}$ is analytic at z_0 and has a zero of order $(k - m)$ at this point;
(2) in the case $k - m < 0$ the function $f(z) = \frac{g(z)}{h(z)}$ has a pole of order $(m - k)$ at z_0.

Proof By the definition of multiplicity of zero, the hypothesis states that $g(z) = (z - z_0)^k g_1(z)$ and $h(z) = (z - z_0)^m h_1(z)$, where the functions $g_1(z)$ and $h_1(z)$ are analytic at the point z_0 and $g_1(z_0) \neq 0$, $h_1(z_0) \neq 0$ (if, for instance, $g(z_0) \neq 0$, then the same formula holds for $k = 0$). Therefore, $f(z)$ is represented in the form

$$f(z) = \frac{g(z)}{h(z)} = (z - z_0)^{k-m} \cdot f_1(z) \,. \tag{3.22}$$

The function $f_1(z) = \frac{g_1(z)}{h_1(z)}$ is analytic at z_0 as the ratio of two analytic functions with the denominator $h_1(z)$ different from zero at z_0: $h_1(z_0) \neq 0$. Also, $f_1(z_0) = \frac{g_1(z_0)}{h_1(z_0)} \neq 0$. If $k - m \geq 0$, then the function $f(z)$ is analytic as the product of two analytic functions and, by the definition, it has a zero of order $(k - m)$ at z_0. If $k - m < 0$, then, by the criterion for multiplicity of a pole (Theorem 3.3), the function $f(z)$ has a pole of order $(m - k)$ at z_0. □

Definition A function $f(z)$ is called *meromorphic* in a domain D if $f(z)$ is analytic in D except possibly at isolated singularities each of which is a pole.

By the definition, any analytic in a domain D function is meromorphic in D (the empty set of singularities is allowed). In particular, any entire function is a meromorphic function on the entire complex plane. Like the set of analytic in D functions, the set of meromorphic functions is closed under the operations of addition, subtraction and multiplication. Unlike analytic functions, quotients of meromorphic functions are also meromorphic, provided that the denominator is not identically zero. Indeed, it follows from Theorem 3.1 and 3.5 that if $g(z) \not\equiv 0$ is a meromorphic function in D, then $\frac{1}{g(z)}$ is also meromorphic in D, and consequently, $\frac{f(z)}{g(z)} = f(z) \cdot \frac{1}{g(z)}$ is a meromorphic function in D.

Example 3.1 Every rational function $f(z) = \frac{a_m z^m + \cdots + a_1 z + a_0}{b_n z^n + \cdots + b_1 z + b_0}$, $b_n \neq 0$ is meromorphic in the entire complex plane and the set of its poles is either a finite set contained in the set of zeros of the denominator polynomial or the empty set.

Example 3.2 The cotangent function $\cot z = \frac{\cos z}{\sin z}$ is meromorphic in the entire complex plane and the set of its poles is the infinite (countable) set $\{k\pi, k \in \mathbb{Z}\}$.

We end this section with an example illustrating the relationship between the form of expansion in the Laurent series at some point and the type of singularity at the same point.

Example Expand the function $f(z) = \frac{1}{z(z+1)}$ in the Laurent series at the points $z = 0$, $z = -1$, $z = \infty$ and find the ring (or the disk) of convergence of the obtained series.

Notice first that $z = 0$ and $z = -1$ are simple poles of $f(z)$, and $z = \infty$ is a removable singularity since $\lim\limits_{z \to \infty} \frac{1}{z(z+1)} = 0$.

To find the Laurent series at the point $z = 0$, let us recall the well-known expansion of the function $\frac{1}{1+z}$ in the power series

$$\frac{1}{1+z} = 1 - z + z^2 - z^3 + \cdots = \sum_{n=0}^{+\infty}(-1)^n z^n .$$

Therefore, for $f(z)$, we obtain

$$\begin{aligned} f(z) &= \frac{1}{z} \cdot \frac{1}{1+z} = \frac{1}{z}\left(1 - z + z^2 - z^3 + \ldots\right) \\ &= \frac{1}{z} - 1 + z - z^2 + \cdots = \sum_{n=-1}^{+\infty} (-1)^{n+1} z^n , \end{aligned} \tag{3.23}$$

which is the Laurent series at the point $z = 0$. The regular part of this series is $f_1(z) = \sum_{n=0}^{+\infty} (-1)^{n+1} z^n$ and the principal part is $f_2(z) = \frac{1}{z}$. The latter contain only one element $\frac{1}{z}$ that corresponds to the fact that $z = 0$ is a simple pole of $f(z)$. Since the power series of the function $\frac{1}{1+z}$ converges in the disk $|z| < 1$, the Laurent series (3.23) converges in the ring $0 < |z| < 1$.

Consider now the point $z = -1$. Following the same reasoning, we obtain

$$f(z) = \frac{1}{z(z+1)} = \frac{1}{z+1} \cdot \frac{-1}{1-(z+1)} = -\frac{1}{z+1}\sum_{n=0}^{+\infty}(z+1)^n = -\sum_{n=-1}^{+\infty}(z+1)^n . \tag{3.24}$$

At the point $z = -1$, this Laurent series of $f(z)$ has the regular part $f_1(z) = -\sum_{n=0}^{+\infty}(z+1)^n$ and the principal part $f_2(z) = \frac{-1}{z+1}$ with only one element, because $z = -1$ is a simple pole of $f(z)$. Employing the fact that the power series of the function $\frac{1}{z} = -\frac{1}{1-(z+1)}$ converges in the disk $|z+1| < 1$, we arrive at the conclusion that the Laurent series (3.24) converges in the ring $0 < |z+1| < 1$.

Finally, the expansion in the Laurent series at $z = \infty$ can be found using a similar technique

$$f(z) = \frac{1}{z(z+1)} = \frac{1}{z^2} \cdot \frac{1}{1+1/z} = \frac{1}{z^2}\sum_{n=0}^{+\infty}(-1)^n z^{-n} = \sum_{n=-2}^{-\infty}(-1)^n z^n . \tag{3.25}$$

Since $z = \infty$ is a removable singularity, the Laurent series (3.25) contains only the regular part $f_1(z) = f(z) = \sum_{n=-2}^{-\infty}(-1)^n z^n$ (the principal part is zero). The series (3.25) converges in the domain $|z| > 1$, because the power series for $\frac{1}{1+1/z}$ with the elements $\frac{1}{z}$ converges when $\left|\frac{1}{z}\right| < 1$.

3.8 Essential Singular Points

Theorem (Criterion for an essential singular point) *Let z_0 be a single-valued isolated singular point of a function $f(z)$. The point z_0 is an essential singularity of $f(z)$*

if and only if the principal part of the Laurent series of $f(z)$ at z_0 has an infinite number of terms.

Proof *Necessity*. Since z_0 is a single-valued isolated singular point (finite or infinite), the function $f(z)$ is analytic in a deleted neighborhood of z_0, that is, $f(z)$ can be expanded in the Laurent series in this neighborhood. Let us suppose, by contradiction, that the principal part of the Laurent series either is absent or contains a finite number of terms. If the principal part is absent, then z_0 is a removable singular point (by the criterion for a removable singular point in Sect. 3.5). If the principal part contains a finite number of terms, then z_0 is a pole (according to the criterion for a pole—see Theorem 3.4 in Sect. 3.7). In either case, we arrive at a contradiction with the Theorem's hypothesis, which means that our supposition is false.

Sufficiency. Using again the method of contradiction, suppose that z_0 is not an essential singular point, although the principal part of the Laurent series possesses an infinite number of terms. Since z_0 is a single-valued isolated singular point, there are two remaining options for z_0—either this point is removable or it is a pole. If z_0 is a removable singular point, then, by the criterion for removable points (Sect. 3.5), the principal part of the Laurent series is zero. If z_0 is a pole, then, by the criterion for poles (Sect. 3.7), the principal part of the Laurent series contains a finite number of terms. In either case, we obtain a contradiction with the Theorem's hypothesis, which means that the supposition is false.

The Theorem is proved. □

Sokhotski–Weierstrass Theorem *If z_0 is an essential singular point of the function $f(z)$, then for any (finite or inifinite) complex number $A \in \overline{\mathbb{C}}$ there exist points in any neighborhood of z_0 at which the function $f(z)$ approximates A with any desired accuracy.*

Proof We prove the Theorem in the case $z_0 \neq \infty$, since the situation with $z_0 = \infty$ can be demonstrated in a similar way and is left to the reader.

We divide the proof into two parts: first, we consider $A = \infty$ and then $A \neq \infty$.

Let us show that $A = \infty$ is a partial limit of the function $f(z)$ as z approaches z_0. We argue by contradiction: suppose that one can find a constant $M > 0$ that bounds all the function values in a neighborhood of z_0, that is, there exist $\delta > 0$ such that for $\forall z,\ 0 < |z - z_0| < \delta$ one has $|f(z)| \leq M$. Then, the function $f(z)$ is analytic in a ring $0 < |z - z_0| < \delta$ (because z_0 is a single-valued isolated singular point) and is bounded there. Therefore, by the Corollary in Sect. 3.5, we conclude that z_0 is a removable singular point of $f(z)$, that contradicts the Theorem's hypothesis.

Now let us show that any $A \in \mathbb{C}, A \neq \infty$ is also a partial limit of $f(z)$ as z approaches z_0. Again we use the method of contradiction: assume that there exists a constant $A_0 \in \mathbb{C}$, which is not a partial limit, that is, there exist $\varepsilon_0 > 0$ and $\delta_0 > 0$ such that for $\forall z,\ 0 < |z - z_0| < \delta_0$ it follows $|f(z) - A_0| \geq \varepsilon_0$. Let us consider the auxiliary function $g(z) = \frac{1}{f(z) - A_0}$. The function $f(z)$, and consequently $f(z) - A_0$, is analytic in the ring $0 < |z - z_0| < \delta_0$ (due to the condition that z_0 is an essential singular point) and $f(z) - A_0 \neq 0$ in this ring according to the supposition that $|f(z) - A_0| \geq \varepsilon_0$. Therefore, the function $g(z)$ is analytic in the ring $0 < |z - z_0| <$

δ_0, which implies (by the definition) that z_0 is a single-valued isolated singular point of $g(z)$. Let us specify the kind of the singularity of $g(z)$ at z_0. By the supposition, in the ring $0 < |z - z_0| < \delta_0$ one has $|f(z) - A_0| \geq \varepsilon_0$, and consequently $|g(z)| \leq \frac{1}{\varepsilon_0}$, that is, $g(z)$ is bounded in this ring, which implies that z_0 is a removable singular point of $g(z)$. Hence, there exist a finite limit of $g(z)$ as z approaches z_0: $g(z_0) = \lim_{z \to z_0} g(z)$. The relationship between $f(z)$ and $g(z)$ can be rewritten in the form $f(z)$: $f(z) = A_0 + \frac{1}{g(z)} = \frac{A_0 g(z)+1}{g(z)}$ and we can conclude that as the ratio of two analytic functions, by Theorem 3.5 of the previous section, $f(z)$ either is analytic or has a pole at z_0. Either case contradicts the Theorem's hypothesis that z_0 is an essential singularity of $f(z)$. Hence, our assumption is false.

This completes the proof of the Theorem. □

Remark The Sokhotski–Weierstrass Theorem reveals an interesting behavior of a function $f(z)$ in a neighborhood of an essential singular point. According to the definition of an essential singular point, there is no limit of $f(z)$ as z approaches z_0. The Sokhotski–Weierstrass Theorem states that any complex number is a partial limit of $f(z)$. Hence, if we take any deleted neighborhood of z_0: $0 < |z - z_0| < r$, where $\forall r > 0$, and denote the set of the values of $f(z)$ in this ring by G: $G = f(0 < |z - z_0| < r)$, then $\overline{G} = \overline{\mathbb{C}}$. In other words, the closure of the image of any deleted neighborhood of an essential singular point is the extended complex plane.

3.9 Residues and the Fundamental Theorem on Residues

Definition Let z_0 be a single-valued isolated singular point or a regular point of a function $f(z)$. The *residue* (abbreviated to *res*) of $f(z)$ at z_0 is defined and denoted as follows:

$$\operatorname*{res}_{z=z_0} f(z) = \frac{1}{2\pi i} \int_{|z-z_0|=\rho} f(z)\, dz\,, \text{ if } z_0 \neq \infty \tag{3.26}$$

and

$$\operatorname*{res}_{z=\infty} f(z) = -\frac{1}{2\pi i} \int_{|z|=\rho} f(z)\, dz, \text{ if } z_0 = \infty\,. \tag{3.27}$$

Remark Since z_0 is a single-valued isolated singular point or a regular point of a function $f(z)$, this function is analytic in a deleted neighborhood of z_0, that is, in the ring $0 < |z - z_0| < r$ if $z_0 \neq \infty$, or in the ring $R < |z| < +\infty$ if $z_0 = \infty$. In the definition of the residue, one uses the integral along a circle of the radius ρ, where ρ is any number such that $0 < \rho < r$ if $z_0 \neq \infty$, or $R < \rho < +\infty$ if $z_0 = \infty$. Let us show that the value of the residue does not depend on the choice of the radius ρ. We consider only the case $z_0 \neq \infty$ since for $z_0 = \infty$ the proof is the same. Take two different radii ρ_1 and ρ_2, $0 < \rho_1 < \rho_2 < r$. Since the function $f(z)$ is analytic in the ring $0 < |z - z_0| < r$, it is analytic in the closed ring $\overline{G} = \{\rho_1 \leq |z - z_0| \leq \rho_2\}$ (see Fig. 3.6). Then, by Goursat's Theorem, we have

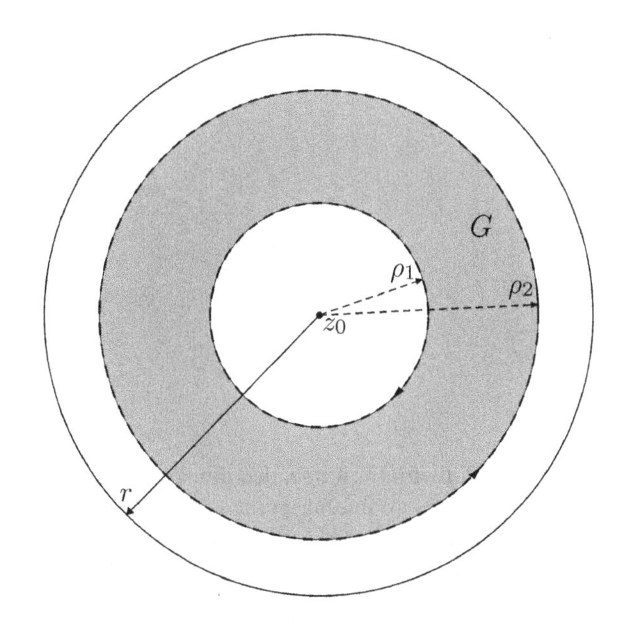

Fig. 3.6 A neighborhood $0 < |z - z_0| < r$ and the auxiliary closed ring $\overline{G} = \{\rho_1 \leq |z - z_0| \leq \rho_2\}$ inside this neighborhood

$$0 = \int_{\partial G} f(z)\,dz = \int_{|z-z_0|=\rho_2} f(z)\,dz - \int_{|z-z_0|=\rho_1} f(z)\,dz\,,$$

and consequently

$$\int_{|z-z_0|=\rho_2} f(z)\,dz = \int_{|z-z_0|=\rho_1} f(z)\,dz\,,$$

which shows that the integral value (and the residue value) does not depend on the radius ρ of a chosen circle.

Theorem (Fundamental Theorem on residues) *Let D be a domain whose boundary consists of a finite number of piecewise smooth curves with folds and does not contain the point ∞. If a function $f(z)$ is analytic in D and continuous in D up to its boundary, except a finite number of single-valued isolated singular points a_k, $k = 1, \ldots, n$ contained in D, then*

$$\int_{\partial D} f(z)\,dz = 2\pi i \cdot \sum_{k=1}^{n} \operatorname*{res}_{z=a_k} f(z) \tag{3.28}$$

(if $\infty \in D$, then the residue at the point ∞ is also included in the above sum of the residues). The last expression is called the residue formula.

Proof We divide the proof into two parts—when $\infty \notin D$ and when $\infty \in D$.

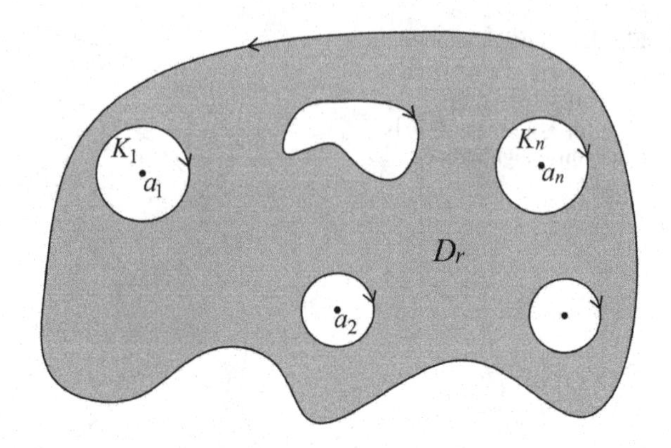

Fig. 3.7 Construction of the auxiliary domain D_r for a bounded domain D

If $\infty \notin D$, then D is a bounded domain. Since the points a_k, $\forall k = 1, \ldots, n$ lie in D, all of them are interior points, and consequently, there is a neighborhood of each a_k contained in D. Since the number of these points a_k is finite, we can choose such a small radius $r > 0$ that all the disks (neighborhoods) $K_k = \{|z - a_k| < r\}, k = 1, \ldots, n$ are contained in D together with their boundaries and there is no intersection between their closures: $\overline{K}_k \cap \overline{K}_j = \varnothing, \forall k \neq j$.

Let us consider the auxiliary domain D_r obtained by excluding the closed disks $\overline{K}_j$ from the domain D: $D_r = D \backslash \left(\overset{n}{\underset{j=1}{\cup}} \overline{K}_j \right)$ (see Fig.3.7). The domain D_r is bounded by a finite number of piecewise smooth curves with folds and the function $f(z)$ is analytic in D_r and continuous in D_r up to its boundary. Therefore, by Goursat's Theorem, we have

$$\int_{\partial D_r} f(z)\, dz = 0 .$$

We rewrite the last integral in a detailed form using each individual part of the boundary of D_r: the boundary ∂D_r consists of the boundary ∂D, whose positive traversal coincide with the positive traversal along ∂D_r, and also of the circles $|z - a_j| = r,\ j = 1, \ldots, n$, which are traversed clockwise to keep the domain D_r on the left (see Fig. 3.7). Considering the circles $|z - a_j| = r$ as individual curves, we should invert the orientation of the traversal to counterclockwise (which is a positive direction for a closed curve), which results in the negative sign for this part of integrals

$$0 = \int_{\partial D_r} f(z)\, dz = \int_{\partial D} f(z)\, dz - \sum_{j=1}^{n} \int_{|z-a_j|=r} f(z)\, dz . \tag{3.29}$$

Since each disk K_j contains only one single-valued isolated singular point a_j, by the definition of the residue, we have $\int_{|z-a_j|=r} f(z)\, dz = 2\pi i \cdot \underset{z=a_j}{\text{res}}\, f(z), j = 1, \ldots, n$, and relation (3.29) can be rewritten in the form (3.28)

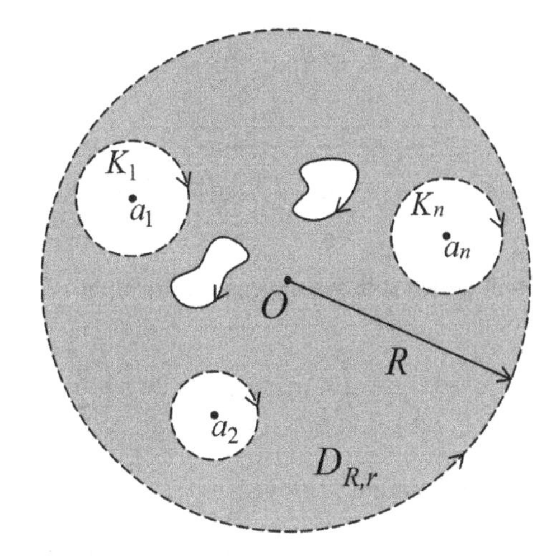

Fig. 3.8 Construction of the auxiliary domain $D_{R,r}$ in the case $\infty \in D$

$$\int_{\partial D} f(z)\,dz = \sum_{j=1}^{n} \int_{|z-a_j|=r} f(z)\,dz = 2\pi i \cdot \sum_{j=1}^{n} \operatorname*{res}_{z=a_j} f(z).$$

Let now $\infty \in D$. In this case, ∞ is an interior point of D, that is, it belongs to D together with its neighborhood. Since there is a finite number of the points a_j, we can choose the radius R of a neighborhood of ∞ such that all the points a_j, $j = 1, \ldots, n$ belong to the disk $|z| < R$. With this choice of $R > 0$, we have $\{|z| \geq R\} \subset D$ and $a_j \in D \setminus \{|z| \geq R\}$, $\forall j = 1, \ldots, n$. In the same way as in the first part of the proof, we find $r > 0$ such that all the disks $K_j = \{|z - a_j| < r\}$, $j = 1, \ldots, n$ together with their boundaries are contained in the domain $D \setminus \{|z| \geq R\}$ and their closures have no common points.

Denote by $D_{R,r}$ the auxiliary domain $D_{R,r} = (D \setminus \{|z| \geq R\}) \setminus \left(\bigcup_{j=1}^{n} \overline{K}_j\right)$ (see Fig. 3.8). The domain $D_{R,r}$ is bounded and its boundary consists of a finite number of a piecewise smooth curves with folds, and the function $f(z)$ is analytic in $D_{R,r}$ and continuous in $D_{R,r}$ up to its boundary. Therefore, by Goursat's Theorem

$$\int_{\partial D_{R,r}} f(z)\,dz = 0.$$

The boundary $\partial D_{R,r}$ consists of the boundary ∂D, whose positive traversal coincide with the positive traversal along $\partial D_{R,r}$, the circle $|z| = R$ traversed counterclockwise and also the circles $|z - a_j| = r$, $j = 1, \ldots, n$ traversed clockwise to keep the domain $D_{R,r}$ on the left (see Fig. 3.8). Considering the circles $|z - a_j| = r$ as individual curves, we invert the orientation of the traversal to counterclockwise and change the sign of this part of integrals to negative. This allows us to rewrite the last integral formula in the form

$$0 = \int_{\partial D_{R,r}} f(z)\, dz = \int_{\partial D} f(z)\, dz + \int_{|z|=R} f(z)\, dz - \sum_{j=1}^{n} \int_{|z-a_j|=r} f(z)\, dz$$

$$= \int_{\partial D} f(z)\, dz - 2\pi i \operatorname*{res}_{z=\infty} f(z) - 2\pi i \sum_{j=1}^{n} \operatorname*{res}_{z=a_j} f(z),$$

from which follows the desired formula

$$\int_{\partial D} f(z)\, dz = 2\pi i \left(\sum_{j=1}^{n} \operatorname*{res}_{z=a_j} f(z) + \operatorname*{res}_{z=\infty} f(z) \right).$$

The Theorem is proved. □

Remark Notice that this remarkable Theorem includes as particular cases Goursat's Theorem, the Cauchy integral formula, and the formula of the derivative of any order of an analytic function.

Corollary *If a function $f(z)$ is analytic in the entire finite complex plane except for a finite number of single-valued isolated singular points a_j, $a_j \in \mathbb{C}$, $j = 1, \ldots, n$, then the sum of all the residues of $f(z)$, including the residue at ∞, is equal zero*

$$\sum_{j=1}^{n} \operatorname*{res}_{z=a_j} f(z) + \operatorname*{res}_{z=\infty} f(z) = 0. \tag{3.30}$$

Proof Since the number of singular points a_j, $j = 1, \ldots, n$ is finite, we can choose such radius R that all these points belong to the disk $K = \{|z| < R\}$. The function $f(z)$ satisfies all the conditions of the Fundamental Theorem on residues in the disk K, and therefore

$$\int_{\partial K} f(z)\, dz = 2\pi i \sum_{j=1}^{n} \operatorname*{res}_{z=a_j} f(z). \tag{3.31}$$

At the same time, using the definition of the residue at ∞, we have

$$\int_{\partial K} f(z)\, dz = \int_{|z|=R} f(z)\, dz = -2\pi i \cdot \operatorname*{res}_{z=\infty} f(z). \tag{3.32}$$

Comparing (3.31) and (3.32), we obtain

$$2\pi i \sum_{j=1}^{n} \operatorname*{res}_{z=a_j} f(z) = -2\pi i \operatorname*{res}_{z=\infty} f(z),$$

which is equivalent to (3.30). □

3.10 Residue Calculus

In this section, we derive formulas for evaluation of residues in the cases of different singular points. Notice that although the residues are defined through the integrals, usually, we evaluate the residues employing another approach in order to find the value of the corresponding integral.

Case 1. Calculation by the definition.
We start with the observation that for any kind of single-valued isolated singular point or regular point z_0 of a function $f(z)$, the residue can be calculated by the definition

$$\operatorname*{res}_{z=z_0} f(z) = \frac{1}{2\pi i} \int_{|z-z_0|=\rho} f(z)\, dz\,, \text{ if } z_0 \neq \infty\,, \tag{3.33}$$

and

$$\operatorname*{res}_{z=\infty} f(z) = -\frac{1}{2\pi i} \int_{|z|=\rho} f(z)\, dz\,, \text{ if } z_0 = \infty\,. \tag{3.34}$$

However, these formulas are rarely employed, since usually the problem is inverse: find integrals through evaluation of the corresponding residues. In this context, other formulas and approaches to evaluation of residues are relevant.

Case 2. Arbitrary single-valued isolated singular point.
If z_0 is a (finite or infinite) single-valued isolated singular point (of any kind) of a function $f(z)$, then

$$\operatorname*{res}_{z=z_0} f(z) = c_{-1}, \text{ if } z_0 \neq \infty, \tag{3.35}$$

and

$$\operatorname*{res}_{z=\infty} f(z) = -c_{-1}, \text{ if } z_0 = \infty, \tag{3.36}$$

where c_{-1} is the coefficient at the term $(z - z_0)^{-1}$ if $z_0 \neq \infty$ or at z^{-1} if $z = \infty$ in the Laurent series of the function $f(z)$ at the point z_0.

First, let us prove formula (3.35). Since z_0 is a single-valued isolated singularity, the function $f(z)$ can be expanded in the Laurent series in the ring $0 < |z - z_0| < r$

$$f(z) = \sum_{n=-\infty}^{+\infty} c_n (z - z_0)^n\,,$$

where

$$c_n = \frac{1}{2\pi i} \int_{|z-z_0|=\rho} \frac{f(z)}{(z - z_0)^{n+1}} dz\,,\ \forall \rho\, :\ 0 < \rho < r\,.$$

In particular, for $n = -1$, we have

$$c_{-1} = \frac{1}{2\pi i} \int_{|z-z_0|=\rho} f(z)\, dz\,,$$

that is

$$\operatorname*{res}_{z=z_0} f(z) = \frac{1}{2\pi i} \int_{|z-z_0|=\rho} f(z)\, dz = c_{-1}, \text{ if } z_0 \neq \infty\,.$$

Now consider $z_0 = \infty$. In this case the function $f(z)$ can be represented in the Laurent series in the ring $R < |z| < +\infty$

$$f(z) = \sum_{n=-\infty}^{+\infty} c_n z^n,$$

where

$$c_n = \frac{1}{2\pi i} \int_{|z|=\rho} \frac{f(z)}{z^{n+1}} dz, \ \forall \rho : R < \rho < +\infty\,.$$

In particular, for $n = -1$, we have

$$c_{-1} = \frac{1}{2\pi i} \int_{|z|=\rho} f(z)\, dz\,,$$

and consequently

$$\operatorname*{res}_{z=\infty} f(z) = -\frac{1}{2\pi i} \int_{|z|=\rho} f(z)\, dz = -c_{-1}\,.$$

Notice that formulas (3.35) and (3.36) are very simple, elegant, and applicable for any kind of single-valued isolated singular points. However, the use of these formulas requires the knowledge of the Laurent series in a neighborhood of z_0 and expansion in the Laurent series is not a simple problem in practice. For this reason, the expressions (3.35) and (3.36) are rarely used, letting space for other ways to find residues.

Case 3. Regular point or removable singularity.
If $z_0 \neq \infty$ is a regular point or removable singularity of $f(z)$, then

$$\operatorname*{res}_{z=z_0} f(z) = 0\,. \tag{3.37}$$

Indeed, if z_0 is a removable singular point, we always can define the function $f(z)$ at z_0 by the formula $f(z_0) = \lim_{z \to z_0} f(z)$, which makes $f(z)$ an analytic function in the entire disk $K = \{|z - z_0| < r\}$. Of course, if z_0 is a regular point of $f(z)$, then the function is analytic in the same disk K by the definition. Employing Goursat's Theorem with the disk K as the domain D and the disk $K_\rho = \{|z - z_0| < \rho\}$, $\forall \rho$: $0 < \rho < r$ as the set G such that $\bar{G} \subset D$, we obtain

$$\operatorname*{res}_{z=z_0} f(z) = \frac{1}{2\pi i}\int_{|z-z_0|=\rho} f(z)\, dz = \frac{1}{2\pi i}\int_{\partial G} f(z)\, dz = 0\,.$$

Formula (3.37) can also be deduced using another reasoning. Since $z_0 \neq \infty$ is a regular point or a removable singularity of $f(z)$, the principal part of the Laurent series at z_0 is absent, that is, $c_{-n} = 0$, $\forall n \in \mathbb{N}$, in particular, $c_{-1} = 0$. Then, using formula (3.35), we get

$$\operatorname*{res}_{z=z_0} f(z) = c_{-1} = 0\,.$$

Notice that the restriction $z_0 \neq \infty$ is important: if ∞ is a regular or removable singular point of $f(z)$, then the residue of $f(z)$ at this point can be different from zero. The following example illustrates this situation.

Example Consider $f(z) = \frac{1}{z}$. This function is analytic in the ring $0 < |z| < +\infty$, that is, 0 and ∞ are single-valued isolated singular points. At the point ∞ we have $\lim\limits_{z\to\infty} f(z) = \lim\limits_{z\to\infty} \frac{1}{z} = 0$, which means that ∞ is a removable singular point and the definition $f(\infty) = \lim\limits_{z\to\infty} f(z) = 0$ determines the analytic function in the domain $|z| > 0$. We can evaluate the residue of $f(z)$ at ∞ by the definition. Using the parametric representation of the circle $|z| = \rho$

$$z = \rho e^{i\varphi}\,,\ 0 \le \varphi \le 2\pi\,,$$

that gives $dz = \rho e^{i\varphi} i d\varphi$, we obtain

$$\operatorname*{res}_{z=\infty} f(z) = -\frac{1}{2\pi i}\int_{|z|=\rho} f(z)\, dz = -\frac{1}{2\pi i}\int_{|z|=\rho} \frac{dz}{z}$$

$$= -\frac{1}{2\pi i}\int_0^{2\pi} \frac{\rho e^{i\varphi} i d\varphi}{\rho e^{i\varphi}} = -\frac{1}{2\pi}\int_0^{2\pi} d\varphi = -1\,.$$

Hence, $z = \infty$ is a removable singular point of the function $f(z) = \frac{1}{z}$, but its residue is not 0: $\operatorname*{res}_{z=\infty} f(z) = -1 \neq 0\,.$

Case 4. Pole of order n.
If $z_0 \neq \infty$ is a pole of order n of a function $f(z)$, then by the criterion for multiplicity of a pole (Theorem 3.3 in Sect. 3.7), we can represent $f(z)$ in the form $f(z) = (z - z_0)^{-n} f_1(z)$, where the function $f_1(z)$ is analytic at the point z_0 and $f_1(z_0) \neq 0$. Then, the residue of $f(z)$ at z_0 can be written as follows:

$$\operatorname*{res}_{z=z_0} f(z) = \frac{1}{2\pi i}\int_{|z-z_0|=\rho} f(z)\, dz = \frac{1}{2\pi i}\int_{|z-z_0|=\rho} \frac{f_1(z)}{(z-z_0)^n}\, dz$$

$$= \frac{1}{(n-1)!}\cdot\frac{(n-1)!}{2\pi i}\int_{|z-z_0|=\rho} \frac{f_1(z)}{(z-z_0)^n} dz\,.$$

Since the function $f_1(z)$ is analytic at z_0, that is, it is analytic in a disk $|z - z_0| < r$, and the radius ρ is chosen to be smaller than r, then the last formula represents the $(n-1)$-th derivative of $f_1(z)$ (see Corollary 2.1 in Sect. 2.4). Using the expression for the derivative and recalling that $f_1(z) = f(z)(z - z_0)^n$, we can rewrite the last formula in the form

$$\operatorname*{res}_{z=z_0} f(z) = \frac{1}{(n-1)!} f_1^{(n-1)}(z_0) = \frac{1}{(n-1)!} \lim_{z\to z_0} f_1^{(n-1)}(z)$$

$$= \frac{1}{(n-1)!} \lim_{z\to z_0} \left(f(z) \cdot (z - z_0)^n\right)^{(n-1)}.$$

Thus, we derive the following formula for the evaluation of the residue at a pole $z_0 \neq \infty$ of order n

$$\operatorname*{res}_{z=z_0} f(z) = \frac{1}{(n-1)!} \lim_{z\to z_0} \left(f(z) \cdot (z - z_0)^n\right)^{(n-1)}. \tag{3.38}$$

Case 4a. If z_0 is a simple pole of $f(z)$, then setting $n = 1$ in formula (3.38), we obtain a simpler expression

$$\operatorname*{res}_{z=z_0} f(z) = \lim_{z\to z_0} \left(f(z) \cdot (z - z_0)\right). \tag{3.39}$$

Case 4b. Suppose a function $f(z)$ is represented in the form $f(z) = \frac{\varphi(z)}{\psi(z)}$, where $\varphi(z)$ and $\psi(z)$ are analytic functions at the point z_0 and $\varphi(z_0) \neq 0$, $\psi(z_0) = 0$, $\psi'(z_0) \neq 0$, that is, z_0 is a simple zero of $\psi(z)$. Then, the Theorem about poles (Theorem 3.5 of Sect. 3.7) states that z_0 is a simple pole of $f(z)$, and consequently, we can employ formula (3.39) to find the residue of $f(z)$ at z_0 (recall that $\psi(z_0) = 0$)

$$\operatorname*{res}_{z=z_0} f(z) = \lim_{z\to z_0} \left(f(z) \cdot (z - z_0)\right) = \lim_{z\to z_0} \frac{\varphi(z)}{\psi(z)} (z - z_0) = \lim_{z\to z_0} \frac{\varphi(z)}{\frac{\psi(z)-\psi(z_0)}{z-z_0}} = \frac{\varphi(z_0)}{\psi'(z_0)}.$$

Hence, we arrive at the very simple formula

$$\operatorname*{res}_{z=z_0} f(z) = \frac{\varphi(z_0)}{\psi'(z_0)}. \tag{3.40}$$

Case 5. Finite number of single-valued isolated singular points.
If a function $f(z)$ is analytic in the entire (finite) complex plane $\mathbb{C}$, except for a finite number of single-valued isolated singular points a_j, $j = 1, \ldots, n$, then

$$\operatorname*{res}_{z=\infty} f(z) = -\sum_{j=1}^{n} \operatorname*{res}_{z=a_j} f(z). \tag{3.41}$$

This is just another representation of formula (3.30) of the previous section.

Case 6. Regularity or removable singularity at ∞, the first formula. If $z = \infty$ is a regular or removable singular point of a function $f(z)$, then

$$\operatorname*{res}_{z=\infty} f(z) = \lim_{z\to\infty} \left(\left(f(\infty) - f(z) \right) \cdot z \right), \tag{3.42}$$

where $f(\infty) = \lim_{z\to\infty} f(z)$.

In fact, by the criterion for a removable singular point (Theorem of Sect. 3.5), the principal part of the Laurent series of $f(z)$ vanishes, that is, the Laurent series of $f(z)$ in the ring $R < |z| < +\infty$ has the form

$$f(z) = \sum_{n=0}^{-\infty} c_n z^n = c_0 + \frac{c_{-1}}{z} + \frac{c_{-2}}{z^2} + \dots, \quad c_0 = f(\infty) = \lim_{z\to\infty} f(z) \tag{3.43}$$

(the last limit exists and is finite by the definition of a removable singular point). According to formula (3.36), one has $\operatorname*{res}_{z=\infty} f(z) = -c_{-1}$. To find the coefficient c_{-1}, we use the following consequence of formula (3.43)

$$\left(f(z) - f(\infty) \right) \cdot z = \left(f(z) - c_0 \right) \cdot z = c_{-1} + \frac{c_{-2}}{z} + \frac{c_{-3}}{z^2} + \dots. \tag{3.44}$$

Since the Laurent series (3.44) converges uniformly, we can pass to the limit term-by-term as $z \to \infty$ and obtain

$$\lim_{z\to\infty} \left(\left(f(z) - f(\infty) \right) \cdot z \right) = c_{-1}.$$

Therefore

$$\operatorname*{res}_{z=\infty} f(z) = -c_{-1} = \lim_{z\to\infty} \left(\left(f(\infty) - f(z) \right) \cdot z \right).$$

Case 6a. If $z = \infty$ is a zero of $f(z)$, then

$$\operatorname*{res}_{z=\infty} f(z) = -\lim_{z\to\infty} z \cdot f(z). \tag{3.45}$$

This formula is an immediate consequence of (3.42), because in this case $f(\infty) = 0$.

Case 6b. If $z = \infty$ is a zero of order k ($k \geq 2$) of a function $f(z)$, then

$$\operatorname*{res}_{z=\infty} f(z) = 0. \tag{3.46}$$

By the definition of a zero of order k, the function $f(z)$ can be represented in the form $f(z) = z^{-k} f_1(z)$, where $f_1(z)$ is an analytic function at ∞ and $f_1(\infty) \neq 0$. Employing formula (3.45) to evaluate the residue of $f(z)$ at ∞, we obtain

$$\operatorname*{res}_{z=\infty} f(z) = -\lim_{z\to\infty} z \cdot f(z) = -\lim_{z\to\infty} z^{1-k} \cdot f_1(z) = 0.$$

The last limit is equal to zero because the function $f_1(z)$ has a finite limit as $z \to \infty$ and $z^{1-k} \to 0$ when $z \to \infty$, since $1 - k < 0$ for $k \geq 2$.

Case 7. Regularity or removable singularity at ∞, the second formula.
Let us derive one more formula of the residue at the regular or removable singular point $z = \infty$. As was shown, in this case the Laurent series of $f(z)$ has the form (3.43) in a ring $R < |z| < +\infty$. Differentiating (3.43) term-by-term (which is possible due to the uniform convergence of the Laurent series), we obtain

$$f'(z) = -\frac{c_{-1}}{z^2} - \frac{2c_{-2}}{z^3} - \frac{3c_{-3}}{z^4} - \dots$$

or

$$z^2 f'(z) = -c_{-1} - \frac{2c_{-2}}{z} - \frac{3c_{-3}}{z^2} - \dots .$$

Taking the limit term-by-term in the last series as $z \to \infty$, we reach the new expression for the residue

$$\operatorname*{res}_{z=\infty} f(z) = -c_{-1} = \lim_{z\to\infty} z^2 \cdot f'(z) . \tag{3.47}$$

Case 8. Pole of order k at ∞.
If $z = \infty$ is a pole of order k of a function $f(z)$, then, by the criterion for multiplicity of a pole, the function $f(z)$ has the Laurent series expansion in the ring $R < |z| < +\infty$

$$f(z) = \sum_{n=-\infty}^{k} c_n z^n = c_k z^k + c_{k-1} z^{k-1} + \dots + c_1 z + c_0 + \frac{c_{-1}}{z} + \frac{c_{-2}}{z^2} + \dots .$$

To determine the residue, we have to find the coefficient c_{-1} of this series. To this end, first we eliminate the summands in the principal part of the Laurent series by differentiating the series term-by-term $k + 1$ times

$$f'(z) = kc_k z^{k-1} + (k-1) c_{k-1} z^{k-2} + \dots + 2c_2 z + c_1 - \frac{c_{-1}}{z^2} - \frac{2c_{-2}}{z^3} - \frac{3c_{-3}}{z^4} - \dots ,$$

$$f''(z) = k(k-1) c_k z^{k-2} + \dots + 2c_2 + \frac{1 \cdot 2 \cdot c_{-1}}{z^3} + \frac{2 \cdot 3 \cdot c_{-2}}{z^4} + \frac{3 \cdot 4 \cdot c_{-3}}{z^5} + \dots ,$$

$$\vdots$$

$$f^{(k)}(z) = k!\, c_k + \frac{(-1)^k k!\, c_{-1}}{z^{k+1}} + \frac{(-1)^k (k+1)!\, c_{-2}}{1!\, z^{k+2}} + \frac{(-1)^k (k+2)!\, c_{-3}}{2!\, z^{k+3}} + \dots ,$$

$$f^{(k+1)}(z) = \frac{(-1)^{k+1}(k+1)!\,c_{-1}}{z^{k+2}} + \frac{(-1)^{k+1}(k+2)!\,c_{-2}}{z^{k+3}} + \frac{(-1)^{k+1}(k+3)!\,c_{-3}}{2!\,z^{k+4}} + \dots .$$

Multiplying the last derivative by z^{k+2} and passing to the limit term-by-term as $z \to \infty$, we obtain

$$\lim_{z\to\infty}\left(f^{(k+1)}(z)\cdot z^{k+2}\right) = \lim_{z\to\infty}\left((-1)^{k+1}(k+1)!\,c_{-1} + \frac{a_2}{z} + \frac{a_3}{z^2} + \dots\right)$$

$$= (-1)^{k+1}(k+1)!\,c_{-1}\,.$$

Therefore

$$\operatorname*{res}_{z=\infty} f(z) = -c_{-1} = \frac{(-1)^k}{(k+1)!}\lim_{z\to\infty}\left(f^{(k+1)}(z)\cdot z^{k+2}\right). \tag{3.48}$$

Notice that if $k = 0$, that is, $z = \infty$ is a removable singular point of $f(z)$, then formula (3.48) is reduced to (3.47).

Consider below an example of the application of different formulas derived in this section.

Example Find singular points and evaluate the residues at single-valued isolated singular points

$$f(z) = \frac{3}{9z^3 - z^5}\,.$$

The singularities of $f(z)$ include the zeros of denominator $9z^3 - z^5 = z^3\left(9 - z^2\right) = 0$, from which we get $z_1 = 0\,,\ z_2 = 3\,,\ z_3 = -3$, and also the point $z_4 = \infty$. The point $z_1 = 0$ is a pole of third order, $z_2 = 3$ and $z_3 = -3$ are simple poles, and $z_4 = \infty$ is a removable singularity, more precisely, ∞ is zero of fifth order.

To evaluate the residue of $f(z)$ at $z_1 = 0$, we use formula (3.38) of this section with $n = 3$

$$\operatorname*{res}_{z=0} f(z) = \frac{1}{2!}\lim_{z\to 0}\left(\frac{3}{z^3\left(9-z^2\right)}\cdot z^3\right)'' = \frac{3}{2}\lim_{z\to 0}\left(\frac{2z}{\left(9-z^2\right)^2}\right)'$$

$$= 3\lim_{z\to 0}\frac{\left(9-z^2\right)^2 + 4z^2\left(9-z^2\right)}{\left(9-z^2\right)^4} = 3\cdot\frac{9}{9^3} = \frac{1}{27}\,.$$

The residues at the simple poles $z_2 = 3$ and $z_3 = -3$ are more easily found using formula (3.40) of this section

$$\operatorname*{res}_{z=3} f(z) = \frac{3}{z^3}\cdot\frac{1}{\left(9-z^2\right)'}\Big|_{z=3} = \frac{3}{27}\cdot\frac{1}{-6} = -\frac{1}{54}\,;$$

$$\operatorname*{res}_{z=-3} f(z) = \frac{3}{z^3} \cdot \frac{1}{\left(9 - z^2\right)'} \Big|_{z=-3} = \frac{3}{-27} \cdot \frac{1}{6} = -\frac{1}{54}.$$

Since $z_4 = \infty$ is zero of order greater than 1, we can apply formula (3.46) of this section that gives immediately $\operatorname*{res}_{z=\infty} f(z) = 0$.

3.11 Application of the Fundamental Theorem on Residues to Improper Integrals. Jordan's Lemma

We start with the definition of the *improper integral of a complex function along a curve*. Let Γ be a continuous rectifiable curve and a function $f(z)$ be defined and continuous either on $\Gamma \subset \mathbb{C}$ except for the point $a \in \Gamma$ (that is, a is a singular point of $f(z)$), or on a curve Γ that contains the point ∞ (in the latter case, Γ itself is not rectifiable, but its part lying in any disk $|z| < R$, $R > 0$ is rectifiable).

Let us consider first the situation when a singular point $a \in \Gamma$ is finite. Take any $r > 0$ and construct the disk $K_r = \{|z - a| < r\}$. Consider the arc γ_r of Γ contained in K_r, which can be described as follows: the first part of γ_r is obtained starting from the point a and traversing Γ in any direction (positive or negative) until the first intersection with ∂K_r, and the second part—starting from a and moving in the opposite direction again until the first intersection with ∂K_r (see Fig. 3.9). These two parts form the connected arc γ_r with the point a in this arc. Denote the remaining part of Γ by $\Gamma_r = \Gamma \backslash \gamma_r$ and notice that Γ_r is not necessarily connected—it can be composed of two separated curves. Since the function $f(z)$ is continuous on Γ_r, the integral $\int_{\Gamma_r} f(z)\,dz$ is defined. If there exists a finite limit

$$\lim_{r\to 0} \int_{\Gamma_r} f(z)\,dz = \int_{\Gamma} f(z)\,dz,$$

then it is called the *improper integral of the function $f(z)$ along the curve Γ*. In this case it is also said that the improper integral converges. Otherwise (if the limit is infinite or does not exists), it is said that the improper integral does not exists or diverges.

In the case when $a = \infty \in \Gamma$, the improper integral is defined in a similar way. Take any $R > 0$ and construct the circle $K_R = \{|z| = R\}$. Starting from the point ∞, we traverse the curve Γ until the first intersection with K_R and denote this part of Γ by γ_R. The remaining part of Γ is denoted by $\Gamma_R = \Gamma \backslash \gamma_R$ (see Fig. 3.10).

Notice that $\int_{\Gamma_R} f(z)\,dz$ exists, because for any R, the curve Γ_R is continuous rectifiable and the function $f(z)$ is continuous on Γ_R. If there exists a finite limit

$$\lim_{R\to\infty} \int_{\Gamma_R} f(z)\,dz = \int_{\Gamma} f(z)\,dz,$$

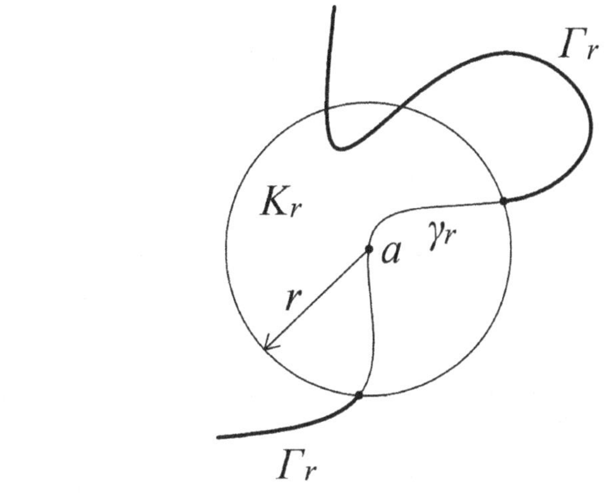

Fig. 3.9 Construction of the arcs γ_r and Γ_r when a singular point $a \neq \infty$

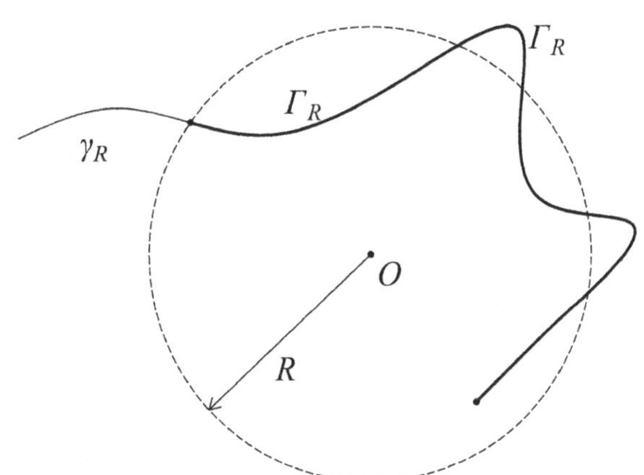

Fig. 3.10 Construction of the arcs γ_R and Γ_R for the singular point $a = \infty$

then it is called the *improper integral of the function $f(z)$ along the curve Γ*. Otherwise, the improper integral does not exists. Equivalently, in the first case, it is said that the improper integral converges and in the second—diverges.

Let now Γ be the boundary of a domain D, $\Gamma = \partial D$ and a be a singular (finite or infinite) point belonging to Γ. Besides the already introduced parts γ_r and Γ_r of Γ, consider also the part of the circle $|z - a| = r$, which is contained in D, and denoted by C_r (see Figs. 3.11 and 3.12 for bounded and infinite domains, respectively). If $a = \infty$, then C_R is the part of the circle $|z| = R$, which lies in D (see Fig. 3.13). In either case (finite or infinite a), the point ∞ does not belong to the boundary of the auxiliary domain D_r or D_R (see Figs. 3.11, 3.12, and 3.13).

We can now formulate the Theorem that generalizes the Fundamental Theorem on residues to the case of improper integrals along a curve.

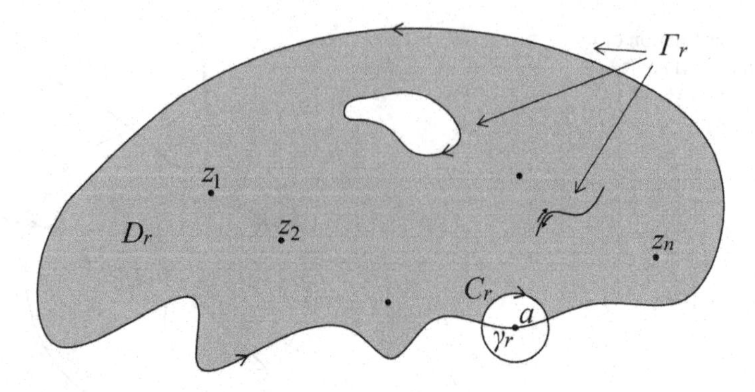

Fig. 3.11 Auxiliary domain D_r in the case of a finite singular point $a \in \partial D$ and a bounded domain D

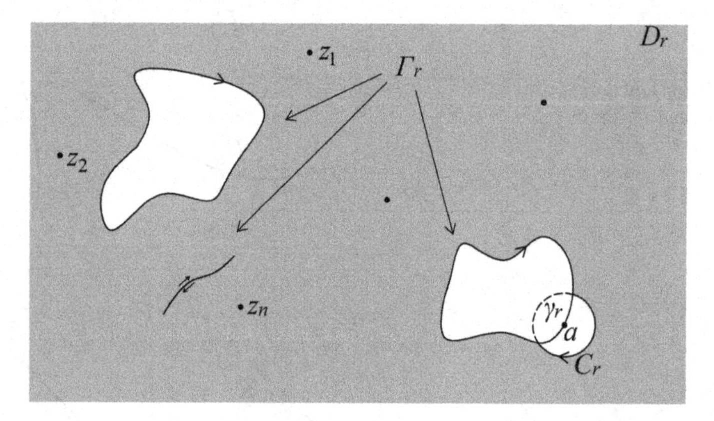

Fig. 3.12 Auxiliary domain D_r in the case of a finite singular point $a \in \partial D$ and an infinite domain D

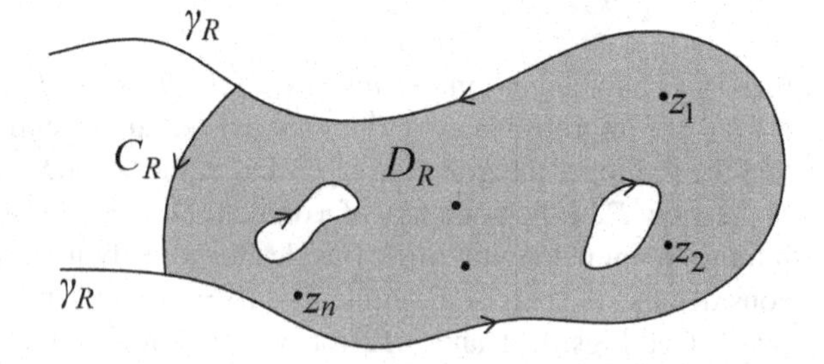

Fig. 3.13 Auxiliary domain D_R in the case of the infinite singular point $a = \infty \in \partial D$ and an unbounded finite domain D

Theorem *Suppose D is a domain whose boundary consists of a finite number of piecewise smooth curves with folds, and the function $f(z)$ is analytic in D and continuous in D up to its boundary, except for a finite number of single-valued isolated singular points z_k, $k = 1, \ldots, n$ contained in D ($z_k \in D$, $\forall k = 1, \ldots, n$) and except for a singular (finite or infinite) point a lying on the boundary of D ($a \in \partial D$). If*

(1) the improper integral $\int_{\partial D} f(z)\,dz$ converges

$$(2) \int_{C_r} f(z)\,dz \underset{r\to 0}{\to} 0 \text{ if } a \neq \infty \text{ (or } \int_{C_R} f(z)\,dz \underset{R\to\infty}{\to} 0 \text{ if } a = \infty\text{)}, \tag{3.49}$$

then

$$\int_{\partial D} f(z)\,dz = 2\pi i \cdot \sum_{k=1}^{n} \operatorname*{res}_{z=z_k} f(z).$$

Proof We present the proof only for the case $a \neq \infty$, since for $a = \infty$ the proof is the same. Take $r > 0$ small enough to ensure that all the points z_k are located out of the closed disk $|z - a| \leq r$ (it is possible since z_k, $k = 1, \ldots, n$ are interior points of D) and denote by $D_r = D \setminus \{|z - a| \leq r\}$ the remaining part of D (Fig. 3.11 illustrates the case of a bounded domain D and Fig. 3.12—an infinite domain). The domain D_r is bounded by a finite number of piecewise smooth curves with folds and $\infty \notin \partial D_r$. The function $f(z)$ is analytic in D_r and continuous in D_r up to its boundary, except for a finite number of single-valued isolated singular points z_k, $k = 1, \ldots, n$. Therefore, by the Fundamental Theorem on residues, we have

$$\int_{\partial D_r} f(z)\,dz = 2\pi i \cdot \sum_{k=1}^{n} \operatorname*{res}_{z=z_k} f(z). \tag{3.50}$$

(If $\infty \in D$ then the right-hand side contains also the residue at ∞.) Notice that the boundary of D_r consists of the curve $\Gamma_r = \partial D \setminus \gamma_r$ and the arc C_r of the circle $|z - a| = r$ (that is, $\partial D_r = \Gamma_r \cup C_r$). Using the definition of the improper integral along the curve (which is convergent), the condition (3.49) and the fact that the integral in (3.50) does not depend on r (if r is sufficiently small), we obtain

$$\int_{\partial D} f(z)\,dz = \lim_{r\to 0} \int_{\Gamma_r} f(z)\,dz = \lim_{r\to 0} \left(\int_{\Gamma_r} f(z)\,dz + \int_{C_r} f(z)\,dz - \int_{C_r} f(z)\,dz \right)$$

$$= \lim_{r\to 0} \left(\int_{\partial D_r} f(z)\,dz - \int_{C_r} f(z)\,dz \right) = \lim_{r\to 0} \left(2\pi i \cdot \sum_{k=1}^{n} \operatorname*{res}_{z=z_k} f(z) - \int_{C_r} f(z)\,dz \right)$$

$$= 2\pi i \cdot \sum_{k=1}^{n} \operatorname*{res}_{z=z_k} f(z).$$

As was mentioned above, in the case $a = \infty$ the proof is the same and the corresponding illustration is given in Fig. 3.13. □

Remark Notice that the condition (2) in the Theorem formulation is essential. To see what conditions the function $f(z)$ should satisfy to guarantee that the condition (2) holds, we evaluate the integral in this condition. If $a \neq \infty$, then denoting $M_f(r) = \sup_{\forall z \in C_r} |f(z)|$, we have

$$\left| \int_{C_r} f(z) \, dz \right| \leq \int_{C_r} |f(z)| \, |dz| \leq M_f(r) \int_{C_r} |dz| = M_f(r) \cdot l_{C_r} \leq 2\pi r \cdot M_f(r).$$

It means that if

$$r \cdot M_f(r) \underset{r \to 0}{\to} 0, \tag{3.51}$$

then the condition (2) is satisfied.

If $a = \infty$, then using $M_f(R) = \sup_{\forall z \in C_R} |f(z)|$, we obtain in a similar manner

$$\left| \int_{C_R} f(z) \, dz \right| \leq \int_{C_R} |f(z)| \, |dz| \leq M_f(R) \cdot \int_{C_R} |dz| \leq 2\pi R \cdot M_f(R)$$

Therefore, the condition

$$R \cdot M_f(R) \underset{R \to \infty}{\to} 0 \tag{3.52}$$

ensures that (2) holds.

The limits (3.51) and (3.52) are very simple restrictions on the function $f(z)$, which guarantee that the condition (2) is satisfied. However, sometimes we need finer restrictions imposed on $f(z)$, which still guarantee that the condition (2) holds. One of these restrictions in the case $a = \infty$ is given in the next Lemma.

Jordan's Lemma *Let C_R be an upper half-circle of a radius R (C_R: $z = Re^{i\varphi}, 0 \leq \varphi \leq \pi$) and a function $f(z)$ be continuous on C_R. Denote $M_f(R) = \sup_{\forall z \in C_R} |f(z)|$. If $M_f(R) \underset{R \to \infty}{\to} 0$, then for any $b > 0$ it follows $\int_{C_R} e^{ibz} f(z) \, dz \underset{R \to \infty}{\to} 0$.*

Proof Since $z \in C_R$ can be represented as $z = Re^{i\varphi}$, $R, \varphi \in \mathbb{R}$, for any $z \in C_R$ and $b > 0$, we have

$$e^{i\,b\,z} = e^{i\,b\,R\,e^{i\varphi}} = e^{i\,b\,R(\cos\varphi + i\sin\varphi)} = e^{-b\,R\sin\varphi} \cdot e^{i\,b\,R\,\cos\varphi},$$

and consequently $\left| e^{ibz} \right| = e^{-bR\sin\varphi}$ since $\left| e^{ibR\cos\varphi} \right| = 1$. To evaluate the integral along C_R, we take into account that on this curve $z = Re^{i\varphi}$ (R is a constant), which implies that $dz = R\,e^{i\varphi} i\,d\varphi$ and $|dz| = Rd\varphi$. Therefore

$$\left|\int_{C_R} e^{ibz} f(z)\, dz\right| \le \int_{C_R} \left|e^{ibz}\right| \cdot |f(z)|\, |dz| = \int_{C_R} e^{-bR\sin\varphi} |f(z)|\, |dz|$$

$$\le M_f(R) \int_{C_R} e^{-bR\sin\varphi}\, |dz| = RM_f(R) \int_0^{\pi} e^{-bR\sin\varphi} d\varphi$$

$$= RM_f(R) \left(\int_0^{\frac{\pi}{2}} e^{-bR\sin\varphi} d\varphi + \int_{\frac{\pi}{2}}^{\pi} e^{-bR\sin\varphi} d\varphi \right) = I\,.$$

In the second integral on the right-hand side we change the variable: $\psi = \pi - \varphi$, which implies $\varphi = \pi - \psi, d\varphi = -d\psi$ and $\sin\varphi = \sin(\pi - \psi) = \sin\psi$. Continuing the evaluation of I, we obtain the two equal integrals:

$$I = RM_f(R) \left(\int_0^{\frac{\pi}{2}} e^{-bR\sin\varphi} d\varphi - \int_{\frac{\pi}{2}}^{0} e^{-bR\sin\psi} d\psi \right)$$

$$= RM_f(R) \left(\int_0^{\frac{\pi}{2}} e^{-bR\sin\varphi} d\varphi + \int_0^{\frac{\pi}{2}} e^{-bR\sin\psi} d\psi \right) = 2R \cdot M_f(R) \int_0^{\frac{\pi}{2}} e^{-bR\sin\varphi} d\varphi = I_1\,.$$

Notice now that for $\varphi \in \left[0, \frac{\pi}{2}\right]$ the inequality $\sin\varphi \ge \frac{2}{\pi}\varphi$ holds (the graph of $\sin\varphi$ is located above the line $t = \frac{2}{\pi}\varphi$ on the segment $\varphi \in \left[0, \frac{\pi}{2}\right]$) which implies $-bR\sin\varphi \le -bR \cdot \frac{2}{\pi}\varphi$. Consequently, since the exponential function $e^x, x \in \mathbb{R}$ is increasing, we have $e^{-bR\sin\varphi} \le e^{-bR\frac{2}{\pi}\varphi}$. Using the last inequality, we can evaluate I_1 as follows:

$$I_1 \le 2RM_f(R) \int_0^{\frac{\pi}{2}} e^{-bR\frac{2}{\pi}\varphi} d\varphi = -\frac{2RM_f(R)\pi}{2bR} \cdot e^{-bR\frac{2}{\pi}\varphi}\Big|_0^{\frac{\pi}{2}}$$

$$= -\frac{\pi M_f(R)}{b} e^{-bR} + \frac{\pi M_f(R)}{b} \underset{R\to\infty}{\to} 0\,.$$

Thus,

$$\left|\int_{C_R} e^{ibz} f(z)\, dz\right| \underset{R\to\infty}{\to} 0\,.$$

This terminates the proof of the Lemma. □

Remark Notice that if C_R is a lower half-circle of the radius R, that is, $C_R : z = Re^{i\varphi}, -\pi \le \varphi \le 0$, then we need to consider the integral $\int_{C_R} e^{-ibz} f(z)\, dz$ for $\forall b > 0$. Similarly, we can reformulate Jordan's Lemma for the case of a right half-circle $C_R : z = Re^{i\varphi}, -\frac{\pi}{2} \le \varphi \le \frac{\pi}{2}$ or a left half-circle $C_R : z = Re^{i\varphi}, \frac{\pi}{2} \le \varphi \le \frac{3\pi}{2}$.

3.12 Main Classes of Integrals Evaluated Through Residues

Class 1. General result

We start with the general result formulated in the Fundamental Theorem on residues. If a domain D and a function $f(z)$ satisfies the conditions of the Fundamental Theorem on residues (Sect. 3.9), then the integral along ∂D can be evaluated using the general formula (3.28), that is

$$\int_{\partial D} f(z)\,dz = 2\pi i \cdot \sum_{k=1}^{n} \operatorname*{res}_{z=z_k} f(z), \tag{3.53}$$

where z_k are single-valued isolated singular points of $f(z)$ contained in D.

Class 2. $\int_{\alpha}^{\alpha+2\pi} f(\cos\varphi, \sin\varphi)\,d\varphi$**, where** α **is an arbitrary real number**

In this case, we change the variable $z = e^{i\varphi}$ that gives the following relations: $dz = e^{i\varphi} i\,d\varphi$, $d\varphi = \frac{dz}{i\,e^{i\varphi}} = \frac{dz}{i\,z}$, $\cos\varphi = \frac{e^{i\varphi}+e^{-i\varphi}}{2} = \frac{z+\frac{1}{z}}{2}$, $\sin\varphi = \frac{e^{i\varphi}-e^{-i\varphi}}{2i} = \frac{z-\frac{1}{z}}{2i}$, and notice that if φ varies on the interval $[\alpha, \alpha+2\pi]$ (where α is an arbitrary real number), then the corresponding point z traverses counterclockwise the entire circle $|z| = 1$. Applying these relations to the given integral, we obtain

$$\int_{\alpha}^{\alpha+2\pi} f(\cos\varphi, \sin\varphi)\,d\varphi = \frac{1}{i}\int_{|z|=1} \frac{1}{z} f\left(\frac{z+\frac{1}{z}}{2}, \frac{z-\frac{1}{z}}{2i}\right) dz \equiv I\,.$$

The circle $|z| = 1$ divides the entire complex plane into the two domains: $|z| < 1$ and $|z| > 1$. If the function $g(z) = \frac{1}{z} f\left(\frac{z+\frac{1}{z}}{2}, \frac{z-\frac{1}{z}}{2i}\right)$ satisfies the conditions of the Fundamental Theorem on residues of Sect. 3.9 or the Theorem of Sect. 3.11 at least in one of these domains, for instance, $g(z)$ has a finite number of single-valued isolated singular points contained in the disk $|z| < 1$, then we have

$$I = 2\pi i \cdot \frac{1}{i}\sum_{k=1}^{n} \operatorname*{res}_{z=z_k, |z_k|<1} g(z) = 2\pi \cdot \sum_{k=1}^{n} \operatorname*{res}_{z=z_k, |z_k|<1} g(z)\,. \tag{3.54}$$

Notice that in this case the counterclockwise traversal along the circle $|z| = 1$ means the positive traversal along the boundary of the domain $D = \{|z| < 1\}$. If we choose the exterior of the unit disk—$D = \{|z| > 1\}$, then counterclockwise traversal along the circle $|z| = 1$ corresponds to the negative traversal along the boundary of this domain. Besides, in the latter case, the residues of $g(z)$ should be calculated both at the finite singular points contained in D and at the point ∞. In this case, we obtain

$$I = -2\pi i \cdot \frac{1}{i}\left(\sum_{k=1}^{n} \operatorname*{res}_{z=z_k,\, |z_k|>1} g(z) + \operatorname*{res}_{z=\infty} g(z)\right) = -2\pi\left(\sum_{k=1}^{n} \operatorname*{res}_{z=z_k,\, |z_k|>1} g(z) + \operatorname*{res}_{z=\infty} g(z)\right). \tag{3.55}$$

Usually one chooses such a domain between $|z| < 1$ and $|z| > 1$ that contains a fewer number of singular points of the function $g(z)$.

Class 3. $\int_{-\infty}^{+\infty} f(x)\,dx$

In this case, we have an improper integral with the singular point being the infinity. The curve—the real axis—divides the complex plane into the two domains: the upper half-plane $\operatorname{Im} z > 0$ and the lower one $\operatorname{Im} z < 0$. Choose, for instance, as the domain D the upper half-plane and consider the function $f(z)$ in this domain. Assume the function $f(z)$ has a finite number of single-valued isolated singular points z_k, $k = 1, \ldots, n$ in the upper half-plane D. If the integral $\int_{-\infty}^{+\infty} f(x)\,dx$ converges and the integral $\int_{C_R} f(z)\,dz$ along the upper half-circle $C_R = \{z = Re^{i\varphi}, 0 \leq \varphi \leq \pi\}$ approaches 0 as R approaches ∞ (that is, $\int_{C_R} f(z)\,dz \underset{R\to\infty}{\to} 0$), then, by the Theorem of Sect. 3.11, we get

$$\int_{-\infty}^{+\infty} f(x)\,dx = 2\pi i \sum_{k=1}^{n} \operatorname*{res}_{z=z_k,\, \operatorname{Im} z_k>0} f(z).$$

If similar conditions are satisfied for the lower half-plane, then

$$\int_{-\infty}^{+\infty} f(x)\,dx = -2\pi i \sum_{k=1}^{n} \operatorname*{res}_{z=z_k,\, \operatorname{Im} z_k<0} f(z)$$

(notice that the motion from $-\infty$ to $+\infty$ along the real axis is negative with respect to the lower half-plane).

Let us consider a specific case of a rational function

$$f(z) = R(z) = \frac{P_n(z)}{P_m(z)} = \frac{a_n z^n + a_{n-1} z^{n-1} + \cdots + a_0}{b_m z^m + b_{m-1} z^{m-1} + \cdots + b_0}. \tag{3.56}$$

To evaluate $|f(z)|$ on the circle $|z| = R$, we transform $f(z)$ as follows:

$$f(z) = \frac{a_n z^n + \cdots + a_0}{b_m z^m + \cdots + b_0} = \frac{a_n}{b_m} \cdot \frac{z^n}{z^m} \cdot \frac{1 + \frac{a_{n-1}}{a_n}\frac{1}{z} + \cdots + \frac{a_0}{a_n}\frac{1}{z^n}}{1 + \frac{b_{m-1}}{b_m}\frac{1}{z} + \cdots + \frac{b_0}{b_m}\frac{1}{z^m}} = \frac{a_n}{b_m} \cdot \frac{1}{z^{m-n}} \cdot g(z). \tag{3.57}$$

Notice that $g(z)$ approaches 1 as $z \to \infty$

$$g(z) = \frac{1 + \frac{a_{n-1}}{a_n}\frac{1}{z} + \cdots + \frac{a_0}{a_n}\frac{1}{z^n}}{1 + \frac{b_{m-1}}{b_m}\frac{1}{z} + \cdots + \frac{b_0}{b_m}\frac{1}{z^m}} \underset{z\to\infty}{\to} 1,$$

that is, for $\forall\varepsilon > 0$ (take $\varepsilon = 1$) there exists R_1 such that for $\forall z$, $|z| > R_1$ it follows $|g(z) - 1| < 1$. From this evaluation, we have immediately that $|g(z)| = |g(z) - 1 + 1| \le |g(z) - 1| + 1 < 2, \forall z, |z| > R_1$. Therefore, for $\forall R$, $R > R_1$, on the circle $|z| = R$, we have

$$|f(z)|_{|z|=R} = \frac{|a_n|}{|b_m|} \cdot \frac{1}{|z|^{m-n}} \cdot |g(z)|_{|z|=R} \le \frac{2|a_n|}{|b_m|} \cdot \frac{1}{R^{m-n}},$$

which implies that

$$M_f(R) = \sup_{|z|=R} |f(z)| \le \frac{2|a_n|}{|b_m|} \cdot \frac{1}{R^{m-n}}. \tag{3.58}$$

The definition (3.56) of the rational function shows that all the singular points of $f(z)$ are zeros of the denominator, and consequently, there is a finite number of singular points and all of them are poles.

Now we impose the following additional conditions:
(1) $P_m(x) \ne 0, \forall x \in \mathbb{R}$, that is, there is no singular point on the real axis, except for the point ∞;
(2) $m - n \ge 2$.
From the expression (3.57), taking into account that $g(x) \to 1$ as $x \to \infty$, we have that

$$f(x) = \frac{a_n}{b_m} \frac{1}{x^{m-n}} g(x) \underset{x\to\infty}{\sim} \frac{a_n}{b_m} \frac{1}{x^{m-n}}.$$

Since $m - n \ge 2$, the integrals $\int_A^{+\infty} \frac{dx}{x^{m-n}}$, $A > 0$ and $\int_{-\infty}^{-B} \frac{dx}{x^{m-n}}$, $B > 0$ converge, and therefore, the integral $\int_{-\infty}^{+\infty} f(x)\,dx$ also converges. Moreover, the condition $m - n \ge 2$ implies that

$$R \cdot M_f(R) \le \frac{2|a_n|}{|b_m|} \cdot \frac{1}{R^{m-n-1}} \underset{R\to\infty}{\to} 0.$$

Hence, all the conditions of the Theorem of Sect. 3.11 are satisfied, and consequently, the integral can be expressed in the form

$$\int_{-\infty}^{+\infty} f(x)\,dx = 2\pi i \sum \underset{z=z_k,\, \mathrm{Im} z_k > 0}{\mathrm{res}} f(z) = -2\pi i \sum \underset{z=z_j,\, \mathrm{Im} z_j < 0}{\mathrm{res}} f(z). \tag{3.59}$$

Class 4. $\int_{-\infty}^{+\infty} e^{ibx} R(x)\,dx$, **where** $b > 0$ **and** $R(z)$ **is a rational function** (3.56) **with real coefficients.**

We impose the following additional conditions on the function $R(z)$:
(1) $P_m(x) \ne 0, \forall x \in \mathbb{R}$, that is, there is no singular point on the real axis, except for the point ∞;
(2) $m - n > 0$.

Let us show how the given integral can be evaluated through residues under the imposed conditions. Notice that the function $F(z) = e^{ibz}R(z)$ has a finite number of singular points and all of them are zeros of denominator of the rational function $R(z)$, that is, all of them are poles of $F(z)$. Since $R(x)$ is a real rational function and $P_m(x) \neq 0$, the function $R(x)$ is continuous on the real axis and has a finite number of extrema, which implies that there exists $A > 0$ such that for all $x > A$ the function $R(x)$ is continuous and monotonic. Besides, the representation (3.57) implies that $R(x) = \frac{a_n}{b_m}\frac{1}{x^{m-n}}g(x) \underset{x\to+\infty}{\to} 0$, since $g(x)$ is a bounded function and $m-n > 0$. Therefore, due to this tendency and monotonicity, the function $R(x)$ approaches zero as $x \to +\infty$ monotonically. Observing also that the primitive of e^{ibx} is bounded—$\left|\int e^{ibx}dx\right| = \left|\frac{1}{bi}e^{ibx}\right| = \frac{1}{b}$, we can see that the conditions of Dirichlet's test are satisfied, and therefore, the integral $\int_A^{+\infty} e^{ibx}R(x)\,dx$, $A > 0$ converges. For the same reasons, the integral $\int_{-\infty}^{-B} e^{ibx}R(x)\,dx$, $B > 0$ also converges. Since there is no other singular points on the real axis, the given integral $\int_{-\infty}^{+\infty} e^{ibx}R(x)\,dx$ converges. Using the inequality (3.58) and the hypothesis that $m-n > 0$, we obtain $M_f(R) \leq \frac{2|a_n|}{|b_m|}\frac{1}{R^{m-n}} \underset{R\to\infty}{\to} 0$. Therefore, Jordan's Lemma ensures that $\int_{C_R} e^{ibz}R(z)\,dz \underset{R\to\infty}{\to} 0$. Thus, all the conditions of the Theorem of Sect. 3.11 are satisfied, and consequently

$$\int_{-\infty}^{+\infty} e^{ibx}R(x)\,dx = 2\pi i \sum \underset{z=z_k,\,\mathrm{Im}z_k>0}{\mathrm{res}} F(z)\,, \ \forall b > 0\,. \tag{3.60}$$

In the case $b < 0$, the domain D should be the lower half-plane and Jordan's Lemma should be applied to the lower half-circle $C_R = \{z = Re^{i\varphi}, \pi \leq \varphi \leq 2\pi\}$. If all the conditions on the rational function $R(z)$ hold, then, taking into account that the traverse along the real axis from $-\infty$ to $+\infty$ corresponds to the negative orientation for the boundary of the lower half-plane, we obtain a similar formula

$$\int_{-\infty}^{+\infty} e^{ibx}R(x)\,dx = -2\pi i \sum \underset{z=z_k,\,\mathrm{Im}z_k<0}{\mathrm{res}} F(z)\,, \ \forall b < 0\,. \tag{3.61}$$

Class 5. $\int_{-\infty}^{+\infty} \cos bx \cdot R(x)\,dx$ **and** $\int_{-\infty}^{+\infty} \sin bx \cdot R(x)\,dx$

Suppose $R(z)$ is a rational function (3.56) with real coefficients that satisfies the two additional conditions of Class 4 functions

(1) $P_m(z) \neq 0$ for $z = x$;

(2) $m - n > 0$.

Notice that $\cos bx = \mathrm{Re}\, e^{ibx}$, $\sin bx = \mathrm{Im}\, e^{ibx}$ and, since $R(x)$ is a real-valued function for $x \in \mathbb{R}$, we have $\cos bx \cdot R(x) = R(x) \cdot \mathrm{Re}\, e^{ibx} = \mathrm{Re}\left(e^{ibx}R(x)\right)$ and similarly $\sin bx \cdot R(x) = \mathrm{Im}\left(e^{ibx}R(x)\right)$. Therefore, taking into account that $R(x)$ satisfies the two additional conditions of the previous Class, we obtain

$$\int_{-\infty}^{+\infty} \cos bx \cdot R(x)\, dx = \operatorname{Re} \int_{-\infty}^{+\infty} e^{ibx} R(x)\, dx = \operatorname{Re} \left\{ 2\pi i \sum \operatorname*{res}_{z=z_k,\, \operatorname{Im} z_k > 0} \left(e^{ibz} R(z) \right) \right\}$$

and

$$\int_{-\infty}^{+\infty} \sin bx \cdot R(x)\, dx = \operatorname{Im} \int_{-\infty}^{+\infty} e^{ibx} R(x)\, dx = \operatorname{Im} \left\{ 2\pi i \sum \operatorname*{res}_{z=z_k,\, \operatorname{Im} z_k > 0} \left(e^{ibz} R(z) \right) \right\}.$$

Class 6. $\int_{-\infty}^{+\infty} e^{iax} f(x)\, dx$ **, where** $a > 0$
We add the following conditions on $f(z)$:

(1) the integral $\int_{-\infty}^{+\infty} e^{iax} f(x)\, dx$ converges;

(2) $M_f(R) = \sup\limits_{|z|=R} |f(z)| \underset{R\to\infty}{\rightarrow} 0$;

(3) $f(z)$ has a finite number of single-valued isolated singularities z_k, $k = 1, \ldots, m$ lying in the upper half-plane and also has a finite number of simple poles x_j, $j = 1, \ldots, p$ located on the real axis.

To represent the given integral through residues, we first construct a domain where the function $F(z) = e^{iaz} f(z)$ satisfies the conditions of the Fundamental Theorem on residues. To this end, choose $R > 0$ large enough to ensure that all the singular points of $f(z)$ belong to the disk $|z| < R$. Also pick $r > 0$ small enough to guarantee that the following three conditions hold: first, each disk $|z - x_j| < r$, $j = 1, \ldots, p$ contains only one singular point x_j, second, each of these disks lies in the large disk $|z| < R$, and third, these disks have no common points. Denote the corresponding large upper half-circle of the radius R centered at the origin by C_R and the small upper half-circles of the radius r centered at x_j by γ_j, $j = 1, \ldots, p$. Consider the domain $D_{R,r}$ bounded by C_R, γ_j, $j = 1, \ldots, p$ and by the corresponding segments of the real axis (see Fig. 3.14). The function $F(z)$ satisfies all the conditions of the Fundamental Theorem on residues in the domain $D_{R,r}$, and therefore

$$\int_{\partial D_{R,r}} e^{iaz} f(z)\, dz = 2\pi i \cdot \sum_{k=1}^{m} \operatorname*{res}_{z=z_k} F(z)\,. \tag{3.62}$$

On the other hand, the integral $\int_{\partial D_{R,r}} e^{iaz} f(z)\, dz$ can be represented as the sum of the following integrals:

$$\begin{aligned} \int_{\partial D_{R,r}} e^{iaz} f(z)\, dz = &\int_{-R}^{x_1 - r} e^{iax} f(x)\, dx + \sum_{j=1}^{p-1} \int_{x_j + r}^{x_{j+1} - r} e^{iax} f(x)\, dx \\ &+ \int_{x_p + r}^{R} e^{iax} f(x)\, dx + \int_{C_R} e^{iaz} f(z)\, dz + \sum_{j=1}^{p} \int_{\gamma_j} e^{iaz} f(z)\, dz\,. \end{aligned} \tag{3.63}$$

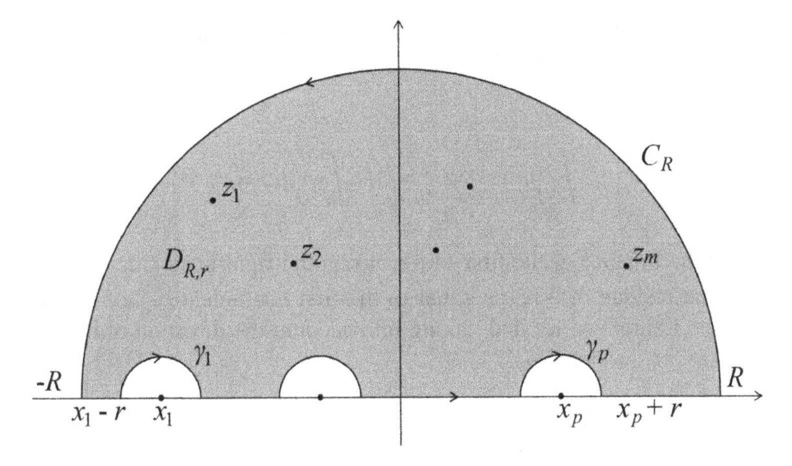

Fig. 3.14 The auxiliary domain $D_{R,r}$

Notice that the first three terms in the last formula converge as $R \to \infty$ and $r \to 0$ because of the convergence of the integral $\int_{-\infty}^{+\infty} e^{iax} f(x)\, dx$

$$\int_{-R}^{x_1-r} e^{i\,ax} f(x)\, dx + \sum_{j=1}^{p-1} \int_{x_j+r}^{x_{j+1}-r} e^{i\,ax} f(x)\, dx + \int_{x_p+r}^{R} e^{i\,ax} f(x)\, dx$$
$$\underset{R\to\infty,\, r\to 0}{\to} \int_{-\infty}^{+\infty} e^{i\,ax} f(x)\, dx\,. \tag{3.64}$$

Since $a > 0$ and $M_f(R) \underset{R\to\infty}{\to} 0$, by Jordan's Lemma, the forth term in (3.63) also converges

$$\int_{C_R} e^{i\,az} f(z)\, dz \underset{R\to\infty}{\to} 0\,. \tag{3.65}$$

Consider now the integrals $\int_{\gamma_j} e^{iaz} f(z)\, dz$. By the hypothesis, the function $f(z)$ has a simple pole at x_j, which implies that $F(z) = e^{i\,a\,z} f(z)$ also has a simple pole at x_j. Then, according to the criterion for multiplicity of a pole (Theorem 3.4 in Sect. 3.7), the function $F(z)$ has the following Laurent series expansion in a deleted neighborhood of x_j:

$$F(z) = \sum_{n=-1}^{+\infty} c_n^{(j)} \left(z - x_j\right)^n = \frac{c_{-1}^{(j)}}{z - x_j} + \sum_{n=0}^{+\infty} c_n^{(j)} \left(z - x_j\right)^n = \frac{c_{-1}^{(j)}}{z - x_j} + g_j(z)\,. \tag{3.66}$$

The function $g_j(z)$—the second summand in the last formula—is analytic at the point x_j as the sum of the convergent power series, and consequently, it is bounded in a neighborhood of this point, that is, there exists M_j such that for $\forall z,\ \left|z - x_j\right| <$

r one has $|g_j(z)| \le M_j$. Using the parameterization of the half-circle γ_j in the form $z - x_j = re^{i\varphi}$, $\varphi \in [0, \pi]$, which implies $z = x_j + re^{i\varphi}$ and $dz = re^{i\varphi} i d\varphi$, we obtain

$$\left| \int_{\gamma_j} g_j(z)\, dz \right| \le \int_{\gamma_j} |g_j(z)|\, |dz| \le M_j \int_0^{\pi} r d\varphi = r\pi M_j \underset{r\to 0}{\to} 0. \tag{3.67}$$

To evaluate the integral of the first summand on the right-hand side of (3.66), we recall that the residue of $F(z)$ is equal to the first coefficient $c_{-1}^{(j)}$ of the negative powers in the Laurent series and, taking into account the direction of the traversal along the curve γ_j, we obtain

$$\int_{\gamma_j} \frac{c_{-1}^{(j)}}{z - x_j} dz = \int_{\pi}^{0} \frac{c_{-1}^{(j)}}{re^{i\varphi}} re^{i\varphi} i d\varphi = i c_{-1}^{(j)} \varphi|_{\pi}^{0} = -i\pi c_{-1}^{(j)} = -i\pi \operatorname*{res}_{z=x_j} F(z). \tag{3.68}$$

Hence, substituting the results (3.67) and (3.68) in the equality (3.66), we get

$$\int_{\gamma_j} e^{iaz} f(z)\, dz = \int_{\gamma_j} F(z)\, dz = \int_{\gamma_j} \frac{c_{-1}^{(j)}}{z - x_j} dz + \int_{\gamma_j} g_j(z)\, dz \underset{r\to 0}{\to} -i\pi \operatorname*{res}_{z=x_j} F(z). \tag{3.69}$$

Finally, recalling that the left-hand sides in (3.62) and (3.63) are equal and do not depend on R and r, passing to the limit on the right-hand side of (3.63) as $R \to \infty, r \to 0$ and employing the results (3.64), (3.65) and (3.69), we obtain

$$2\pi i \cdot \sum_{k=1}^{m} \operatorname*{res}_{z=z_k} F(z) = \int_{-\infty}^{+\infty} e^{iax} f(x)\, dx - i\pi \sum_{j=1}^{p} \operatorname*{res}_{z=x_j} F(z).$$

Solving the last equation with respect to the given integral, we arrive at the formula

$$\int_{-\infty}^{+\infty} e^{iax} f(x)\, dx = 2\pi i \cdot \left\{ \sum_{k=1}^{m} \operatorname*{res}_{z=z_k} \left(e^{iaz} f(z)\right) + \frac{1}{2} \sum_{j=1}^{p} \operatorname*{res}_{z=x_j} \left(e^{iaz} f(z)\right) \right\}. \tag{3.70}$$

In particular, if $f(x) = R(x) = \frac{P_n(x)}{P_q(x)}$, where $q - n > 0$ and the polynomial $P_q(x)$ has only simple zeros on the real axis (which means that $f(z)$ has a finite number of simple poles on the real axis), then all the conditions imposed on the function $f(z)$ are satisfied, and consequently, the given integral can be evaluated by formula (3.70).

Consider two examples of the evaluation of integrals, which employ the theory of residues and the formulas deduced in this section.

Example 3.1 Evaluate the integral by applying the theory of residues

$$\oint_{|z|=1} \frac{e^z}{z^2\left(z^2+25\right)}dz.$$

We start with finding singularities of the function $f(z) = \frac{e^z}{z^2(z^2+25)}$. Solving the equation $z^2\left(z^2+25\right)=0$, we have three singular points: $z_1 = 0$, $z_2 = 5i$, $z_3 = -5i$. One more singular point is $z_4 = \infty$. The circle $|z| = 1$ divides the entire plane into the two domains: $|z| < 1$ and $|z| > 1$, for each of which $|z| = 1$ is a boundary. To evaluate the integral, we can choose any of these two domains. In such cases, to simplify the calculations, the preference is usually given to such a domain that contains a fewer number of singular points and/or such that contains singular points with an easier evaluation of residues. In our example, we choose the domain $D = \{|z| < 1\}$, since it contains only one singular point $z_1 = 0$, which is a pole of second order of $f(z)$. Notice that the positive (counterclockwise) traversal along the circle $|z| = 1$ coincides with the positive orientation of the boundary of the chosen domain $D = \{|z| < 1\}$. Using the Fundamental Theorem on residues (Sect. 3.9) and formula (3.38) of Sect. 3.10 (with $n = 2$), we obtain

$$\oint_{|z|=1} \frac{e^z dz}{z^2\left(z^2+25\right)} = \oint_{\partial D} f(z)\,dz = 2\pi i \cdot \operatorname*{res}_{z=0} f(z) = 2\pi i \lim_{z\to 0}\left(\frac{e^z}{z^2\left(z^2+25\right)}\cdot z^2\right)'$$

$$= 2\pi i \lim_{z\to 0} \frac{e^z\left(z^2+25\right)-2ze^z}{\left(z^2+25\right)^2} = \frac{2\pi i}{25}.$$

Example 3.2 Evaluate the integral by applying the theory of residues

$$\int_0^{+\infty} \frac{5x\,\sin 5x}{x^4+3x^2+2}dx.$$

Notice first that the integrand $\frac{5x\,\sin 5x}{x^4+3x^2+2}$ in the improper integral is even, and consequently

$$\int_0^{+\infty} \frac{5x\sin 5x}{x^4+3x^2+2}dx = \frac{1}{2}\int_{-\infty}^{+\infty} \frac{5x\,\sin 5x}{x^4+3x^2+2}dx.$$

Besides, for $x\in\mathbb{R}$ one has $\sin 5x = \operatorname{Im} e^{5ix}$, and since $\frac{5x}{x^4+3x^2+2}\in\mathbb{R}$, we can represent the original integral in the form

$$\int_0^{+\infty} \frac{5x\,\sin 5x}{x^4+3x^2+2}dx = \frac{1}{2}\operatorname{Im}\int_{-\infty}^{+\infty} \frac{5x\,e^{5ix}}{x^4+3x^2+2}dx.$$

Consider the function

$$f(z) = \frac{5z\,e^{5iz}}{z^4+3z^2+2} = e^{5iz}R(z)\,,\ \text{where } R(z) = \frac{5z}{z^4+3z^2+2}.$$

The zeros of the denominator are singular points of $f(z)$

$$z^4 + 3z^2 + 2 = 0 \Rightarrow z^2 = -1\,,\; z^2 = -2 \Rightarrow z_1 = i\,,\; z_2 = -i\,,\; z_3 = i\sqrt{2}\,,\; z_4 = -i\sqrt{2}\,,$$

all of which are simple poles. The function $f(z)$ satisfies the conditions of the functions in Class 4 of this section, and using formula (3.40) in Sect. 3.10, we obtain

$$\int_0^{+\infty} \frac{5x\,\sin 5x}{x^4+3x^2+2}dx = \frac{1}{2}\,\mathrm{Im}\int_{-\infty}^{+\infty} \frac{5x\,e^{5ix}}{x^4+3x^2+2}dx = \frac{1}{2}\,\mathrm{Im}\left\{2\pi i\left(\operatorname*{res}_{z=i} f(z) + \operatorname*{res}_{z=i\sqrt{2}} f(z)\right)\right\}$$

$$= \frac{1}{2}\,\mathrm{Im}\left\{2\pi i\left(\left.\frac{5z\,e^{5iz}}{4z^3+6z}\right|_{z=i} + \left.\frac{5z\,e^{5iz}}{4z^3+6z}\right|_{z=i\sqrt{2}}\right)\right\}$$

$$= \frac{1}{2}\,\mathrm{Im}\left\{2\pi i\left(\frac{5\,e^{-5}}{-4+6} + \frac{5\,e^{-5\sqrt{2}}}{-8+6}\right)\right\} = \frac{5\pi}{2}\left(e^{-5} - e^{-5\sqrt{2}}\right).$$

Notice that the original integrand (which is a real-valued function of a real variable) has no primitive expressed in the terms of elementary functions, and for this reason, an evaluation of the integral through the techniques of Real Analysis is either very difficult or even impossible. At the same time, the theory of residues allows us to calculate such integrals in a rather simple way.

3.13 The Argument Principle

Theorem (The Argument Principle) *Let D be a bounded domain whose boundary consists of a finite number of piecewise smooth curves with folds. If a function $f(z)$ is analytic in D, continuous in D up to its boundary, except for a finite number of poles contained in D, and $f(z) \neq 0$, $\forall z \in \partial D$, then*

$$\frac{1}{2\pi i}\int_{\partial D} \frac{f'(z)}{f(z)}dz = N_f(D) - P_f(D)\,, \tag{3.71}$$

where $N_f(D)$ and $P_f(D)$ are, respectively, the numbers of zeros and of poles of $f(z)$ in the domain D counted according to their multiplicities.

Proof First, we show that the function $f(z)$ can have at most a finite number of zeros in the domain D. Suppose, by contradiction, that the set E of the zeros of $f(z)$ in D contains an infinite number of points. Denote the poles of $f(z)$ in D by b_j, $j = 1, \ldots, p$, $b_j \in D$ and the remaining part of D by $\tilde{D} = D \setminus \bigcup_{j=1}^{p} \{b_j\}$. The function $f(z)$ is analytic in the domain $\tilde{D}$ (since all the poles were removed) and the set E is contained in $\tilde{D}$: $E \subset \tilde{D}$. By supposition, the set E is infinite, which implies that it has at least one limit point z_0. This point lies either in $\tilde{D}$ or on the boundary $\partial\tilde{D}$.

In the first case (if $z_0 \in \tilde{D}$), we can see that the hypotheses of the Theorem on uniqueness of an analytic function are satisfied (see Sect. 2.8): $f(z)$ is analytic in $\tilde{D}$ and equal to zero on the set $E \subset \tilde{D}$, which has the limit point z_0 contained in $\tilde{D}$. Therefore, this Theorem ensures that $f(z) \equiv 0$ in $\tilde{D}$ and, according to the definition of the function on the boundary of a domain, we have $f(z) = 0,\ \forall z \in \partial\tilde{D}$, which implies that $f(z) = 0, \forall z \in \partial D$, but the last equality contradicts the Theorem's hypothesis.

In the second case (if $z_0 \in \partial\tilde{D}$), it follows that $z_0 \in \partial D$ (since z_0 may not coincide with any point b_j). Since z_0 is a limit point of E, we can select in E a sequence of points $\{z_n\}$ that converges to z_0 as $n \to \infty$. By the construction, $f(z_n) = 0,\ \forall z_n \in E$, and then, by the definition of a function on the boundary of a domain, we get $f(z_0) = \lim_{n\to\infty} f(z_n) = 0$, that contradicts the Theorem's hypothesis because $z_0 \in \partial D$. Thus, in either case, we arrive at a contradiction, which means that our supposition is false and the function $f(z)$ can have at most a finite number of zeros in the domain D.

Denote the zeros of $f(z)$ in D by $a_k, k = 1, \ldots, m$ and their multiplicities by $\alpha_k, k = 1, \ldots, m$, respectively. The poles of $f(z)$ in D were already denoted by b_j, while we denote their corresponding multiplicities by $\beta_j,\ j = 1, \ldots, p$. Consider now the function $g(z) = \frac{f'(z)}{f(z)}$. This function may have singularities only at the zeros and poles of $f(z)$, that is, at the points where the denominator equals zero and at the singular points of $f(z)$. Notice that all these points are single-valued isolated singularities of the function $g(z)$ and investigate the behavior of $g(z)$ at these points.

First consider a zero a_k of a multiplicity α_k. By the definition, the function $f(z)$ has the following representation: $f(z) = (z - a_k)^{\alpha_k} f_k(z)$, where the function $f_k(z)$ is analytic at the point a_k and $f_k(a_k) \neq 0$. Then for $g(z)$, we have

$$g(z) = \frac{f'(z)}{f(z)} = \frac{\alpha_k (z - a_k)^{\alpha_k - 1} f_k(z) + (z - a_k)^{\alpha_k} f_k'(z)}{(z - a_k)^{\alpha_k} f_k(z)} = \frac{\alpha_k}{z - a_k} + \frac{f_k'(z)}{f_k(z)}\,. \qquad (3.72)$$

The function $\frac{f_k'(z)}{f_k(z)}$ is analytic at the point a_k as the ratio of two analytic functions with the denominator different from 0 at the point a_k. Therefore, the function $\frac{f_k'(z)}{f_k(z)}$ can be expanded in the power series in a neighborhood of a_k, and this expansion turns the right-hand side in (3.72) into the Laurent series of $g(z)$ in some deleted neighborhood of a_k. Moreover, formula (3.72) reveals that the principal part of this Laurent series consists of only one term $\frac{\alpha_k}{z-a_k}$, which is the first negative power. By the criterion for multiplicity of a pole, this means that the point a_k is a simple pole of the function $g(z)$ and

$$\operatorname*{res}_{z=a_k} g(z) = \alpha_k\,. \qquad (3.73)$$

Now take a pole b_j of multiplicity β_j. By the criterion for multiplicity of a pole, the function $f(z)$ can be expressed as $f(z) = \left(z - b_j\right)^{-\beta_j} f_j(z)$, where $f_j(z)$ is an analytic function at the point b_j and $f_j\left(b_j\right) \neq 0$. Then for $g(z)$, we obtain

$$g(z)=\frac{f'(z)}{f(z)}=\frac{-\beta_j\left(z-b_j\right)^{-\beta_j-1}f_j(z)+\left(z-b_j\right)^{-\beta_j}f_j'(z)}{\left(z-b_j\right)^{-\beta_j}f_j(z)}=\frac{-\beta_j}{z-b_j}+\frac{f_j'(z)}{f_j(z)}. \tag{3.74}$$

The function $\frac{f_j'(z)}{f_j(z)}$ is analytic at b_j as the ratio of two analytic functions with the denominator different from 0 at the point b_j, and consequently, the function $\frac{f_j'(z)}{f_j(z)}$ has the power series expansion in a neighborhood of b_j. Bringing this expansion in the right-hand side of (3.74), we obtain the representation of $g(z)$ in the Laurent series in some deleted neighborhood of b_j. The principal part of this series contains the only term $\frac{-\beta_j}{z-b_j}$, which is the first negative power. Therefore, by the criterion for multiplicity of a pole, we conclude that b_j is a simple pole of $g(z)$ and

$$\operatorname*{res}_{z=b_j} g(z)=-\beta_j. \tag{3.75}$$

Thus, the function $g(z)=\frac{f'(z)}{f(z)}$ is analytic in the domain D and continuous in D up to its boundary, except for a finite number of simple poles $a_k,\ k=1,\ldots,m$ and $b_j,\ j=1,\ldots,p$ contained in D. Then, applying the Fundamental Theorem on residues and using the results (3.73) and (3.75), we arrive at the main formula (3.71)

$$\begin{aligned}\frac{1}{2\pi i}\int_{\partial D}\frac{f'(z)}{f(z)}dz&=\frac{1}{2\pi i}\int_{\partial D}g(z)\,dz=\frac{1}{2\pi i}2\pi i\left(\sum_{k=1}^{m}\operatorname*{res}_{z=a_k}g(z)+\sum_{j=1}^{p}\operatorname*{res}_{z=b_j}g(z)\right)\\&=\sum_{k=1}^{m}\alpha_k-\sum_{j=1}^{p}\beta_j=N_f(D)-P_f(D).\end{aligned} \tag{3.76}$$

Notice that the sum $\sum_{k=1}^{m}\alpha_k=N_f(D)$ contains all the zeros of $f(z)$ including their multiplicities, and the sum $\sum_{j=1}^{p}\beta_j=P_f(D)$ contains all the poles with their multiplicities.

The Theorem is proved. □

Remark The main formula (3.71) can be expressed in another way. It was shown in Sect. 2.6 that $\int_\Gamma\frac{dt}{t}=\int_1^z\frac{dt}{t}=\ln|z|+i\arg z+2k\pi i$, that is, the integral of $\frac{1}{t}$ along a closed curve represents the variation of the function $\ln t$: $\oint_\Gamma\frac{dt}{t}=2k\pi i=\underset{\Gamma}{var}\ln t$. Therefore, the integral $\int_{\partial D}\frac{f'(z)}{f(z)}dz=\int_{\partial D}\frac{df(z)}{f(z)}$ represents the variation of the function $\ln f(z)$ when the point z traverses the boundary ∂D. Then, we have

$$\begin{aligned}N_f(D)-P_f(D)&=\frac{1}{2\pi i}\int_{\partial D}\frac{f'(z)}{f(z)}dz=\frac{1}{2\pi i}\underset{z\in\partial D}{var}\ln f(z)\\&=\frac{1}{2\pi i}\left(\underset{z\in\partial D}{var}\ln|f(z)|+i\underset{z\in\partial D}{var}\arg f(z)\right).\end{aligned}$$

Notice that the function $|f(z)|$ (and consequently $\ln|f(z)|$) returns to its initial value after a circuit along a closed curve. It means that the value of the function $\ln|f(z)|$

does not change after a circuit along the closed curve is completed: $\underset{z\in\partial D}{var} \ln |f(z)| = 0$. The only quantity that changes is $\arg f(z)$. Hence, the main formula takes the form

$$N_f(D) - P_f(D) = \frac{1}{2\pi i} \underset{z\in\partial D}{var} \ln f(z) = \frac{1}{2\pi} \underset{z\in\partial D}{var} \arg f(z). \tag{3.77}$$

3.14 Rouché's Theorem

Rouché's Theorem *Suppose that D is a bounded domain whose boundary consists of a finite number of piecewise smooth curves with folds, and that functions $f(z)$ and $F(z)$ are analytic in D and continuous in D up to its boundary. If the inequality $|f(z)| < |F(z)|$ is satisfied on the boundary ∂D (that is, for $\forall z \in \partial D$), then $N_{f+F}(D) = N_F(D)$, that is, the number of zeros of the sum $F(z)+f(z)$ in the domain D coincides with the number of zeros of the function that has larger absolute values on the boundary of D.*

Proof The condition $|F(z)| > |f(z)|, \forall z \in \partial D$ ensures that $F(z)$ is different from zero at any point on the boundary of D: $F(z) \neq 0, \forall z \in \partial D$. Similarly, the inequality $|F(z) + f(z)| \geq |F(z)| - |f(z)| > 0, \forall z \in \partial D$ guarantees that $f(z) + F(z)$ is also different from zero on ∂D: $F(z) + f(z) \neq 0, \forall z \in \partial D$. Therefore, we can apply the Argument Principle to the functions $F(z)$ and $F(z) + f(z)$. Since the functions $f(z)$ and $F(z)$ are analytic in D by the Theorem's hypothesis, we have that $P_F(D) = 0$ and $P_{F+f}(D) = 0$ (there is no pole). Then, the application of the Argument Principle to the function $F(z) + f(z)$ gives (see formula (3.77) in the previous section)

$$N_{F+f}(D) = \frac{1}{2\pi} \underset{z\in\partial D}{var} \arg (F(z) + f(z)) = \frac{1}{2\pi} \underset{z\in\partial D}{var} \arg \left(F(z) \left(1 + \frac{f(z)}{F(z)} \right) \right)$$

$$= \frac{1}{2\pi} \underset{z\in\partial D}{var} \arg F(z) + \frac{1}{2\pi} \underset{z\in\partial D}{var} \arg \left(1 + \frac{f(z)}{F(z)} \right) = N_F(D) + \frac{1}{2\pi} \underset{z\in\partial D}{var} \arg \left(1 + \frac{f(z)}{F(z)} \right). \tag{3.78}$$

Notice that the function $1 + \frac{f(z)}{F(z)}$ is continuous on ∂D, because $F(z) \neq 0, \forall z \in \partial D$. When the point z traverses all the boundary ∂D, the corresponding point $t = 1 + \frac{f(z)}{F(z)}$ traverses a finite number of closed curves on the complex plane $\mathbb{T}$. The condition $|f(z)| < |F(z)|, \forall z \in \partial D$ implies that $|t - 1| = \left|\frac{f(z)}{F(z)}\right| < 1, \forall z \in \partial D$, and consequently, all these curves in the plane $\mathbb{T}$ are contained in the disk $|t - 1| < 1$. Therefore, under the traversal along each closed curve in the plane $\mathbb{T}$, the vector $t = 1 + \frac{f(z)}{F(z)}$ does not complete circuits around the origin, that is, the value of $\arg t$ at the final point of any circuit along the closed curve coincides with its value at the starting point (see Fig. 3.15). This reasoning is valid for each component of the boundary ∂D, that justifies the formula $\underset{z\in\partial D}{var} \arg \left(1 + \frac{f(z)}{F(z)}\right) = 0$. Hence, the equality (3.78) takes the form $N_{F+f}(D) = N_F(D)$. □

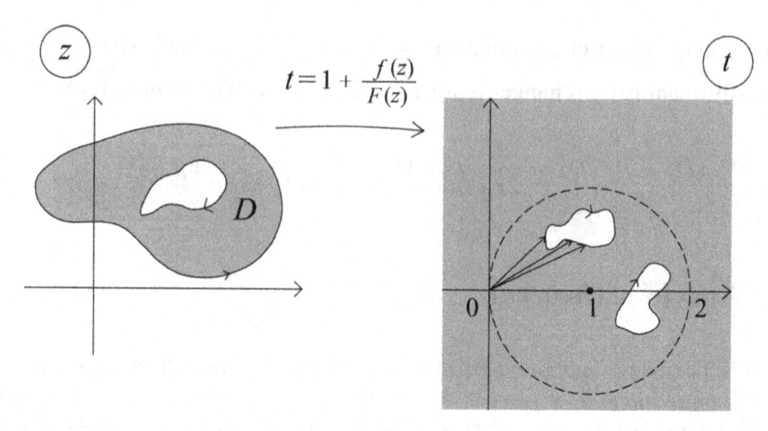

Fig. 3.15 The domain D and its image under the transformation $t = 1 + \frac{f(z)}{F(z)}$; all the boundary components lie inside the disk $|t - 1| < 1$

We end this section with some typical applications of Rouché's Theorem. An immediate application consists of the evaluation of the roots of specific functions like shown in the example below.

Example Find the number of roots of the given equation inside the disk $|z| < 1$: $z^9 + 2z^7 - 2z^2 + 9z - 1 = 0$.

To solve this problem, we apply Rouché's Theorem to the function $f(z) = z^9 + 2z^7 - 2z^2 + 9z - 1$ considered in the domain $D = \{|z| < 1\}$. Let us represent $f(z)$ in the form

$$f(z) = f_1(z) + f_2(z) , \text{ where } f_1(z) = 9z, \ f_2(z) = z^9 + 2z^7 - 2z^2 - 1 .$$

Notice that both functions $f_1(z)$ and $f_2(z)$ are analytic in the entire finite complex plane (in particular, in the closed domain $\overline{D} = \{|z| \leq 1\}$) and

$$|f_1(z)|_{\partial D} = 9 , \ |f_2(z)|_{\partial D} \leq \left(|z|^9 + 2|z|^7 + 2|z|^2 + 1\right)_{|z|=1} = 6 ,$$

that is, $|f_1(z)|_{\partial D} > |f_2(z)|_{\partial D}$. Notice also that $f_1(z) = 9z$ has the only zero in the domain D—the point $z = 0$. Therefore, using Rouché's Theorem, we have $N_f(D) = N_{f_1}(D) = 1$, that is, the function $f(z)$ has the only zero in the domain D, which means that the given equation has the only root in the disk $|z| < 1$.

The use of Rouché's Theorem leads to a quite simple proof of the *fundamental theorem of Algebra.*

The Fundamental Theorem of Algebra *Every polynomial of degree n (with complex or real coefficients) has exactly n zeros in the set of complex numbers (taking into account their multiplicity).*

Proof A polynomial of degree n has the form

$$P_n(z) = c_n z^n + c_{n-1} z^{n-1} + \ldots + c_1 z + c_0, \text{ where } c_n \neq 0.$$

Take $F(z) = c_n z^n$, $f(z) = c_{n-1} z^{n-1} + \cdots + c_1 z + c_0$ and $P_n(z) = F(z) + f(z)$. Choose the disk $|z| < R$ as the domain D, with $R = 1 + \frac{|c_{n-1}|}{|c_n|} + \cdots + \frac{|c_1|}{|c_n|} + \frac{|c_0|}{|c_n|}$. The functions $F(z) = c_n z^n$ and $f(z) = c_{n-1} z^{n-1} + \cdots + c_1 z + c_0$ are analytic in the entire (finite) complex plane, and consequently, they are analytic on the closed disk $|z| \le R$. The absolute values of these functions on the boundary of the disk $|z| \le R$ (that is, on the circle $|z| = R \ge 1$) satisfy the inequality $|f(z)|_{|z|=R} < |F(z)|_{|z|=R}$, that can be seen from the following evaluation:

$$|f(z)|_{|z|=R} = \left| c_{n-1} z^{n-1} + c_{n-2} z^{n-2} + \cdots + c_1 z + c_0 \right|_{|z|=R}$$

$$\le \left\{ |c_{n-1}| \cdot |z|^{n-1} + \cdots + |c_1| \cdot |z| + |c_0| \right\}_{|z|=R} = |c_{n-1}| R^{n-1} + |c_{n-2}| R^{n-2} + \cdots + |c_0|$$

$$= |c_n| R^{n-1} \left\{ \frac{|c_{n-1}|}{|c_n|} + \frac{|c_{n-2}|}{|c_n|} \frac{1}{R} + \cdots + \frac{|c_1|}{|c_n|} \frac{1}{R^{n-2}} + \frac{|c_0|}{|c_n|} \frac{1}{R^{n-1}} \right\}$$

$$\le |c_n| R^{n-1} \left\{ \frac{|c_{n-1}|}{|c_n|} + \frac{|c_{n-2}|}{|c_n|} + \cdots + \frac{|c_1|}{|c_n|} + \frac{|c_0|}{|c_n|} \right\}$$

$$< |c_n| R^{n-1} \left\{ 1 + \frac{|c_{n-1}|}{|c_n|} + \frac{|c_{n-2}|}{|c_n|} + \cdots + \frac{|c_0|}{|c_n|} \right\}$$

$$= |c_n| \cdot R^{n-1} \cdot R = |c_n| \cdot R^n = \left| c_n z^n \right|_{|z|=R} = |F(z)|_{|z|=R}.$$

Thus, all the conditions of Rouché's Theorem hold, and therefore, $N_{P_n(z)} = N_{F+f} = N_F$. The function $F(z) = c_n z^n$ has zero of the multiplicity n at the point $z = 0$ and does not possess other zeros, that is, $N_F = n$. Then $N_{P_n(z)} = n$, which completes the proof of the theorem. □

The investigation of the existence of fixed points in a suitably chosen domain can be quite simple.

The number of fixed points. If $h(z)$ is an analytic function in the unit disk $K = \{|z| < 1\}$ and continuous on $\overline{K}$, and $h(\partial K) \subset \{|w| < 1\}$, then $h(z)$ has exactly one fixed point in K. In fact, taking $f(z) = h(z)$ and $F(z) = -z$ in the conditions of Rouché's Theorem, we obtain $|f(z)| < 1 = |F(z)|$ on the boundary $|z| = 1$. Therefore, $f(z) + F(z) = h(z) - z$ and $F(z) = -z$ have the same number of zeros in K, that is, there is exactly one $c \in K$ such that $h(c) = c$.

One can extract information about the zeros of a function from the knowledge of the zeros of its Taylor polynomials.

The number of roots through Taylor polynomials. If $f(z)$ is an analytic function in the disk $K_R = \{|z| < R\}$, then using the Taylor theorem, we can write $f(z) = T_n(z) + z^{n+1}h_n(z)$, where $T_n(z)$ is the Taylor polynomial of degree n and $h_n(z)$ is an analytic function in K_R representing the corresponding residual. For some suitably chosen $r < R$ and n, it may happen that we can prove the inequality $r^{n+1}|h_n(z)| < |T_n(z)|$ on the circle $|z| = r$. Then, $f(z)$ has the same number of zeros in the disk $K_r = \{|z| < r\}$ as $T_n(z)$, and this number can be determined by (approximate) solution of the polynomial equation $T_n(z) = 0$.

One more theoretical application of Rouché's Theorem is the result on univalence of a limit function usually attributed to Hurwitz. We start with the following preliminary statement:

Theorem *If the functions $\{f_n(z)\}$ are analytic and different from 0 in a domain D, and if $\{f_n(z)\}$ converges normally to $f(z)$, then $f(z)$ is either identically zero or never equal to zero in D. (Recall that normal convergence means uniform convergence on every compact subset of D.)*

Proof Suppose that $f(z)$ is not identically zero. We show that in this case $f(z) \neq 0$, $\forall z \in D$. We argue by contradiction. Suppose that there exists at least one zero of $f(z)$ in the domain D: $f(z_0) = 0$, $z_0 \in D$. Notice that this zero is isolated, for if not, then z_0 is a limit point of zeros in D, and by Theorem of the uniqueness of an analytic function (Sect. 2.8), $f(z) \equiv 0$ in D, that contradicts the condition of the chosen case.

Since z_0 is an isolated zero of $f(z)$, there exists $r > 0$ such that in the closed disk $\overline{K} = \{|z - z_0| \leq r\} \subset D$ there is no other zeros of $f(z)$. In particular, $f(z) \neq 0$, $\forall z \in \partial K \equiv C$. Since the circle C is a compact set, the function $|f(z)|$ attains the minimum value on C and $\inf\limits_{z \in C} |f(z)| = m > 0$ because $f(z) \neq 0$, $\forall z \in C$. The condition of normality of $\{f_n(z)\}$ implies that $\forall \epsilon > 0$ (take $\epsilon = m$) there exists N such that $\forall n > N$ and $\forall z \in C$ it follows $|f_n(z) - f(z)| < \epsilon = m$. Representing $f_n(z)$ in the form $f_n(z) = (f_n(z) - f(z)) + f(z)$ and noting that $|f_n(z) - f(z)| < m$ and $|f(z)| \geq m$, $\forall z \in C$, we conclude that all the functions $f_n(z)$, $n > N$, have the same number of zeros in K as the function $f(z)$ with largest module (according to Rouché's Theorem). Therefore, each function $f_n(z)$, $n > N$, has one zero inside the disk K and consequently at least one zero in D, in contradiction to the Theorem conditions. □

Using this result, we get almost trivial proof of Hurwitz's Theorem.

Hurwitz's Theorem *Let $\{f_n(z)\}$ be a sequence of univalent analytic functions defined on a domain D. If $\{f_n(z)\}$ converges normally on D, then the limit function $f(z)$ is either constant or univalent.*

Proof Suppose that $f(z)$ is not constant. Choose a point $z_1 \in D$ and consider the sequence $h_n(z) = f_n(z) - f_n(z_1)$. Since each $h_n(z)$ is univalent by the Theorem conditions, we have $h_n(z) \neq 0$ for all $z \in D\backslash\{z_1\}$. The normal convergence of $\{f_n(z)\}$ on D implies the normal convergence of $h_n(z)$ on $D\backslash\{z_1\}$, and consequently, by the preceding Theorem, the limit function $h(z) = f(z) - f(z_1)$ is either identically zero or never zero in $D\backslash\{z_1\}$. However, according to the supposition, $h(z)$ is not constant, that eliminates the former option. Hence, $h(z) \neq 0$, $\forall z \in D\backslash\{z_1\}$, and consequently $f(z) \neq f(z_1)$, $\forall z \neq z_1$. Since z_1 is an arbitrary point, we have proved that $f(z)$ is univalent. □

Remark Sometimes the preceding Theorem is also referred to as Hurwitz's Theorem.

Exercises

1. Find singular points and classify them:

(1) $\frac{1}{4z-z^3}$;

(2) $\frac{z^4}{1+z^4}$;

(3) $\frac{z+3}{z(z^2+25)^2}$;

(4) $\frac{z^5}{(z+1)^3}$;

(5) $\frac{e^z}{9+z^2}$;

(6) $\frac{9+z^2}{e^z}$;

(7) $\frac{1}{z^3(4-\sin z)}$;

(8) $e^{\frac{z}{1-z}}$;

(9) $e^{z-\frac{1}{z}}$;

(10) $\frac{1}{\cos z}$;

(11) $\frac{\sin z}{z^3}$;

(12) $\sin\frac{1}{1+z}$;

(13) $\frac{1}{\sin z+\sin a}$;

(14) $e^{-z}\cos\frac{1}{z}$;

(15) $\frac{1-\cos z}{z^2}$;

(16) $\frac{1-\cos z}{\sin^2 z}$;

(17) $\frac{1}{z^2-1}\cos\frac{z\pi}{z+1}$;

(18) $\frac{1}{e^z-1}-\frac{1}{\sin z}$.

2. Expand the function in the Laurent series at the given points and find the ring (or the disk) of convergence of the obtained series:

(1) $\frac{1}{z-5}$; $z=0$, $z=5$, $z=\infty$;

(2) $\frac{2}{(z^2+1)^2}$; $z=0$, $z=i$, $z=\infty$;

(3) $z^3e^{\frac{1}{z}}$; $z=0$, $z=\infty$;

(4) $z^2\sin\frac{1}{z+i}$; $z=-i$;

(5) $\cos\frac{z^2+2z}{(z+1)^2}$; $z=-1$;

(6) $e^{z+\frac{1}{z}}$; $z=0$, $z=\infty$.

3. Find the set of convergence of the Laurent series:

(1) $\sum_{n=-\infty}^{+\infty}2^{-|n|}z^n$;

(2) $\sum_{n=-\infty}^{+\infty}\frac{z^{3n}}{3^n+1}$;

(3) $\sum_{n=-\infty}^{+\infty}\frac{(z-1)^n}{\cosh an}$, $a>0$;

(4) $\sum_{n=-\infty}^{+\infty} \frac{z^n}{n^2+1}$;

(5) $\sum_{n=-\infty}^{+\infty} 2^{-n^2} (z+1)^n$;

(6) $\sum_{n=-\infty}^{+\infty} 2^{-n^2} (z+1)^{n^2}$;

(7) $\sum_{n=-\infty}^{+\infty} 2^n z^n$.

4. Find singular points and evaluate the residues at the single-valued isolated singularities:

(1) $\frac{z^3}{(z^2+4)^2}$;

(2) $\frac{z^5}{(z^2+4)^2}$;

(3) $\frac{e^z}{z^3(z^2+25)}$;

(4) $\frac{1}{\sin z}$;

(5) $\cos \frac{z}{z-1}$;

(6) $z^3 \cos \frac{1}{z+1}$;

(7) $e^{z+\frac{1}{z}}$;

(8) $\frac{1}{z+z^3}$;

(9) $\frac{z^2}{1+z^4}$;

(10) $\frac{z^2}{(1+z)^3}$;

(11) $\frac{1}{(z^2+1)^3}$;

(12) $\frac{z^{2n}}{(z-1)^n}$, $\forall n \in \mathbb{N}$;

(13) $\frac{1}{e^z+1}$;

(14) $\frac{\sin z\pi}{(z-1)^3}$.

5. Let $z = a \neq \infty$ be an essential singularity of $f(z)$ and a pole of $g(z)$. Prove that a is essential singularity of $h(z) = f(z)g(z)$.
Hint: use a proof by contradiction.
6. Assume that $z = a$ is a single-valued isolated singularity of $f(z)$ and the function $f(z)$ satisfies the condition $|f(z)| < M \cdot |z-a|^{-m}$ in a deleted neighborhood of a, where M and m are positive constants. Demonstrate that $z = a$ cannot be an essential singularity of $f(z)$.
7. Let $z = a \neq \infty$ be a pole of order k of a function $f(z)$. Determine what kind of point is $z = a$ for $f^{(n)}(z)$.
8. Suppose $f(z)$ and $g(z)$ have a pole of order m and n, respectively, at $z = \infty$. Prove that the composite function $F(z) = f(g(z))$ has a pole of order mn at $z = \infty$.
9. Let the Laurent series $\sum_{n=-\infty}^{+\infty} c_n(z-a)^n$ be convergent in the closed ring $\overline{D} = \{0 < r \leq |z-a| \leq R\}$. Demonstrate that $|c_n| \leq M(r^{-n} + R^{-n})$, $\forall n \in \mathbb{Z}$, where M is a constant independent of n.
10. Let $z = 0$ and $z = \infty$ be single-valued isolated singularities (or regular points) of a function $f(z)$. Prove that if $f(z)$ is an even function, then $\operatorname*{res}_{z=0} f(z) = \operatorname*{res}_{z=\infty} f(z) = 0$.
11. Let $z = a$ and $z = -a$ be single-valued isolated singularities of $f(z)$. Demonstrate that:

(1) if $f(z)$ is an even function, then $\operatorname*{res}_{z=a} f(z) = -\operatorname*{res}_{z=-a} f(z)$;

(2) if $f(z)$ is an odd function, then $\operatorname*{res}_{z=a} f(z) = \operatorname*{res}_{z=-a} f(z)$.

12. Suppose that $f(z)$ and $g(z)$ are analytic at $z = a \neq \infty$ and have a zero of order m at this point. Demonstrate that:

(1) $\operatorname*{res}_{z=a} \left(\frac{f(z)}{g(z)} \cdot \frac{1}{z-a}\right) = \frac{f^{(m)}(a)}{g^{(m)}(a)}$;

(2) $\operatorname*{res}_{z=a} \left(\frac{f(z)}{g(z)} \cdot \frac{1}{(z-a)^2}\right) = \frac{1}{m+1} \frac{f^{(m)}(a)}{g^{(m)}(a)} \left(\frac{f^{(m+1)}(a)}{f^{(m)}(a)} - \frac{g^{(m+1)}(a)}{g^{(m)}(a)}\right)$.

13. Assume that a function $g(z)$ is analytic at a, $g'(a) \neq 0$, while a function $f(z)$ has a simple pole at $b = g(a)$ and $\operatorname*{res}_{z=b} f(z) = B$. Find $\operatorname*{res}_{z=a} f(g(z))$.
14. Assume that $f(z)$ and $g(z)$ are analytic at $z = a \neq \infty$, $f(a) \neq 0$ and $g(z)$ has a zero of order 2 at a. Find $\operatorname*{res}_{z=a} \frac{f(z)}{g(z)}$.
Hint: $\operatorname*{res}_{z=a} \frac{f(z)}{g(z)} = \frac{2f'(a)}{g''(a)} - \frac{2}{3} \frac{f(a)g'''(a)}{(g''(a))^2}$.
15. Suppose that $z = \infty$ is a single-valued isolated singularity of $f(z)$, that is, in a deleted neighborhood of ∞ the function $f(z)$ has a representation in the

Laurent series: $f(z) = \sum_{n=-\infty}^{+\infty} c_n z^n$. Find $\operatorname*{res}_{z=\infty} f^2(z)$ if:

(1) $z = \infty$ is a removable singularity;

(2) $z = \infty$ is a pole of order k;

(3) $z = \infty$ is an essential singularity.

16. Find $\operatorname*{res}_{z=a} \left(g(z)\frac{f'(z)}{f(z)}\right)$ if $g(z)$ is analytic at $z = a \neq \infty$ and $f(z)$:

(1) has a zero of order m at $z = a$;

(2) has a pole of order m at $z = a$.
Hint: (1) $\operatorname*{res}_{z=a} \left(g(z)\frac{f'(z)}{f(z)}\right) = mg(a)$; (2) $\operatorname*{res}_{z=a} \left(g(z)\frac{f'(z)}{f(z)}\right) = -mg(a)$.

17. Suppose that D is a finite domain, $g(z)$ is analytic in D and continuous in $\overline{D}$, and $f(z)$ is analytic in D and continuous in $\overline{D}$ except at a finite number of poles $b_j \in D$, $j = 1, \dots, m$ of order k_j, respectively, and additionally $f(z) \neq 0$, $\forall z \in \partial D$. Show that if $f(z)$ has a finite number of zeros $a_k \in D, k = 1, \dots, n$ of order l_k, respectively, then

$$\int_{\partial D} g(z)\frac{f'(z)}{f(z)}dz = 2\pi i \sum_{k=1}^{n} l_k g(a_k) - 2\pi i \sum_{j=1}^{m} k_j g(b_j).$$

Hint: make use of Exercise 16.

18*. Let $f(z)$ be an entire function. Prove that any analytic branch of the function $\ln \frac{z-b}{z-a}$, which is analytic in a neighborhood of ∞, satisfies the formula $\operatorname*{res}_{z=\infty} \left(f(z) \ln \frac{z-b}{z-a}\right) = \int_a^b f(z)dz$.
Hint: find the Laurent series expansion at ∞ of an analytic branch of $f(z) \ln \frac{z-b}{z-a}$.

19. Evaluate integrals using the theory of residues:

(1) $\oint_{x^2+y^2=2x} \frac{dz}{z^4+1}$;

(2) $\oint_{|z-2|=1} \frac{(z-1)dz}{(z+1)(z-2)^2}$;

(3) $\oint_{|z|=2} \frac{zdz}{(z-5)(z^5-1)}$;

(4) $\oint_{|z|=3} \frac{z^3dz}{z^4-1}$;

(5) $\oint_{|z|=2} \sin \frac{1}{z} dz$;

(6) $\oint_{|z|=2} \sin^2 \frac{1}{z} dz$;

(7) $\oint_{|z|=3} \left(z^2 - 2\right) \sin \frac{1}{z-2} dz$;

(8) $\oint_{|z|=1} z^n e^{\frac{3}{z}} dz, \ \forall n \in \mathbb{Z}$;

(9) $\oint_{|z|=4} \frac{z}{z+3} e^{\frac{1}{3z}} dz$;

(10) $\oint_{|z|=2} \frac{dz}{z^3\left(z^{10}-2\right)}$;

(11) $\oint_{|z|=3} \frac{z^2}{(z-1)(z-2)} \sin^2 \frac{1}{z} dz$;

(12) $\oint_{|z|=5} \left(z^2 + 1\right) e^{\frac{1}{z-1}} dz$;

(13) $\oint_{|z|=2} \left(z^2 + 5z + 3\right) \sin \frac{1}{z} dz$;

(14) $\int_0^{2\pi} \frac{d\varphi}{(a+b\cos\varphi)^2}$, $a > b > 0$;

(15) $\int_{-\pi}^{\pi} \frac{1+4\cos^2\varphi}{(17-8\cos\varphi)^2} d\varphi$;

(16) $\int_0^{2\pi} \frac{2+3i\sin 2\varphi}{15-8i\sin\varphi} d\varphi$;

(17*) $\int_0^{2\pi} e^{\cos\varphi} \cos\left(n\varphi - \sin\varphi\right) d\varphi, \ \forall n \in \mathbb{Z}$;

(18) $\int_0^{2\pi} \frac{\cos^2\varphi}{13+12\cos\varphi} d\varphi$;

(19) $\int_0^{\pi} \frac{\cos^4\varphi}{1+\sin^2\varphi} d\varphi$;

20*) $\int_{-\pi}^{\pi} \frac{\sin n\varphi \, d\varphi}{1-2a\sin\varphi+a^2}$, $-1 < a < 1, \ a \neq 0, \ \forall n \in \mathbb{N}$;

(21) $\int_{-\infty}^{+\infty} \frac{x\,dx}{\left(x^2+4x+13\right)^2}$;

(22) $\int_0^{+\infty} \frac{dx}{(x^2+1)^n}$, $\forall n \in \mathbb{N}$;

(23) $\int_0^{+\infty} \frac{x^2+7}{x^4+10x^2+9} dx$;

(24) $\int_{-\infty}^{+\infty} \frac{x^2-3x+1}{(x^2+4x+8)^2} dx$;

(25) $\int_0^{+\infty} \frac{x^2+9}{x^6+1} dx$;

(26) $\int_{-\infty}^{+\infty} \frac{(x-1)e^{ix}}{x^2-2x+2} dx$;

(27) $\int_{-\infty}^{+\infty} \frac{e^{ix} dx}{(x^2+4ix-5)^2}$;

(28) $\int_{-\infty}^{+\infty} \frac{(x+1)e^{-3ix}}{x^2-2x+5} dx$;

(29) $\int_{-\infty}^{+\infty} \frac{x \cos x}{x^2+4x+20} dx$;

(30) $\int_{-\infty}^{+\infty} \frac{(x+1) \sin 3x}{(x^2+6x+10)^2} dx$;

(31) $\int_{-\infty}^{+\infty} \frac{\sin 2x \, dx}{(x^2-7x+10)(x^2+1)}$;

(32) $\int_{-\infty}^{+\infty} \frac{\cos 3x}{(x-1)(x^2+1)^2} dx$.

20*. Let $f(z)$ be analytic in a ring $R < |z| < +\infty$ and $\operatorname{res}_{z=\infty} f(z) = 0$. Prove that in this ring there exists an analytic primitive of the function $f(z)$.
Hint: show that the integral $\int_a^z f(\zeta) d\zeta$, $R < |a|, |z| < +\infty$ does not depend on the form of curve lying in the given ring and connecting the points a and z.

21. Find the number of roots of the given equation in the disk $|z| < 1$ using Rouché's Theorem:

(1) $3z^6 - z^4 + 2z^2 + z - 8 = 0$;

(2) $z^5 - 6z^4 + z^3 - 2z + 1 = 0$;

(3) $2z^4 - 5z + 2 = 0$;

(4) $z^7 - 5z^4 + z^2 - 2 = 0$;

(5) $z^8 - 4z^5 + z^2 - 1 = 0$.

22. Find the number of roots of the equation $z^4 + 7z - 1 = 0$:

    (1) in the disk $|z| < 1$;

    (2) in the ring $1 < |z| < 2$.

23. Show that the equation $ze^{\lambda - z} = 1$, where $\lambda > 1$, has exactly one root in the disk $|z| < 1$ and this root is real.
24. Show that the equation $z = \lambda - e^{-z}$, where $\lambda > 1$, has the only root in the right half-plane $\operatorname{Re} z > 0$ and this root is real.
25. Verify whether the statement of Rouché's theorem will be true if the inequality $|f(z)| < |F(z)|$ on the boundary of a domain D is replaced by the non-strict version $|f(z)| \leq |F(z)|$.
    Hint: find a counterexample.
26. Let $D = \{|z| < 1\}$. Consider the equation $e^z = az^n$. Prove that:

    (1) if $|a| > e$, then the equation has n roots in D;

    (2) if $|a| < e^{-1}$, then the equation has no roots in D.

27*. Prove that the equation $z \sin z = 1$ has a countable set of roots in the complex plane and all these roots are real.
    Hint: find the number of real roots of the equation $x \sin x = 1$ in the segment $[-R, R]$ and compare with the number of roots of the given equation in the square $\overline{D} = \{x \in [-R, R], y \in [-R, R]\}$, where $R = \left(n + \frac{1}{2}\right)\pi$, $\forall n \in \mathbb{N}$.

Chapter 4
Conformal Mappings. Elementary Functions

In this chapter, we examine in detail the properties of elementary functions. We employ results discussed in the preceding parts of the book as well as some new concepts and theorems that are introduced in the first sections of this chapter. We begin with the geometric interpretation of derivative and the notion of conformal mapping. The fundamental principles of conformal mappings are formulated and briefly discussed, while their proofs are postponed to Chap. 5. Then, the third section is concerned with the description of symmetric points, which is another geometric property used in the analysis of elementary functions.

The remaining part of this chapter is devoted to an extensive study of the properties (including conformal ones) of a large number of analytic functions important both in theory and applications. The list of the functions contains linear and fractional linear functions, power and root functions, exponential and logarithmic functions, and four basic trigonometric functions. The analysis is focused on the three groups of the properties: analytical, geometric, and conformal. The first group includes the investigation of analyticity and domains of univalence, determination and classification of singularities, and finding the fixed points of a transformation. Among geometric features we consider the relations between nets of curves (in the first place, coordinate curves), transformations of standard domains, and relevant symmetries. Finally, the specification of the domains of conformality allows us to find inverse functions, including the cases of multi-valued functions, and construct the corresponding Riemann surfaces. Some elementary and also a bit more sophisticated functions are given to work out as exercises at the end of this chapter, but the majority of the essential elementary functions are analyzed in detail in the text. The theoretical analysis of each function is illustrated by solution of specific analytic and geometric problems.

A. Bourchtein and L. Bourchtein, *Complex Analysis*, Hindustan Publishing Corporation,
https://doi.org/10.1007/978-981-15-9219-5_4

4.1 Geometric Interpretation of the Absolute Value and Argument of Derivative. Concept of Conformal Mapping

Let $w = f(z)$ be an analytic function in a domain D and G be the image of D under $f(z)$: $G = f(D)$. (Later, we will show that G is also a domain.)

Definition 4.1 A function $f(z)$ is called *univalent (injective) in a domain* D, if for $\forall z_1, z_2 \in D,\ z_1 \neq z_2$ it follows that $f(z_1) \neq f(z_2)$, that is, different points of a domain D are sent to different points of the complex plane w that contains the image G. In this case, $f(z)$ is said to be *one-to-one correspondence* between D and G, or $f(z)$ is a *bijective function (mapping)*.

Definition 4.2 A function $f(z)$ is called *univalent at a point* $z_0 \in D$ if it is univalent in a neighborhood of this point.

Let us investigate the geometric meaning of the derivative of an analytic function. Suppose $f'(z_0) \neq 0$ at some point $z_0 \in D$. Trace a smooth curve γ in the domain D through the point z_0, that is, $\gamma:\ z = z(t),\ t \in [a, b]$ and the point z_0 corresponds to some value of the parameter t: $z_0 = z(t_0)$, $t_0 \in [a, b]$. By the definition of a smooth curve, there exists a tangent line continuously turning while moving along γ, or, in analytic terms, there exists the continuous function $z'(t)$ such that $z'(t) \neq 0,\ \forall t \in [a, b]$, in particular, $z'(t_0) \neq 0$. Denote by Γ the image of the curve γ under the mapping f: $\Gamma = f(\gamma),\ \Gamma:\ w(t) = f(z(t)),\ \forall t \in [a, b]$. Since $z(t)$ is a continuously differentiable function and $f(z)$ is an analytic function, by the properties of a composite function, the function $w(t)$ is also continuously differentiable on $[a, b]$ and $w'(t_0) = f'(z_0) \cdot z'(t_0) \neq 0$. Therefore, there exists a tangent line to Γ at the point $w_0 = w(t_0) = f(z(t_0)) = f(z_0)$. Hence, a smooth curve γ in the domain D is mapped by the analytic function $f(z)$ onto a piecewise smooth curve Γ in the plane w (this curve is smooth at the point $w_0 = f(z_0)$).

Consider again the equality

$$w'(t_0) = f'(z_0) \cdot z'(t_0)\,. \tag{4.1}$$

Denote by φ the (oriented) angle between the positive direction of the real axis in the z-plane and the tangent vector $z'(t_0)$, and by ψ the (oriented) angle between the positive direction of the real axis in the w-plane and the tangent vector $w'(t_0)$ (see Fig. 4.1). (Notice that we use the oriented angles, where the change of the order of the two vectors results in the change of the sign of the angle between them.)

Then, the equality (4.1) can be expressed in the form

$$\left|w'(t_0)\right| \cdot e^{i\psi} = \left|f'(z_0)\right| \cdot e^{i \arg f'(z_0)} \cdot \left|z'(t_0)\right| \cdot e^{i\varphi},$$

and consequently

$$\left|w'(t_0)\right| = \left|f'(z_0)\right| \cdot \left|z'(t_0)\right| \tag{4.2}$$

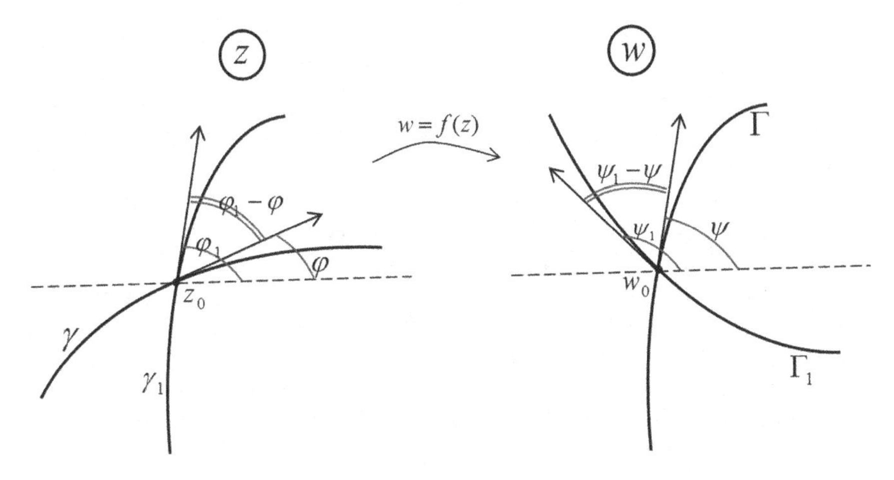

Fig. 4.1 Preserving the angles between the curves under a conformal mapping

and

$$\psi = \varphi + \arg f'(z_0). \tag{4.3}$$

Trace another smooth curve γ_1 in D passing through z_0 and denote by Γ_1 the image of γ_1 under the mapping f: $\Gamma_1 = f(\gamma_1)$. Denote by φ_1 the (oriented) angle between the positive direction of the real axis in the z-plane and the tangent vector to γ_1 at the point z_0, and by ψ_1 the (oriented) angle between the positive direction of the real axis in the w-plane and the tangent vector to the curve Γ_1 at the corresponding point $w_0 = f(z_0)$ (see Fig. 4.1). Just as before, we obtain the following relationship between the angles of the curves γ_1 and Γ_1:

$$\psi_1 = \varphi_1 + \arg f'(z_0). \tag{4.4}$$

It follows from (4.3) and (4.4) that

$$\psi_1 - \psi = \varphi_1 - \varphi. \tag{4.5}$$

Let us analyze the geometric meaning of the obtained relations. The equalities (4.3), (4.4), and (4.5) contain information regarding the angles. The equality (4.5) shows that the angle between the curves γ and γ_1 in the z-plane is equal to the angle between their images Γ and Γ_1 in the w-plane (recall that by definition the angle between two curves is the angle between the tangents to these curves at a chosen point). The equalities (4.3) and (4.4) state that the direction of the tangent vector at the image point w_0 is obtained by rotating the tangent vector at the original point z_0 by the angle $\arg f'(z_0)$. Therefore, we can make the following two conclusions on the *geometric interpretation of the argument of the non-zero derivative of an analytic function*:

(1) *the angles between curves are preserved under the mapping by an analytic function with non-zero derivative*;
(2) *the argument of the non-zero derivative of an analytic function at a point z_0 represents the angle of rotation of any vector emanating from z_0* under the f-mapping of the z-plane into the w-plane.

Consider now the Eq. (4.2), which is true for any smooth curve γ passing through the point z_0. It shows that the length of the tangent vector and also the length of any vector starting from z_0 is changed by the same factor under f-mapping of the z-plane into the w-plane, namely, the length of each such vector is amplified $\left|f'(z_0)\right|$ times. Hence, we arrive at the following *geometric interpretation of the absolute value of the non-zero derivative of an analytic function*:
the absolute value of the derivative of an analytic function at a point z_0 is a linear extension factor applied to any vector emanating from z_0 under its f-mapping from the z-plane into the w-plane.

Definition 4.3 We say that a *mapping is conformal of the first kind* at a chosen point if it preserves the oriented angle between any two arcs (that is, both the absolute value and direction of the angle are preserved) and keeps the same linear extension factor at a given point (that is, the extension factor does not depend on the direction).

Definition 4.4 We say that a mapping $w = f(z)$ is *conformal in a domain D*, if it is conformal at every point of D and $f(z)$ is univalent on D.

The provided above reasoning shows that an analytic at a point $z_0 \neq \infty$ function $w = f(z)$ with non-zero derivative $f'(z_0) \neq 0$ represents a conformal mapping of the first kind at the point z_0.

Another group of the mappings with similar properties is the class of conformal mappings of the second kind. The only difference between the two types of conformal mappings is that the latter changes the direction of the angles between curves. One example of such a mapping is provided below.

Example Consider the function $w = \bar{z}$. For convenience of illustration, let us represent the points z and w on the same plane. The mapping $w = \bar{z}$ is a symmetry with respect to the real axis: if $z = x + iy$, then $w = \bar{z} = x - iy$. Notice that the linear extension factor of this mapping is equal to 1 at every point of the plane (that is, the absolute values of the vectors do not change) and the angles preserve their absolute values, but change their orientation as shown in Fig. 4.2.

Definition 4.5 We say that a *mapping is conformal of the second kind* at a chosen point if it preserves the absolute value of the angles between any two arcs, but changes its direction, while keeping the same linear extension factor at a given point (that is, the extension factor does not depend on the direction).

Theorem *If $w = f(z) = \overline{g(z)}$, where $g(z)$ is an analytic function at a point z_0 and $g'(z_0) \neq 0$, then the function $f(z)$ is a conformal mapping of the second kind at z_0.*

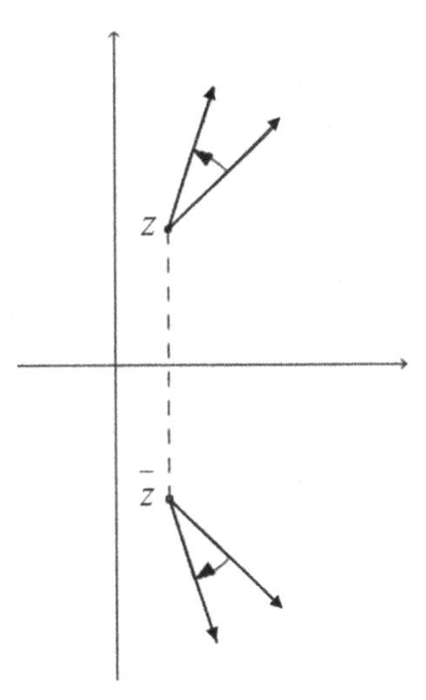

Fig. 4.2 Geometric representation of the transformation $w = \bar{z}$

Proof Since $g(z)$ is an analytic function with non-zero derivative at z_0, then $g(z)$ is a conformal mapping of the first kind at the same point. It means that under mapping $g(z)$ the length of any vector emanating from z_0 is extended by the same factor and the oriented angles between any two vectors are preserved. Denoting $\zeta = g(z)$, we have $w = \bar{\zeta}$. The example considered above shows that the passage from the ζ-plane to the w-plane consists geometrically in the inversion of the orientation of the angles between any two vectors, while keeping the absolute values of both the angles and the vectors. Therefore, the composition of the mappings $w = \bar{\zeta} = \overline{g(z)}$ results, by the definition, in a conformal mapping of the second type: it changes the absolute value of any vector starting from z_0 by the same factor and preserves the absolute values of angles between any two vectors, but inverts the angle direction. □

In what follows we will focus on the conformal mappings of the first kind and, for brevity, will call them simply conformal mappings.

The concept of conformal mapping can be extended to the case of the point $z_0 = \infty$ and also when $f(z_0) = \infty$. In the last case, we assume that $f(z)$ is analytic in a deleted neighborhood of z_0 and $\lim\limits_{z \to z_0} f(z) = \infty$, that is, z_0 is a pole of the function $f(z)$.

Definition 4.6 We say that a function $w = f(z)$ is a *conformal mapping at the point* $z_0 = \infty$, if the function $g(t) = f\left(\frac{1}{t}\right)$ is analytic at the point $t_0 = 0$ and $g'(0) \neq 0$, that is, the function $g(t)$ is a conformal mapping at $t_0 = 0$.

Definition 4.7 Let z_0 be a *pole* of a function $f(z)$. We say that the function $f(z)$ is *a conformal mapping at the point* z_0, if the function $g(z) = \frac{1}{f(z)}$ is analytic at the point z_0 and $g'(z_0) \neq 0$, that is, $g(z)$ is a conformal mapping at z_0.

Remark 4.1 Notice that in the last definition the function $g(z)$ has a simple zero at z_0, since $g(z_0) = 0$ and at the same time $g'(z_0) \neq 0$. Consequently, the function $f(z)$ has a simple pole at z_0.

Remark 4.2 The given definitions allow us to conclude that a transformation $f(z)$ is conformal in a domain D if the function $f(z)$ is analytic with non-zero derivative in D, except, possibly, at one simple pole, and $f(z)$ is univalent in D.

Hereinafter we will use the concept of a non-degenerate domain.

Definition 4.8 A domain D is called degenerate if $D = \overline{\mathbb{C}}$ or $D = \overline{\mathbb{C}} \setminus \{a\}$, where a is a finite or infinite point. Otherwise a domain is non-degenerate. It is seen that any non-degenerate domain is a domain with more than one boundary point.

The following statements are *the fundamental principles of conformal mappings.*

1. **The principle of domain preservation**. The image of any domain in $\overline{\mathbb{C}}$ under a mapping by an analytic non-constant function is a domain.

2. **The principle of boundary correspondence**. Suppose D and G are simply connected domains whose boundaries $\Gamma = \partial D$ and $\Upsilon = \partial G$ are piecewise smooth curves with folds, and suppose a function $w = f(z)$ is analytic in the domain D and continuous on D up to its boundary, except, possibly, at one simple pole contained in D. If under the mapping f one positive circuit along Γ induces one positive circuit along Υ, then $w = f(z)$ is a univalent (and bijective) mapping of D onto G.

3. **The criterion for local univalence**. If a function $f(z)$ is analytic at a point $z_0 \neq \infty$, then $f(z)$ is univalent at z_0 if and only if $f'(z_0) \neq 0$.

4. **The Riemann Mapping Theorem**. For an arbitrary simply connected non-degenerate domain D, there exists an analytic function $w = f(z)$ defined on D that maps D conformally and univalently onto the open unit disk $|w| < 1$.

If additionally $f(z_0) = w_0$ and $\arg f'(z_0) = \alpha$, where z_0 and w_0 are given points ($z_0 \in D$ and $|w_0| < 1$) and α is a given real number, then this function $w = f(z)$ is unique.

It follows immediately from the Riemann Mapping Theorem that *any simply connected non-degenerate domain can be mapped conformally and univalently on any other simply connected non-degenerate domain.* Indeed, if D and G are two simply connected non-degenerate domains, then both of them can be mapped onto the open unit disk: $w = f(z),\ f : D \to |w| < 1$ and $w = g(\zeta),\ g : G \to |w| < 1$. Since the function $g(\zeta)$ is analytic and univalent in G, the mapping $w = g(\zeta)$ is bijective. Therefore, there exists the inverse (analytic and univalent) mapping $\zeta = g^{-1}(w) = h(w)$ such that $h : \{|w| < 1\} \to G$. Hence, the composite function $\zeta = h(f(z))$ that maps the domain D onto G is analytic and univalent.

In other words, all simply connected domains with more than one boundary point are conformally equivalent, and the mapping between two such domains is uniquely defined if the correspondence between two chosen points and between the directions at these points is established.

To make the presented results more accessible and emphasize their importance, we systematically illustrate their application to the analysis of various complex functions carried out in the principal part of this chapter. The proofs of these statements are postponed to the next chapter.

4.2 Linear Function

A *linear function has the form* $w = f(z) = az + b$, $a \neq 0$. Consider some properties of this function.

1. Evidently, the function $w = f(z) = az + b$ is analytic in the entire (finite) complex plane $\mathbb{C}$ (its derivative is $f'(z) = a$). In the neighborhood of the point $z = \infty$ we can represent this functions in the form $f(z) = z\left(a + \frac{b}{z}\right)$, where $f_1(z) = a + \frac{b}{z}$ is analytic at $z = \infty$ and $f_1(\infty) = \lim\limits_{z\to\infty} f_1(z) = a \neq 0$. Therefore, by the criterion for multiplicity of poles (Theorem 3.3 in Sect. 3.7), the point ∞ is a simple pole of $f(z)$.

2. The function $w = f(z) = az + b$ establishes a one-to-one correspondence between the points of the z-plane and w-plane: the function $f(z)$ associates with each $z \in \mathbb{C}_z$ a single value $w = az + b$, and conversely, for any $w \in \mathbb{C}_w$ there exists the only $z = \frac{w-b}{a}$ $(a \neq 0)$ corresponding to this w. Since $f(z) = az + b$ is analytic and univalent on the entire plane $\mathbb{C}_z$, it maps conformally the plane $\mathbb{C}_z$ onto the plane $\mathbb{C}_w$. Since the singular point $z = \infty$ is a simple pole, the mapping is also conformal at $z = \infty$.

3. Let us find fixed points of this mapping. (Recall that z_0 is a fixed point of a function $w = f(z)$ if it is mapped to the same point—$w_0 = z_0$—that is, z_0 is a solution of the equation $f(z) = z$.) Notice first that $f(\infty) = \infty$, which means that ∞ is a fixed point of the linear function. Let us analyze the equation $az + b = z$ to check whether there are other fixed points. The following three options may occur

(1) if $a = 1, b = 0$, then $w \equiv z$, $\forall z \in \overline{\mathbb{C}}$, that is, all the points of the complex plane are fixed and the mapping is the identity;

(2) if $a = 1, b \neq 0$, then the equation $z + b = z$ has no solutions in $\mathbb{C}$, and consequently, the only fixed point is ∞;

(3) if $a \neq 1$, the only solution of the equation $az + b = z$ is $z = \frac{b}{1-a}$, which is a finite fixed point.

4. We analyze now what kind of mapping is performed by a linear function. To this end, we represent a in the exponential form $a = |a| \cdot e^{i \arg a} = |a| \cdot e^{i\alpha}$ and express a given function as a chain of simpler transformations:

(1) $z_1 = |a|\, z$;

(2) $z_2 = z_1 e^{i\alpha}$;

(3) $w = z_2 + b$.

Let us study each of these elementary transformations. For any point $z = |z| \cdot e^{i \arg z} = |z| \cdot e^{i\varphi}$ the first mapping gives $z_1 = |z_1| \cdot e^{i \arg z_1} = |a| \cdot z = |a| \cdot |z| \cdot e^{i\varphi}$,

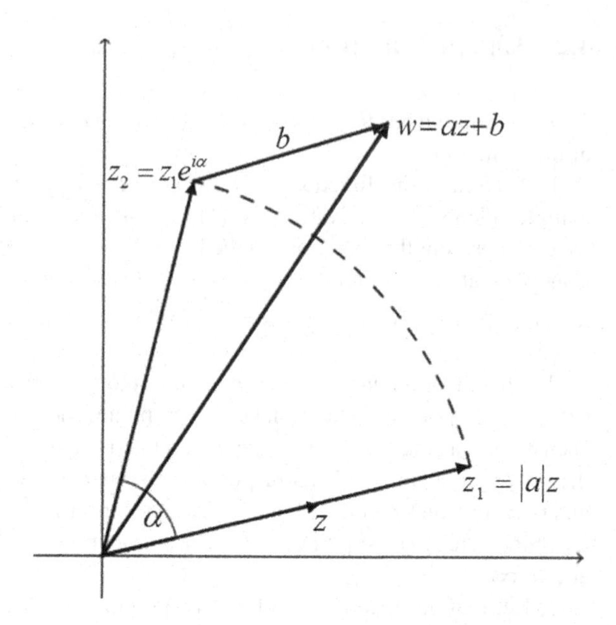

Fig. 4.3 Geometric meaning of linear function: extension, rotation, and translation

that is, the length of each vector increases $|a|$ times ($|z_1| = |a| \cdot |z|$), while the direction of each vector does not change since the argument is preserved ($\arg z_1 = \varphi = \arg z$) (see Fig. 4.3). For this reason, the mapping $z_1 = |a| \cdot z$ is called dilation or contraction. The second transformation $z_2 = |z_2| \cdot e^{i \arg z_2} = z_1 \cdot e^{i\alpha} = |z_1| \cdot e^{i(\arg z_1 + \alpha)}$ preserves the vector lengths ($|z_2| = |z_1|$) and rotates all the vectors by the angle $\arg a = \alpha$ (that is, $\arg z_2 = \arg z_1 + \alpha$). Accordingly, the mapping $z_2 = z_1 e^{i\alpha}$ is called rotation (see Fig. 4.3). The last transformation $w = z_2 + b$ is referred to as parallel translation (or simply translation), because it shifts all the vectors z_2 by the same vector b (see Fig. 4.3). Thus, a linear function consists of the three successive simpler transformations: dilation/contraction, rotation, and translation.

5. The principle of boundary correspondence suggests that a natural way to find the image of a chosen domain under a given mapping is by starting with the image of its boundary. In other words, it is convenient to determine first the images of different curves. Let us pick the simplest curves—straight lines and circles—and find their images under the transformation by a linear function. Recall that the standard general equation of a circle can be written as follows:

$$k\left(x^2 + y^2\right) + bx + cy + d = 0, \tag{4.6}$$

where k, b, c, d are real parameters. In the singular case $k = 0$, the equation of a circle turns into the equation of a straight line, that is, a straight line can be considered as a particular case of a circle with the infinite radius.

To make a connection with complex functions, we rewrite the Eq. (4.6) in the complex form by using the relations $z = x + iy$, $\bar{z} = x - iy$ and $x = \frac{1}{2}(z + \bar{z})$, $y = \frac{1}{2i}(z - \bar{z})$, $x^2 + y^2 = z \cdot \bar{z}$:

$$kz \cdot \bar{z} + \frac{b}{2}(z + \bar{z}) - \frac{ci}{2}(z - \bar{z}) + d = 0\,.$$

Denoting $A = k \in \mathbb{R}$, $D = d \in \mathbb{R}$, $B = \frac{1}{2}(b - ic)$, we have

$$A\,z \cdot \bar{z} + B\,z + \bar{B}\,\bar{z} + D = 0\,. \tag{4.7}$$

If $k = 0$ (and consequently $A = 0$), then we have the complex equation of a straight line:

$$Bz + \bar{B}\bar{z} + D = 0\,. \tag{4.8}$$

Using the same splitting of a linear function into the three transformations, we start with the first mapping—$z_1 = |a|\,z$—that leads to substitution of the relation $z = \frac{z_1}{|a|}$ in Eqs. (4.7) and (4.8):

$$\frac{A}{|a|^2}z_1 \cdot \bar{z}_1 + \frac{B}{|a|}z_1 + \frac{\bar{B}}{|a|}\bar{z}_1 + D = 0$$

and

$$\frac{B}{|a|}z_1 + \frac{\bar{B}}{|a|}\bar{z}_1 + D = 0\,.$$

Denoting $A_1 = \frac{A}{|a|^2} \in \mathbb{R}$, $D_1 = D \in \mathbb{R}$, and $B_1 = \frac{B}{|a|} \in \mathbb{C}$, we obtain the complex equation of a circle

$$A_1 z_1 \cdot \bar{z}_1 + B_1 z_1 + \bar{B}_1 \bar{z}_1 + D_1 = 0 \tag{4.9}$$

and the complex equation of a straight line

$$B_1 z_1 + \bar{B}_1 \bar{z}_1 + D_1 = 0 \tag{4.10}$$

in the z_1-plane. Notice that again the second equation is a particular case of first in the case $A_1 = 0$ that corresponds to $A = 0$. Hence, a circle in the z-plane is transformed into a circle in the z_1-plane and a straight line is transformed into a straight line. Looking at this property from geometric point of view it appears to be rather evident, since the first transformation results in dilation/contraction of any original figure.

Using a geometric approach, it is easy to conclude that the second and third transformations also map circles into circles (and lines into lines), since they represent, respectively, rotation and translation of any figure in the z_1-plane. We leave to the reader the algebraic verification of this property.

Thus, a linear function transforms any circle into another circle and any straight line into another straight line. This property is called the *circular property* of a linear function.

Consider an example of a linear function that performs the required conformal mapping.

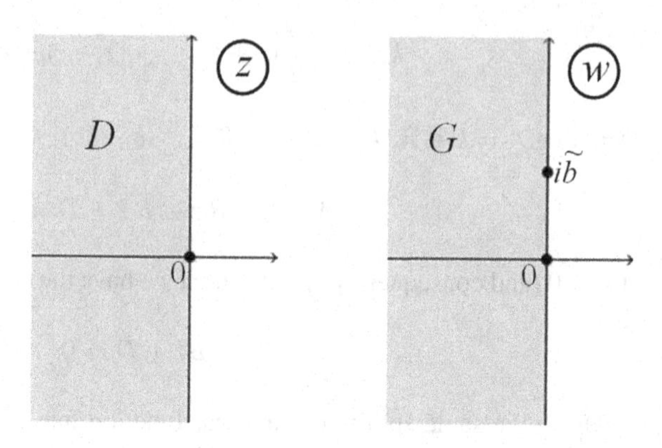

Fig. 4.4 The domains of Example

Example Find a general form of the linear function that transforms the left half-plane into itself.

In the definition of a linear function $w = f(z) = az + b$, we have to specify the coefficients a and b in such a way that this function maps the left half-plane onto itself. Using the principle of boundary correspondence, we note that the point $z = 0$ is carried into a point of the imaginary axis in the w-plane (see Fig. 4.4), that is, $f(0) = b = i\tilde{b}$, where $\tilde{b} \in \mathbb{R}$. At the same time, the point $w = 0$ corresponds to a point at the imaginary axis in the z-plane ($z = iy$): $a \cdot iy + i\tilde{b} = 0$, which implies that $a \in \mathbb{R}$.

On the boundary of the left half-plane D we have $iv = ayi + i\tilde{b}$, that is, $v = ay + \tilde{b}$. When the point z traverses the boundary of D in the positive direction, the corresponding point $w = f(z)$ traverses the boundary of G also in the positive direction. This means that v increases with respect to y, and consequently $a > 0$. Thus, a general form of the linear function that performs the required transformation is $f(z) = az + i\tilde{b}$, where $\tilde{b} \in \mathbb{R}$, $a > 0$.

4.3 Symmetric Points

Definition The *points* z_1 *and* z_2 *are called symmetric with respect to a circle* $\gamma = \{|z - a| = r\}$ if they belong to the same ray emanating from the center of the circle γ and the product of their distances to the point a (center of γ) is equal to the square of the radius.

Notice that this definition is a generalization of the symmetry of two points with respect to a straight line: the points z_1 and z_2 are symmetric with respect to a straight line l if they belong to the same straight line L perpendicular to l and their distances to the line l are equal.

Given a circle $\gamma = \{|z - a| = r\}$ and a point $z_1 \in \mathbb{C}$, let us find the formula of the point z_2 symmetric to z_1 with respect to γ. By the definition, z_1 and z_2 lie on the same ray starting from the point a, and consequently, the vectors $z_1 - a$ and $z_2 - a$ form the same angle with the positive direction of the x-axis:

$$\arg(z_1 - a) = \arg(z_2 - a)\,. \tag{4.11}$$

The second condition is about the product of the lengths of the vectors $z_1 - a$ and $z_2 - a$:

$$|z_1 - a| \cdot |z_2 - a| = r^2. \tag{4.12}$$

Employing (4.11) and (4.12), we get

$$z_2 - a = |z_2 - a| \cdot e^{i \arg(z_2 - a)} = \frac{r^2}{|z_1 - a|} \cdot e^{i \arg(z_1 - a)} = \frac{r^2}{|z_1 - a| \cdot e^{-i \arg(z_1 - a)}} = \frac{r^2}{\overline{z_1 - a}}\,.$$

Therefore,

$$z_2 = a + \frac{r^2}{\overline{z_1 - a}}, \tag{4.13}$$

which is the formula for symmetric point z_2.

There are two special cases included in the last formula:

(1) if $z_1 = a + re^{i\varphi}$, $\forall \varphi \in \mathbb{R}$, then $z_2 = a + \frac{r^2}{re^{-i\varphi}} = z_1$, or expressing it geometrically, if z_1 belongs to the circle γ, then the point z_2, symmetric to z_1 with respect to γ, coincides with z_1;

(2) if $z_1 \to a$, then $z_2 \to \infty$, that is, the points a (center of the circle) and ∞ are symmetric with respect to γ.
Notice that these two special symmetries also follow directly from the definition.

Let us specify what kind of transformation is defined by formula (4.13). Since the function $\frac{1}{z-a}$ is analytic in the entire complex plane, except at the point a, and $\left(\frac{1}{z-a}\right)' = -\frac{1}{(z-a)^2} \neq 0$, $\forall z \in \mathbb{C} \backslash \{a\}$, the Theorem of Sect. 4.1 states that the function $\zeta = a + \frac{r^2}{\overline{z-a}}$ is a conformal mapping of the second kind. Hence, the symmetry with respect to a circle is a conformal mapping of the second kind.

Now, we derive a simple criterion for the symmetry of points with respect to a circle. Recall first the following theorem from Geometry: given a circle L and a point A lying out of the disk enclosed by L, if we trace the two line segments from A—the secant to L passing through points B and C and the tangent to L at a point D—then the product of the lengths of the segments AB and AC is equal to the square of the length of the segment AD: $|AB| \cdot |AC| = |AD|^2$. Figure 4.5 shows that the last equality follows from the similarity of the triangles ACD and ABD, which is guaranteed by the common angle A and the equality between the angles BCD and BDA (both are equal to a half of the arc BD).

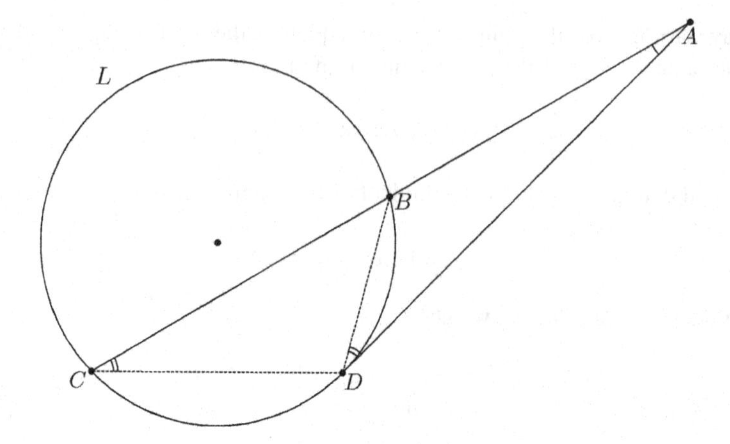

Fig. 4.5 Secant AC and tangent AD lines from the point A to the circle L

Theorem (Criterion for symmetry of points with respect to a circle) *The points z_1 and z_2 are symmetric with respect to a circle γ if and only if the set of circles, passing through the points z_1 and z_2, is orthogonal to γ.*

Proof *Necessity*. Suppose the points z_1 and z_2 are symmetric with respect to a circle γ. Consider the set of the circles $L = \{l\}$ passing through z_1 and z_2 and pick any circle l from this set L. This circle l intercepts γ in two points (since z_1 and z_2 are located on different sides of γ). Denote one of these points of intersection by z_0 (see Fig. 4.6). Since $z_0 \in \gamma$, then $|z_0 - a| = r$. By the definition of symmetric points (using the equality (4.12)), we obtain $|z_1 - a| \cdot |z_2 - a| = r^2 = |z_0 - a|^2$. At the same time, the points z_1 and z_2 belong to the same ray, that is, they lie on the same secant line emanating from the centerpoint a of γ and crossing the circle l. Then, from the above mentioned theorem of Geometry, it follows that the vector $z_0 - a$ is tangent to the circle l. Since $z_0 - a$ is a radius-vector of the circle γ, it is perpendicular to the tangent line of γ at the point z_0. Hence, the tangent lines to the circles l and γ are perpendicular at z_0 (the point of the intersection of the circles), that is, l and γ are orthogonal. Since l is an arbitrary circle of the set L, the proof of the first part is completed.

Sufficiency. Consider a set of the circles $L = \{l\}$ passing through the points z_1 and z_2 and orthogonal to the circle γ. Since any circle $l \in L$ is orthogonal to γ, the points z_1 and z_2 are located on different sides of the circle γ. In the set L, there exists the circle l_0 of the infinite radius, that is, the straight line that contains z_1 and z_2 and passes through the centerpoint of γ, since l_0 is orthogonal to γ. Therefore, the points z_1 and z_2 lie on the same ray emanating from the centerpoint of γ, but on different sides of γ. Any circle $l \in L, l \neq l_0$ intercepts γ in two points (since, by the condition, l is orthogonal to γ). Pick any of these circles $l \in L, l \neq l_0$ and denote one of the intersection points by z_0 (see Fig. 4.6). The orthogonality between l and γ means that their tangents at z_0 are perpendicular. Since the radius-vector and tangent

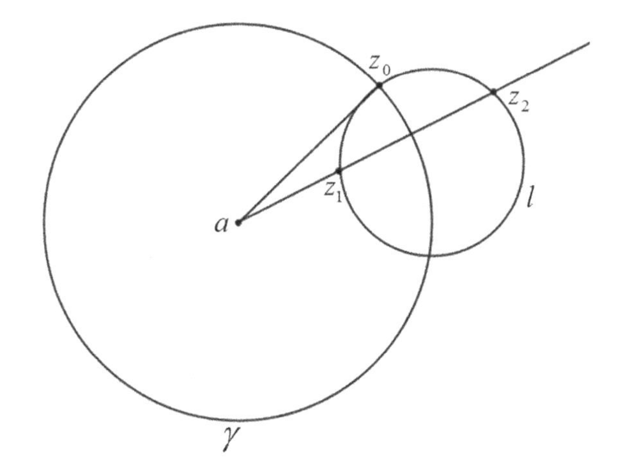

Fig. 4.6 Symmetry of the points z_1 and z_2 with respect to a circle γ

line of a circle are perpendicular at the contact point, we conclude that the vector $z_0 - a$ is tangent to the circle l. Therefore, by the theorem of Geometry, we have $|z_1 - a| \cdot |z_2 - a| = |z_0 - a|^2 = r^2$. Thus, the points z_1 and z_2 lie on the same ray emanating from the center of the circle γ and the product of their distances to the centerpoint of γ is equal to the square of the radius, which means that z_1 and z_2 are symmetric with respect to γ.

This terminates the proof of the Theorem. □

4.4 Function $w = \frac{1}{z}$

In this section, we study the *function* $w = \frac{1}{z}$.

1. The function $w = \frac{1}{z}$ is analytic in the entire (finite) complex plane $\mathbb{C}$ except at the point $z = 0$. Since $\lim\limits_{z\to\infty} f(z) = \lim\limits_{z\to\infty} \frac{1}{z} = 0$, the point $z = \infty$ is a removable singularity. If we define $f(\infty) = \lim\limits_{z\to\infty} f(z) = 0$, then the function $f(z) = \frac{1}{z}$ becomes analytic at ∞. To reveal a behavior of $f(z) = \frac{1}{z}$ at the point $z = 0$, first notice that the origin is a single-valued isolated singular point. Calculating the limit $\lim\limits_{z\to 0} f(z) = \lim\limits_{z\to 0} \frac{1}{z} = \infty$, we see that $z = 0$ is a pole, and representing the function in the form $f(z) = \frac{1}{z} = z^{-1} \cdot 1$, we deduce that $z = 0$ is a simple pole.

2. If we define the function $\frac{1}{z}$ at the point $z = 0$ as $f(0) = \infty$, then $w = \frac{1}{z}$ establishes a one-to-one correspondence between the extended complex planes $\overline{\mathbb{C}}_z$ and $\overline{\mathbb{C}}_w$: with every point $z \in \overline{\mathbb{C}}_z$ the function associates a single point $w = \frac{1}{z}$ and the converse is also true. Thus, the function $w = \frac{1}{z}$ is a conformal mapping from the extended complex plane $\overline{\mathbb{C}}_z$ onto the extended complex plane $\overline{\mathbb{C}}_w$.

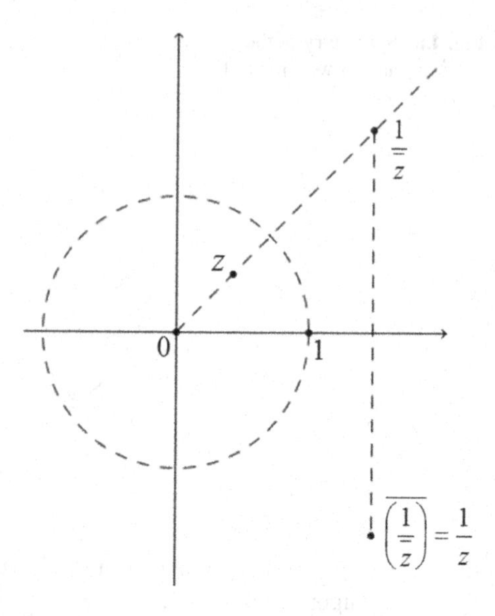

Fig. 4.7 The successive symmetries of $w = \frac{1}{z}$: with respect to the unit circle and with respect to the real axis

3. The fixed points of the transformation $w = \frac{1}{z}$ are found by solving the equation $\frac{1}{z} = z$, or equivalently, $z^2 = 1$ whose roots are $z = 1$ and $z = -1$.

4. To analyze the geometric transformations made by the function $w = \frac{1}{z}$, we represent this function through the two already known functions: $\zeta = \frac{1}{\bar{z}}$ and $w = \bar{\zeta}$. The former is the symmetry with respect to the circle $|z| = 1$ (see formula (4.13) in Sect. 4.3), while the latter is the symmetry with respect to the real axis. Therefore, the original function $w = \overline{\left(\frac{1}{\bar{z}}\right)} = \frac{1}{z}$ is the successive application of the two symmetries (see Fig. 4.7).

Let us find the images of the circles under the transformation $w = \frac{1}{z}$. In Sect. 4.2, we have introduced the equation of a circle in the complex form:

$$Az \cdot \bar{z} + Bz + \bar{B}\bar{z} + D = 0, \tag{4.14}$$

where A and D are real numbers. In the singular case $A = 0$, the Eq. (4.14) represents a straight line. Substituting $z = \frac{1}{w}$ in (4.14), we obtain

$$A \cdot \frac{1}{w} \cdot \frac{1}{\bar{w}} + \frac{B}{w} + \frac{\bar{B}}{\bar{w}} + D = 0$$

or

$$A + B\bar{w} + \bar{B}w + Dw\bar{w} = 0\,. \tag{4.15}$$

The last relation is the equation of a circle in the w-plane. Therefore, under the function $w = \frac{1}{z}$, a circle in the z-plane is transformed into a circle in the w-plane, that is, $w = \frac{1}{z}$ has a *circular property* (notice that the notion of a circle includes a straight line as a particular case).

Consider some specific cases.

(1) If we choose in the z-plane a straight line that passes through the origin, that is, $A = 0$ and $D = 0$ in (4.14), then in the w-plane we obtain the straight line $B\bar{w} + \bar{B}w = 0$ that also passes through the origin. Therefore, any straight line passing through the origin is mapped onto a straight line passing through the origin under the transformation $w = \frac{1}{z}$.
(2) Taking a straight line that does not pass through the origin in the z-plane ($A = 0$, $D \neq 0$ in (4.14)), we obtain a circle that passes through the origin in the w-plane.
(3) The image of a circle passing through the origin in the z-plane (it corresponds to $A \neq 0$, $D = 0$ in (4.14)) is a straight line that does not contain the origin in the w-plane.
(4) Taking a circle that does not pass through the origin in the z-plane ($A \neq 0, D \neq 0$ in (4.14)), we obtain a circle that also does not pass through the origin in the w-plane.

The following example illustrates the application of the circular property of $w = \frac{1}{z}$ to a family of circles.

Example Find the images of the family of circles $x^2 + y^2 = by$, $\forall b \in \mathbb{R}$ under the transformation $w = \frac{1}{z}$.

Using the algebraic forms of z and w—$z = x + iy$, $w = u + iv$—we can rewrite the function $w = \frac{1}{z}$ in the form

$$u + iv = \frac{1}{x + iy} = \frac{x - iy}{x^2 + y^2},$$

or

$$\begin{cases} u = \dfrac{x}{x^2 + y^2} \\ v = -\dfrac{y}{x^2 + y^2} \end{cases}. \tag{4.16}$$

Considering the family of the circles $x^2 + y^2 = by$, $b \neq 0$ (in the canonical form $x^2 + \left(y - \frac{b}{2}\right)^2 = \frac{b^2}{4}$), each of which has the centerpoint located on the imaginary axis, the radius $\frac{b}{2}$ and is tangent to the real axis at the origin (see Fig. 4.8), the analysis of (4.16) reveals that $v = -1/b$ is a constant, while $u = \frac{x}{by}$ takes all the real values when $x \in [-\frac{b}{2}, \frac{b}{2}]$, that is, in the w-plane we have a family of the straight lines parallel to the real axis (but not the real axis itself, since $v \neq 0$).

In the singular case $b = 0$, the circle $x^2 + y^2 = by$ shrinks to the point $z = 0$, which the transformation $w = \frac{1}{z}$ carries to ∞. On the other hand, there is no circle of the considered family that is mapped to the real axis, since $v = -\frac{y}{x^2+y^2} \neq 0$ if $y \neq 0$. Thus, in the w-plane, we obtain the family of the straight lines parallel to the real axis, except for the real axis itself.

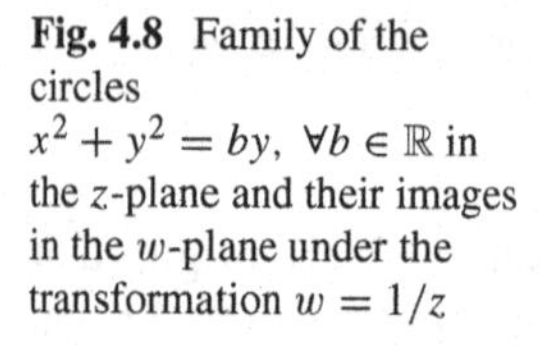

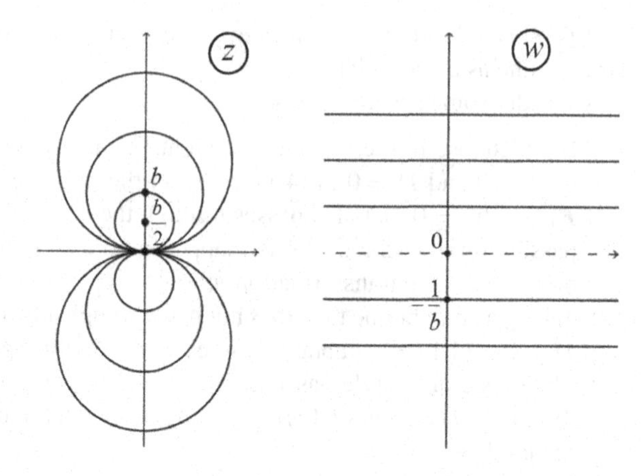

Fig. 4.8 Family of the circles $x^2 + y^2 = by,\ \forall b \in \mathbb{R}$ in the z-plane and their images in the w-plane under the transformation $w = 1/z$

4.5 Fractional Linear Transformation

The *fractional linear transformation* (also known as *bilinear or homographic or Möbius transformation*) has the form

$$w = f(z) = \frac{az + b}{cz + d}, \tag{4.17}$$

where a, b, c, d are complex numbers such that $ad - bc \neq 0$. Notice that if $ad - bc = 0$, that is, $\frac{a}{c} = \frac{b}{d} = k$, then $w = \frac{az+b}{cz+d} = \frac{ckz+dk}{cz+d} = k,\ \forall z \in \mathbb{C}$ and the function is reduced to a constant $w \equiv k$. In what follows we will disregard this trivial case, considering that $ad - bc \neq 0$. Also we usually suppose that $c \neq 0$, since otherwise the given function is simplified to the form of a linear function already discussed in Sect. 4.2. Let us study the properties of the fractional linear transformation.

1. The function (4.17) is analytic in the entire (finite) complex plane except at the point $z = -\frac{d}{c}$ (when the denominator is zero). The two singular points $z = -\frac{d}{c}$ and $z = \infty$ are single-valued and isolated. Let us see what kind of singularity the function has at these two points. Since $\lim\limits_{z\to\infty} f(z) = \lim\limits_{z\to\infty} \frac{az+b}{cz+d} = \frac{a}{c}$, the point ∞ is a removable singularity and the function can be defined at ∞ using the limit: $f(\infty) = \lim\limits_{z\to\infty} f(z) = \frac{a}{c}$. At the point $z = -\frac{d}{c}$ we have $\lim\limits_{z\to -\frac{d}{c}} f(z) = \infty$, which means that this point is a pole. Applying the Theorem 3.5 in Sect. 3.7 on the ratio of two analytic functions, we see that the point $z = -\frac{d}{c}$ is a simple pole. Therefore, the given function is analytic in the domain $\overline{\mathbb{C}} \setminus \left\{-\frac{d}{c}\right\}$ and the point $z = -\frac{d}{c}$ is its simple pole.

2. To investigate whether the function (4.17) is univalent, we solve the last relation with respect to z and obtain $z = \frac{-dw+b}{cw-a} = \frac{a_1 w+b_1}{c_1 w+d_1}$, where $a_1 d_1 - b_1 c_1 = (-d)(-a) - bc = ad - bc \neq 0$, that is, the inverse function is also fractional linear and non-degenerate (since $a_1 d_1 - b_1 c_1 \neq 0$). Therefore, we deduce that a single

value $w=\frac{az+b}{cz+d}$ is associated with each complex number $z\in\overline{\mathbb{C}}_z$ and the converse is also true—a single point $z=\frac{-dw+b}{cw-a}$ corresponds to each complex number $w\in\overline{\mathbb{C}}_w$. Hence, the fractional linear transformation provides a one-to-one correspondence between the extended z- and w-planes. Since, additionally, this function is analytic in $\overline{\mathbb{C}}\setminus\left\{-\frac{d}{c}\right\}$ and the point $z=-\frac{d}{c}$ is its only simple pole, the fractional linear transformation maps conformally the extended complex plane $\overline{\mathbb{C}}_z$ onto the extended complex plane $\overline{\mathbb{C}}_w$.

It can be shown that the converse statement is also true: if a function $w=f(z)$ is analytic on the entire extended z-plane, except for one singular point z_0 (finite or infinite), and univalent on the same extended plane, then this function is fractional linear. The proof can be made as follows. Notice first that z_0 is a single-valued isolated singular point of $f(z)$, because the function is analytic in a deleted neighborhood of z_0. This singularity is not removable, since otherwise it would exist a finite limit $\lim\limits_{z\to z_0}f(z)=A$ and, defining $f(z_0)=A$, the resulting function $f(z)$ would be analytic in the entire extended complex z-plane. Then, Corollary 3.2 to Liouville's Theorem (Sect. 3.6) implies that $f(z)$ would be a constant on the extended z-plane that contradicts the condition of univalence of $f(z)$. The point z_0 cannot be also an essential point, since the Sokhotski–Weierstrass Theorem (Sect. 3.8) states that $f(z)$ is not univalent function in any deleted neighborhood of an essential point, which contradicts the univalence of $f(z)$. Therefore, z_0 is a pole of $f(z)$ and the condition of univalence requires that this pole should be simple (see Remark 4.1 to Definition 4.7 in Sect. 4.1). Then, the principal part of the Laurent series of $f(z)$ contains only one term. For $z_0\neq\infty$ the principal part has the form $\frac{A}{z-z_0}$, $A\neq 0$, while for $z_0=\infty$ the principal part is Az, $A\neq 0$. Consider an auxiliary function $g(z)=f(z)-\frac{A}{z-z_0}$ if $z_0\neq\infty$ or $g(z)=f(z)-Az$ if $z_0=\infty$. This function is analytic on the extended complex z-plane, and consequently (by Corollary 3.2 to Liouville's Theorem in Sect. 3.6) $g(z)=B$, $B=constant$. Hence, $f(z)=\frac{A}{z-z_0}+B$ if $z_0\neq\infty$ or $f(z)=Az+B$ if $z_0=\infty$, that is, $f(z)$ is a fractional linear function if z_0 is a finite point or a linear function (which is a particular case of a fractional linear function) if z_0 is an infinite point.

3. Next, we find the fixed points of the fractional linear transformation by solving the equation $\frac{az+b}{cz+d}=z$ or, equivalently, $cz^2+(d-a)\,z-b=0$. Since $c\neq 0$, we have

$$z_{1,2}=\frac{(a-d)\pm\sqrt{(a-d)^2+4bc}}{2c}\,. \tag{4.18}$$

It may occur two situations:

(1) if $(a-d)^2+4bc\neq 0$, then $f(z)$ has two different fixed points z_1 and z_2;

(2) if $(a-d)^2+4bc=0$, then $z_1=z_2$ and $f(z)$ has the only (double) fixed point $z_1=z_2=\frac{a-d}{2c}$.

4. Consider now a composition of fractional linear transformations $\zeta = \frac{az+b}{cz+d}, ad - bc \neq 0$ and $w = \frac{a_1\zeta+b_1}{c_1\zeta+d_1}, a_1d_1 - b_1c_1 \neq 0$. The composed function can be written in the form

$$w(z) = w(\zeta(z)) = \frac{a_1\frac{az+b}{cz+d} + b_1}{c_1\frac{az+b}{cz+d} + d_1} = \frac{a_1az + a_1b + b_1cz + b_1d}{ac_1z + bc_1 + cd_1z + dd_1}$$

$$= \frac{(a_1a + b_1c)z + (a_1b + b_1d)}{(ac_1 + cd_1)z + (bc_1 + dd_1)} = \frac{a_2z + b_2}{c_2z + d_2},$$

where

$$a_2d_2 - b_2c_2 = (aa_1 + b_1c)(bc_1 + dd_1) - (a_1b + b_1d)(ac_1 + cd_1)$$

$$= aa_1bc_1 + bb_1cc_1 + aa_1dd_1 + b_1cdd_1 - aa_1bc_1 - ab_1c_1d - a_1bcd_1 - b_1cdd_1$$

$$= (ad - bc)(a_1d_1 - b_1c_1) \neq 0.$$

Hence, a composition of fractional linear transformations is again a fractional linear transformation. Recall that it was already shown that the inverse function to (4.17) is also a fractional linear function $z = \frac{-dw+b}{cw-a}$.

5. To evaluate the geometric properties of the function (4.17), we assume that $c \neq 0$ (the function is not reduced to a linear function) and express the function formula as follows:

$$w = \frac{az+b}{cz+d} = \frac{1}{c} \cdot \frac{acz + bc + ad - ad}{cz+d} = \frac{1}{c}\left(\frac{a(cz+d)}{cz+d} + \frac{bc-ad}{cz+d}\right) = \frac{a}{c} + \frac{1}{c} \cdot \frac{bc-ad}{cz+d}.$$

Denoting $\frac{1}{c}(bc - ad) = a_1$ and $\frac{a}{c} = b_1$, we have

$$w = \frac{az+b}{cz+d} = \frac{a_1}{cz+d} + b_1. \tag{4.19}$$

The last expression shows that a fractional linear transformation can be represented as the composition of the three functions studied previously:

$$z_1 = cz + d, \quad z_2 = \frac{1}{z_1}, \quad w = a_1z_2 + b_1.$$

The first and third functions are linear acting as dilation/contraction, rotation and translation, and the second function is a conformal mapping that transforms circles into other circles. Since each of these three transformations possesses a circular property, a fractional linear transformation also has this property, that is, it maps any circle (including a straight line) onto another circle (or a straight line).

6. The next geometric feature is the invariance of symmetric points under a fractional linear transformation.

Theorem *If the points z_1 and z_2 are symmetric with respect to a circle γ, then their images w_1 and w_2 under a fractional linear transformation $w = \frac{az+b}{cz+d}$ are symmetric with respect to the circle Γ which is the image of γ under $w = \frac{az+b}{cz+d}$.*

Proof It was already shown that a fractional linear transformation possesses the circular property, and consequently, the image of a circle γ is another circle Γ. Trace a set of circles $\{l\}$ passing through the points z_1 and z_2. By the criterion for symmetry of the points with respect to a circle, this set is orthogonal to γ (see Theorem in Sect. 4.3). The image $L = f(l)$ of each circle l is again a circle, according to the circular property. Therefore, the image of the set $\{l\}$ is the set of circles $\{L\}$ in the w-plane passing through the points w_1 and w_2: $w_1, w_2 \in L, \forall L \in \{L\}$. Since each l is orthogonal to γ and a fractional linear transformation is conformal (implying that angles are preserved), we deduce that each L is orthogonal to Γ. Hence, the set of circles $\{L\}$ passes through the points w_1 and w_2, and is orthogonal to Γ. Applying one more time the criterion for symmetry of points (Sect. 4.3), we conclude that w_1 and w_2 are symmetric with respect to the circle Γ. □

7. Invariant of a fractional linear transformation.

Let us show that any fractional linear transformation (4.17) satisfies the property

$$\frac{w-w_1}{w-w_2} : \frac{w_3-w_1}{w_3-w_2} = \frac{z-z_1}{z-z_2} : \frac{z_3-z_1}{z_3-z_2}, \tag{4.20}$$

where z, z_1, z_2, z_3 are four arbitrary different points in the z-plane and w, w_1, w_2, w_3 are the corresponding (different) images of these points in the w-plane. The expression $\frac{z-z_1}{z-z_2} : \frac{z_3-z_1}{z_3-z_2}$ is called the cross ratio of the four points z, z_1, z_2, z_3, and the equality (4.20) states that the cross ratio of the four points is invariant under a fractional linear transformation. For now we assume that all the eight points involved are different from ∞, but later we will withdraw this restriction.

The desired property is easily proved by substituting the expressions

$$w = \frac{az+b}{cz+d}, \quad w_1 = \frac{az_1+b}{cz_1+d}, \quad w_2 = \frac{az_2+b}{cz_2+d}, \quad w_3 = \frac{az_3+b}{cz_3+d}$$

(the images of the points z, z_1, z_2, z_3 under a fractional linear function $w = \frac{az+b}{cz+d}$, $ad - bc \neq 0$) in the left-hand side of (4.20) and performing simplifications:

$$\frac{w-w_1}{w-w_2} : \frac{w_3-w_1}{w_3-w_2} = \frac{\frac{az+b}{cz+d} - \frac{az_1+b}{cz_1+d}}{\frac{az+b}{cz+d} - \frac{az_2+b}{cz_2+d}} : \frac{\frac{az_3+b}{cz_3+d} - \frac{az_1+b}{cz_1+d}}{\frac{az_3+b}{cz_3+d} - \frac{az_2+b}{cz_2+d}}$$

$$= \frac{(aczz_1 + bcz_1 + adz + bd - aczz_1 - adz_1 - bcz - bd)\,(cz+d)\,(cz_2+d)}{(aczz_2 + bcz_2 + adz + bd - aczz_2 - adz_2 - bcz - bd)\,(cz+d)\,(cz_1+d)}$$

$$: \frac{(acz_1z_3 + bcz_1 + adz_3 + bd - acz_1z_3 - adz_1 - bcz_3 - bd)\,(cz_3+d)\,(cz_2+d)}{(acz_2z_3 + bcz_2 + adz_3 + bd - acz_2z_3 - adz_2 - bcz_3 - bd)\,(cz_3+d)\,(cz_1+d)}$$

$$= \frac{(ad - bc)(z - z_1)(cz_2 + d)}{(ad - bc)(z - z_2)(cz_1 + d)} : \frac{(ad - bc)(z_3 - z_1)(cz_2 + d)}{(ad - bc)(z_3 - z_2)(cz_1 + d)} = \frac{z - z_1}{z - z_2} : \frac{z_3 - z_1}{z_3 - z_2}.$$

Hence, a fractional linear transformation satisfies the invariant property (4.20).

Let us see what happens if one of the points z, z_1, z_2, z_3 or w, w_1, w_2, w_3 is infinity, for instance, let $z_2 = \infty$. We consider the right-hand side of (4.20) and take the limit as $z_2 \to \infty$:

$$\lim_{z_2 \to \infty} \frac{z - z_1}{z - z_2} : \frac{z_3 - z_1}{z_3 - z_2} = \lim_{z_2 \to \infty} \frac{z - z_1}{z_3 - z_1} : \frac{z - z_2}{z_3 - z_2} = \lim_{z_2 \to \infty} \frac{z - z_1}{z_3 - z_1} : \frac{z_2 \left(\frac{z}{z_2} - 1 \right)}{z_2 \left(\frac{z_3}{z_2} - 1 \right)}$$

$$= \frac{z - z_1}{z_3 - z_1} : \frac{(-1)}{(-1)} = \frac{z - z_1}{z_3 - z_1} : \frac{1}{1} = \frac{z - z_1}{1} : \frac{z_3 - z_1}{1}.$$

Therefore, if one of the points z, z_1, z_2, z_3 is infinity, then the differences containing this point in (4.20) should be changed to 1. The same is true if one of the points w, w_1, w_2, w_3 is infinity.

Another application of the formula (4.20) is obtained if we consider the point z be a variable, and consequently, w be the function value determined by (4.20). In this case, it is easily seen that the functional relation (4.20) between z and w is identical to (4.17) in the definition of a fractional linear transformation, and rewriting (4.20) in the form $w = \frac{az+b}{cz+d}$ we immediately represent the coefficients a, b, c, d in the terms of the six remaining points z_1, z_2, z_3 and w_1, w_2, w_3. Therefore, there exists exactly one fractional linear transformation that carries three different arbitrarily given points z_1, z_2, z_3 into three different preassigned images w_1, w_2, w_3.

8. The next problem is to find a fractional linear transformation that maps the upper half-plane $\operatorname{Im} z > 0$ onto the (open) unit disk in such a way that a given point α of the upper half-plane is carried to the origin $w = 0$ (the centerpoint of the disk).

The existence of a conformal mapping with these properties is guaranteed by the Riemann Mapping Theorem. The principle of boundary correspondence indicates that we should start construction of such transformation by finding the relationship between the boundaries of the given domains. The boundary of the upper half-plane is the real axis (straight line) and that of the unit disk is the unit circle. Therefore, the straight line should be mapped onto the circle. Since fractional linear transformations possess the circular property, we may expect that required transformation can be found among fractional linear functions (4.17). For convenience, we rewrite (4.17) in a slightly modified form:

$$w = \frac{az + b}{cz + d} = \frac{a}{c} \cdot \frac{z + \frac{b}{a}}{z + \frac{d}{c}}. \tag{4.21}$$

Since the transformation we are looking for carries the point $z = \alpha$ into $w = 0$, we have $\alpha + \frac{b}{a} = 0$, that is, $\frac{b}{a} = -\alpha$. The point $\bar{\alpha}$ is symmetric to the point α with respect to the real axis (a singular case of a circle), and the point ∞ is symmetric to 0 with respect the unit circle $|w| = 1$. Since a fractional linear transformation preserves the symmetry of two points, we deduce that the point $z = \bar{\alpha}$ should be mapped to the point ∞, that is, $\bar{\alpha} + \frac{d}{c} = 0$ and consequently $\frac{d}{c} = -\bar{\alpha}$. Bringing these values in (4.21), we obtain

$$w = \frac{a}{c} \cdot \frac{z - \alpha}{z - \bar{\alpha}} . \tag{4.22}$$

Since the real axis is mapped onto the unit circle centered at the origin, for any $z = x$ we have $|w| = 1$, and then (4.22) gives

$$1 = |w|_{z=x} = \left|\frac{a}{c}\right| \cdot \left|\frac{x - \alpha}{x - \bar{\alpha}}\right| = \left|\frac{a}{c}\right| \cdot \left|\frac{x - \alpha}{\overline{x - \alpha}}\right| = \left|\frac{a}{c}\right| .$$

(In the last formula, we have used the fact that x is a real number and consequently $x = \bar{x}$ and $x - \bar{\alpha} = \overline{x - \alpha}$.) From the relation $\left|\frac{a}{c}\right| = 1$, it follows that $\frac{a}{c} = e^{i\gamma}$, $\forall \gamma \in \mathbb{R}$ and (4.22) can be specified as follows:

$$w = e^{i\gamma} \frac{z - \alpha}{z - \bar{\alpha}} .$$

The last function is the required fractional linear transformation.

9. Consider one more problem: find a transformation that maps the unit disk $|z| < 1$ in the z-plane onto the unit disk $|w| < 1$ in the w-plane in such a way that a given point α in the unit disk $|z| < 1$ is sent to the center of the unit disk $|w| < 1$.

The Riemann Mapping Theorem guarantees the existence of such a conformal mapping. According to the principle of boundary correspondence, we start with the relationship between the two unit circles—$|z| = 1$ and $|w| = 1$. Since fractional linear transformations have the circular property, we may try to find the required mapping among them. First, we rewrite a fractional linear function (4.17) in the same way as in the preceding problem:

$$w = \frac{az + b}{cz + d} = \frac{a}{c} \cdot \frac{z + \frac{b}{a}}{z + \frac{d}{c}} . \tag{4.23}$$

Since the point $z = \alpha$ is mapped to the point $w = 0$, we have $\alpha + \frac{b}{a} = 0$, that is, $\frac{b}{a} = -\alpha$. The point $\frac{1}{\bar{\alpha}}$ is symmetric to α with respect to the circle $|z| = 1$, and the point ∞ is symmetric to 0 with respect to the circle $|w| = 1$. Then, the point $z = \frac{1}{\bar{\alpha}}$ should be mapped to $w = \infty$, since fractional linear mappings preserve symmetry between two points. Therefore, $\frac{1}{\bar{\alpha}} + \frac{d}{c} = 0$, that is, $\frac{d}{c} = -\frac{1}{\bar{\alpha}}$. Bringing the specified values of $\frac{b}{a}$ and $\frac{d}{c}$ in (4.23), we get

$$w = \frac{a}{c} \cdot \frac{z + \frac{b}{a}}{z + \frac{d}{c}} = \frac{a}{c} \cdot \frac{z - \alpha}{z - \frac{1}{\bar{\alpha}}} = -\frac{a\bar{\alpha}}{c} \cdot \frac{z - \alpha}{1 - \bar{\alpha}z} . \tag{4.24}$$

Since the circle $|z| = 1$ is mapped onto the circle $|w| = 1$, we deduce that $|w| = 1$ for any point $z = e^{i\varphi}$, $\forall\varphi \in \mathbb{R}$. Substituting $z = e^{i\varphi}$ in (4.24), we obtain

$$1 = |w|_{z=e^{i\varphi}} = \left|-\frac{a\bar{\alpha}}{c}\right| \cdot \left|\frac{e^{i\varphi} - \alpha}{1 - \bar{\alpha}e^{i\varphi}}\right| = \left|-\frac{a\bar{\alpha}}{c}\right| \cdot \frac{1}{|e^{i\varphi}|} \cdot \left|\frac{e^{i\varphi} - \alpha}{e^{-i\varphi} - \bar{\alpha}}\right| = \left|-\frac{a\bar{\alpha}}{c}\right|$$

(notice that $e^{-i\varphi} - \bar{\alpha} = \overline{e^{i\varphi} - \alpha}$ and $\left|e^{i\varphi} - \alpha\right| = \left|e^{-i\varphi} - \bar{\alpha}\right|$, $\forall\varphi \in \mathbb{R}$). The condition $\left|-\frac{a\bar{\alpha}}{c}\right| = 1$ means that $-\frac{a\bar{\alpha}}{c} = e^{i\gamma}$, $\forall\gamma \in \mathbb{R}$, and, therefore, the required function takes the form of the fractional linear transformation:

$$w = e^{i\gamma}\frac{z - \alpha}{1 - \bar{\alpha}z}\,.$$

Consider some examples, which illustrate the application of different properties of a fractional linear function.

Example 4.1 Find the image of the domain $\begin{cases} |z| < 2 \\ \operatorname{Im} z < 0 \end{cases}$ under the transformation $w = \frac{iz+1}{z-4i}$.

The given domain D is the lower half-disk centered at the origin and of the radius 2, which is shown in Fig. 4.9.

To find the image of D under the transformation $w = f(z) = \frac{iz+1}{z-4i}$ one could write down the parametric equations of the curves, which compose the boundary of D, and analyze their images. However, we will employ another method of solution.

The function $w = f(z) = \frac{iz+1}{z-4i}$ is a fractional linear transformation, which carries any circle in the z-plane to a circle in the w-plane (recall that a straight line is considered to be a circle of an infinite radius). Since the boundary ∂D consists of the parts of the two circles, $f(z)$ carries them to parts of some circles. To construct each of these two parts, it is sufficient to find three points for each of these two circles. Consider first the interval $[-2,\ 2]$ of the real axis in the z-plane (the first part of ∂D) and choose the three points -2, 0, and 2 in this interval. The corresponding three points in the w-plane are the following:

$$w_1 = f(-2) = \frac{-2i+1}{-2-4i} = -\frac{1}{2} \cdot \frac{1-2i}{1+2i} = -\frac{1}{2} \cdot \frac{1-4i-4}{5} = \frac{3}{10} + \frac{4}{10}i,$$

$$w_2 = f(2) = \frac{2i+1}{2-4i} = \frac{1}{2} \cdot \frac{1+2i}{1-2i} = \frac{1}{2} \cdot \frac{1+4i-4}{5} = -\frac{3}{10} + \frac{4}{10}i,$$

$$w_3 = f(0) = \frac{1}{-4i} = \frac{1}{4}i\,.$$

Using these three points in the w-plane, we can specify the arc of the circle which is the image of the interval $[-2,\ 2]$ (Fig. 4.10). Indeed, since w_1 and w_2 are symmetric with respect to the imaginary axis, it is clear that the centerpoint of the required circle lies on this axis. Denoting this centerpoint by $w_0' = iv_0'$, we have

$$\left|w_1 - w_0'\right| = \left|w_2 - w_0'\right| = \left|w_3 - w_0'\right| = r_1,$$

from which it follows that

$$\left|\frac{1}{4} - v_0'\right| = \sqrt{\frac{9}{100} + \left(\frac{4}{10} - v_0'\right)^2},$$

and consequently

$$\frac{1}{16} - \frac{1}{2}v_0' + v_0'^2 = \frac{25}{100} - \frac{4}{5}v_0' + v_0'^2 \Rightarrow \frac{3}{10}v_0' = \frac{3}{16} \Rightarrow v_0' = \frac{5}{8}; \; r_1 = \left|\frac{1}{4} - \frac{5}{8}\right| = \frac{3}{8}.$$

Thus, the circle that passes through $w_1, \; w_2, \; w_3$ satisfies the equation $\left|w - \frac{5i}{8}\right| = \frac{3}{8}$.

Similarly, we can find the image of the lower half-circle of the radius 2 (see Figs. 4.9 and 4.10). The corresponding images of the endpoints $z = -2$ and $z = 2$ were already found $w_1 = f(-2)$ and $w_2 = f(2)$. Choosing one more point on this half-circle, for instance, $z = -2i$, we get

$$w_4 = f(-2i) = \frac{i(-2i) + 1}{-2i - 4i} = \frac{3}{-6i} = \frac{i}{2}.$$

Now, we can specify the equation of the circle in the w-plane that passes through the points $w_1 = f(-2), \; w_2 = f(2)$, and $w_4 = f(-2i)$ in the same way as for the preceding circle. Denoting the centerpoint of this circle by $w_0'' = iv_0''$, we have

$$\left|w_1 - w_0''\right| = \left|w_4 - w_0''\right| = r_2,$$

and consequently

$$\left|\frac{1}{2} - v_0''\right| = \sqrt{\frac{9}{100} + \left(\frac{4}{10} - v_0''\right)^2},$$

which results in

$$\frac{1}{4} - v_0'' + v_0''^2 = \frac{25}{100} - \frac{4}{5}v_0'' + v_0''^2 \Rightarrow v_0'' = 0, \; r_2 = \frac{1}{2}.$$

Therefore, the equation of the second circle is $|w| = \frac{1}{2}$ (see Fig. 4.10).

The constructed arcs of the circles divide the entire w-plane into two domains. To specify the required domain, we can pick $\forall z \in D$ and find the corresponding $w = f(z) \in f(D) = G$. Take, for simplicity, the point $z = -i \in D$; $f(-i) = \frac{2}{-5i} = \frac{2i}{5}$. Hence, the image of D is the domain G shown in Fig. 4.10.

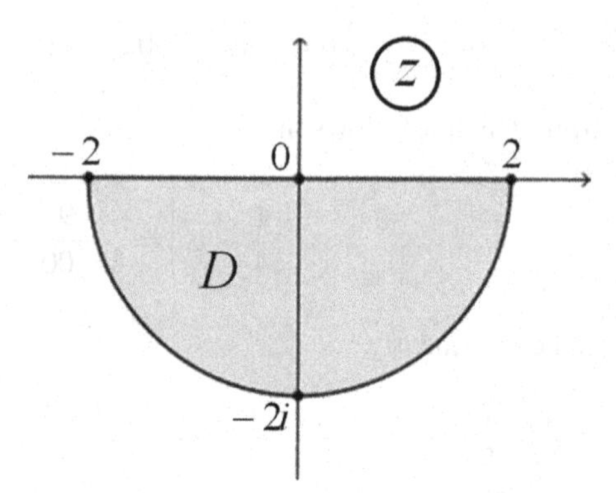

Fig. 4.9 The domain D of Example 4.1

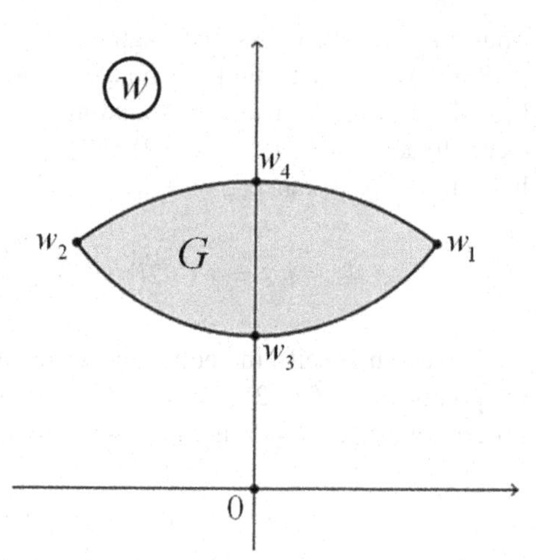

Fig. 4.10 The image of the domain D under the transformation of Example 4.1

Example 4.2 Transform the eccentric ring $\begin{cases} |z-i\,| > 2 \\ |z+i\,| < 5 \end{cases}$ into the concentric ring $1 < |w| < R$ and find the value of R. The original eccentric and final concentric rings are shown in Fig. 4.11.

Notice first that the points $w = 0$ and $w = \infty$ are symmetric with respect to both circles $|w| = 1$ and $|w| = R$ in the w-plane. This suggests a method for finding the required transformation: first determine the points z_1 and z_2, which are symmetric with respect to both circles $\gamma_1 = \{|z-i\,| = 2\}$ and $\gamma_2 = \{|z+i\,| = 5\}$ in the z-plane, and then construct a fractional linear function that carries one of the points z_1 or z_2 to 0 and another one to ∞.

Therefore, let us start finding the points symmetric with respect to γ_1 and γ_2, which is accomplished by solving the system:

$$\begin{cases} \bar{z}_1 + i = \dfrac{4}{z_2 - i} \\ \bar{z}_1 - i = \dfrac{25}{z_2 + i} \end{cases}.$$

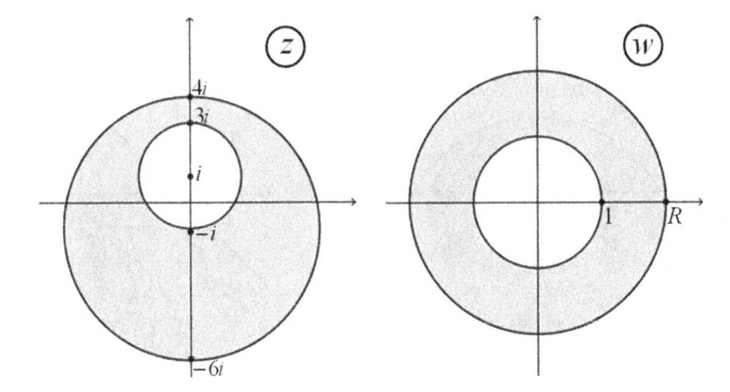

Fig. 4.11 Example 4.2: The original eccentric ring in the z-plane and the final concentric ring in the w-plane

Subtracting the second equation from the first, we get the relation $2i = \frac{4}{z_2-i} - \frac{25}{z_2+i} = \frac{-21z_2+29i}{z_2^2+1}$, which is reduced to the quadratic equation $2z_2^2 - 21iz_2 - 27 = 0$. The roots of the last equation are the required symmetric points in the z-plane:

$$z_{1,2} = \frac{21i \pm \sqrt{-441+216}}{4} = \frac{21i \pm 15i}{4} \Rightarrow z_1 = \frac{3}{2}i,\ z_2 = 9i\ .$$

Consider now the following fractional linear function:

$$\zeta = \frac{z-z_1}{z-z_2} = \frac{z-3i/2}{z-9i}\ . \tag{4.25}$$

According to the properties of the fractional linear functions, (4.25) transforms the given eccentric ring into a concentric ring. To find the radii of the circles that bound this ring, let us pick some points in γ_1 and γ_2 and find their images. For instance, choosing $a = -i \in \gamma_1$ and $b = -6i \in \gamma_2$ we have

$$\zeta(a) = \frac{-i-3i/2}{-i-9i} = \frac{5/2}{10} = \frac{1}{4},\quad \zeta(b) = \frac{-6i-3i/2}{-6i-9i} = \frac{15/2}{15} = \frac{1}{2}\ .$$

We see that the function (4.25) maps the original ring onto the ring $\frac{1}{4} < |\zeta| < \frac{1}{2}$. To obtain the final ring, we have to amplify the last ring four times:

$$w = 4\zeta = 4\frac{z-3i/2}{z-9i}\ .$$

Thus, the last function maps conformally the eccentric original ring onto the concentric ring $1 < |w| < 2$ with the larger radius $R = 2$.

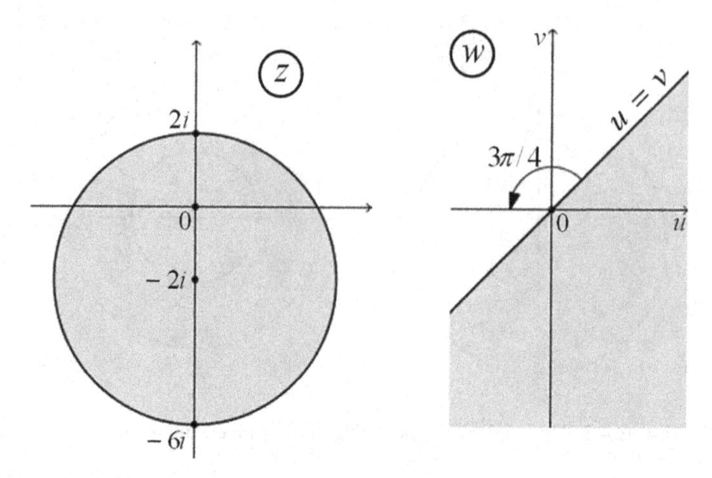

Fig. 4.12 Example 4.3: The original domain in the z-plane and the final domain in the w-plane

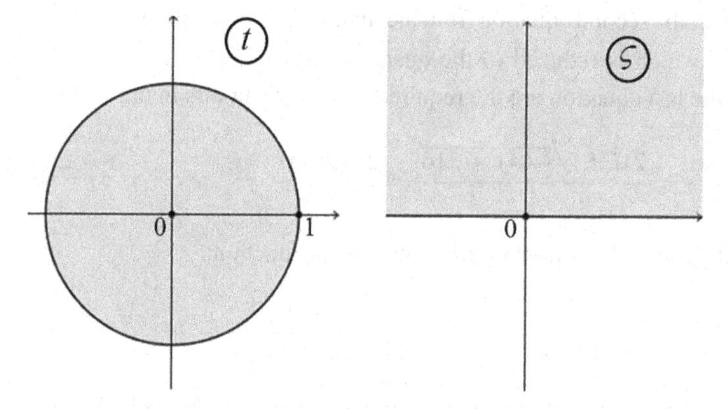

Fig. 4.13 Example 4.3: The auxiliary domains—the unit disk and the upper half-plane

Example 4.3 Map the disk $|z + 2i\,| < 4$ in the z-plane onto the half-plane $u > v$ in the w-plane ($w = u + iv$) in such a way that $w\,(-2i) = 3$ and $w\,(-6i) = 0$ (see Fig. 4.12).

The conformal transformation which maps the upper half-plane onto the unit disk is already known (see item 8 in this section). For this reason, we first transform the given domains into the auxiliary domains, which have the form required in that conformal transformation: the function $t = \frac{1}{4}\,(z + 2i)$ transforms the disk $|z + 2i\,| < 4$ into the unit disk in the t-plane and the function $\zeta = we^{\frac{3}{4}\pi i}$ transforms the half-plane $u > v$ into the upper half-plane in the ζ-plane (Fig. 4.13).

Now, we can apply the above-mentioned transformation $t = e^{i\gamma}\frac{\zeta-\alpha}{\zeta-\bar{\alpha}}$ to map conformally the upper half-plane onto the unit disk. Returning to the original variables, we obtain

$$\frac{1}{4}(z+2i) = e^{i\gamma}\frac{we^{3\pi i/4}-\alpha}{we^{3\pi i/4}-\bar{\alpha}}. \tag{4.26}$$

To specify the parameters α and γ, we employ the two additional conditions: $w(-2i) = 3$ and $w(-6i) = 0$. This leads to the following system

$$\begin{cases} e^{i\gamma}\dfrac{3e^{3\pi i/4}-\alpha}{3e^{3\pi i/4}-\bar{\alpha}} = \dfrac{1}{4}(-2i+2i) = 0, \\ e^{i\gamma}\dfrac{\alpha}{\bar{\alpha}} = \dfrac{1}{4}(-6i+2i) = -i. \end{cases}$$

The first equation gives $\alpha = 3e^{3\pi i/4}$ and substituting this value in the second we get $e^{i\gamma}\frac{3e^{3\pi i/4}}{3e^{-3\pi i/4}} = -i$ or, simplifying, $e^{i\gamma} = -i\,e^{-3\pi i/2} = 1$. Using these values of α and γ in (4.26), we obtain

$$\frac{1}{4}(z+2i) = \frac{we^{3\pi i/4}-3e^{3\pi i/4}}{we^{3\pi i/4}-3e^{-3\pi i/4}} = \frac{w-3}{w-3i},$$

and finally, solving the last equation for w we find the required transformation:

$$w = \frac{3iz-18}{z-4+2i}.$$

Example 4.4 Find a fractional liner function that has the double fixed point $z = 5$ and carries ∞ to i.

Recall that z_0 is a fixed point of $f(z)$ if $f(z_0) = z_0$. To find the fixed points of a fractional liner function $w = \frac{az+b}{cz+d}$, we have to solve the equation $\frac{az+b}{cz+d} = z$ which can be transformed to

$$cz^2 + (d-a)z - b = 0. \tag{4.27}$$

Since, in this exercise, $z = 5$ is a double fixed point, we conclude that $c \neq 0$ and that $z = 5$ is a double root of (4.27), that is, $\left(cz^2 + (d-a)z - b\right)' = 2cz + (d-a) = 0$ for $z = 5$. Therefore,

$$\begin{cases} 25c + 5(d-a) - b = 0, \\ 10c + (d-a) = 0. \end{cases}$$

The relations among coefficients in this system can be expressed in the following way: the second equation can be rewritten as $d - a = -10c$ and, using this relation in the first equation, we get $25c - 50c - b = 0$ or $b = -25c$.

Employing the second condition of the exercise—∞ is carried into i—we have

$$w(\infty) = \lim_{z\to\infty}\frac{az+b}{cz+d} = \frac{a}{c} = i,$$

that is, $a = ci$ and consequently $d = a - 10c = ci - 10c = c(-10+i)$. Substituting the specified values of a, b, and d in the general form of fractional liner function and recalling that $c \neq 0$, we obtain

$$w = \frac{ciz - 25c}{cz + c(-10 + i)} = \frac{iz - 25}{z - 10 + i}.$$

Notice that the relations among the coefficients a, b, c, d of the required fractional linear function can be obtained in a different manner. Since $z = 5$ is a double fixed point, the discriminant of the quadratic Eq. (4.27) is 0—$(d - a)^2 + 4bc = 0$— and consequently, the roots are found by the formula $z_1 = z_2 = \frac{a-d}{2c} = 5$. The latter gives $a - d = 10c$ and using this in the former we get $100c^2 + 4bc = 0$ or $b = -25c$ (since $c \neq 0$). Therefore, we arrive at the same relations as above.

4.6 Functions $w = z^n$ and $z = \sqrt[n]{w}$

We start with the *power function* $w = z^n$, $n \in \mathbb{N}$ and make the restriction that $n \neq 1$, since for $n = 1$ the function is linear.

1. The function $w = f(z) = z^n$ is analytic in the entire (finite) complex plane $\mathbb{C}$. The point $z = \infty$ is a pole of order n.

2. The derivative $f'(z) = nz^{n-1}$ exists in the entire (finite) complex plane and is different from 0 for any point except for $z = 0$. Then, the function $w = z^n$ represents a conformal mapping at any $z \neq 0$. Since the derivative vanishes at the origin, the transformation is not conformal at the point $z = 0$. Let us verify what conditions are not satisfied in the definition of a conformal mapping. Take two vectors z_1 and z_2, which emanate from the origin and form the angles φ_1 and φ_2, respectively, with the positive direction of the real axis: $z_1 = r_1 e^{i\varphi_1}$, $z_2 = r_2 e^{i\varphi_2}$. The angle between the vectors z_1 and z_2 is equal to $\varphi_2 - \varphi_1$. The corresponding vectors under the transformation $w = z^n$ are $w_1 = \rho_1 e^{i\theta_1} = r_1^n e^{in\varphi_1}$, $w_2 = \rho_2 e^{i\theta_2} = r_2^n e^{in\varphi_2}$, that gives the corresponding angles $\theta_1 = n\varphi_1$, $\theta_2 = n\varphi_2$ and the angle between w_1 and w_2 is equal $\theta_2 - \theta_1 = n(\varphi_2 - \varphi_1)$. Therefore, the mapping from the z-plane onto the w-plane under the transformation $w = z^n$ does not preserve the angles at the origin: all the angles between the vectors emanating from the origin are increased n times. (Recall that the angles are preserved under the transformation $w = z^n$ at any point $z \neq 0$, since this transformation is conformal at any point different from zero.) Notice that the function is univalent at any point $z \in \mathbb{C}\setminus\{0\}$ by the criterion for local univalence, since this function is analytic on the entire complex plane and $f'(z) \neq 0$, $\forall z \in \mathbb{C}\setminus\{0\}$.

3. Let us see if the function $w = z^n$ has the global univalence. Recall that for any fixed $w \neq 0$ there are exactly n different solutions of the equation $z^n = w$:

$$z_k = \sqrt[n]{|w|} \cdot e^{i\frac{\arg w + 2k\pi}{n}}, \quad k = 0, 1, \ldots, n-1. \tag{4.28}$$

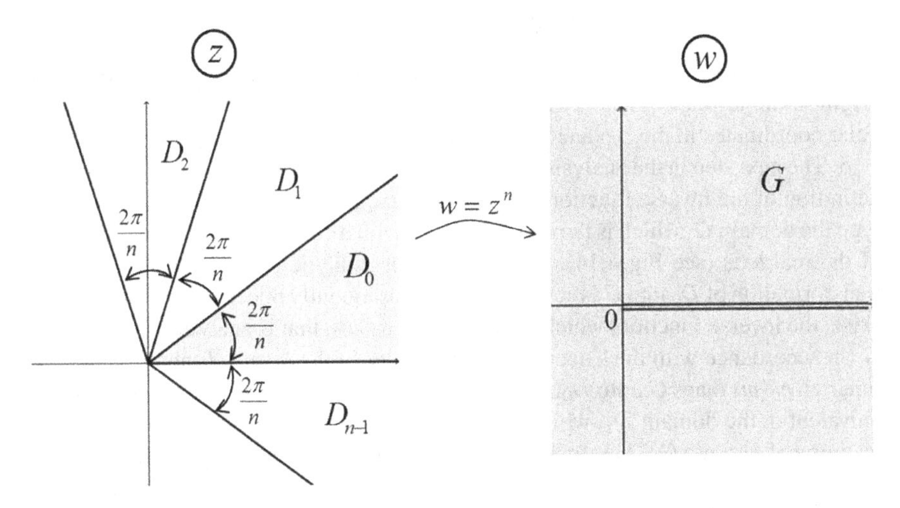

Fig. 4.14 Domains of univalence of the function $w = z^n$

This means that different points z_k and z_j, $k \neq j$ of the z-plane are carried to the same point in the w-plane. Hence, $w = z^n$ is not univalent on $\mathbb{C}\backslash\{0\}$. Therefore, the natural problem which now arises is to find the domains of univalence, that is, the domains where the function $w = z^n$ is univalent. It is clear that such a domain may not contain the points z_k of the form (4.28) with different indices k. Then, we obtain the sectors of the central angles $\frac{2\pi}{n}$, each of which is a domain of univalence of the given function, and the number of these sectors is n. The system of the sectors can be rotated in any angle around the origin, but it is convenient to take as the first sector D_0 that whose angle varies from 0 to $\frac{2\pi}{n}$; the next sector D_1 has the angle varying between $\frac{2\pi}{n}$ and $\frac{4\pi}{n}$, and so on (see Fig. 4.14).

Let us find the image of each of these sectors in the w-plane. Since the angle between any two vectors emanating from the origin increases n times under the transformation $w = z^n$, we deduce that the sector D_0 is mapped onto the entire w-plane with a cut along the positive part of the real axis, which we denote by G. The next sector D_1 is also mapped onto the same domain G, and so on. Therefore, each sector D_k, $k = 0, 1, \ldots, n-1$ is mapped on the domain G under the transformation $w = z^n$ (see Fig. 4.14).

4. Now, we consider the net of the polar coordinate curves in the z-plane, that is, the circles centered at the origin and the rays starting from the origin, and find their images under the transformation $w = z^n$. The equation of such a circle is $z = re^{i\varphi}$, where r is a constant and φ varies from 0 to 2π. Then $w = \rho e^{i\theta} = r^n e^{in\varphi} = z^n$, that is, $\rho = r^n$ is a constant and θ varies from 0 to $2n\pi$. Therefore, a circle in the z-plane is carried to the circle in the w-plane. Notice, that when the point z makes one loop along the circle $|z| = r$, the corresponding point $w = z^n$ circuits n times along the circle $|w| = \rho = r^n$.

For the rays with the initial point at the origin in the z-plane, we have $z = re^{i\varphi}$, where φ is a constant and r varies from 0 to ∞. Then $w = \rho e^{i\theta} = r^n e^{in\varphi} = z^n$, that is, the vectors w have the same angle $\theta = n\varphi$ with the positive direction of the real axis in the w-plane and their radius $\rho = r^n$ varies from 0 to ∞. Therefore, a ray emanating from the origin in the z-plane is carried to the ray emanating from the origin in the w-plane. Hence, the power function $w = z^n$ transforms the net of the polar coordinates in the z-plane into the net of the polar coordinates in the w-plane.

5. The next step in the analysis of the domains of univalence of $w = z^n$ leads to the definition of the inverse function. Indeed, the function $w = z^n$ maps the domain D_0 onto the domain G, which is the entire w-plane with a cut along the positive direction of the real axis (see Fig. 4.14). Since $w = z^n$ is analytic and univalent in D_0, the transformation of D_0 into G is conformal and consequently bijective. Therefore, there exists the inverse function, which we denote by $g_0(w)$, that is analytic in the domain G (in accordance with the Theorem on analyticity of the inverse function), and this function $g_0(w)$ maps G onto D_0. In the same way, the function $w = z^n$ is analytic and univalent in the domain D_1, which implies that $w = z^n$ is a conformal and bijective mapping of D_1 onto G. Therefore, there exists the inverse function $g_1(w)$ from G onto D_1, which is analytic and univalent. Similarly, the function $w = z^n$ is analytic and univalent in any sector D_k, $k = 0, 1, \ldots, n-1$, which implies that it is a conformal and bijective mapping of D_k onto G. Consequently, for each fixed k, there exists the inverse function $g_k(w)$ from G onto D_k, which is analytic and univalent. Thus, we obtain n different analytic and univalent functions $g_k(w)$, $k = 0, 1, \ldots, n-1$, each of which is inverse to the original function $w = z^n$. Therefore, the *inverse function* $z = \sqrt[n]{w}$ *is an analytic multi-valued function* whose analytic branches (elements) are the functions $g_k(w)$.

Let us see how these analytic branches $g_k(w)$, $k = 0, 1, \ldots, n-1$ are connected with each other. Consider an arbitrary simple closed curve Γ in the w-plane, which has no circuit around the origin (that is, Γ does not intersect the positive part of the real axis). Denote the initial (and at the same time the final) point of the curve Γ by w_0: $w_0 = |w_0| \cdot e^{i\theta_0}$. The function $z = g_0(w)$ maps Γ onto a simple closed curve γ_0 contained in D_0, whose the initial/end point is $z_0 = \sqrt[n]{|w_0|} \cdot e^{i\frac{\theta_0}{n}}$. The next function

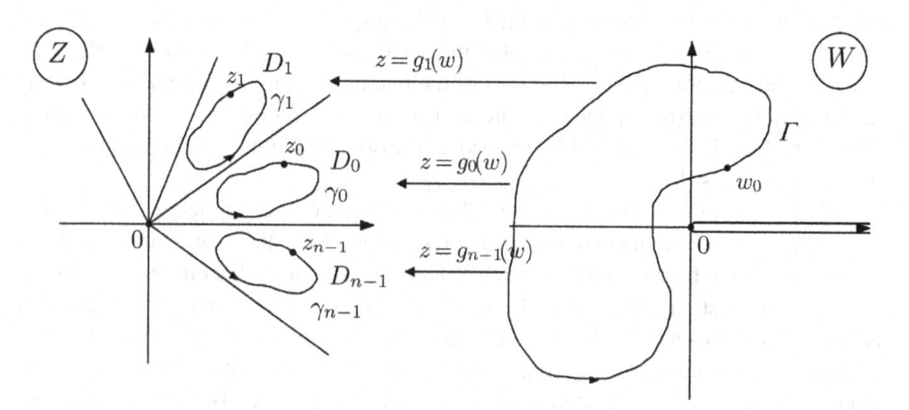

Fig. 4.15 A closed curve Γ in the w-plane with no circuit around the origin and its images $\gamma_0, \gamma_1, \ldots, \gamma_{n-1}$ in the z-plane under $z = g_k(w) = \sqrt[n]{w}$, $k = 0, 1, \ldots, n-1$

$z = g_1(w)$ carries the same curve Γ to a simple closed curve γ_1 contained in D_1 with the initial/end point $z_1 = \sqrt[n]{|w_0|} \cdot e^{i\left(\frac{\theta_0}{n}+\frac{2\pi}{n}\right)}$. And the same happens with other functions $z = g_k(w)$ (see Fig. 4.15).

Consider now a simple closed curve Γ in the w-plane, which makes one circuit in the counterclockwise direction around the origin starting and ending at a point $w_0 = |w_0| \cdot e^{i\theta_0}$. Pick any value of the function $z = \sqrt[n]{w}$ at the point w_0, for instance, $z_0 = \sqrt[n]{|w_0|} \cdot e^{i\frac{\theta_0}{n}}$, which corresponds to $g_0(w_0)$. After one circuit along Γ we arrive at the same point w_0, but with the changed argument $\theta_0 + 2\pi$: $w_0 = |w_0| \cdot e^{i(\theta_0+2\pi)}$. With continuous traversal along Γ in the w-plane, starting at $w_0 = |w_0| \cdot e^{i\theta_0}$ and ending at $w_0 = |w_0| \cdot e^{i(\theta_0+2\pi)}$, there is associated continuous path in the z-plane along the curve γ_0 lying in D_0 from $z_0 = \sqrt[n]{|w_0|} \cdot e^{i\frac{\theta_0}{n}}$ to $z_1 = \sqrt[n]{|w_0|} \cdot e^{i\left(\frac{\theta_0}{n}+\frac{2\pi}{n}\right)}$, where $z_1 = g_1(w_0)$.

Making one more circuit along Γ we move from $w_0 = |w_0| \cdot e^{i(\theta_0+2\pi)}$ to $w_0 = |w_0| \cdot e^{i(\theta_0+4\pi)}$ and, correspondingly, we proceed along γ_1 in the z-plane from $z_1 = g_1(w_0)$ to $z_2 = g_2(w_0) = \sqrt[n]{|w_0|} \cdot e^{i\left(\frac{\theta_0}{n}+\frac{4\pi}{n}\right)}$. A similar situation occurs when we make the next circuits along Γ (see Fig. 4.16).

The nth circuit around the origin we start with the point $w_0 = |w_0| \cdot e^{i(\theta_0+2(n-1)\pi)}$, which gives the corresponding value $z_{n-1} = g_{n-1}(w_0) = \sqrt[n]{|w_0|} \cdot e^{i\left(\frac{\theta_0}{n}+\frac{2\pi}{n}(n-1)\right)}$, and terminate with the point $w_0 = |w_0| \cdot e^{i(\theta_0+2n\pi)}$, which is mapped to the point $\sqrt[n]{|w_0|} \cdot e^{i\left(\frac{\theta_0}{n}+\frac{2\pi}{n}(n-1)+\frac{2\pi}{n}\right)} = \sqrt[n]{|w_0|} \cdot e^{i\left(\frac{\theta_0}{n}+\frac{2\pi}{n}\cdot n\right)} = \sqrt[n]{|w_0|} \cdot e^{i\frac{\theta_0}{n}} = g_0(w_0)$, that is, we arrive at the initial value of the function $g_0(w)$ (see Fig. 4.16). Hence, after n circuits around the origin along Γ we return to the original value of the function $z = \sqrt[n]{w}$ at the point w_0: $\sqrt[n]{w_0} = z_0 = g_0(w_0)$, which means that the origin $w = 0$ is the branch point of the finite order. Thus, the function $\sqrt[n]{w}$ is a multi-valued (n-valued) analytic function

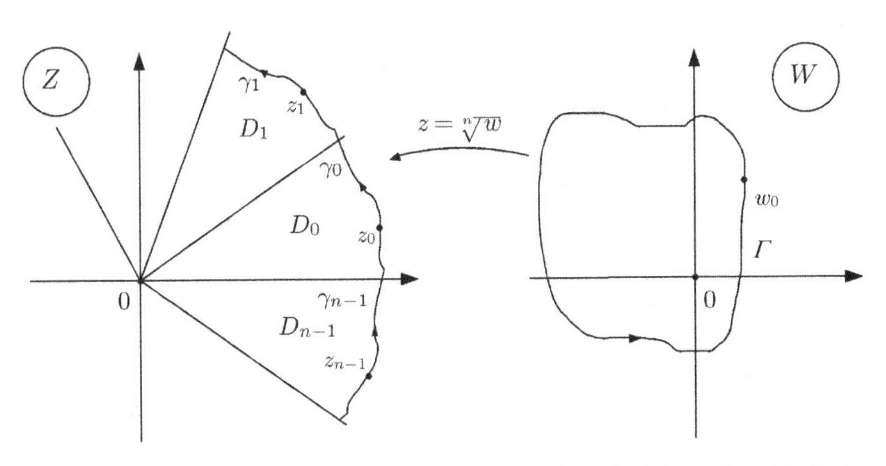

Fig. 4.16 A closed curve Γ in the w-plane with a single circuit around the origin and its images $\gamma_0, \gamma_1, \ldots, \gamma_{n-1}$ in the z-plane under $z = \sqrt[n]{w}$

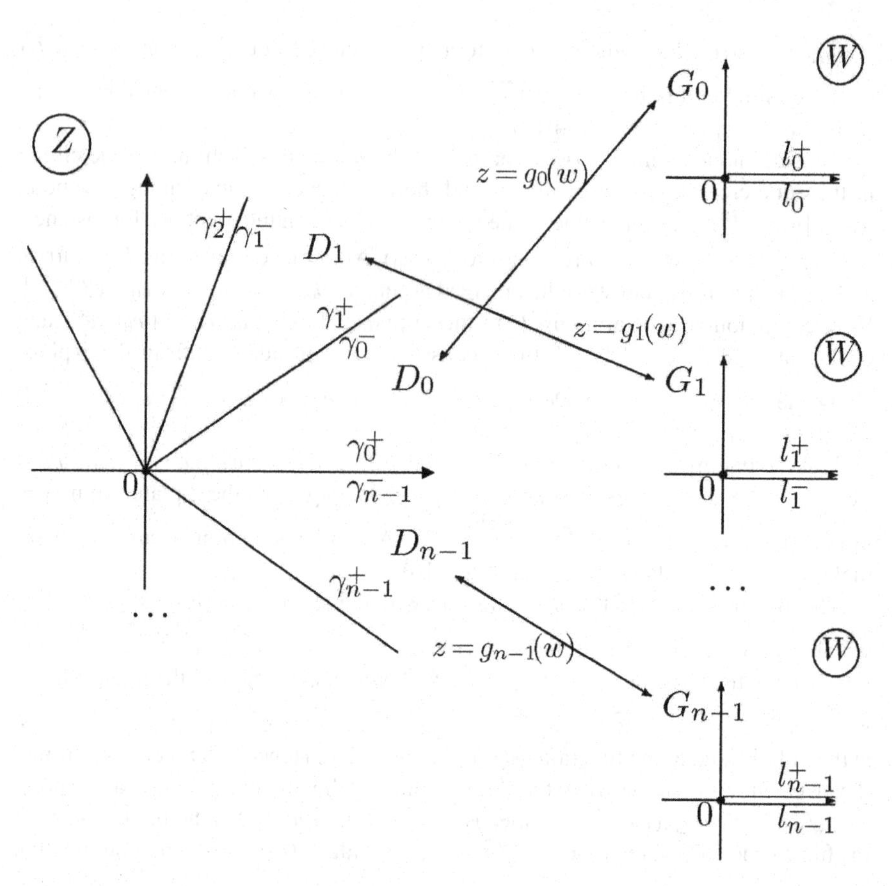

Fig. 4.17 Copies of the domain G and one-to-one correspondence between domains G_k and D_k under $z = g_k(w)$, $k = 0, 1, \ldots, n-1$

in the domain $\mathbb{C}\backslash\{0\}$. It is understood to be a *multi-valued function composed of a set of n analytic branches (elements)* $g_k(w)$, $k = 0, 1, \ldots, n-1$.

6. Let us construct the Riemann surface for the multi-valued (n-valued) analytic function $\sqrt[n]{w}$. We take n copies (sheets) of the domains G, $G = \mathbb{C}\backslash\{[0, \infty)\}$ and denote them by $G_0, G_1, \ldots, G_{n-1}$. Recall that the function (branch) $g_k(w)$, $k = 0, 1, \ldots, n-1$, transforms the domain G_k into the sector D_k bijectively. Denote the upper and lower edges of the cut along the positive part of the real axis in the domain G_k by l_k^+ and l_k^-, $k = 0, 1, \ldots, n-1$, respectively, and denote the corresponding rays of the borders of the domains D_k by γ_k^+ and γ_k^- (see Fig. 4.17).

Take a point $w_0 \neq 0$ in the w-plane and fix the value of the branch $g_0(w)$ at this point (w_0 belongs to the domain G_0 and the values of $g_0(w)$ belong to D_0). As was discussed above, going counterclockwise around the origin and completing one loop along a simple closed curve Γ, we pass from the branch $g_0(w)$ to the branch $g_1(w)$, that is, from the domain G_0 to G_1. Therefore, we glue together the domains D_0 and

D_1 along their common part (the ray γ_0^- of D_0 coincides with the ray γ_1^+ of D_1), and consequently, on the w-plane we paste together the domains G_0 and G_1 along the corresponding rays: glue the lower edge of the cut l_0^- of the sheet G_0 with the upper edge of the cut l_1^+ of the sheet G_1 (see Fig. 4.17).

Performing the next circuit around the origin, we pass from the branch $g_1(w)$ to the branch $g_2(w)$, that is, from the domain D_1 to D_2 (respectively, from the sheet G_1 to G_2). Then we join the sectors D_1 and D_2 along their common ray (γ_1^- of D_1, which coincides with γ_2^+ of D_2) and, correspondingly, we paste together the domains G_1 and G_2 in the w-plane by gluing the lower edge of the cut l_1^- of the sheet G_1 with the upper edge of the cut l_2^+ of G_2 (see Fig. 4.17).

Continuing this process, we finally make the nth circuit around the origin and return from the last branch $g_{n-1}(w)$ to the first one $g_0(w)$, and consequently, we join the sectors D_{n-1} and D_0 along their common ray (γ_{n-1}^- of D_{n-1} is the same as γ_0^+ of D_0). Accordingly, in the w-plane, we connect the sheets G_{n-1} and G_0 by pasting together the lower edge of the cut l_{n-1}^- with the upper edge of the cut l_0^+ (see Fig. 4.17). The obtained surface in the w-plane composed of n sheets $G_0, G_1, \ldots, G_{n-1}$ is the Riemann surface of the n-valued function $\sqrt[n]{w}$.

To visualize the constructed Riemann surface, we apply the following standard procedure. We take n replicas of the w-plane with a cut along the positive part of the real axis and place them one upon another in the prescribed order $G_0, G_1, \ldots, G_{n-1}$ from the bottom to the top. Then, we connect the lowermost sheet G_0 with the second sheet G_1 by pulling the lower (negative) edge of G_0 upward and the upper (positive) edge of G_1 downward and gluing them together. In the same way, we connect the second sheet G_1 with the third G_2 by gluing the lower edge of G_1 with the upper edge of G_2, and so on, until we reach the sheet G_{n-2} and connect its lower edge with the upper edge of the uppermost sheet G_{n-1}. When each pair of adjacent sheets is connected in this manner, it remains only to connect the lower edge of the uppermost sheet G_{n-1} with the upper edge of the lowermost sheet G_0. Since the last connection actually penetrates all the preceding connections, it is impossible to build a correct paper model of this Riemann surface. Therefore, we have two options of a visualization: either leave the replicas G_{n-1} and G_0 unconnected, reminding that analytically they are connected, or make its connection in some artificial way, bearing in mind that in a real geometric model it is impossible. The illustration of these two options of the visualization of the Riemann surface for $n = 2$ is shown in Fig. 4.18. One more visualization for $n = 3$ is presented in Fig. 4.19.

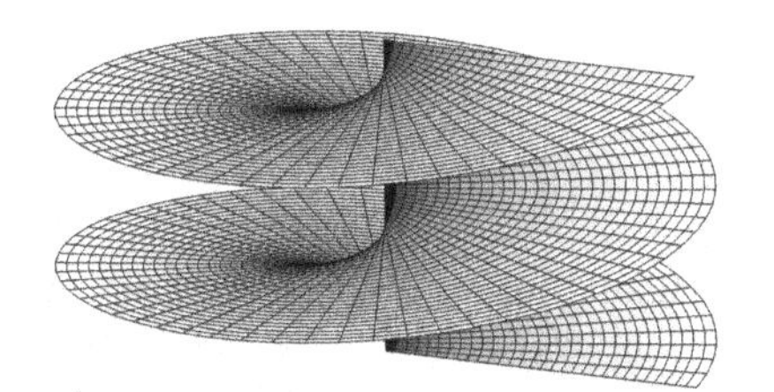
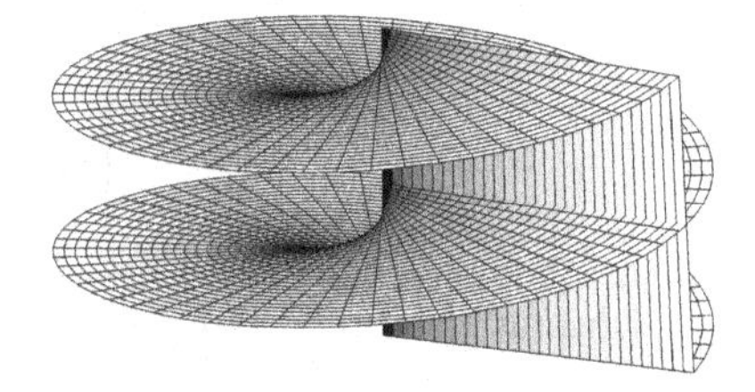

Fig. 4.18 The Riemann surface for $z = \sqrt{w}$ without (at the left) and with (at the right) the final connection of G_0 and G_1 through the cuts l_1^- and l_0^+

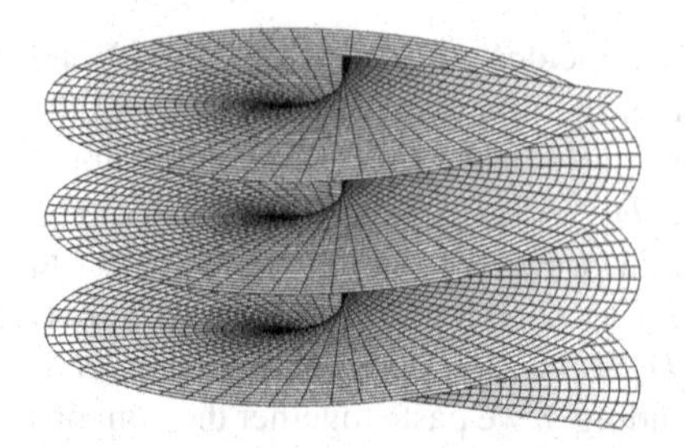
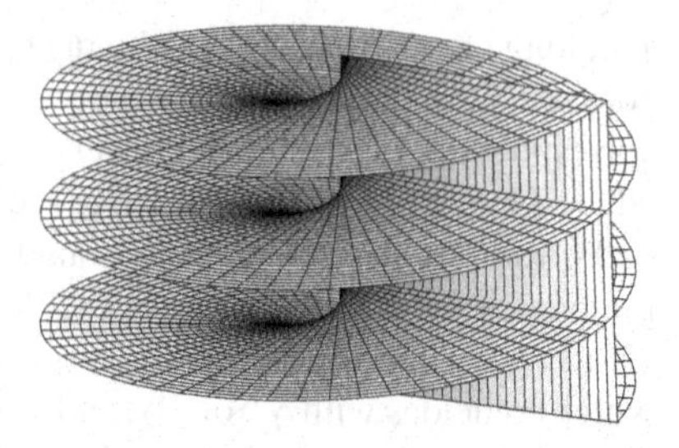

Fig. 4.19 The Riemann surface for $z=\sqrt[3]{w}$ without (at the left) and with (at the right) the final connection of G_0 and G_2 through the cuts l_2^- and l_0^+

Example Construct a conformal transformation of the domain $D = \begin{cases} |z| > 1 \\ |z-1| < \sqrt{2} \end{cases}$ into the upper half-plane. The original domain D is shown in Fig. 4.20.

Remark Before to find a solution to this Example, it is pertinent to make a general note regarding the problems of this type, which frequently appear in the theory of conformal mappings. The formulation of the above Example does not allow us to immediately find out the function that performs the required mapping. Even though the target domain is a standard one (usually the upper half-plane or the unit disk are considered to be standard simply connected domains), there is no known elementary function that satisfies the mapping conditions. This is a typical situation in the construction of conformal mappings and a common approach employed for solution of these problems is the following: one searches for a chain of mappings, transforming gradually the original domain to a form more similar to that of standard domains.

In this specific Example, the boundary of the original domain consists of the two arcs of circles, which gives rise to the following plan: first, find the points of intersection of these arcs and use the fractional linear function to carry one of these points into 0 and another into ∞; under this transformation, the circle arcs are transformed into the rays emanating from the origin, which form an angular sector (due to the circular property of a fractional linear function); then using rotation and change of the angle (that is, applying a power function), this sector can be transformed into the sector of the angle π, i.e., into the half-plane. The solution presented below illustrates the implementation of this procedure in detail.

We start with finding the points of intersection of the two circles whose arcs form the boundary of the domain D:

$$\begin{cases} |z| = 1, \\ |z-1| = \sqrt{2}. \end{cases}$$

The first equation can be rewritten in the exponential form $z = e^{i\varphi}$, $\varphi \in [-\pi, \pi]$ and substituting this expression in the second equations we obtain

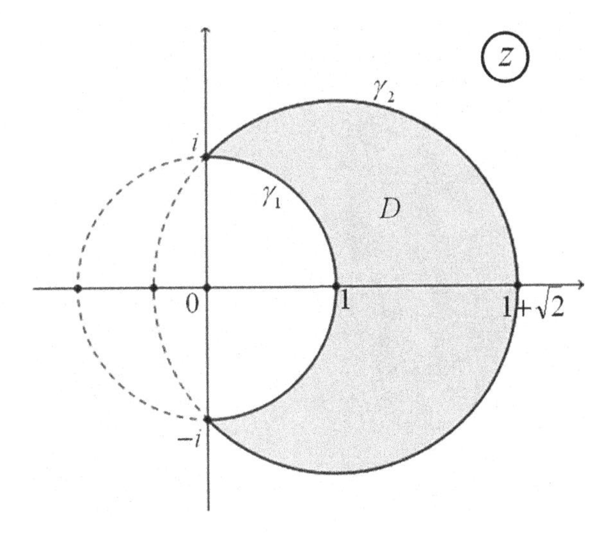

Fig. 4.20 The original domain D of Example

$$\left|e^{i\varphi} - 1\right| = |(\cos\varphi - 1) + i\,\sin\varphi| = \sqrt{(\cos\varphi - 1)^2 + \sin^2\varphi} = \sqrt{2 - 2\cos\varphi} = \sqrt{2}\,.$$

Therefore, $\cos\varphi = 0$, that is, $\varphi = \pm\frac{\pi}{2}$ and we have the two intersection points $z' = e^{i\pi/2} = i$ and $z'' = e^{-i\pi/2} = -i$ (see Fig. 4.20).

Next, we transform conformally the domain D into a simpler auxiliary domain by mapping conformally the arcs γ_1 and γ_2 of the two circles, which form the boundary of D, in the parts of straight lines. A fractional linear function is suitable for this task, and setting additional conditions that the point z' is carried into ∞ and z'' into 0, we arrive at the following function:

$$z_1 = \frac{z+i}{z-i}\,. \tag{4.29}$$

Let us determine the form of the auxiliary domain D_1 in the z_1-plane. It follows from the construction of (4.29) that the arcs γ_1 and γ_2 are transformed into the rays emanating from the origin $z_1 = 0$. To know the exact position of these rays, we need to find the image of one more point for each of the arcs. For instance, choosing $z = 1 \in \gamma_1$ and $z = 1 + \sqrt{2} \in \gamma_2$, we get (see Fig. 4.21)

$$z_1\,(1) = \frac{1+i}{1-i} = \frac{(1+i)^2}{2} = i,$$

$$z_1\left(1+\sqrt{2}\right) = \frac{1+\sqrt{2}+i}{1+\sqrt{2}-i} = \frac{\left(1+\sqrt{2}\right)^2 - 1 + 2i\left(1+\sqrt{2}\right)}{\left(1+\sqrt{2}\right)^2 + 1}$$

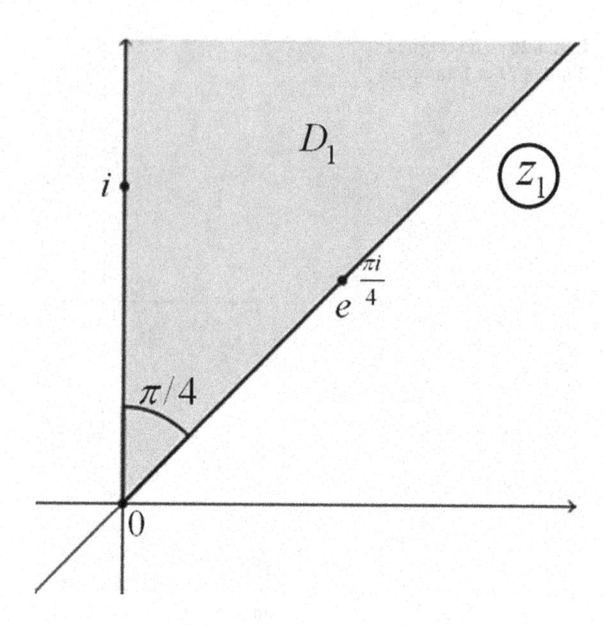

Fig. 4.21 The domain D_1 in the z_1-plane obtained from the domain D under the transformation (4.29)

$$= \frac{\left(2+2\sqrt{2}\right)(1+i)}{4+2\sqrt{2}} = \frac{1}{\sqrt{2}} + \frac{i}{\sqrt{2}} = e^{i\frac{\pi}{4}}.$$

The obtained rays divide the entire plane into the two infinite sectors. Due to conformality of a fractional linear transformation, it preserves the angles and, therefore, we have to choose the sector with the acute angle at the origin, since the corresponding angle in D at the point $z = -i$ is acute. Notice that the angle of this sector is equal to $\pi/4$.

Finally, to reach the upper half-plane, we rotate clockwise the sector D_1 by the angle $\pi/4$ about the origin

$$z_2 = z_1 e^{-i\pi/4} \tag{4.30}$$

and use the power function to increase four times the angle at the origin:

$$w = z_2^4 . \tag{4.31}$$

Joining all the applied transformations (4.29), (4.30) and (4.31), we obtain the function that maps conformally the original domain D onto the upper half-plane:

$$w = \left(e^{-i\frac{\pi}{4}} \frac{z+i}{z-i}\right)^4 = e^{-i\pi} \left(\frac{z+i}{z-i}\right)^4 = -\left(\frac{z+i}{z-i}\right)^4 .$$

4.7 Joukowski Function

In this section, we study the properties of the *Joukowski function*

$$w = f(z) = \frac{1}{2}\left(z + \frac{1}{z}\right). \tag{4.32}$$

1. The function (4.32) is analytic in the entire complex plane, except at the points 0 and ∞, which are single-valued isolated singularities. Since $\lim\limits_{z\to 0} f(z) = \lim\limits_{z\to 0} \frac{1}{2}\left(z+\frac{1}{z}\right) = \infty$ and $\lim\limits_{z\to\infty} f(z) = \lim\limits_{z\to\infty} \frac{1}{2}\left(z+\frac{1}{z}\right) = \infty$, both points 0 and ∞ are poles. Representing the function (4.32) in a neighborhood of 0 in the form $f(z) = \frac{1}{2}\left(z+\frac{1}{z}\right) = z^{-1}\cdot\frac{1}{2}\left(z^2+1\right)$, we see that $z=0$ is a simple pole by the criterion for multiplicity of a pole (see Theorem 3.3 in Sect. 3.7). Similarly, rewriting the function in a neighborhood of ∞ in the form $f(z) = \frac{1}{2}\left(z+\frac{1}{z}\right) = z\cdot\frac{1}{2}\left(1+\frac{1}{z^2}\right)$, we clarify that $z=\infty$ is also a simple pole. Therefore, the Joukowski function is analytic in the domain $\mathbb{C}\setminus\{0\}$ and possesses the two simple poles 0 and ∞.

2. In order to verify the local univalence of the function (4.32), we find its derivative in the domain $\mathbb{C}\setminus\{0\}$: $f'(z) = \frac{1}{2}\left(1-\frac{1}{z^2}\right)$. The derivative is zero only at the points $z=\pm 1$. Therefore, the function (4.32) is locally univalent at every point of $\mathbb{C}\setminus\{0\}$, except at $z=1$ and $z=-1$. Since $z=0$ and $z=\infty$ are simple poles, the function is also univalent at these two points. Thus, the obtained results on analyticity and local univalence imply that the Joukowski function represents a conformal mapping at any point of $\overline{\mathbb{C}}$ except at the points $z=1$ and $z=-1$, where the transformation is not conformal.

3. Let us pass to the study of the global univalence of the function. Pick two different points $z_1, z_2 \in \mathbb{C}_z$, $z_1 \neq z_2$ and verify if they have the same image: $f(z_1) = \frac{1}{2}\left(z_1+\frac{1}{z_1}\right) = \frac{1}{2}\left(z_2+\frac{1}{z_2}\right) = f(z_2)$. The last equation can be rewritten in the form $(z_1-z_2)+\left(\frac{1}{z_1}-\frac{1}{z_2}\right)=0$ or $(z_1-z_2)(1-z_1z_2)=0$. Since $z_1\neq z_2$, the solution is

$$z_1\cdot z_2 = 1. \tag{4.33}$$

Using the exponential form of z_1 and z_2: $z_1 = r_1e^{i\varphi_1}$, $z_2 = r_2e^{i\varphi_2}$, we have $z_1\cdot z_2 = r_1\cdot r_2\cdot e^{i(\varphi_1+\varphi_2)} = 1$ and consequently

$$\begin{cases} r_1r_2 = 1, \\ \varphi_1+\varphi_2 = 2k\pi, \quad \forall k\in\mathbb{Z}. \end{cases} \tag{4.34}$$

Thus, if two different points z_1 and z_2 satisfy relation (4.33), or equivalently, the system (4.34), then they are carried to the same point in the w-plane. Therefore, any domain of univalence of the function (4.32) may not contain a pair of points satisfying the condition (4.33), that is, at least one of the equalities in (4.34) should be violated for any two points in a domain of univalence. Consider first the domains in which the first condition of (4.34) does not hold: $r_1 \cdot r_2 \neq 1$. The simplest domains of this type are $|z| < 1$ and $|z| > 1$: in the domain $|z| < 1$, for any pair of points z_1, z_2 we have $r_1 r_2 < 1$, and in the domain $|z| > 1$ any two points satisfy the inequality $r_1 r_2 > 1$. Another option to obtain a domain of univalence is to violate the second equation in (4.34): $\varphi_1 + \varphi_2 \neq 2k\pi, \forall k \in \mathbb{Z}$. The simplest domains of this type are the upper half-plane $\mathrm{Im}\, z > 0$ and the lower half-plane $\mathrm{Im}\, z < 0$. Thus, the simplest domains of univalence of the Joukowski function are $|z| < 1$, $|z| > 1$, $\mathrm{Im}\, z > 0$, and $\mathrm{Im}\, z < 0$.

4. To find the fixed points of the function (4.32), we have to solve the equation $\frac{1}{2}\left(z + \frac{1}{z}\right) = z$, which is transformed to $z^2 - 1 = 0$ giving the two roots $z = 1$ and $z = -1$. Besides, the infinity point is mapped to itself: $f(\infty) = \lim_{z\to\infty} f(z) = \infty$. Therefore, the points 1, -1, and ∞ are the fixed points under the Joukowski function.

5. Let us find the images of the net of the polar coordinate curves (that is, the circles centered at the origin and the rays starting from the origin) under the transformation (4.32) (see Fig. 4.22). To this end, we employ the exponential representation of z—$z = re^{i\varphi}$—in the function definition

$$w = u + iv = \frac{1}{2}\left(z + \frac{1}{z}\right) = \frac{1}{2}\left(re^{i\varphi} + \frac{1}{re^{i\varphi}}\right) = \frac{1}{2}\left(re^{i\varphi} + \frac{1}{r}e^{-i\varphi}\right)$$

$$= \frac{1}{2}\left(r(\cos\varphi + i\sin\varphi) + \frac{1}{r}(\cos\varphi - i\sin\varphi)\right) = \frac{1}{2}\left(r + \frac{1}{r}\right)\cos\varphi + \frac{i}{2}\left(r - \frac{1}{r}\right)\sin\varphi,$$

which gives

$$\begin{cases} u = \dfrac{1}{2}\left(r + \dfrac{1}{r}\right)\cos\varphi, \\ v = \dfrac{1}{2}\left(r - \dfrac{1}{r}\right)\sin\varphi. \end{cases} \tag{4.35}$$

Consider first the system of the circles $z = re^{i\varphi}$, where r is an arbitrary positive constant and φ varies from 0 to 2π. For any fixed $r \neq 1$, we can rewrite the Eq. (4.35) in the form $\cos\varphi = \frac{u}{\frac{1}{2}(r+\frac{1}{r})}$ and $\sin\varphi = \frac{v}{\frac{1}{2}(r-\frac{1}{r})}$, and, according to the main trigonometric identity, we obtain

$$\cos^2\varphi + \sin^2\varphi = \frac{u^2}{\left(\frac{1}{2}\left(r + \frac{1}{r}\right)\right)^2} + \frac{v^2}{\left(\frac{1}{2}\left(r - \frac{1}{r}\right)\right)^2} = 1, \tag{4.36}$$

that is, we arrive at the equation of ellipse centered at the origin in the w-plane. Hence, the transformation (4.32) carries any circle $|z| = r, r \neq 1$ in the z-plane to the ellipse (4.36) in the w-plane. All these ellipses have the major semi-axis $a = \frac{1}{2}\left(r + \frac{1}{r}\right)$ on the u-axis, the minor semi-axis $b = \frac{1}{2}\left|r - \frac{1}{r}\right|$ on the v-axis and the foci located on the u-axis at the points $(\pm c, 0)$, where $c^2 = a^2 - b^2 == \frac{1}{4}\left(r + \frac{1}{r}\right)^2 - \frac{1}{4}\left(r - \frac{1}{r}\right)^2 = 1$, that is, $c = 1$ (see Fig. 4.22). The circle $r = 1$ is considered separately. In this case, (4.35) takes the form

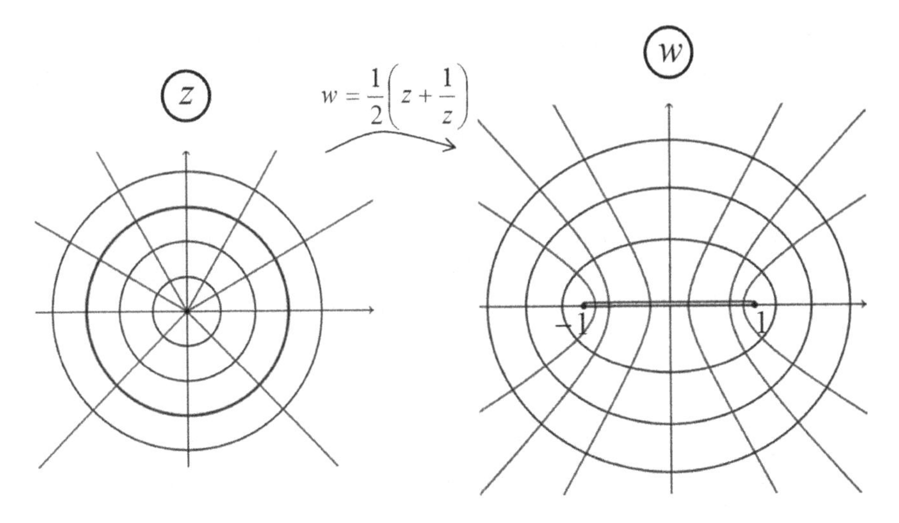

Fig. 4.22 The images of the net of the polar coordinate curves under the Joukowski transformation

$$\begin{cases} u = \cos\varphi, \\ v = 0\,. \end{cases} \tag{4.37}$$

Since φ varies from 0 to 2π, the Eq. (4.37) represents the segment $[-1, 1]$ on the u-axis of the w-plane, which is traversed twice. This segment can be understood as a degenerate ellipse, because for $r \to 1$ the semi-axes a and b have the following limit values: $a = \frac{1}{2}\left(r + \frac{1}{r}\right) \to 1, b = \frac{1}{2}\left|r - \frac{1}{r}\right| \to 0$ (see Fig. 4.22).

For the rays emanating from the origin in the z-plane, the angle φ is a constant (in the interval $[0, 2\pi]$) and r varies from 0 to ∞. Supposing, for a moment, that $\cos\varphi \neq 0$ and $\sin\varphi \neq 0$, we can rewrite Eq. (4.35) in the form $\frac{u}{\cos\varphi} = \frac{1}{2}\left(r + \frac{1}{r}\right)$ and $\frac{v}{\sin\varphi} = \frac{1}{2}\left(r - \frac{1}{r}\right)$. Taking the squares of the two last expressions and subtracting the second result from the first, we obtain

$$\frac{u^2}{\cos^2\varphi} - \frac{v^2}{\sin^2\varphi} = \frac{1}{4}\left(r + \frac{1}{r}\right)^2 - \frac{1}{4}\left(r - \frac{1}{r}\right)^2 = 1\,.$$

Thus, we arrive at the equation of a hyperbola in the w-plane

$$\frac{u^2}{\cos^2\varphi} - \frac{v^2}{\sin^2\varphi} = 1 \tag{4.38}$$

with the real semi-axis $a_1 = |\cos\varphi|$ and the imaginary semi-axis $b_1 = |\sin\varphi|$. The corresponding foci are located on the u-axis and has the coordinates $(\pm c_1, 0)$, where $c_1^2 = a_1^2 + b_1^2 = \cos^2\varphi + \sin^2\varphi = 1$, that is, $c_1 = 1$ (see Fig. 4.22).

Consider now the rays defined by the conditions $\cos\varphi = 0$ and $\sin\varphi = 0$. In this case, we have the angles $\varphi = k \cdot \frac{\pi}{2}$, $k = 0, 1, 2, 3$. If $\varphi = 0$, then (4.35) takes the form

$$\begin{cases} u = \frac{1}{2}\left(r + \frac{1}{r}\right), \\ v = 0. \end{cases} \tag{4.39}$$

Since r changes from 0 to $+\infty$, the last system represents the ray $[1, +\infty)$ in the w-plane, which is traversed twice. The second ray in the z-plane is defined by the condition $\varphi = \frac{\pi}{2}$, which leads to the following specification of (4.35):

$$\begin{cases} u = 0, \\ v = \frac{1}{2}\left(r - \frac{1}{r}\right). \end{cases} \tag{4.40}$$

For r varying from 0 to $+\infty$, (4.40) represents all the imaginary axis of the w-plane, which is traversed in the direction from the bottom to the top. Similarly, for the ray $\varphi = \pi$ in the z-plane, we obtain the ray $(-\infty, -1]$ in the w-plane and for $\varphi = \frac{3}{2}\pi$ in the z-plane we get the imaginary axis of the w-plane traversed in the direction from the top to the bottom. Notice that the rays $[1, +\infty)$, $(-\infty, -1]$ and the imaginary axis we can understand as degenerate hyperbolas.

Let us return to the rays $\varphi = constant$, $\varphi \neq k \cdot \frac{\pi}{2}$ in the z-plane and take a closer look on the correspondence between these rays and different parts of the hyperbolas. If a ray is located in the first quadrant, that is, $0 < \varphi < \frac{\pi}{2}$, then $\cos\varphi > 0$ and $\sin\varphi > 0$, and the systems (4.35) reveal that the image of this ray is the right side of the hyperbola (4.38) (since $u > 0$), which is traversed from the lower to upper part. If we pick a ray in the fourth quadrant of the z-plane ($\cos\varphi > 0$ and $\sin\varphi < 0$), then this ray is also mapped onto the right side of the hyperbola (4.38), but the direction of the traversal is from the top to the bottom. The rays of the second and third quadrants are mapped on the left side of the hyperbola (4.38).

Recall that the sets of the circles and the rays are orthogonal to each other in the z-plane and the Joukowski function is a conformal mapping at any point $z \neq \pm 1$, meaning that the angles between any two curves in the z-plane are preserved by their images in the w-plane (except at the points $z = \pm 1$). Therefore, the families of the ellipses and hyperbolas are orthogonal to each other in the w-plane, except at the points $w = \pm 1$. Thus, we can summarize that under the Joukowski transformation the net of the polar coordinate curves in the z-plane is mapped onto the two sets of the confocal ellipses and the confocal hyperbolas in the w-plane, which are orthogonal to each other (see Fig. 4.22).

Let us clarify what happens at the point $z = 1$. Take a circle and a ray passing through the point $z = 1$: $\gamma = \{|z| = 1\}$ and $l = \{\varphi = 0\}$. Notice that γ and l are orthogonal to each other. It was already shown that the unit circle is mapped onto the segment $\Gamma = [-1, 1]$ of the real axis, and the ray $\varphi = 0$ is mapped to the ray $L = [1, +\infty)$ under the Joukowski transformation. The angle between Γ and L is equal to π, which means that the angle at the point $z = 1$ is not preserved but doubled. It is easy to see that at the point $z = -1$ the situation is analogous.

6. Let us consider the possibility to construct an inverse function to (4.32). Choose any of the domains of univalence of the Joukowski function, say $D_1 = \{|z| < 1\}$. The boundary of D_1—the unit circle $|z| = 1$—is transformed into the segment $[-1,\ 1]$ of the real axis in the w-plane. Since the Joukowski function is analytic in D_1 except at the simple pole $z = 0$ and is univalent in D_1, it maps bijectively the domain D_1 onto the entire w-plane with the cut along the segment $[-1,\ 1]$. We denote the last domain by G: $G = \overline{\mathbb{C}} \setminus \{[-1,\ 1]\}$. Therefore, there exists the inverse function $g_1(w)$, which is analytic in G and maps G onto D_1 bijectively. Similarly, considering the domain $D_2 = \{|z| > 1\}$, we deduce that there exists the inverse function $g_2(w)$, which is analytic in G and transforms G into D_2 bijectively.

Consider now the definition of the function (4.32) as the equation for the unknown z. Transforming this equation to the form $z^2 - 2zw + 1 = 0$ and solving the last with respect to z, we have

$$z = w \pm \sqrt{w^2 - 1}\,. \tag{4.41}$$

Therefore, for any point of the domain G, the choice of an analytic branch of the inverse function ($g_1(w)$ or $g_2(w)$) is defined by the choice of the sign of the root in (4.41). Pick, for instance, the point $w = 2$. If we choose the minus sign, that is, $z_1 = 2 - \sqrt{3}$ at this point, then this is the value of the function $g_1(w)$; but if we take $z_2 = 2 + \sqrt{3}$, corresponding to the plus sign in (4.41), then we get the value of $g_2(w)$. Therefore, the inverse to the Joukowski function is a multi-valued (more precisely, two-valued) analytic function. The points $w = 1$ and $w = -1$ are the branch points of the inverse function, since the rotation around any of these points changes the analytic branch of the function. Performing one rotation about $w = 1$, we change the branch $g_1(w)$ by $g_2(w)$; making one more circuit around $w = 1$, we return from the analytic branch $g_2(w)$ to $g_1(w)$. A similar situation occurs under rotation about $w = -1$.

For other domains of univalence, the images may be different. For instance, the domain $D_3 = \{\forall z \in \mathbb{C} :\ \operatorname{Im} z > 0\}$ (as well as $D_4 = \{\forall z \in \mathbb{C} :\ \operatorname{Im} z < 0\}$) is mapped onto the entire w-plane with the cuts along the rays $(-\infty,\ -1]$ and $[1,\ +\infty)$, since the ray $\varphi = 0$ (the right side of the boundary $y = 0$) is carried to the ray $[1,\ +\infty)$ in the w-plane and the ray $\varphi = \pi$ (the left side of the boundary $y = 0$) is carried to the ray $(-\infty,\ -1]$. Let us denote this domain in the w-plane by G_1. Then, for any point of the domain G_1, the choice of an analytic branch of the inverse function is defined by the choice of the sign of the root in (4.41). Taking, for instance, $w = 0$ and choosing the corresponding value $z(0) = i$, we deal with the analytic branch $g_3(w)$, which maps bijectively G_1 onto D_3; but if we pick the corresponding value $z(0) = -i$, then we work with the analytic branch $g_4(w)$, which maps bijectively G_1 onto D_4.

In the example below, we analyze the transformation of a given domain under the Joukowski function.

Example 4.1 Find the image of the domain $D = \{r < |z| < 1,\ \operatorname{Im} z > 0\}$ under the Joukowski function.

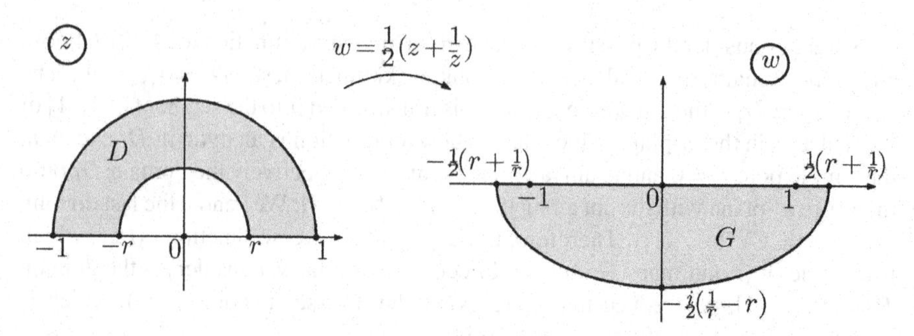

Fig. 4.23 The original domain D of Example 4.1 and its image under the Joukowski transformation

The original domain D is shown on the left side of Fig. 4.23. To find its image, we analyze the transformation of the boundary of D under the Joukowski function $w = \frac{1}{2}\left(z + \frac{1}{z}\right)$. The half-circle $\begin{cases} |z| = 1 \\ \operatorname{Im} z \geq 0 \end{cases}$ is carried to the interval $[-1,\ 1]$. The intervals of the real axis $[-1,\ -r]$ and $[r,\ 1]$ are mapped onto the intervals of the real axis $\left[-\frac{1}{2}\left(r + \frac{1}{r}\right),\ -1\right]$ and $\left[1,\ \frac{1}{2}\left(r + \frac{1}{r}\right)\right]$, respectively. Finally, the half-circle $\begin{cases} |z| = r < 1 \\ \operatorname{Im} z \geq 0 \end{cases}$ is carried to the lower half-ellipse $\frac{u^2}{a^2} + \frac{v^2}{b^2} = 1$ with the semi-axes $a = \frac{1}{2}\left(r + \frac{1}{r}\right)$ and $b = \frac{1}{2}\left|\frac{1}{r} - r\right|$. Thus, the image of D is the lower half-ellipse shown on the right side of Fig. 4.23.

The next example illustrates an application of different elementary functions, including the Joukowski function, to construction of the required conformal mapping.

Example 4.2 Construct a conformal mapping of a given domain D onto the upper half-plane: the original domain D is a circular sector $\begin{cases} |z| < 4 \\ \pi < \arg z < 3\pi/2 \end{cases}$ with the cuts $\begin{cases} 0 \leq |z| \leq 1 \\ \arg z = 5\pi/4 \end{cases}$ and $\begin{cases} 2 \leq |z| \leq 4 \\ \arg z = 5\pi/4 \end{cases}$.

The original domain D is shown on the left side of Fig. 4.24. Like in Example of Sect. 4.6, let us construct a chain of transformations choosing appropriate elementary functions. Notice that various elementary functions handle "nicely" domains, whose boundary consists of segments of the real axis and arcs of the unit circle. Therefore, the primitive idea is to employ rotations, contractions, and change of angles in order to obtain such image of the original domain, which possesses these simple boundaries. It is important to keep in mind that if any of employed functions is not univalent, we need to use only such domains which are contained in the domain of univalence of the non-univalent function.

Following this general scheme, at the first stage, we make the rotation of the given sector D by the angle π around the origin and the subsequent contraction by the factor $1/4$ in order to obtain the arc of the circle of the radius 1 (see the right side of Fig. 4.24):

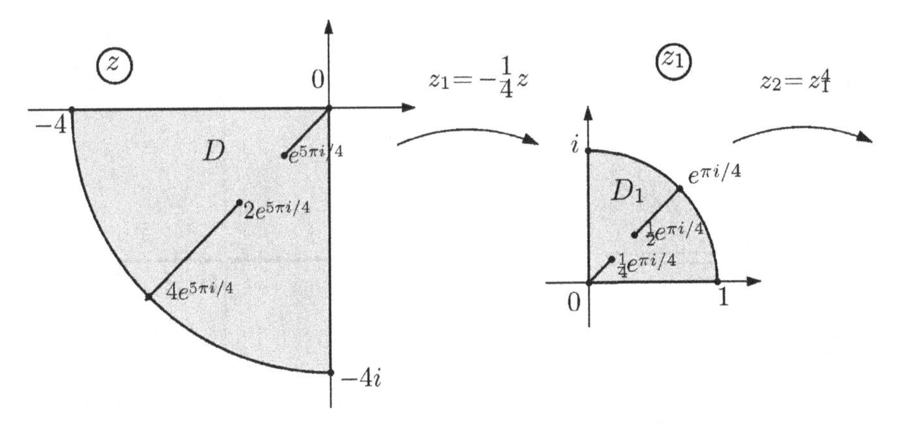

Fig. 4.24 Example 4.2: the original domain D and its image D_1 under the transformation (4.42)

$$z_1 = \frac{1}{4} z e^{i\pi} = -\frac{1}{4} z \,. \tag{4.42}$$

At the next stage, we use the power function to magnify the angles at the origin 4 times:

$$z_2 = z_1^4 \,. \tag{4.43}$$

Then, we obtain the unit disk D_2 with the cuts shown on the left side of Fig. 4.25. Since the boundary of the last domain in the z_2-plane is composed of the unit circle and the parts of the real axis, it is suitable to apply the Joukowski function

$$z_3 = \frac{1}{2}\left(z_2 + \frac{1}{z_2}\right), \tag{4.44}$$

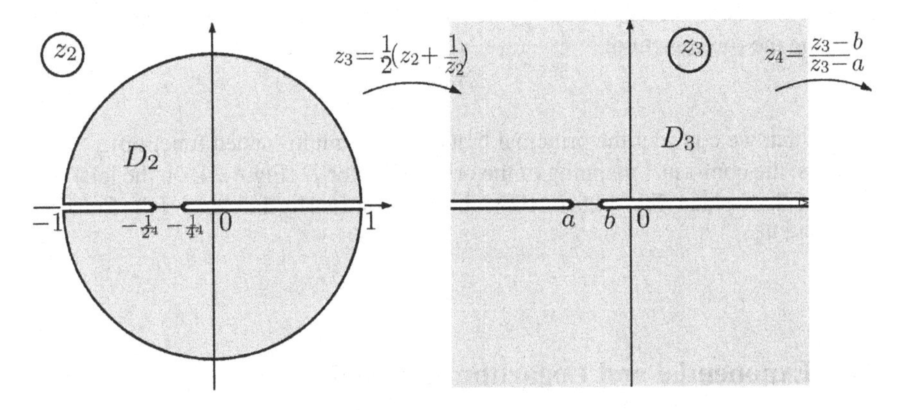

Fig. 4.25 Example 4.2: the unit disk with the cuts obtained from D_1 in the z_1-plane by applying the power function (4.43), and the entire plane with the cuts along the real axis obtained by using the Joukowski function (4.44)

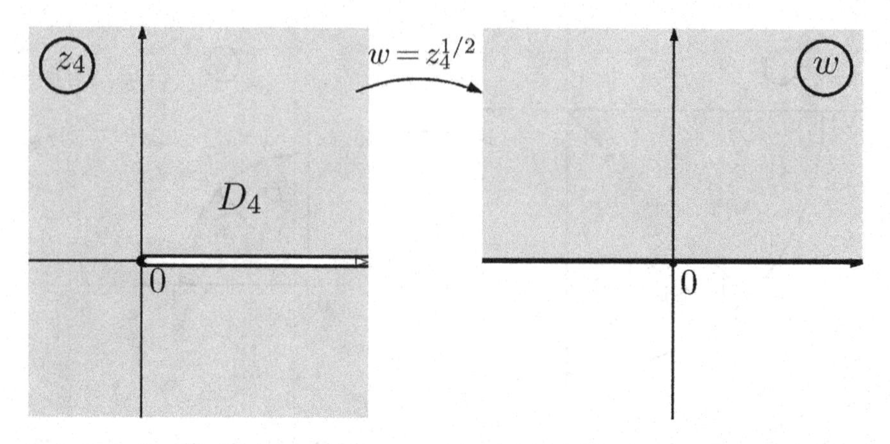

Fig. 4.26 Example 4.2: the domain in the z_4-plane obtained by applying the fractional linear function (4.45), and the upper half-plane in the w-plane obtained by using the root function (4.46)

which transforms D_2 into the plane with the cuts along the real axis shown on the right side of Fig. 4.25: $(-\infty, a]$ and $[b, +\infty)$, where $a = -\frac{1}{2}\left(\frac{1}{4^4} + 4^4\right) = -\left(2^7 + 2^{-9}\right)$, $b = -\frac{1}{2}\left(\frac{1}{2^4} + 2^4\right) = -\left(2^3 + 2^{-5}\right)$. Notice that the domain D_2 in the z_2-plane (Fig. 4.25) is contained in one of the domains of univalence of the Joukowski function, and consequently, the transformation of the disk with the cuts in the z_2-plane into the plane with the specified cuts D_3 in the z_3-plane (Fig. 4.25) is univalent (bijective) under the Joukowski function.

To transform the two cuts into the unique ray along the real axis starting at the origin (see the left side of Fig. 4.26), we use the fractional linear function

$$z_4 = \frac{z_3 - b}{z_3 - a} . \tag{4.45}$$

Finally, to obtain the upper half-plane, we halve the angles around the origin by applying the root function:

$$w = z_4^{1/2} \tag{4.46}$$

(notice that we consider the principal branch of this multi-valued function).

Thus, the conformal mapping of the original sector D (Fig. 4.24, on the left) onto the upper half-plane (Fig. 4.26, on the right) is performed by the chain of the functions (4.42)–(4.46).

4.8 Exponential and Logarithmic Functions

In Sect. 2.9, we have defined the *exponential function* e^z as the analytic continuation of the real function e^x from the real axis onto the whole complex plane:

$$e^z = \sum_{n=0}^{\infty} \frac{z^n}{n!} . \tag{4.47}$$

Let us study the properties of this function.

1. We have already seen that e^z is an entire function (analytic function in the entire complex plane). The unique singular point of this function is ∞, which is a single-valued isolated singularity. Since the power series (4.47) is, at the same time, the Laurent series of e^z in a neighborhood of infinity whose principal part contains an infinite number of terms, by the criterion for an essential point (see Sect. 3.8), ∞ is an essential singular point of the exponential function.

2. In Sect. 2.9, we have already shown the equality $e^{z_1+z_2} = e^{z_1} \cdot e^{z_2}, \forall z_1, z_2 \in \mathbb{C}$. In particular, $e^z = e^{x+iy} = e^x \cdot e^{iy} = e^x (\cos y + i \sin y)$, and consequently $|e^z| = e^x$, $\arg e^z = y + 2k\pi, \forall k \in \mathbb{Z}$. The derivative of the exponential function $(e^z)' = e^z$ is different from 0 on the entire plane $\mathbb{C}$, because $|(e^z)'| = |e^z| = e^x \neq 0, \forall x \in \mathbb{R}$. Therefore, $w = e^z$ is univalent at any point $z \in \mathbb{C}$ and, together with analyticity, it implies that $w = e^z$ is a conformal transformation at every point of the finite complex plane.

3. Using the representation

$$e^z = e^{x+iy} = e^x (\cos y + i \sin y) , \tag{4.48}$$

we can evaluate $e^{z+2k\pi i}$ for $\forall k \in \mathbb{Z}$ as follows:

$$e^{z+2k\pi i} = e^{x+i(y+2k\pi)} = e^x (\cos (y + 2k\pi) + i \sin (y + 2k\pi)) = e^x (\cos y + i \sin y) = e^x e^{iy} = e^z ,$$

since $\cos y$ and $\sin y$ are periodic functions with the period 2π. Therefore, e^z is a periodic function with the period $2\pi i$.

4. The periodicity of e^z implies that this function is not univalent in the entire complex plane $\mathbb{C}$, although it is univalent at every point $z \in \mathbb{C}$. For instance, the function $w = e^z$ carries the different points z and $z + 2k\pi i$, $\forall k \in \mathbb{Z}$, $k \neq 0$ of the z-plane to the same point in the w-plane, since $e^{z+2k\pi i} = e^z, \forall k \in \mathbb{Z}, \forall z \in \mathbb{C}$. Let us find the domains of univalence. If a point z_0 belongs to a domain of univalence, then no other point of the form $z_0 + 2k\pi i$, where $\forall k \in \mathbb{Z} \backslash \{0\}$, can belong to the same domain. Therefore, the domains of univalence are strips such that any vertical section of a strip has the length less than or equal to 2π. It is more convenient to choose the infinite horizontal strips of width 2π whose boundaries are parallel to the real axis. There are an infinite number of these strips D_k, $\forall k \in \mathbb{Z}$, which divide the entire plane $\mathbb{C}_z$. Usually the principal strip is chosen in the form

$$D_0 = \{\forall z \in \mathbb{C} : \ -\infty < x < +\infty ; \ -\pi < y < \pi\}$$

or

$$D_0' = \{\forall z \in \mathbb{C} : \ -\infty < x < +\infty ; \ 0 < y < 2\pi\} .$$

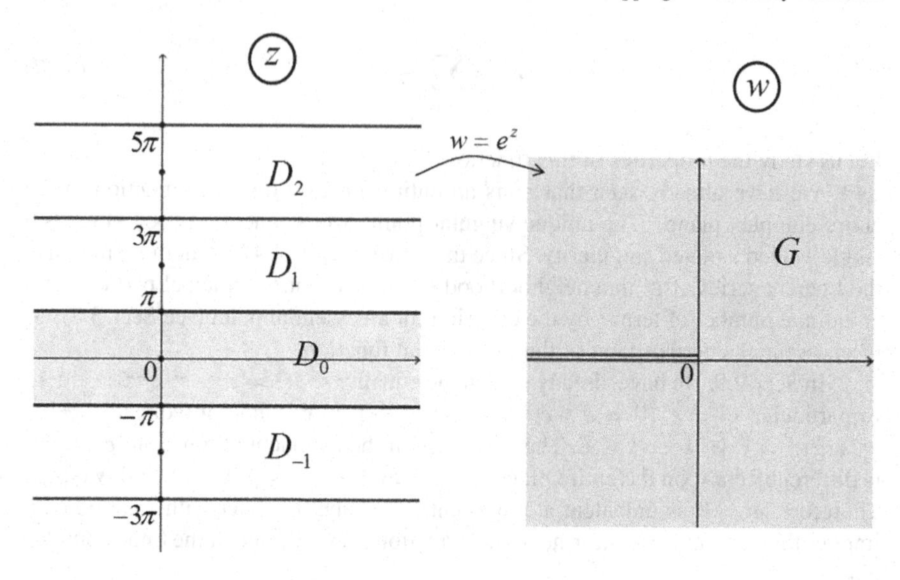

Fig. 4.27 Domains of univalence of the function $w = e^z$

The next strips D_k or $D_k^{'}$, $\forall k \in \mathbb{Z}$ are obtained using parallel translations of the principal strip D_0 or $D_0^{'}$ by $2k\pi$, $k \in \mathbb{Z}$ along the imaginary axis: the translations upward when $k > 0$ and downward when $k < 0$ (see Fig. 4.27).

5. Let us find the images of the net of the Cartesian coordinate lines (that is, the straight lines $x = constant$ and $y = constant$) under the exponential function. Representing w in the form $w = \rho e^{i\theta}$, we obtain from the definition of the function $w = e^z = e^{x+iy} = e^x e^{iy}$ that

$$\begin{cases} \rho = e^x, \\ \theta = y + 2k\pi, \ \forall k \in \mathbb{Z}. \end{cases} \tag{4.49}$$

Considering first the straight lines $x = c$, $c = constant$, $\forall c \in \mathbb{R}$, from (4.49) we see that $\rho = e^c = c_1 > 0$, that is, $\rho = constant$. This means that the straight line $x = c$ in the z-plane is mapped onto the circle of the radius $\rho = c_1$ $(c_1 = e^c)$ in the w-plane (see Fig. 4.28). Next, taking the straight lines $y = c, c = constant, \forall c \in \mathbb{R}$, we deduce using (4.49) that $\theta = constant$. Therefore, the straight line $y = c$ in the z-plane is mapped onto the ray emanating from the origin and having the angle $y = c$ with the real axis. Notice that $y = c$ and $y = c + 2k\pi$, $\forall k \in \mathbb{Z}$ are mapped onto the same ray in the w-plane. Thus, under the transformation $w = e^z$, the Cartesian net in the z-plane is mapped onto the polar net in the w-plane (see Fig. 4.28).

6. We now return to the domains of univalence of $w = f(z) = e^z$. Take, for instance, the domain D_0 and consider its boundary. We have just shown that the straight lines $y = constant$ are mapped onto rays emanating from the origin, in particular, the line $y = -\pi$ is mapped onto the ray $\theta = -\pi$ and the line $y = \pi$ onto

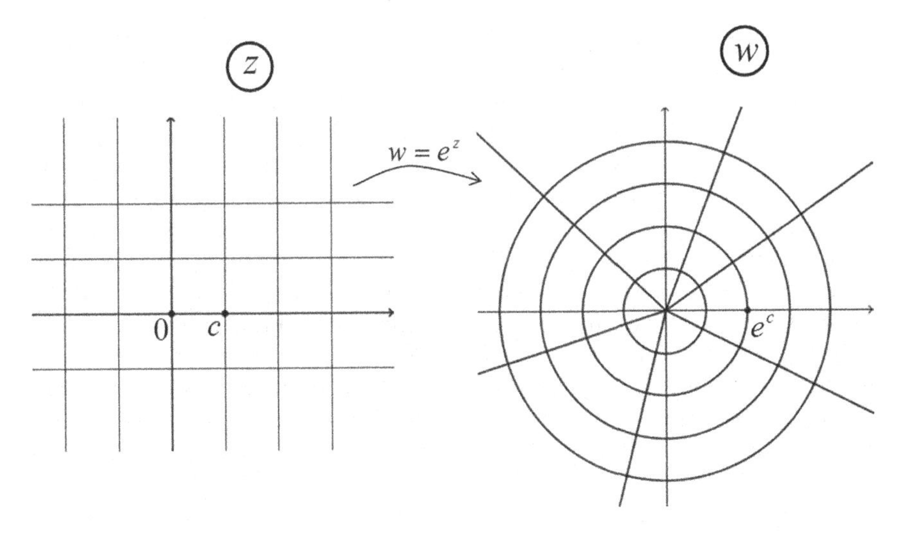

Fig. 4.28 The images of the Cartesian coordinate lines under the exponential function

the ray $\theta = \pi$. Then, the image G of D_0 is the entire w-plane with a cut along the ray $(-\infty,\ 0]$ (see Fig. 4.27). Since the function $w = e^z$ is analytic and univalent in D_0, it maps D_0 on G bijectively. Therefore, there exists the inverse function $g_0(w)$, which is analytic in G and maps G onto D_0 bijectively. On the other hand, $w = e^z$ maps the strip $D_1 = \{\forall z \in \mathbb{C} : \pi < y < 3\pi\}$ again onto G by $w = e^z$ analytically and univalently, that is, bijectively. Consequently, there exists the inverse function $g_1(w)$, which is analytic in G and maps G onto D_1 bijectively. In the same way, each strip D_k ($\forall k \in \mathbb{Z}$) is mapped by $w = e^z$ onto G analytically and univalently, which means bijectively. Therefore, there exists the inverse function $g_k(w)$, which is analytic in G and transforms G into D_k bijectively. The functions $g_k(w)$, $\forall k \in \mathbb{Z}$ are the analytic branches (elements) of the analytic function inverse to $w = e^z$.

Pick an arbitrary $w \neq 0$ in the w-plane and find all the solutions of the equation $e^z = w$. Representing z in the algebraic form and w in the exponential—$z = x + iy$; $w = \rho\, e^{i\theta}$—we rewrite the equation $e^z = w$ in the following form: $e^{x+iy} = e^x \cdot e^{iy} = \rho e^{i\theta}$. Consequently, $e^x = \rho$, $y = \theta + 2k\pi$ or, equivalently, $x = \ln \rho$, $y = \theta + 2k\pi$, $\forall k \in \mathbb{Z}$. Then, we obtain the solutions

$$z = x + iy = \ln \rho + i\theta + 2k\pi i = \ln |w| + i \arg w + 2k\pi i = \ln w, \ \forall k \in \mathbb{Z}\,. \tag{4.50}$$

In Sect. 2.6, we have derived the same representation for the function $\ln w$ (here we have chosen the principal value of the argument of w to be defined on the interval $(-\pi,\ \pi]$). Thus, we conclude that the *logarithmic function is inverse to the exponential*. Besides, the relation (4.50) clearly shows that the function $\ln w$ is multi-valued analytic. If in (4.50) we set $k = 0$, then we obtain the principal analytic branch $g_0(w)$ of the logarithmic function; for $k = 1$ we have the analytic branch $g_1(w)$, and so on.

Let us see how these different analytic branches of the logarithmic function are interconnected. Take an arbitrary point $w_0 = |w_0| \cdot e^{i\,\theta_0}$, $-\pi < \theta_0 \le \pi$, $w_0 \ne 0$ in the w-plane and choose the value of some of analytic branches at this point, say $g_0\,(w)$: $z_0 = g_0\,(w_0) = \ln|w_0| + i\theta_0$ (the point z_0 belongs to the domain D_0). Consider a simple closed curve L that makes one circuit around the origin starting and ending at the point w_0. One traversal of L in the positive (counterclockwise) direction reflects in increasing $\arg w_0 = \theta_0$ by 2π, and, therefore, we arrive at the value of the function $g_1(w)$ at the point w_0: $\ln|w_0| + i\theta_0 + 2\pi i = g_1\,(w_0) = z_1$ (the point z_1 lies in the domain D_1). Hence, performing one circuit in the positive direction around the origin, we pass from the branch $g_0(w)$ to the branch $g_1(w)$. Making one more loop along L in the positive direction we pass from $g_1(w)$ to $g_2(w)$, and so on. If we traverse L clockwise, then we switch from the branch $g_0(w)$ to the branch $g_{-1}(w)$, next from $g_{-1}(w)$ to $g_{-2}(w)$, and so on. Therefore, a rotation about $w = 0$ changes the analytic branches of the logarithmic function. Continuing rotation, with

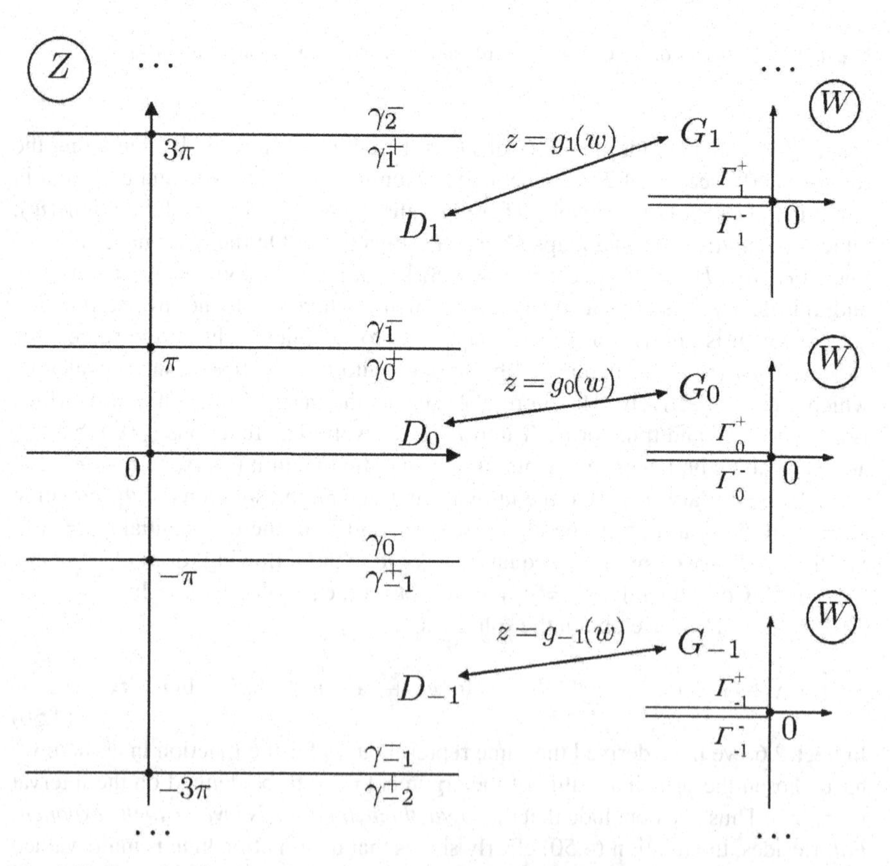

Fig. 4.29 Domains G_k, $\forall k \in \mathbb{Z}$ in the w-plane and one-to-one correspondence between G_k and D_k, $\forall k \in \mathbb{Z}$

each circuit we obtain a new branch and in this manner we pass through an infinite number of different analytic branches of the logarithmic function $g_k(w)$, $\forall k \in \mathbb{Z}$. Thus, the *logarithmic function is a multi-valued (countable-valued) analytic function* in the domain $\mathbb{C}\backslash\{0\}$. The points $w = 0$ and $w = \infty$ are the branch points of infinite order (or logarithmic branch points) and the functions $g_k(w)$, $\forall k \in \mathbb{Z}$ are the analytic branches (elements) of $\ln w$ in the domain $G = \mathbb{C}\backslash\{(-\infty, 0]\}$.

7. Now construct the Riemann surface of the multi-valued analytic function $z = \ln w$. To this end, we take a countable set of the copies of the domain G in the w-plane and denote these domains by G_k, $\forall k \in \mathbb{Z}$. The function $z = g_k(w)$ maps each domain G_k onto D_k bijectively. Denote the upper and lower edges of the cut in each G_k by Γ_k^+ and Γ_k^-, respectively, and denote the corresponding straight lines of the boundary of each D_k by γ_k^+ and γ_k^- (see Fig. 4.29).

Take an arbitrary point $w_0 \neq 0$ in the w-plane and fix the value of the analytic branch $g_0(w)$ at this point. As we have seen, after one counterclockwise circuit around $w = 0$ along a simple closed curve L, we change the branch $g_0(w)$ by $g_1(w)$, that is, move from the domain D_0 to D_1 (correspondingly, from G_0 to G_1). Therefore, we have to paste the domains D_0 and D_1 along their common line $\left(\gamma_0^+ = \gamma_1^-\right)$, and consequently, we connect the replicas G_0 and G_1 of the w-plane along the corresponding rays: the ray Γ_0^+ is glued with the ray Γ_1^-. At the next rotation along L, we change from $g_1(w)$ to $g_2(w)$, that is, we pass from D_1 to D_2 and, correspondingly, from G_1 to G_2 in the w-plane. Then, we paste the domains D_1 and D_2 by the common line $\left(\gamma_1^+ = \gamma_2^-\right)$, and consequently, we glue together the sheets G_1 and G_2 along the corresponding rays Γ_1^+ and Γ_2^-. Continuing this process infinitely, for each integer k, we connect the domains D_k and D_{k+1} along their common line and, correspondingly, glue together the sheets G_k and G_{k+1} along the corresponding rays Γ_k^+ and Γ_{k+1}^-

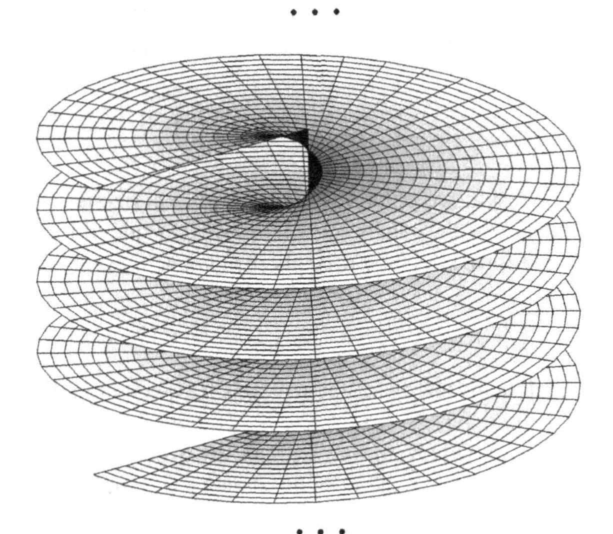

Fig. 4.30 The Riemann surface for $z = \ln w$

in the w-plane. The result of this infinite process is the entire z-plane and a surface consisting of infinite number of sheets of the w-plane. The latter surface is the Riemann surface of the multi-valued analytic function $\ln w$ (see Fig. 4.30).

Consider now an example of the application of the exponential function.

Example Construct a conformal mapping of the given domain $D = \begin{cases} |z+4i| > 4 \\ \operatorname{Im} z < 0, \ \operatorname{Re} z < 0 \end{cases}$ onto the upper half-plane.

The illustration of the domain D is given on the left side of Fig. 4.31. The boundary of the domain D consists of the arcs of the three circles (considering that a straight line is a circle with infinite radius) and all these circles pass through the origin. To "rectify" these arcs, we can use the fractional linear function which brings the origin to infinity:

$$z_1 = \frac{1}{z} . \tag{4.51}$$

Let us find the images of each of the three arcs under the transformation (4.51). The interval $\gamma_1 = (-\infty, 0]$ is transformed into itself. The part of the imaginary axis $\gamma_3 = (-i\infty, -8i]$ is carried into the interval $[0, i/8]$. Finally, the arc γ_2 of the circle $|z+4i| = 4$ is mapped onto a ray emanating from the point $i/8$. To specify the location of this ray, we pick one more point of the arc γ_2, for instance, $a = -4 - 4i \in \gamma_2$ and obtain

$$z_1(a) = \frac{1}{-4-4i} = -\frac{1}{4} \cdot \frac{1-i}{2} = -\frac{1}{8} + \frac{i}{8} .$$

This means that the image of γ_2 under (4.51) is the ray parallel to the real axis. Hence, we obtain the semi-infinite horizontal strip in the z_1-plane shown in the middle part of Fig. 4.31.

Next we slightly modify the obtained strip performing successively the rotation by π, parallel translation and stretching expressed by the formula

$$z_2 = \left(z_1 e^{i\pi} + \frac{i}{8} \right) \cdot 8\pi = -8\pi z_1 + i\pi . \tag{4.52}$$

As a result, we obtain the semi-infinite horizontal strip of width π in the positive direction of the x-axis shown on the right side of Fig. 4.31.

Now, we are ready to apply the exponential function

$$z_3 = e^{z_2} , \tag{4.53}$$

which maps the ray of the real axis $[0, +\infty)$ onto the ray $[1, +\infty)$, the ray parallel to the real axis $z_2 = x_2 + i\pi$, $x_2 \geq 0$ onto the ray $(-\infty, -1]$, and the interval of the imaginary axis from 0 to $i\pi$ onto the upper unit half-circle. In this way, we obtain the domain D_3 in the z_3-plane shown in Fig. 4.32, on the left.

Finally, the Joukowski function

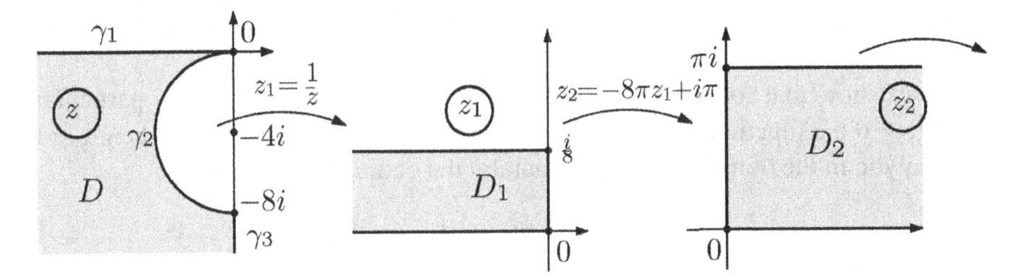

Fig. 4.31 Example: the original domain D (on the left), its image D_1 under the transformation (4.51) (in the middle) and the domain D_2 obtained as the successive rotation, translation, and stretching (4.52) of D_1 (on the right)

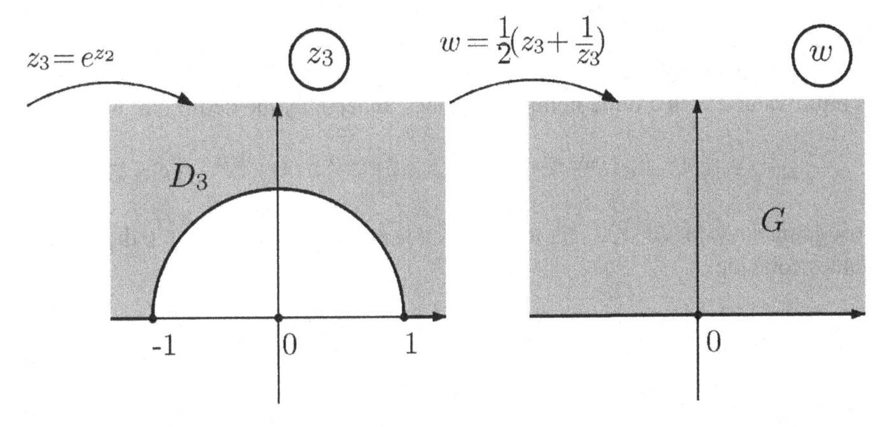

Fig. 4.32 Example: the image D_3 of the half-strip D_2 under the exponential function (4.53) (on the left), and the upper half-plane G obtained by application of the Joukowski function (4.54) to D_3 (on the right)

$$w = \frac{1}{2}\left(z_3 + \frac{1}{z_3}\right) \tag{4.54}$$

transforms the domain D_3 into the upper half-plane G (see Fig. 4.32, on the right). Hence, the composition of the functions (4.51)–(4.54) performs the required conformal transformation.

4.9 Function $w = z^\alpha$

The *function $w = z^\alpha$ is defined by the formula $w = z^\alpha = e^{\alpha \ln z}$*, $\forall \alpha \in \mathbb{C}$. Therefore, by the definition, this function is a composition of the already studied functions: the logarithmic function $z_1 = \ln z$, the linear function $z_2 = \alpha z$ and the exponential function $w = e^{z_2}$. The function $z_1 = \ln z$ is multi-valued analytic, and consequently,

$w = z^\alpha$ is (generally) multi-valued analytic. The points $z = 0$ and $z = \infty$ are the logarithmic branch points.

Let us show that for $\alpha \in \mathbb{Z}$ the function $w = z^\alpha$ is single-valued; in particular, when $\alpha \geq 0$ the function is analytic in the entire complex plane $\mathbb{C}$, and when $\alpha < 0$ it is analytic in the domain $\overline{\mathbb{C}} \setminus \{0\}$. In fact, by the definition, we have

$$w = z^\alpha = e^{\alpha \ln z} = e^{\alpha(\ln|z| + i \arg z + 2k\pi i)} = e^{\alpha(\ln|z| + i \arg z)} \cdot e^{2\alpha k \pi i}$$

and if α is an integer, then so is $k \cdot \alpha$ and consequently $e^{2\alpha k\pi i} = 1$, $\forall k \in \mathbb{Z}$. Therefore, for α integer and $\alpha \geq 0$ the function $w = z^\alpha$ has a unique value at any point $z \in \mathbb{C}$, and for α integer and $\alpha < 0$ it takes a single value at any point $z \in \overline{\mathbb{C}} \setminus \{0\}$. Notice that if $\alpha \in \mathbb{N}$ the function has the form of the power function z^n, and if $\alpha < 0$, $\alpha \in \mathbb{Z}$ $(\alpha = -n, n \in \mathbb{N})$ the function takes the form $z^{-n} = \frac{1}{z^n}$.

If α is a rational number $\left(\alpha = \frac{m}{n}, m \in \mathbb{Z}, n \in \mathbb{N}\right)$, then the function $w = z^\alpha = z^{\frac{m}{n}}$ is multi-valued with a finite number of values. Indeed, by the definition, we have

$$z^{\frac{m}{n}} = e^{\frac{m}{n} \ln z} = e^{\frac{m}{n}(\ln|z| + i \arg z + 2k\pi i)} = e^{\frac{m}{n} \ln|z| + i \frac{m}{n} \arg z + i \frac{2mk\pi}{n}}, \quad \forall k \in \mathbb{Z}.$$

Assigning the values $0, 1, 2, \ldots$ to k, we notice that for $k = n$ and $k = 0$ the function values coincide:

$$\left(z^{\frac{m}{n}}\right)_{k=n} = e^{\frac{m}{n} \ln|z| + i \frac{m}{n} \arg z + i \frac{2mn\pi}{n}} = e^{\frac{m}{n} \ln|z| + i \frac{m}{n} \arg z} = \left(z^{\frac{m}{n}}\right)_{k=0}.$$

The same occurs for $k = n + 1$ and $k = 1$: $\left(z^{\frac{m}{n}}\right)_{k=n+1} = \left(z^{\frac{m}{n}}\right)_{k=1}$, for $k = -1$ and $k = n - 1$: $\left(z^{\frac{m}{n}}\right)_{k=-1} = \left(z^{\frac{m}{n}}\right)_{k=n-1}$, and so on. Hence, the function $w = z^{\frac{m}{n}}$ takes exactly n different values at each $z \neq 0$ if $\alpha = \frac{m}{n}$.

If α is an irrational or complex number, then $w = z^\alpha$ is a multi-valued analytic function with an infinite number of values. Indeed, for irrational α, we have

$$z^\alpha = e^{\alpha \ln|z| + i\alpha \cdot \arg z + 2k\pi\alpha i}, \quad \forall k \in \mathbb{Z},$$

where $k \cdot \alpha$ is not an integer for any integer k different from 0. Then, we deduce that $(z^\alpha)_k \neq (z^\alpha)_0$, $\forall k \in \mathbb{Z}$, $k \neq 0$. Analogously, $(z^\alpha)_k \neq (z^\alpha)_1$, $\forall k \in \mathbb{Z}$, $k \neq 1$, and so on. Hence, taking $\forall k \in \mathbb{Z}$, $k \neq k_0$, we have

$$(z^\alpha)_k = e^{\alpha \ln|z| + i\alpha \cdot \arg z + 2k\pi\alpha i} \neq e^{\alpha \ln|z| + i\alpha \cdot \arg z + 2k_0\pi\alpha i} = (z^\alpha)_{k_0}.$$

If α is a complex number, $\alpha = \alpha_1 + i\,\alpha_2$ $(\alpha_2 \neq 0)$, then

$$z^\alpha = e^{(\alpha_1 + i\alpha_2)\cdot(\ln|z| + i \arg z + 2k\pi i)} = e^{(\alpha_1 \ln|z| - \alpha_2 \arg z - 2k\pi\alpha_2) + i(\alpha_2 \ln|z| + \alpha_1 \arg z + 2k\pi\alpha_1)},$$

that is, assigning different integers to k we obtain different values of the function. In this case, the change of k alters both the argument and absolute value of the function $w = z^\alpha$.

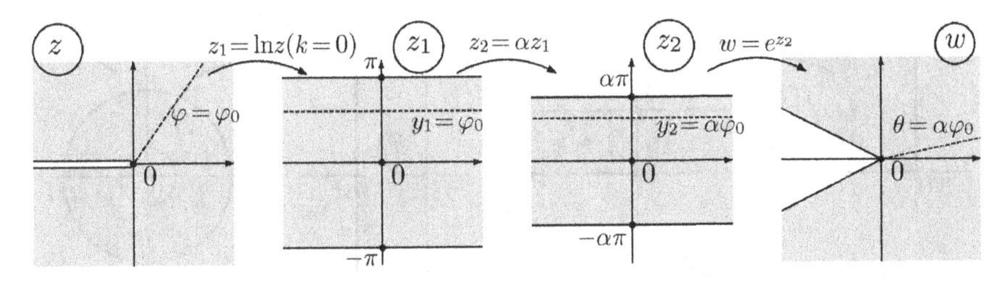

Fig. 4.33 Transformation of rays under the function $w = z^\alpha$, $\alpha \in (0, 1)$

Thus, $w = z^\alpha$ is an analytic single-valued function for $\forall\alpha \in \mathbb{Z}$, it takes a finite number of values if α is rational and infinite number of values if α is irrational or complex with non-zero imaginary part.

Let us see what kind of mapping is performed under the function $w = z^\alpha$ when α is real positive. To this end, choose some single-valued analytic branch of the logarithmic function, say the branch corresponding to $k = 0$ in the z-plane with a cut along the negative part of the real axis. Take a ray $\varphi = \varphi_0$, $-\pi < \varphi_0 \le \pi$ in the z-plane and consider the function $w = z^\alpha = e^{\alpha \ln z}$ as the composition of the known transformations: $z_1 = \ln z$, $z_2 = \alpha z_1$, and $w = e^{z_2}$. For the first function—$z_1 = \ln z = \ln |z| + i \arg z + 2k\pi i$—we choose the analytic branch corresponding to $k = 0$ and obtain the straight line $y_1 = \varphi_0$, $-\pi < \varphi_0 \le \pi$ in the z_1-plane, where $z_1 = x_1 + iy_1$. On the next step, this straight line is transformed into another straight line $y_2 = \alpha\varphi_0$ in the z_2-plane, $z_2 = x_2 + iy_2$. Finally, the last line is mapped onto the ray $\theta = \alpha\varphi_0$ in the w-plane, $w = \rho e^{i\theta}$. Hence, the function $w = z^\alpha$, $\alpha > 0$, changes the angles of the vectors emanating from the origin α times. Figure 4.33 illustrates these transformations in the case $\alpha \in (0, 1)$.

Example 4.1 Let us calculate the number i^i. By definition, $i^i = e^{i \ln i}$. Representing the number i in the exponential form $i = 1 \cdot e^{i\frac{\pi}{2}}$, we have $i^i = e^{i \ln i} = e^{i\left(\ln 1 + i\frac{\pi}{2} + 2k\pi i\right)} = e^{-\frac{\pi}{2} - 2k\pi}$, $\forall k \in \mathbb{Z}$. Therefore, we obtain a countable set of the positive real values.

Example 4.2 Let us find the image of the first quadrant $D = \{z : 0 < \arg z < \frac{\pi}{2}\}$ under the function $w = z^i$ and verify if this mapping is one-to-one.

According to the definition, the function $w = z^i = e^{i \ln z}$ is represented as a composition of the three functions: the (multi-valued) logarithmic function $z_1 = \ln z$, the linear function $z_2 = iz_1 = e^{i\pi/2} z_1$, and the exponential function $w = e^{z_2}$. In what follows we use the standard notations $z = x + iy = re^{i\varphi}$, $z_1 = x_1 + iy_1$, $z_2 = x_2 + iy_2$, $w = u + iv$.

In the given domain D, one can choose a single-valued analytic branch of $\ln z$ by fixing the variation of the argument: $\varphi = \arg z \in (-\pi, \pi]$. Then, the first function takes the form $z_1 = \ln z = \ln |z| + i\varphi$, which gives the relations $x_1 = \ln |z|$ and $y_1 = \varphi$. Therefore, the ray $\varphi = 0$ is carried into the line $y_1 = 0$ and the ray $\varphi = \frac{\pi}{2}$ – into the line $y_1 = \frac{\pi}{2}$ (see Fig. 4.34). Since $|z|$ varies from 0 to $+\infty$, the values of $\ln |z|$ range from $-\infty$ to $+\infty$, that is, the corresponding point z_1 moves along the entire

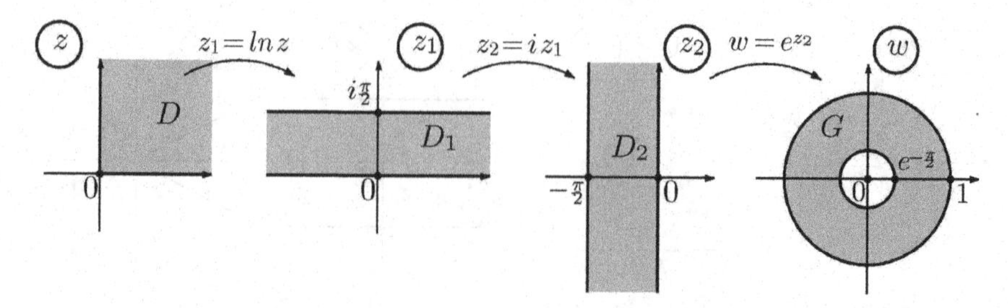

Fig. 4.34 Transformation of the domain $D = \{0 < \arg z < \frac{\pi}{2}\}$ by the function $w = z^i$

lines $y_1 = 0$ and $y_1 = \frac{\pi}{2}$. Hence, in the z_1-plane, we obtain the infinite horizontal strip D_1 between the lines $y_1 = 0$ and $y_1 = \frac{\pi}{2}$ shown in Fig. 4.34. For the chosen single-valued branch of $\ln z$, the mapping of D onto D_1 is bijective.

The second function $z_2 = e^{i\pi/2} z_1$ rotates the strip D_1 counterclockwise by the angle $\frac{\pi}{2}$ about the origin, and consequently, in the z_2-plane we get the vertical strip D_2 between the lines $x_2 = 0$ and $x_2 = -\frac{\pi}{2}$: $D_2 = \{z_2 : -\frac{\pi}{2} < x_2 < 0\}$ (see Fig. 4.34).

Finally, the function $w = e^{z_2} = e^{x_2} e^{iy_2}$ carries the line $x_2 = 0$ into the unit circle $|w| = 1$, and the line $x_2 = -\frac{\pi}{2}$—into the circle $|w| = e^{-\pi/2}$. In this way, the vertical strip D_2 is transformed into the ring $G = \{e^{-\pi/2} < |w| < 1\}$ (see Fig. 4.34). Notice that the last mapping is not univalent (and not bijective), because different points of D_2 such as $z_2 = x_2 + iy_2 + 2k\pi i$, $\forall k \in \mathbb{Z}$ are carried to the same point of G: $w = e^{x_2+iy_2+2k\pi i} = e^{x_2+iy_2}$, $\forall k \in \mathbb{Z}$.

Thus, the function $w = z^i$ maps the first quadrant D onto the ring G. This transformation is single-valued since we have chosen a single-valued analytic branch of $\ln z$, but it is not univalent. If we choose another analytic branch of $\ln z$, we obtain another ring in the w-plane.

4.10 Functions $w = \cos z$ and $w = \sin z$

In Sect. 2.9, we have introduced the *trigonometric functions* $\sin z$ *and* $\cos z$ as analytic continuations of the functions $\sin x$ and $\cos x$ from the real axis to the entire complex plane:

$$\sin z = \sum_{n=0}^{+\infty} (-1)^n \frac{z^{2n+1}}{(2n+1)!}, \quad \cos z = \sum_{n=0}^{+\infty} (-1)^n \frac{z^{2n}}{(2n)!}. \tag{4.55}$$

In this section, we study the properties of these two functions.

1. We have already shown that $w = \sin z$ and $w = \cos z$ are entire functions. The infinity point is a single-valued isolated singularity for both functions. The power series (4.55) of the functions $\sin z$ and $\cos z$ can be considered as the Laurent series in a neighborhood of $z = \infty$. Since the principal parts of these Laurent series contain an

infinite number of terms, the point $z = \infty$ is an essential singularity of the functions $\sin z$ and $\cos z$ (according to the criterion for an essential singular point in Sect. 3.8).

2. Also, we have proved the Euler formulas for any complex number:

$$\cos z = \frac{e^{iz} + e^{-iz}}{2}, \quad \sin z = \frac{e^{iz} - e^{-iz}}{2i}. \tag{4.56}$$

Let us derive the lower and upper bounds for $|\cos z|$ and $|\sin z|$. We start with the simple inequalities following from the properties of absolute values:

$$\begin{aligned} \frac{1}{2}\left|\left|e^{iz}\right| - \left|e^{-iz}\right|\right| &\leq |\cos z| \leq \frac{1}{2}\left(\left|e^{iz}\right| + \left|e^{-iz}\right|\right), \\ \frac{1}{2}\left|\left|e^{iz}\right| - \left|e^{-iz}\right|\right| &\leq |\sin z| \leq \frac{1}{2}\left(\left|e^{iz}\right| + \left|e^{-iz}\right|\right). \end{aligned} \tag{4.57}$$

Representing z in the algebraic form $z = x + iy$, we have

$$\left|e^{iz}\right| = \left|e^{ix-y}\right| = e^{-y}, \ \left|e^{-iz}\right| = \left|e^{-ix+y}\right| = |e^{-ix}| \cdot |e^{y}| = e^{y}.$$

Notice also that for $y \geq 0$ one has $e^y \geq e^{-y}$, and consequently

$$\left|\left|e^{iz}\right| - \left|e^{-iz}\right|\right| = \left|e^{y} - e^{-y}\right| = e^{y} - e^{-y} = e^{|y|} - e^{-|y|}$$

(since $|y| = y$ when $y \geq 0$). Otherwise, if $y < 0$, then $e^y < e^{-y}$ and

$$\left|\left|e^{iz}\right| - \left|e^{-iz}\right|\right| = \left|e^{-y} - e^{y}\right| = e^{-y} - e^{y} = e^{|y|} - e^{-|y|}$$

(since $|y| = -y$ when $y < 0$). Therefore, the chains of the inequalities in (4.57) take the following form:

$$\begin{aligned} \frac{1}{2}\left(e^{|y|} - e^{-|y|}\right) &\leq |\cos z| \leq \frac{1}{2}\left(e^{|y|} + e^{-|y|}\right), \\ \frac{1}{2}\left(e^{|y|} - e^{-|y|}\right) &\leq |\sin z| \leq \frac{1}{2}\left(e^{|y|} + e^{-|y|}\right). \end{aligned} \tag{4.58}$$

The left-hand sides of these inequalities show that $\cos z$ and $\sin z$ can be zero only if $e^{|y|} - e^{-|y|} = 0$, which means if $y = 0$, that is, only for the real numbers. However, if z is a real number ($z = x$), the zeros of $\sin x$ and $\cos x$ are well-known:

$$\sin x = 0 \text{ if } x = k\pi, \ \forall k \in \mathbb{Z}; \ \cos x = 0 \text{ if } x = \frac{\pi}{2} + n\pi, \ \forall n \in \mathbb{Z}.$$

These are all zeros of $\sin z$ and $\cos z$, and all of them lie on the real axis, that is, these functions do not vanish outside the real axis.

If $y \to \infty$, then the left-hand sides in (4.58) growth infinitely: $\frac{1}{2}\left(e^{|y|} - e^{-|y|}\right) \to \infty$, and consequently $|\cos z| \to \infty$ and $|\sin z| \to \infty$ as $y \to \infty$, which means that the functions $\sin z$ and $\cos z$ are not bounded on the complex plane (although they are bounded for real z: $|\sin x| \le 1$ and $|\cos x| \le 1$, $\forall x \in \mathbb{R}$). Moreover, from (4.58) we obtain the growth rate of $|\cos z|$ and $|\sin z|$ as $y \to \infty$. Indeed, dividing all the sides of the inequalities in (4.58) by $\frac{e^{|y|}}{2}$, we have

$$1 - e^{-2|y|} \le \frac{|\cos z|}{\frac{1}{2}e^{|y|}} \le 1 + e^{-2|y|}, \; 1 - e^{-2|y|} \le \frac{|\sin z|}{\frac{1}{2}e^{|y|}} \le 1 + e^{-2|y|}. \qquad (4.59)$$

Applying the limit to the inequalities in (4.59) as $y \to \pm\infty$, we have (using the squeeze theorem) that $\lim\limits_{y\to\infty} \frac{|\cos z|}{\frac{1}{2}e^{|y|}} = 1$ and $\lim\limits_{y\to\infty} \frac{|\sin z|}{\frac{1}{2}e^{|y|}} = 1$, that is, $|\cos z|$ and $|\sin z|$ are equivalent to the function $\frac{1}{2}e^{|y|}$ as $y \to \infty$, which means that both functions increase with the rate of the exponential function.

3. The derivatives of $\sin z$ and $\cos z$ can be calculated using the Euler formulas:

$$(\cos z)' = \left(\frac{e^{iz} + e^{-iz}}{2}\right)' = \frac{ie^{iz} - ie^{-iz}}{2} = -\frac{e^{iz} - e^{-iz}}{2i} = -\sin z,$$

$$(\sin z)' = \left(\frac{e^{iz} - e^{-iz}}{2i}\right)' = \frac{ie^{iz} + ie^{-iz}}{2i} = \frac{e^{iz} + e^{-iz}}{2} = \cos z.$$

Taking into account the above results on zeros of the trigonometric functions, we deduce that the derivative of $\cos z$ vanishes at the points $z_k = k\pi$, $\forall k \in \mathbb{Z}$, and the derivative of $\sin z$ vanishes at $z_n = \frac{\pi}{2} + n\pi$, $\forall n \in \mathbb{Z}$. Therefore, the transformation $\cos z$ is neither univalent nor conformal at the points $z_k = k\pi$, $\forall k \in \mathbb{Z}$; the same is true for $\sin z$ at the points $z_n = \frac{\pi}{2} + n\pi$, $\forall n \in \mathbb{Z}$. At other points of the complex plain $\mathbb{C}$ both transformations are univalent and conformal.

4. Recalling that the function e^z has the period $2\pi i$, we can use the Euler formulas (4.56) to deduce that $\sin z$ and $\cos z$ has the period 2π:

$$\cos(z + 2k\pi) = \frac{e^{iz+2k\pi i} + e^{-iz-2k\pi i}}{2} = \frac{e^{iz} + e^{-iz}}{2} = \cos z, \; \forall k \in \mathbb{Z}, \; \forall z \in \mathbb{C}$$

and

$$\sin(z + 2k\pi) = \frac{e^{iz+2k\pi i} - e^{-iz-2k\pi i}}{2i} = \frac{e^{iz} - e^{-iz}}{2i} = \sin z, \; \forall k \in \mathbb{Z}, \; \forall z \in \mathbb{C}.$$

The periodicity implies that $\sin z$ and $\cos z$ are not univalent on the complex plane even after elimination of the zeros of the derivatives, because these functions map different points z_0 and $z_0 + 2k\pi$ ($\forall k \in \mathbb{Z}$, $k \ne 0$) of the z-plane into the same point of the w-plane.

5. To reveal the transformation performed by the function $w = \cos z$, it is useful to represent it as the composition of the already studied functions:

$$z_1 = iz = e^{i\frac{\pi}{2}} z\,, \quad z_2 = e^{z_1}\,, \quad w = \frac{1}{2}\left(z_2 + \frac{1}{z_2}\right) = \cos z\,. \tag{4.60}$$

Therefore, $w = \cos z$ is composed by the rotation, exponential function and Joukowski function. Similarly, $w = \sin z$ can be represented as the composition of the same functions with additional rotation:

$$z_1 = iz = e^{i\frac{\pi}{2}} z,\ z_2 = e^{z_1},\ z_3 = \frac{1}{i} z_2 = -iz_2 = e^{-i\frac{\pi}{2}} z_2\,,\ w = \sin z = \frac{1}{2}\left(z_3 + \frac{1}{z_3}\right). \tag{4.61}$$

Knowing how works each of the involved transformations, we can analyze the transformations $w = \cos z$ and $w = \sin z$.

First we find the domains of univalence of $w = \cos z$ using (4.60). The function $z_1 = e^{i\frac{\pi}{2}} z = iz$ is a bijective mapping between the extended complex z-plane and the extended complex z_1-plane, keeping, therefore, the univalence for the entire extended complex plane. The second function $z_2 = e^{z_1}$ is not univalent on the entire complex z_1-plane and we should choose one of its domains of univalence, for example, a strip $D_0 = \{\forall z_1 = x_1 + iy_1 \in \mathbb{C} : \forall x_1 \in \mathbb{R}, -\pi < y_1 < \pi\}$. Then, $z_2 = e^{z_1}$ transforms this strip into the z_2-plane with a cut along the negative part of the real axis, which we denote by G_0. The third function $w = \frac{1}{2}\left(z_2 + \frac{1}{z_2}\right)$ is not univalent on G_0. The simplest domains of the univalence of the Joukowski function are the upper and lower half-plane and also the interior and exterior of the unit disk: $G_1 = \{\forall z_2 \in \mathbb{C} :\ \operatorname{Im} z_2 > 0\}$, $G_2 = \{\forall z_2 \in \mathbb{C} :\ \operatorname{Im} z_2 < 0\}$, $G_3 = \{\forall z_2 \in \mathbb{C} :\ |z_2| < 1\}$ and $G_4 = \{\forall z_2 \in \mathbb{C} :\ |z_2| > 1\}$.

The need to reduce the domain G_0 in order to establish the univalent mapping under the Joukowski function implies the correspondent shrinking of D_0, whatever domain of univalence of the Joukowski function is chosen. If we take the upper half-plane G_1, which is mapped by the Joukowski function onto the entire w-plane with the cuts along the rays $(-\infty, -1]$ and $[1, +\infty)$, then the domain in the z_1-plane, which the function $z_2 = e^{z_1}$ maps univalently on G_1, is a strip $D_1 = \{\forall z_1 \in \mathbb{C} :\ \forall x_1 \in \mathbb{R},\ 0 < y_1 < \pi\}$. If we choose the lower half-plane G_2 as the domain of univalence of the Joukowski function, then in the z_1-plane we need to use the strip $D_2 = \{\forall z_1 \in \mathbb{C} :\ \forall x_1 \in \mathbb{R},\ -\pi < y_1 < 0\}$. The illustration of these two cases is provided in Fig. 4.35.

Choosing in G_0 the domains of univalence G_3 or G_4 of the Joukowski function, we have the interior of the unit disk with a cut along the segment $[-1, 0]$ or the exterior of the unit disk with a cut along $(-\infty, -1]$, which the Joukowski function maps onto the w-plane with a cut along the ray $(-\infty, 1]$ (the circle $|z_2| = 1$ is carried to the segment $[-1, 1]$, and the segment $[-1, 0]$ or the ray $(-\infty, -1]$ are carried to the ray $(-\infty, -1]$). If we take the domain G_3, then the domain in the z_1-plane, which the function $z_2 = e^{z_1}$ maps univalently on G_3, is a half-strip $D_3 = \{\forall z_1 \in \mathbb{C} :\ -\infty < x_1 < 0;\ -\pi < y_1 < \pi\}$. If we pick the domain G_4, then the corresponding domain in the z_1-plane is a half-strip $D_4 = \{\forall z_1 \in \mathbb{C} :\ 0 < x_1 < +\infty;\ -\pi < y_1 < \pi\}$. The illustration of these two choices is given in Fig. 4.36.

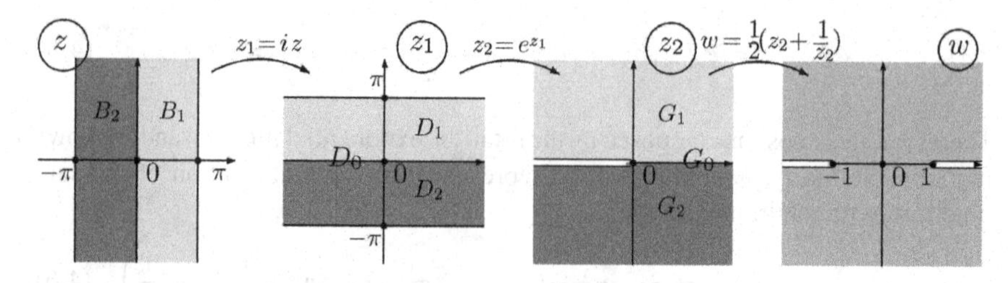

Fig. 4.35 Construction of the domains of univalence B_1 and B_2 for $w = \cos z$ in the cases $\operatorname{Im} z_2 > 0$ and $\operatorname{Im} z_2 < 0$ in the z_2-plane

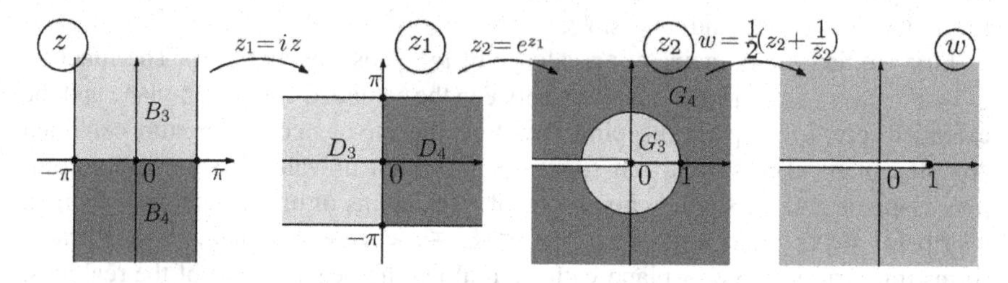

Fig. 4.36 Construction of the domains of univalence B_3 and B_4 for $w = \cos z$ in the cases $|z_2| < 1$ and $|z_2| > 1$ in the z_2-plane

Returning back to the original z-plane, we see that the relation $z = z_1 \cdot e^{-i\frac{\pi}{2}}$ means that the domains of univalence in the z-plane are obtained from the domains of univalence in the z_1-plane rotating the latter clockwise by the angle $\frac{\pi}{2}$. Therefore, we obtain the following simple examples of the domains of univalence of the function $w = \cos z$ (see Figs. 4.35 and 4.36):

$$B_1 = \{\forall z \in \mathbb{C}_z : 0 < x < \pi; -\infty < y < +\infty\};$$

$$B_2 = \{\forall z \in \mathbb{C}_z : -\pi < x < 0; -\infty < y < +\infty\};$$

$$B_3 = \{\forall z \in \mathbb{C}_z : -\pi < x < \pi; 0 < y < +\infty\};$$

$$B_4 = \{\forall z \in \mathbb{C}_z : -\pi < x < \pi; -\infty < y < 0\}.$$

Notice that any domain obtained by translation of B_1, B_2, B_3 or B_4 by 2π along the real axis is also a domain of univalence of $w = \cos z$.

The procedure of construction of domains of univalence for the function $w = \sin z$ is absolutely analogous and is left to the reader.

6. Let us show that the image of the functions $w = \cos z$ and $w = \sin z$ is the entire complex plane $\mathbb{C}$. Indeed, for any complex number $w \in \mathbb{C}$, we can solve the equation $w = \cos z$ in the following manner. Represent $\cos z$ by the Euler formula (4.56) $w = \cos z = \frac{e^{iz}+e^{-iz}}{2}$ and denote $e^{iz} = t$. Then, we have a quadratic equation

with respect to t: $\frac{t+\frac{1}{t}}{2} = w$, that is, $t^2 - 2tw + 1 = 0$. For any complex w, this equation has two roots $t_{1,2} = w \pm \sqrt{w^2 - 1}$ (which coincide for $w = \pm 1$), that is,

$$e^{iz} = w \pm \sqrt{w^2 - 1}\,. \tag{4.62}$$

Since $w \pm \sqrt{w^2 - 1} \neq 0$ for any $w \in \mathbb{C}$, the Eq. (4.62) always has a solution, moreover, the number of these solutions is infinite (countable):

$$iz_k = \ln\left(w \pm \sqrt{w^2 - 1}\right) = \ln\left|w \pm \sqrt{w^2 - 1}\right| + i \arg\left(w \pm \sqrt{w^2 - 1}\right) + 2k\pi i, \ \forall k \in \mathbb{Z}.$$

Consequently, for any complex w, the solutions to the equation $w = \cos z$ are

$$z_k = -i \ln\left|w \pm \sqrt{w^2 - 1}\right| + \arg\left(w \pm \sqrt{w^2 - 1}\right) + 2k\pi, \ \forall k \in \mathbb{Z}.$$

The equation $\sin z = w$ is solved in a similar way for any fixed $w \in \mathbb{C}$. This solution is left to the reader.

Example 4.1 Solve the equation $\cos z = 7$.

We represent $\cos z$ by the Euler formula $\cos z = \frac{e^{iz}+e^{-iz}}{2}$ and, using an auxiliary variable $t = e^{iz}$, rewrite the given equation in the form $t^2 - 14t + 1 = 0$. Solving the last equation $t_{1,2} = 7 \pm \sqrt{49 - 1} = 7 \pm 4\sqrt{3}$, we return to the variable z: $e^{iz} = 7 \pm 4\sqrt{3}$. Representing the number $7 \pm 4\sqrt{3}$ in the exponential form $7 \pm 4\sqrt{3} = \left(7 \pm 4\sqrt{3}\right) \cdot e^{i0}$, we rewrite the exponential equation as $e^{iz} = \left(7 \pm 4\sqrt{3}\right) \cdot e^{i0}$, which implies that $iz_k = \ln\left(7 \pm 4\sqrt{3}\right) + i \cdot 0 + 2k\pi i$ or, equivalently, $z_k = 2k\pi - i \ln\left(7 \pm 4\sqrt{3}\right)$, $\forall k \in \mathbb{Z}$.

Example 4.2 Find the image of the half-strip $D = \{z = x + iy : 0 < x < \pi, \ y < 0\}$ under the function $w = \sin z$ and verify if this mapping is one-to-one.

First, we represent the function $w = \sin z$ as a chain of simpler transformations: $w = \sin z = \frac{e^{iz}-e^{-iz}}{2i} = \frac{1}{2}\left(\frac{e^{iz}}{i} + \frac{i}{e^{iz}}\right)$. Accordingly, we have the chain of the functions (4.61): $z_1 = iz = e^{i\pi/2}z$; $z_2 = e^{z_1}$; $z_3 = e^{-i\pi/2}z_2$; $w = \frac{1}{2}\left(z_3 + \frac{1}{z_3}\right)$, that is, the function $w = \sin z$ is composed of the linear functions (rotations), exponential function and Joukowski function. Since $w = \sin z$ is analytic in the entire complex plane (and consequently single-valued), this mapping is one-to-one if each of the involved transformations is univalent. The function $z_1 = iz = e^{i\pi/2}z$ performs a counterclockwise rotation of the given domain D about the origin by the angle $\frac{\pi}{2}$ (see Fig. 4.37). The resulting half-strip D_1 lies in the domain of univalence of the subsequent exponential function $z_2 = e^{z_1}$. This function transforms the ray $[0, +\infty)$ of the real axis into the ray $[1, +\infty)$ of the real axis, the ray $z_1 = x_1 + i\pi$, $x_1 \in [0, +\infty)$—into the ray $(-\infty, -1]$ of the real axis, and the segment $[0, i\pi]$ of the imaginary axis —into the upper half-circle $|z_2| = 1$. Then, in the z_2-plane we get the domain D_2, representing the upper half-plane with the removed upper unit half-disk (see Fig. 4.37).

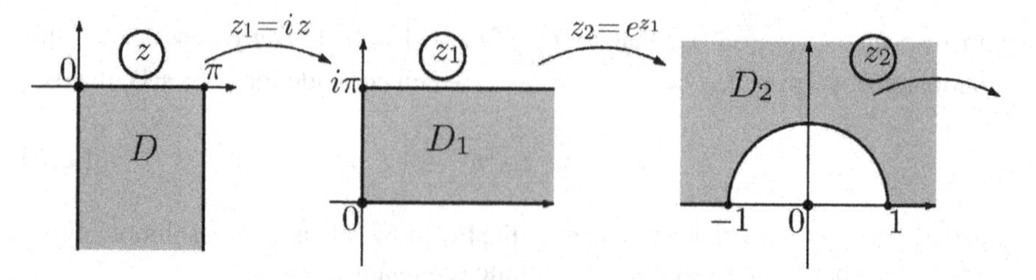

Fig. 4.37 Example 4.2: transformation of the half-strip $D = \{0 < x < \pi,\, y < 0\}$ by the function $w = \sin z$—the first two functions

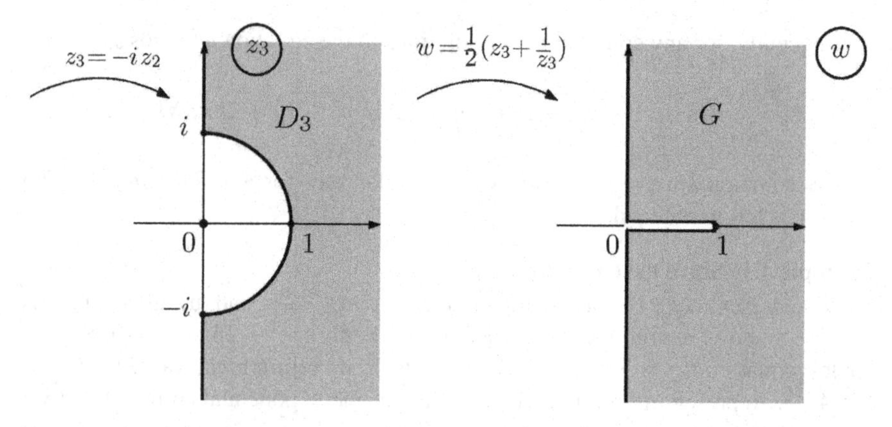

Fig. 4.38 Example 4.2: transformation of the half-strip $D = \{0 < x < \pi,\, y < 0\}$ by the function $w = \sin z$—the last two functions

The next function $z_3 = e^{-i\pi/2} z_2$ is a clockwise rotation about the origin by the angle $\frac{\pi}{2}$ (see Fig. 4.38). The obtained domain D_3 is contained in the domain of univalence of the Joukowski function. The last function carries the ray $[i, +i\infty)$ of the imaginary axis into the ray $[0, +i\infty)$, the ray $(-i\infty, -i]$—into the negative part of the imaginary axis, and the right unit half-circle $|z_3| = 1$, $\operatorname{Re} z_3 \geq 0$—into the double segment (fold) $[0, 1]$ of the real axis (see Fig. 4.38).

Therefore, the function $w = \sin z$ maps the given half-strip D onto the right half-plane with the fold (cut) along the segment $[0, 1]$. This transformation is one-to-one, because each elementary mapping involved in the constructed chain is one-to-one.

4.11 Functions tan *z* and cot *z*

The *trigonometric functions* tan *z and* cot *z* are defined by the formulas:

$$\tan z = \frac{\sin z}{\cos z}, \quad \cot z = \frac{\cos z}{\sin z}. \tag{4.63}$$

Consider some properties of these functions.

1. Since $\sin z$ and $\cos z$ are the entire functions, the function $\tan z$ is analytic in $\mathbb{C}$ except at the points where $\cos z$ is equal to zero, that is, $\tan z$ is analytic in the domain $D_1 = \mathbb{C} \backslash \left\{ z_k = \frac{\pi}{2} + k\pi,\ \forall k \in \mathbb{Z} \right\}$. All the points $z_k = \frac{\pi}{2} + k\pi, \forall k \in \mathbb{Z}$ are simple zeros of $\cos z$, and consequently, they are simple poles of $\tan z$. Notice that $z_k = \frac{\pi}{2} + k\pi \to \infty$ as $k \to \infty$, that is, the poles of $\tan z$ are found in any deleted neighborhood of the point ∞, which means that ∞ is a singular point, which is a limit point of the poles of $\tan z$. Similarly, the function $\cot z$ is analytic in the domain $D_2 = \mathbb{C} \backslash \{z_n = n\pi,\ \forall n \in \mathbb{Z}\}$, the points $z_n = n\pi, \forall n \in \mathbb{Z}$ are simple poles and the point ∞ is a singular point, which is a limit point of the poles of $\cot z$.

2. Using the rules of differentiation and the derivatives of $\cos z$ and $\sin z$, we obtain the derivatives of $\tan z$ and $\cot z$:

$$(\tan z)' = \frac{\cos z \cdot \cos z + \sin z \cdot \sin z}{\cos^2 z} = \frac{1}{\cos^2 z} \tag{4.64}$$

and

$$(\cot z)' = \frac{-\sin z \cdot \sin z - \cos z \cdot \cos z}{\sin^2 z} = -\frac{1}{\sin^2 z} . \tag{4.65}$$

It follows from (4.64) that $(\tan z)' \neq 0$ at any point $z \in D_1$ and $(\cot z)' \neq 0$ at any point $z \in D_2$. Therefore, the function $w = \tan z$ is univalent and represents a conformal transformation at any point of D_1. Since the points $z_k = \frac{\pi}{2} + k\pi, \forall k \in \mathbb{Z}$ are simple poles of $\tan z$, this function is univalent and represents a conformal transformation at these points as well. Thus, the function $w = \tan z$ is univalent and represents a conformal mapping at any point of the (finite) complex plane $\mathbb{C}$. The same results follow from (4.65) for the function $w = \cot z$.

3. Rewriting the definitions (4.63) with the use of the Euler formulas (4.56)

$$\tan z = \frac{\sin z}{\cos z} = \frac{\frac{1}{2i}\left(e^{iz} - e^{-iz}\right)}{\frac{1}{2}\left(e^{iz} + e^{-iz}\right)} = -i\frac{e^{iz} - e^{-iz}}{e^{iz} + e^{-iz}} = -i\frac{e^{2iz} - 1}{e^{2iz} + 1}, \tag{4.66}$$

$$\cot z = \frac{\cos z}{\sin z} = \frac{\frac{1}{2}\left(e^{iz} + e^{-iz}\right)}{\frac{1}{2i}\left(e^{iz} - e^{-iz}\right)} = i\frac{e^{iz} + e^{-iz}}{e^{iz} - e^{-iz}} = i\frac{e^{2iz} + 1}{e^{2iz} - 1}, \tag{4.67}$$

we can represent $w = \tan z$ and $w = \cot z$ as a composition of the known functions:

$$z_1 = 2iz = 2e^{i\frac{\pi}{2}}z,\ z_2 = e^{z_1},\ w = \tan z = -i\frac{z_2 - 1}{z_2 + 1}, \tag{4.68}$$

$$z_1 = 2iz = 2e^{i\frac{\pi}{2}}z,\ z_2 = e^{z_1},\ w = \cot z = i\frac{z_2 + 1}{z_2 - 1}. \tag{4.69}$$

Hence, the functions $w = \tan z$ and $w = \cot z$ are composed of simpler functions: linear, exponential and fractional linear.

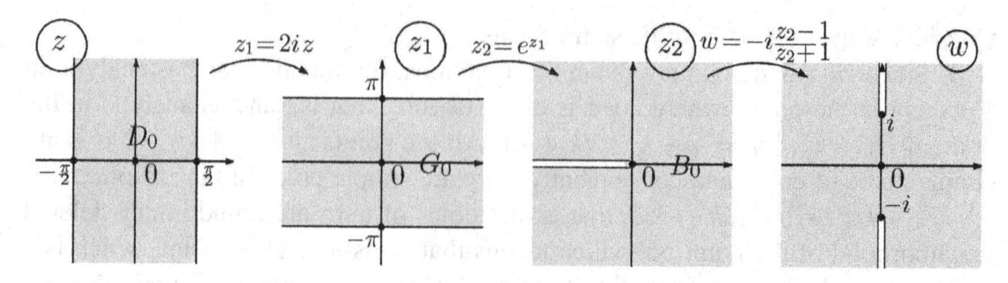

Fig. 4.39 Construction of the domains of univalence for $w = \tan z$

4. Employing (4.68) and (4.69), we can find the domains of univalence of the functions $w = \tan z$ and $w = \cot z$. Let us specify this procedure for $w = \tan z$. The function $w = -i\frac{z_2-1}{z_2+1}$ performs a bijective transformation from the extended complex z_2-plane onto the extended complex w-plane. The preceding function $z_2 = e^{z_1}$ is not univalent on the entire z_1-plane, but we can choose one of its domains of univalence, for instance, the horizontal strip $G_0 = \{\forall z_1 \in \mathbb{C}_{z_1} : \forall x_1 \in \mathbb{R}\,;\ -\pi < y_1 < \pi\}$. This strip is mapped by the function $z_2 = e^{z_1}$ onto the entire complex z_2-plane with a cut along the negative part of the real axis (denote this domain by B_0). Accordingly, the function $w = -i\frac{z_2-1}{z_2+1}$ transforms B_0 into the entire complex w-plane with the cuts along the rays $(-i\,\infty\,,\ -i]$ and $[i\ ,\ +i\,\infty)$ (see Fig. 4.39). Returning to the original z-plane, the relation $z = \frac{1}{2}e^{-i\frac{\pi}{2}}z_1$ shows that the horizontal strip G_0 in the z_1-plane corresponds to the vertical strip $D_0 = \{\forall z \in \mathbb{C} : -\frac{\pi}{2} < x < \frac{\pi}{2}\,;\ \forall y \in \mathbb{R}\}$ in the z-plane. The last strip is one of the domains of univalence of the function $w = \tan z$ (see Fig. 4.39). Notice that any domain obtained by translation of D_0 along the real axis is also a domain of univalence of the function $w = \tan z$.

5. The functions $\tan z$ and $\cot z$ are π-periodic that follows from the elementary manipulations:

$$\tan\,(z+\pi) = \frac{\sin\,(z+\pi)}{\cos\,(z+\pi)} = \frac{-\sin z}{-\cos z} = \tan\, z,$$

$$\cot\,(z+\pi) = \frac{\cos\,(z+\pi)}{\sin\,(z+\pi)} = \frac{-\cos z}{-\sin z} = \cot\, z\,.$$

6. The image of $f(z) = \tan z$ can be found by studying the solutions of the equation $\tan z = w$, where w is any complex number. Employing the exponential representation (4.66)

$$\tan\, z = -i\frac{e^{2i\,z}-1}{e^{2i\,z}+1}$$

and denoting $e^{2i\,z} = t$, we can reduce the equation $\tan z = w$ to the form $-i\frac{t-1}{t+1} = w$, or equivalently $\frac{t-1}{t+1} = i\,w$. Rewriting this equation as $t - 1 = i\,wt + i\,w$ and solving the last linear equation in t, we obtain $t = \frac{-w+i}{w+i}$. Then, the corresponding expression for z is

$$e^{2i\,z} = \frac{-w+i}{w+i}\,. \tag{4.70}$$

Since $e^{2i\,z} \neq 0$ and $e^{2i\,z} \neq \infty$ for any complex number z, the Eq. (4.70) has no roots if w equals i or $-i$. Therefore, the Eq. (4.70) has the following countable set of solutions for any given $w \neq \pm i$:

$$2iz_k = \ln\frac{-w+i}{w+i} = \ln\left|\frac{-w+i}{w+i}\right| + i\arg\frac{-w+i}{w+i} + 2k\pi i,\, \forall k \in \mathbb{Z},$$

or isolating z, we have

$$z_k = \frac{1}{2}\arg\frac{-w+i}{w+i} + k\pi - \frac{i}{2}\ln\left|\frac{-w+i}{w+i}\right|,\quad \forall k \in \mathbb{Z}\,. \tag{4.71}$$

Thus, the equation $\tan z = w$ has a countable set of solutions (4.71) for any complex number w, except for i and $-i$, in other words, the image of the function $f(z) = \tan z$ is the entire complex plane with the points i and $-i$ excluded.

Similarly, the function $w = \cot z$ maps the entire z-plane onto the entire w-plane without the points i and $-i$.

Exercises

1. Find a general form of a linear function that transforms

   (1) the upper half-plane into itself;

   (2) the upper half-plane into the lower half-plane;

   (3) the upper half-plane into the left half-plane.

2. Find symmetric points to the point $5 + 2i$ with respect to the circles:

   (1) $|z| = 1\,;$

   (2) $|z - 3i| = 7\,.$

3. Find a curve symmetric with respect to the circle $|z| = 2$ to the given curves:

   (1) $|z| = 1\,;$

   (2) $|z - 2i| = 2\,;$

(3) $x = 4$.

4. Find the images of the given curves under the transformation $w = \frac{1}{z}$:

(1) the family of the circles $x^2 + y^2 = ax, \ \forall a \in \mathbb{R}$;

(2) the family of the straight lines $y = x + a, \ \forall a \in \mathbb{R}$;

(3) the family of the straight lines $y = kx, \ \forall k \in \mathbb{R}$.

5. Find the image of the given domains under the indicated transformation:

(1) $\begin{cases} x < 0 \\ y > 0 \end{cases}; \ w = \dfrac{z - 3i}{z + 3i}$;

(2) strip $0 < x < 5$; $w = \dfrac{z - 5}{z}$;

(3) strip $0 < x < 5$; $w = \dfrac{z - 5}{z + 1}$.

6. Show that any fractional linear function $w = \frac{az+b}{cz+d}$ satisfies the differential equation $2w'w''' = 3\left(w''\right)^2$.
7. Find a fractional linear function $w(z)$ that carries the points $z = i, \ 1 + i, \ \infty$ to the corresponding points $w = i, \ 1, \ 1 + i$.
8. Find a fractional linear function that carries the points $z = 1, \ 2, \ 3$ to the corresponding points $w = 1, \ i, \ 2 + i$.
9. Find a fractional linear function that satisfies the following conditions:

(1) the points 2 and $-i$ are fixed and the point $-\frac{1}{2} + \frac{1}{2}i$ is carried to -2;

(2) $-3i$ is a double fixed point and 2 is carried to ∞.

10. Prove that if a fractional linear transformation has two finite fixed points, then the product of its derivatives at these points is equal to 1.
11. Show that the inverse image of the family of concentric circles $|w| = c$, $0 < c < +\infty$ under the function $w = \frac{z - z_1}{z - z_2}$ is a family of circles. For each c find the radius and centerpoint of the corresponding circle in the z-plane.

    Hint: use the equality $|z|^2 = z \cdot \bar{z}$.
12*. Demonstrate the following criterion of symmetry of points with respect to a circle: two points z_1 and z_2 are symmetric with respect to the circle $C = \{|z -$

$a| = R\}$ if and only if $\frac{|z-z_1|}{|z-z_2|} = c \equiv constant > 0$ for $\forall z \in C$.
Hint: make use of the result of Exercise 11.

13. For the given function $w = z^2$, find

(1) the images of the straight lines $x = c,\ y = c,\ \forall c \in \mathbb{R}\ (z = x + iy)$;

(2) the inverse images of the straight lines $u = c,\ v = c,\ \forall c \in \mathbb{R}\ (w = u + iv)$.

14. Find the images of the given domains under the Joukowski function:

(1) $|z| < r,\ r < 1$;

(2) $|z| > r,\ r > 1$;

(3) $|z| > 1$;

(4) $\operatorname{Im} z < 0$;

(5) $|z| < 1,\ \operatorname{Im} z > 0$;

(6) $|z| > 1,\ \operatorname{Im} z < 0$;

(7) $1 < |z| < r,\ \operatorname{Im} z > 0$;

(8) $1/r < |z| < r,\ \operatorname{Im} z > 0,\ \operatorname{Re} z > 0\ (r > 1)$.

15. Find the images of the given sets under the transformation $w = e^z$:

(1) the Cartesian net: $x = c;\quad y = c\quad (\forall c \in \mathbb{R})$;

(2) the straight lines $y = ax + b$;

(3) the strip $\frac{\pi}{4} < y < \frac{3}{2}\pi$;

(4) the strip $0 < y < 2\pi$;

(5) the half-strip $0 < y < \alpha,\ \ x > 0\ \ (\alpha \leq 2\pi)$;

(6) the half-strip $0 < y < \alpha,\ \ x < 0\ \ (\alpha \leq 2\pi)$.

16. Find the images of the given sets under the function $w = z^i$:

(1) the polar net: $r = c > 0,\ \ \varphi = c\ \ (z = re^{i\varphi})$;

(2) the lower half-plane $\operatorname{Im} z < 0$; is this mapping one-to-one?

(3*) the half-ring $1 < r < a, -\pi < \varphi < 0$ $(z = re^{i\varphi})$; state conditions under which this mapping is one-to-one.

17. Find the images of the given sets under the function $w = \cos z$:

(1) the Cartesian net: $x = c,\ \ y = c\ \ (\forall c \in \mathbb{R})$;

(2) the half-strip: $0 < x < \pi,\ \ y < 0$;

(3) the half-strip: $0 < x < \frac{\pi}{2},\ \ y > 0$;

(4) the half-strip: $-\frac{\pi}{2} < x < \frac{\pi}{2},\ \ y > 0$;

(5) the strip $0 < x < \pi$;

(6) the rectangle $0 < x < \pi,\ \ -1 < y < 1$.

18. Find the images of the given sets under the function $w = \sin z$:

(1) the Cartesian net: $x = c,\ \ y = c\ \ (\forall c \in \mathbb{R})$;

(2) the strip $-\pi < y < 0$;

(3) the strip $-\frac{\pi}{2} < x < \frac{\pi}{2}$;

(4) the half-strip: $-\frac{\pi}{2} < x < \frac{\pi}{2},\ y > 0$;

(5) the half-strip: $-\frac{\pi}{6} < x < \frac{\pi}{6},\ y > 0$;

(6) the rectangle $-\frac{\pi}{2} < x < \frac{\pi}{2}$, $-2 < y < 0$.

Verify if these mappings are univalent.

19. Transform the upper half-plane $\operatorname{Im} z > 0$ into the unit disk under the following conditions:

(1) $w(2i) = 0$, $\arg w'(2i) = \pi$;

(2) $w(1+i) = 0$, $\arg w'(1+i) = -\frac{\pi}{4}$.

20. Transform the disk $|z| < 3$ into the half-plane $\operatorname{Re} w < 0$ in such a way that $w(0) = -2$ and $\arg w'(0) = -\frac{\pi}{4}$.
21. Transform the disk $|z| < 1$ into the disk $|w| < 1$ in such a way that

(1) $w\left(\frac{i}{4}\right) = 0$, $\arg w'\left(\frac{i}{4}\right) = \frac{\pi}{4}$;

(2) $w\left(-\frac{1}{3}\right) = 0$, $\arg w'\left(-\frac{1}{3}\right) = -\frac{\pi}{2}$.

22. Transform the disk $|z-1| < 2$ into the disk $|w+4i| < 4$ in such a way that $w(1) = i$, $\arg w'(1) = \pi$.
23. Construct a conformal mapping of a given domain onto the upper half-plane:

(1) $\begin{cases} |z-1| < \sqrt{2} \\ \operatorname{Im} z > 0 \end{cases}$;

(2) $\begin{cases} |z| < 1 \\ |z-1| > \sqrt{2} \end{cases}$;

(3) $\begin{cases} |z| > 2 \\ |z-2| > 2 \end{cases}$;

(4) the sector $\frac{2}{3}\pi < \arg z < \pi$ with the cuts $\begin{cases} 2\pi/3 \le \arg z \le 3\pi/4 \\ |z| = 1/2 \end{cases}$ and $\begin{cases} 5\pi/6 \le \arg z \le \pi \\ |z| = 1/2 \end{cases}$;

(5) the sector $-\frac{\pi}{2} < \arg z < -\frac{\pi}{4}$ with the cut $\begin{cases} -\pi/3 \le \arg z \le -\pi/4 \\ |z| = 5 \end{cases}$;

(6) the sector $\begin{cases} |z| > 1/2 \\ 3\pi/4 < \arg z < \pi \end{cases}$ with the cuts $\begin{cases} 1/2 \le |z| \le 1 \\ \arg z = 7\pi/8 \end{cases}$ and $\begin{cases} |z| \ge 2 \\ \arg z = 7\pi/8 \end{cases}$;

(7) $\begin{cases} |z-2| < 2 \\ |z-3| > 1 \end{cases}$;

(8) $\begin{cases} |z+i| > 1 \\ \operatorname{Im} z > -2 \end{cases}$;

(9) $\begin{cases} |z| > 3 \\ |z+4i| > 1 \end{cases}$;

(10) $\begin{cases} |z| > 1 \\ |z+i| < 2 \\ \operatorname{Re} z > 0 \end{cases}$;

(11) $\begin{cases} |z-2| > 2 \\ |z+2| > 2 \\ \operatorname{Im} z < 0 \end{cases}$;

(12) $\begin{cases} |z-2| > 2 \\ |z+2| > 2 \end{cases}$ with the cut $\begin{cases} -\infty \le x \le -4 \\ y = 0 \end{cases}$;

(13) $\begin{cases} |z-2| > 2 \\ |z+2| > 2 \end{cases}$ with the cut $\begin{cases} 4 \le x \le 8 \\ y = 0 \end{cases}$;

(14) the half-strip $\begin{cases} 0 < x < \pi \\ y > 0 \end{cases}$ with the cut $\begin{cases} x = \pi/2 \\ 0 \le y \le 2 \end{cases}$.

24. Employing the function $w = z^2$ and its inverse, construct the conformal mappings of the given domains onto the upper half-plane:

(1) the interior of the right part of the hyperbola $x^2 - y^2 = a^2$;

(2) the exterior of the parabola $y^2 = 2px,\ \ p > 0\ \ (z = x + iy)$.
Hint: use the results of Exercise 13.

25. Find the image of the given domain under the indicated function:

(1) $|z| < 1;\ w = \frac{z}{z^2+1}$;

(2) $\begin{cases} |z| < 1 \\ \operatorname{Im} z < 0 \end{cases};\ w = \frac{1}{z^2+1}$;

(3) $0 < \arg z < \frac{\pi}{n};\ w = \frac{1}{2}\left(z^n + \frac{1}{z^n}\right)$.

26. Transform the given eccentric ring into the concentric ring $1 < |w| < R$ and find the value of R:

(1) $\begin{cases} |z-5|>2 \\ \operatorname{Re} z>0 \end{cases}$;

(2) $\begin{cases} |z-3|>9 \\ |z-8|<16 \end{cases}$;

(3) $\begin{cases} |z-3i|>1 \\ |z-4|>2 \end{cases}$.

27*. Suppose that $f(z)=\sum_{n=0}^{+\infty} c_n z^n$ is an analytic function in the disk $D=\{|z|\leq R\}$ and maps univalently the disk D onto a set G with the area $A(G)$. Prove that

(1) $A(G)=\pi \sum_{n=1}^{+\infty} n|c_n|^2 R^{2n}$;

(2) $A(G)\geq A(D)=\pi R^2$ under the additional condition $f'(0)=1$.
Investigate for which functions the last evaluation turns into equality.
Hint: apply Green's formula for calculation of the area and use the polar coordinates in the z-plane.

28*. Let $f(z)$ be analytic and univalent in the disk $|z|<1$, $f(0)=0$. Show that the multi-valued function $f_p(z)=\sqrt[p]{f(z^p)}$, $p\in\mathbb{N}$, $p>1$, is split in p analytic and univalent functions in $|z|<1$.

29. Let $f(z)$ be analytic at the origin and $f(0)=0$. The radius of univalence r_u of the function $f(z)$ is the maximum radius of the disk centered at the origin in which $f(z)$ is univalent. Find the radius of univalence for the following functions:

(1) $f(z)=\frac{z}{1-z}$;

(2) $f(z)=z+z^2$;

(3) $f(z)=\frac{z}{(1-z)^2}$;

(4) $f(z)=e^z-1$.
Hints: (1) $r_u=+\infty$; (2) $r_u=\frac{1}{2}$; (3) $r_u=1$; (4) $r_u=\pi$.

30*. Let $f(z)$ be analytic at the origin and $f(0)=0$. The radius of convexity r_c of the function $f(z)$ is the maximum radius of the disk centered at the origin, which $f(z)$ maps univalently onto a convex domain. Show that $f(z)$ transforms the circle $|z|=r$ into a convex curve if and only if the inequality $\frac{\partial}{\partial\varphi}\left(\frac{\pi}{2}+\varphi+\arg f'(z)\right)=1+\operatorname{Re}\left(\frac{zf''(z)}{f'(z)}\right)\geq 0$ is satisfied for $\forall\varphi$ $(z=re^{i\varphi})$.

Hint: analyze the properties of the function $\frac{f''(z)}{f'(z)}$ and use the Cauchy–Riemann conditions in the polar coordinates.

31*. Find the radius of convexity for the following functions:

(1) $f(z) = \frac{z}{1-z}$;

(2) $f(z) = z + z^2$;

(3) $f(z) = \frac{z}{(1-z)^2}$;

(4) $f(z) = e^z - 1$.

Hints: (1) $r_c = 1$; (2) $r_c = \frac{1}{4}$; (3) $r_c = 2 - \sqrt{3}$; (4) $r_c = 1$.

Chapter 5
Fundamental Principles of Conformal Mappings. Transformations of Polygons

In this chapter, we provide the proofs of the main results on conformal mappings. The analyzed fundamental principles include the criterion for local univalence (Sect. 5.1), the principle of domain preservation (Sect. 5.2), the principle of boundary correspondence (Sect. 5.3), and the Riemann mapping theorem (Sect. 5.4). Unlike other parts of the text, the subjects considered here are almost exclusively theoretical. Even some more specific problems, such as a conformal mapping between two rings (Sect. 5.4) and construction of conformal mappings between half-planes, rectangles and polygons (Sects. 5.6 and 5.7), have rather theoretical significance than represent illustrative examples.

The main reason behind this form of exposition is that the entire Chap. 4 was reserved for different examples and illustrations of the theoretical principles of conformal mappings considered here: starting with essential geometric properties of conformal mappings and proceeding with examples of application of conformal mappings to study of elementary functions. Another reason is an attempt to focus attention on the sequences of logical arguments in the presented proofs, some of which are rather intricate, at least for the inexperienced reader. This is especially true for the proof of the Riemann mapping theorem on the possibility to transform conformally any simply connected domain with at least two boundary points into the open unit disk.

After demonstrating the four fundamental principles, in Sect. 5.5, we consider the Riemann–Schwarz symmetry principle as one of the important forms of performing an analytic continuation. The two final sections (5.6 and 5.7) are devoted to construction of conformal mappings between the domains of theoretical and practical importance—half-planes, rectangles, and polygons.

A. Bourchtein and L. Bourchtein, *Complex Analysis*, Hindustan Publishing Corporation,
https://doi.org/10.1007/978-981-15-9219-5_5

5.1 The Criterion for Local Univalence

Let us recall the definitions of a univalent function in a domain and at a point given in Sect. 4.1.

Definition 5.1 A function $f(z)$ is called *univalent in a domain* D, if for $\forall z_1, z_2 \in D$, $z_1 \neq z_2$ it follows that $f(z_1) \neq f(z_2)$.

Definition 5.2 A function $f(z)$ is called *univalent at a point* $z_0 \in D$ if it is univalent in a neighborhood of this point.

Evidently, a function $w = f(z)$ univalent in a domain D is univalent at every point of this domain. The converse statement is not true. For example, the function $w = e^z$ (studied in Sect. 4.8) is univalent at every point of the complex plane $\mathbb{C}$, although it is not univalent in the entire domain $\mathbb{C}$, because this function maps different points z and $z + 2k\pi i$, $\forall k \in \mathbb{Z}$ $(k \neq 0)$ of the z-plane into the same point of the w-plane: $e^z = e^{z+2k\pi i}$, $\forall k \in \mathbb{Z}$.

Remark Notice that a composition of univalent functions is again a univalent function. Indeed, suppose $w = f(z)$ is a univalent function in a domain D, whose image under $f(z)$ is G: $G = f(D)$, and $\zeta = g(w)$ is a univalent function in G. Consider the composite function $h(z) = g(f(z))$ and pick two different points $z_1 \neq z_2$ in D. Since G is the image of D under $f(z)$, the corresponding points $w_1 = f(z_1)$ and $w_2 = f(z_2)$ lie in G, and $w_1 \neq w_2$ due to the univalence of $w = f(z)$ in D. Then, the univalence of $\zeta = g(w)$ in G implies that $\zeta_1 = g(w_1) \neq g(w_2) = \zeta_2$. Since $\zeta_1 = g(w_1) = g(f(z_1)) = h(z_1)$ and $\zeta_2 = g(w_2) = g(f(z_2)) = h(z_2)$, we obtain that for any two points z_1 and z_2 $(z_1 \neq z_2)$ in D it follows that $h(z_1) \neq h(z_2)$. Hence, the composite function $\zeta = h(z) = g(f(z))$ is univalent in the domain D.

Theorem (The criterion for local univalence) *An analytic at a point $z_0 \neq \infty$ function $f(z)$ is univalent at z_0 if and only if $f'(z_0) \neq 0$.*

Proof *Necessity.* Since $w = f(z)$ is analytic and univalent at a point $z_0 \neq \infty$, it means that there exists a neighborhood of z_0 where $f(z)$ is univalent and has an expansion in the power series, that is, there exists $r > 0$ such that the series $f(z) = \sum_{n=0}^{+\infty} c_n (z - z_0)^n$ converges in the disk $K = \{|z - z_0| < r\}$ and the function $w = f(z)$ is univalent in K. Let us show that $f'(z_0) \neq 0$.

Suppose, by contradiction, that $f'(z_0) = 0$. First, we show that there exists ρ, $0 < \rho < r$, such that $f'(z) \neq 0$ at all points of the ring $0 < |z - z_0| \leq \rho$. If it is not so, then for each ρ, $0 < \rho < r$, there exists a point z_ρ, $0 < |z_\rho - z_0| < \rho$, such that $f'(z_\rho) = 0$. In particular, we can take a sequence of the numbers $\rho_n = \frac{1}{n}$ and for each $n \in \mathbb{N}$ there exists a point z_n such that $0 < |z_n - z_0| < \frac{1}{n}$ and $f'(z_n) = 0$. Therefore, we construct the sequence of the points $\{z_n\}_{n=1}^{+\infty}$ such that $z_n \in K$, $f'(z_n) = 0$, $\forall n \in \mathbb{N}$, this sequence converges: $z_n \underset{n\to\infty}{\to} z_0$ (since $0 < |z_n - z_0| < \frac{1}{n} \underset{n\to\infty}{\to} 0$), and the limit point z_0 of this sequence belongs to the disk K. Then, by the Theorem of the uniqueness of an analytic function (see Sect. 2.8), we have that $f'(z) \equiv 0$ in the disk K, that is, $f(z) \equiv constant$ in K, that contradicts the condition of univalence of the function $w = f(z)$ in K. Hence, there exists ρ, $0 < \rho < r$, such that $f'(z) \neq 0$ at every point of the ring $0 < |z - z_0| \leq \rho$. Denote by K_ρ the corresponding disk $K_\rho = \{|z - z_0| < \rho\}$ and evaluate $|f(z) - f(z_0)|$ for an arbitrary $z \in \partial K_\rho = \{|z - z_0| = \rho\}$. By the hypothesis of the Theorem, the function $w = f(z)$ is univalent in the disk K, which means that $f(z) \neq f(z_0) = w_0$ for any $z \in \{0 < |z - z_0| \leq \rho\}$ and, in particular, $f(z) \neq f(z_0) = w_0, \forall z \in \partial K_\rho$. The function $f(z) - f(z_0)$ is continuous on the circle $|z - z_0| = \rho$, which is a compact set. Therefore, by the Weierstrass theorem, the function $|f(z) - f(z_0)|$ attains its infimum on $|z - z_0| = \rho$, that is, the infimum of the function $|f(z) - f(z_0)|$ on the circle $|z - z_0| = \rho$ is the value of this function at some point of the circle. Since $|f(z) - f(z_0)| \neq 0$ at every point of the circle $|z - z_0| = \rho$, we have $\inf\limits_{|z-z_0|=\rho} |f(z) - f(z_0)| = m > 0$.

Consider now the disk centered at the point w_0 with the radius m in the w-plane: $|w - w_0| < m$, and take an arbitrary point $w_1 \neq w_0$ in this disk. Represent the function $f(z) - w_1$ as the sum of two functions:

$$f(z) - w_1 = (f(z) - w_0) + (w_0 - w_1).$$

In the last expression, both functions on the right-hand side are analytic in the closed disk $\overline{K}_\rho = \{|z - z_0| \leq \rho\}$. On the boundary of this disk we have $|f(z) - w_0| \geq m$, $\forall z \in \partial K_\rho$, and, together with the inequality $|w_0 - w_1| = |w_1 - w_0| < m$, it gives $|f(z) - w_0| > |w_0 - w_1|, \forall z \in \partial K_\rho$. Then, by Rouché's Theorem (Sect. 3.14), the sum $f(z) - w_1$ of these functions has as many zeros in the disk K_ρ as the function with the larger absolute values on the boundary of K_ρ, that is:

$$N_{f(z)-w_1}\left(K_\rho\right) = N_{f(z)-w_0}\left(K_\rho\right). \tag{5.1}$$

Since $(f(z) - w_0)_{z=z_0} = (f(z) - f(z_0))_{z=z_0} = 0$ and $(f(z) - w_0)'_{z=z_0} = f'(z_0) = 0$, by the criterion for multiplicity of zeros (Theorem 3.2 in Sect. 3.7), the function $f(z) - w_0$ has zero of multiplicity at least two at the point z_0 (and consequently in the disk K_ρ), that is, $N_{f(z)-w_0}\left(K_\rho\right) \geq 2$. Then, from the equality (5.1), it follows that

$$N_{f(z)-w_1}\left(K_\rho\right) \geq 2. \tag{5.2}$$

Consider the two possible cases: the function $f(z) - w_1$ has the unique zero in K_ρ or it has at least two zeros there. If $f(z) - w_1$ has only one zero z_1 in the disk K_ρ, it is clear that $z_1 \neq z_0$, because $w_1 \neq w_0 = f(z_0)$, and, due to the condition (5.2), z_1 is a zero of multiplicity at least two, that is, $(f(z) - w_1)_{z=z_1} = 0$ and $(f(z) - w_1)'_{z=z_1} = f'(z_1) = 0$. However, the last condition contradicts the construction of the disk K_ρ, and consequently, the function $f(z) - w_1$ can possess only simple zeros in the disk K_ρ. Therefore, the condition (5.2) implies that there exist at least two different zeros z_1 and z_2 in $K_\rho \subset K$ such that $f(z_1) = w_1 = f(z_2)$, but this contradicts the univalence of the function $f(z)$ in the disk K. Thus, the necessity is proved.

Sufficiency. Since the function is analytic at z_0, it can be expanded in a power series convergent in a neighborhood of z_0: $f(z) = \sum_{n=0}^{+\infty} c_n (z - z_0)^n$ in the disk $K_r = \{|z - z_0| < r\}$. Besides, $f'(z_0) \neq 0$, which means that $c_1 = f'(z_0) \neq 0$. We need to prove that $f(z)$ is univalent at z_0. From the Theorem on infinite differentiabilitiy of an analytic function (see Sect. 2.1), it follows that $f'(z)$ is analytic in the disk K_r and consequently continuous in K_r. Since $f'(z_0) \neq 0$ and $f'(z)$ is continuous at z_0, there exists a neighborhood of z_0 where $f'(z) \neq 0$, that is, there exists ρ, $0 < \rho \leq r$, such that $f'(z) \neq 0$, $\forall z \in K_\rho = \{|z - z_0| < \rho\}$. Using the term-by-term rule of differentiation of a power series in the disk of convergence, we get

$$f'(z) = \sum_{n=1}^{+\infty} n c_n (z - z_0)^{n-1} = c_1 + \sum_{n=2}^{+\infty} n c_n (z - z_0)^{n-1}, \ \forall z \in K_\rho .$$

Notice that the series $\sum_{n=2}^{+\infty} n c_n (z - z_0)^{n-1}$ converges absolutely (and uniformly) in K_ρ, that is, $\sum_{n=2}^{+\infty} n |c_n| |z - z_0|^{n-1}$ converges in the disk K_ρ. Therefore, we can calculate the limit of the last series as z approaches z_0 term by term:

$$\lim_{z \to z_0} \sum_{n=2}^{+\infty} \left| n c_n (z - z_0)^{n-1} \right| = \sum_{n=2}^{+\infty} \lim_{z \to z_0} n |c_n| |z - z_0|^{n-1} = 0 .$$

By the definition of a limit, for $\forall \varepsilon > 0$ (take $\varepsilon = \frac{|c_1|}{2}$) there exists δ, $0 < \delta \leq \rho$ (we can always choose δ smaller than ρ) such that for $\forall z \in K_\delta = \{|z - z_0| < \delta\}$ it follows $\sum_{n=2}^{+\infty} n |c_n| |z - z_0|^{n-1} < \frac{|c_1|}{2}$. In particular, at the points of the circle $|z - z_0| = r_0 = \frac{\delta}{2}$, we have

$$\sum_{n=2}^{+\infty} n |c_n| |z - z_0|^{n-1} \Bigg|_{|z - z_0| = r_0} = \sum_{n=2}^{+\infty} n |c_n| r_0^{n-1} < \frac{|c_1|}{2} . \tag{5.3}$$

Let us show that $f(z)$ is univalent in the disk $K_{r_0} = \{|z - z_0| < r_0\}$. To this end, pick any two different points $z_1, z_2 \in K_{r_0}$ and evaluate the difference $f(z_1) - f(z_2)$:

$$|f(z_1) - f(z_2)| = \left| \sum_{n=0}^{+\infty} c_n (z_1 - z_0)^n - \sum_{n=0}^{+\infty} c_n (z_2 - z_0)^n \right| = \left| \sum_{n=1}^{+\infty} c_n \left((z_1 - z_0)^n - (z_2 - z_0)^n \right) \right|$$

$$= |c_1 (z_1 - z_0 - z_2 + z_0) + \sum_{n=2}^{+\infty} c_n \left((z_1 - z_0)^n - (z_2 - z_0)^n \right) |$$

$$= |c_1 (z_1 - z_2) + \sum_{n=2}^{+\infty} c_n (z_1 - z_0 - z_2 + z_0) \left((z_1 - z_0)^{n-1} \right.$$

$$\left. + (z_1 - z_0)^{n-2} (z_2 - z_0) + \ldots + (z_2 - z_0)^{n-1} \right) |$$

$$= |z_1 - z_2| \left| c_1 + \sum_{n=2}^{+\infty} c_n \left((z_1 - z_0)^{n-1} + (z_1 - z_0)^{n-2} (z_2 - z_0) + \ldots + (z_2 - z_0)^{n-1} \right) \right| \equiv S\,.$$

Using the inequalities $|a \pm b| \geq |a| - |b|$ and $|a \pm b| \leq |a| + |b|$, as well the fact that $|z_1 - z_0| < r_0$ and $|z_2 - z_0| < r_0$ (since z_1 and z_2 lie in the disk K_{r_0}) and the inequality (5.3), we can proceed with the evaluation of $|f(z_1) - f(z_2)|$ as follows:

$$|f(z_1) - f(z_2)| \equiv S$$

$$\geq |z_1 - z_2| \left\{ |c_1| - \left| \sum_{n=2}^{+\infty} c_n \left((z_1 - z_0)^{n-1} + (z_1 - z_0)^{n-2} (z_2 - z_0) + \ldots + (z_2 - z_0)^{n-1} \right) \right| \right\}$$

$$\geq |z_1 - z_2| \left\{ |c_1| - \sum_{n=2}^{+\infty} |c_n| \left(|z_1 - z_0|^{n-1} + |z_1 - z_0|^{n-2} |z_2 - z_0| + \ldots + |z_2 - z_0|^{n-1} \right) \right\}$$

$$> |z_1 - z_2| \left\{ |c_1| - \sum_{n=2}^{+\infty} |c_n| \left(r_0^{n-1} + r_0^{n-2} r_0 + \ldots + r_0^{n-1} \right) \right\}$$

$$= |z_1 - z_2| \left\{ |c_1| - \sum_{n=2}^{+\infty} n\, |c_n|\, r_0^{n-1} \right\} > |z_1 - z_2| \left\{ |c_1| - \frac{|c_1|}{2} \right\} = |z_1 - z_2| \frac{|c_1|}{2}\,.$$

Thus, for any two different points in the disk K_{r_0} we obtain that $|f(z_1) - f(z_2)| > |z_1 - z_2| \frac{|c_1|}{2} > 0$, that is, $f(z_1) \neq f(z_2)$, which means that the function $f(z)$ is univalent in the disk K_{r_0}, and consequently, it is univalent at the point z_0.

This completes the proof of the Theorem. □

Using Remark at the beginning of this section, we can extend the result of the Theorem to the case when $z_0 = \infty$ or z_0 is a pole of the function $f(z)$.

Corollary 5.1 *Suppose $w = f(z)$ is an analytic function at the point ∞, that is, it can be expanded in the series $f(z) = \sum_{n=0}^{+\infty} \frac{a_n}{z^n}$, which converges in a neighborhood of ∞ ($|z| > R$). The function $w = f(z)$ is univalent at ∞ if and only if $a_1 \neq 0$ (in other words, $\operatorname*{res}_{z=\infty} f(z) = -a_1 \neq 0$).*

Proof Let us introduce a new independent variable $t = \frac{1}{z}$ and a new function $g(t) = f\left(\frac{1}{t}\right) = \sum_{n=0}^{+\infty} a_n t^n$ with the last series converging in a neighborhood of the point $t = 0$: $|t| < \frac{1}{R}$. Then, $g(t)$ is analytic at $t = 0$ and, according to the above Theorem, $g(t)$ is univalent at $t = 0$ if and only if $g'(0) = a_1 \neq 0$. Notice that the function $z = \frac{1}{t}$ transforms bijectively the entire extended complex plane $\overline{\mathbb{C}}_t$ into the entire extended complex plane $\overline{\mathbb{C}}_z$, that is, $z = \frac{1}{t}$ is a univalent function. Therefore, $f(z) = g\left(\frac{1}{z}\right)$ is univalent as the composition of two univalent functions if and only if $g(t)$ is univalent. This means that the function $f(z)$ is univalent at $z = \infty$ if and only if $a_1 \neq 0$. □

Corollary 5.2 *Let z_0 be a pole of a function $f(z)$. The function $f(z)$ is univalent at the point z_0 if and only if z_0 is a simple pole of $f(z)$.*

Proof Consider the function $\zeta = \frac{1}{w} = \frac{1}{f(z)} = g(z)$. Since z_0 is a pole of $w = f(z)$, the function $g(z)$ is analytic at z_0 and $g(z_0) = 0$ (see Sect. 3.7), that is, z_0 is a zero of $g(z)$. By the Theorem of this section, $g(z)$ is univalent at z_0 if and only if $g'(z_0) \neq 0$, or, in other words, z_0 is a simple zero of $g(z)$. Since the function $w = \frac{1}{\zeta}$ establishes one-to-one correspondence between the extended complex planes $\overline{\mathbb{C}}_\zeta$ and $\overline{\mathbb{C}}_w$, the function $f(z) = \frac{1}{g(z)}$ is univalent if and only if $g(z)$ is univalent. The existence of a simple zero z_0 of $g(z)$ is equivalent to the fact that $f(z)$ has a simple pole at z_0. Thus, the function $f(z)$ is univalent at z_0 if and only if z_0 is a simple pole of $f(z)$. □

Notice that if $z = \infty$ is a pole of a function $f(z)$, then, using the same arguments, we can deduce that $f(z)$ is univalent at this point if and only if $z = \infty$ is a simple pole of $f(z)$.

Corollary 5.3 *There exists only one function that transforms conformally the entire extended complex plane into the entire extended complex plane, and this function is either linear or fractional linear.*

A similar statement was considered in Property 2 of fractional linear transformations in Sect. 4.5, but for the sake of completeness, we provide here this important consequence of the criterion for local univalence.

Proof Notice first that the linear function $w = Az + B,\ A \neq 0$ (see Sect. 4.2) is a singular case of the fractional linear function $w = \frac{az+b}{cz+d},\ ad - bc \neq 0$ when $c = 0$ (see Sect. 4.5). In Sects. 4.2 and 4.5, we have already shown that the linear and fractional linear functions establish one-to-one correspondence between the points of the extended complex planes $\overline{\mathbb{C}}_z$ and $\overline{\mathbb{C}}_w$ (see Property 2 in Sects. 4.2 and 4.5, respectively).

Now, we consider the inverse problem: find a function that maps conformally the extended complex plane $\overline{\mathbb{C}}_z$ onto the extended plane $\overline{\mathbb{C}}_w$. Under such a transformation, there exists a point $z = a$, which corresponds to the point $w = \infty$, and this point is unique due to univalence of $f(z)$. Let us analyze the two possible situations: $a \neq \infty$ and $a = \infty$. If $a \neq \infty$, then a is a singular point of $f(z)$, because $f(a) = \infty$. This cannot be an essential singularity, since otherwise the Sokhotski–Weierstrass Theorem (in Sect. 3.8) ensures that $f(z)$ is not univalent in any deleted neighborhood of $z = a$. The condition $f(a) = \infty$ implies that $z = a$ is a pole of $f(z)$ and the univalence of $f(z)$ guarantees that this pole is simple (see Corollary 5.2). Then $f(z)$ can be written in the form $f(z) = \frac{d}{z-a} + g(z),\ d \neq 0$. Since $g(z) = f(z) - \frac{d}{z-a}$ is analytic on the extended complex plane $\overline{\mathbb{C}}_z$, applying Corollary 3.2 of Liouville's Theorem (see Sect. 3.6), we conclude that $g(z) \equiv constant,\ \forall z \in \overline{\mathbb{C}}_z$, which implies that $f(z) = c + \frac{d}{z-a}$, that is, $f(z)$ is a fractional linear function.

In the second case, if $a = \infty$ and consequently $f(\infty) = \infty$, using the same line of reasoning we can deduce that $a = \infty$ is a simple pole of $f(z)$ and this function takes the form of a linear function: $f(z) = dz + b$, $d \neq 0$. □

5.2 The Principle of Domain Preservation

Theorem (The principle of domain preservation) *Any analytic non-constant function maps a domain onto a domain.*

Proof Let $w = f(z)$ be an analytic non-constant function in a domain D and G be the image of D under the transformation f: $G = f(D)$. To show that G is a domain we need to establish that G is an open and linearly connected set. We will show these two properties separately.

(1) We first prove that G is an open set, that is, all points of this set are interior. Take an arbitrary point w_0 of G. Since G is the image of the domain D under the transformation f, the point w_0 is the value of the function $f(z)$ at some (may be not unique) point $z_0 \in D$: $w_0 = f(z_0)$. Let us show that there exists such a neighborhood of z_0 (contained in the domain D) in which $f(z)$ is different from w_0 except at the point z_0: there exists $r > 0$ such that for $\forall z,\ 0 < |z - z_0| < r$ (where $\{|z - z_0| < r\} \subset D$) it follows that $f(z) \neq w_0$. Assume, for contradiction, that for any $r > 0$ there exists a point $z_r \neq z_0$, $0 < |z_r - z_0| < r$, such that $f(z_r) = w_0$. Then, we can construct a sequence $r_n = \frac{1}{n}$ such that for any $n \in \mathbb{N}$ there exists a point z_n, $0 < |z_n - z_0| < \frac{1}{n}$ such that $f(z_n) = w_0$. Therefore, we have a sequence of the points $\{z_n\}_{n=1}^{+\infty}$, $z_n \in D$, such that $f(z_n) = w_0$, $z_n \underset{n\to\infty}{\to} z_0$ (since $0 < |z_n - z_0| < \frac{1}{n} \underset{n\to\infty}{\to} 0$) and the limit point z_0 of z_n belongs to the domain D. Hence, by the Theorem of the uniqueness of an analytic function (see Sect. 2.8), we deduce that $f(z) \equiv w_0$ on the domain D, but this contradicts the hypothesis of the Theorem. Therefore, there exists $r > 0$ such that for $\forall z$, $0 < |z - z_0| < r$, we have $f(z) \neq w_0$. In particular, $f(z) \neq w_0$ at the points of the circle $|z - z_0| = \frac{r}{2}$. Denote $m = \inf\limits_{|z-z_0|=\frac{r}{2}} |f(z) - w_0|$. Since $f(z)$ is analytic in D, the function $|f(z) - w_0|$ is continuous in D, and consequently, by the Weierstrass theorem, it attains its infimum on the compact set $|z - z_0| = \frac{r}{2}$, that is, the number m is the value of the function $|f(z) - w_0|$ at some point of the circle $|z - z_0| = \frac{r}{2}$. Since $f(z) - w_0 \neq 0$ at any point of this circle, it follows that $m > 0$.

Consider now the disk $K_{w_0} = \{|w - w_0| < m\}$ in the plane $\mathbb{C}_w$ and show that this disk is contained in G. The centerpoint $w_0 \in G$ according to our choice and we need to show that all the remaining points of K_{w_0} are located also in G. Pick any w_1 in the disk K_{w_0}, $w_1 \neq w_0$, and represent the difference $f(z) - w_1$ as the sum of two functions:

$$f(z) - w_1 = (f(z) - w_0) + (w_0 - w_1)\,.$$

In the last expression, both functions on the right-hand side are analytic in the domain D and consequently analytic in the closed disk $|z - z_0| \leq \frac{r}{2}$. On the boundary of this disk we have $|f(w) - w_0|_{|z-z_0|=\frac{r}{2}} \geq \inf_{|z-z_0|=\frac{r}{2}} |f(z) - w_0| = m$ and $|w_0 - w_1| = |w_1 - w_0| < m$, which leads to $|f(w) - w_0| > |w_0 - w_1|$, $\forall z \in \{|z - z_0| = \frac{r}{2}\}$. The function $f(z) - w_0$ has a zero at least of the first order at the point z_0, because $f(z_0) = w_0$ (recall that by the construction $f(z) - w_0 \neq 0$ at other points of the disk $|z - z_0| \leq \frac{r}{2}$). Then, by Rouché's Theorem, we have

$$N_{f(z)-w_1}\left(|z - z_0| \leq \frac{r}{2}\right) = N_{f(z)-w_0}\left(|z - z_0| \leq \frac{r}{2}\right) \geq 1 .$$

Thus, the function $f(z) - w_1$ possesses at least one zero in $|z - z_0| \leq \frac{r}{2}$, that is, there exists at least one point z_1 belonging to the disk $|z - z_0| \leq \frac{r}{2}$ (and consequently belonging to the domain D) such that $f(z_1) - w_1 = 0$, or $f(z_1) = w_1$. This means that w_1 is the value of the function $f(z)$ at some point of D, that is, $w_1 \in G$. Since w_1 is an arbitrary point of the disk K_{w_0}, it follows that $K_{w_0} \subset G$, and, therefore, the point w_0 belongs to G together with some neighborhood, which means that w_0 is an interior point of G. Since w_0 is an arbitrary point of G, we conclude that G is an open set.

(2) To show that G is a linearly connected set let us take two arbitrary points w_1 and w_2, $w_1 \neq w_2$ in the set G and construct a continuous curve lying in G that joins these points. Since $w_1, w_2 \in G$ and G is the image of the domain D under $w = f(z)$, there exist the points z_1 and z_2 in D such that $w_1 = f(z_1)$ and $w_2 = f(z_2)$ (if there are many such points we pick one pair of them). Notice that $z_1 \neq z_2$, since for $z_1 = z_2$ one would have $w_1 = f(z_1) = f(z_2) = w_2$ that contradicts the choice of w_1 and w_2. Since D is a domain, we can connect the points z_1 and z_2 by a continuous curve contained in D. Denote this curve by γ: $z = z(t)$, $t \in [a, b]$, $z(a) = z_1$, $z(b) = z_2$, where the function $z = z(t)$ is continuous on the interval $[a, b]$ and $z(t) \in D$, $\forall t \in [a, b]$. Consider the image Γ of the curve γ under the transformation $f : \Gamma = f(\gamma)$. The points of Γ are defined by the function $f(z(t)) = w(t)$, $t \in [a, b]$, and consequently, Γ is a curve. Since $z = z(t)$ is a continuous function on the interval $[a, b]$ and $f(z)$ is an analytic (and consequently continuous) function in D, the composite function $w(t) = f(z(t))$ is continuous on $[a, b]$. Besides, Γ is contained in G as the image of $\gamma \subset D$ under $f(z)$ and it joins the points w_1 and w_2: $w(a) = f(z(a)) = f(z_1) = w_1$ and $w(b) = f(z(b)) = f(z_2) = w_2$. Therefore, we find the continuous curve Γ contained in G, which connects the points w_1 and w_2. Since w_1 and w_2 are arbitrary points of G, we conclude that G is linearly connected. Thus, G is an open and linearly connected set, that is, $G = f(D)$ is a domain.

The Theorem is proved. □

Corollary 5.1 (The maximum modulus principle) *If a non-constant function $w = f(z)$ is analytic in a domain D, then $|f(z)|$ cannot attain its maximum at an interior point of D.*

Proof Arguing by contradiction, assume that there exists a point $z_0 \in D$ such that $|f(z_0)| = \sup_{\forall z \in D} |f(z)|$. Denote by G the image of D under the transformation f: $G = f(D)$, and consider the point $w_0 = f(z_0)$ belonging to G. By the preceding

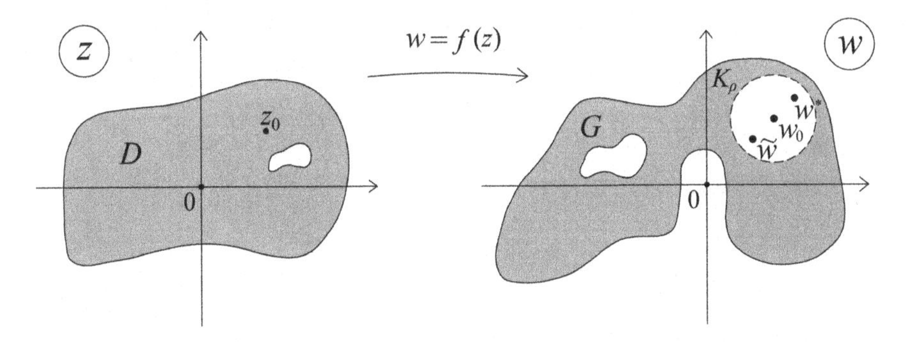

Fig. 5.1 The disk $K_\rho = \{|w - w_0| < \rho\} \subset G = f(D)$, $w_0 = f(z_0)$, with points w^* and $\tilde{w}$: $|w^*| > |w_0|$ and $|\tilde{w}| < |w_0|$

Theorem, G is a domain, and consequently, w_0 is an interior point of G, that is, w_0 is contained in G together with its neighborhood: there exists $\rho > 0$ such that the open disk $K_\rho = \{|w - w_0| < \rho\}$ is contained in G (see Fig. 5.1). Evidently, in this disk there are points more distant from the origin than the point w_0, that is, there exists $w^* \in K_\rho$ (and corresponding $z^* \in D$: $f(z^*) = w^*$) such that $|f(z^*)| = |w^*| > |w_0| = |f(z_0)|$. However, this contradicts the assumption that $|f(z_0)|$ is the maximum value. □

Remark Using the same arguments, we can deduce that a non-constant function $f(z)$ analytic in a domain D has the real and imaginary parts $\operatorname{Re} f(z)$ and $\operatorname{Im} f(z)$ that cannot attain their maxima and minima at interior points of D. Indeed, if we take any point z_0 in D, the corresponding point $w_0 = f(z_0)$ is an interior point in the domain $G = f(D)$. Tracing the horizontal diameter in the disk $K_\rho = \{|w - w_0| < \rho\} \subset G$ through the centerpoint w_0, we see that there are points of this diameter such that $\operatorname{Re} w = \operatorname{Re} f(z) > \operatorname{Re} f(z_0) = \operatorname{Re} w_0$ and $\operatorname{Re} w = \operatorname{Re} f(z) < \operatorname{Re} f(z_0) = \operatorname{Re} w_0$ (see Fig. 5.2). The same is true with respect to the vertical diameter and $\operatorname{Im} f(z)$.

Corollary 5.2 *If a non-constant function $w = f(z)$ is analytic in a domain D and $f(z) \neq 0, \forall z \in D$, then $|f(z)|$ cannot attain its minimum at an interior point of D.*

Proof If there exists at least one point in the domain D at which $|f(z_0)| = 0$, then this is the minimum of the function $|f(z)|$. However, by the conditions of the Corollary, $f(z) \neq 0$ at any point of D, which means that the point $w = 0$ does not belong to the domain $G = f(D)$. Using the same line of reasoning as in Corollary 5.1, we can deduce that for $\forall w_0 \in G$, the value $|w_0|$ cannot be a minimum of the function $|f(z)|$, because in the disk $K_\rho = \{|w - w_0| < \rho\} \subset G$ there are points which are closer to the origin than w_0 (see Fig. 5.1). □

Corollary 5.3 (The maximum modulus principle for a closed domain) *Let D be a bounded domain whose boundary consists of a finite number of piecewise smooth*

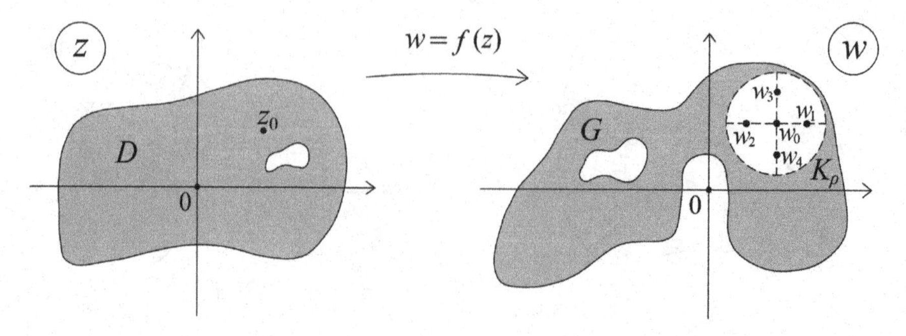

Fig. 5.2 The disk $K_\rho = \{|w - w_0| < \rho\} \subset G = f(D)$, $w_0 = f(z_0)$, with points w_1, w_2: $\text{Re}w_1 > \text{Re}w_0$, $\text{Re}\, w_2 < \text{Re}\, w_0$ and points w_3, w_4: $\text{Im}\, w_3 > \text{Im}\, w_0$, $\text{Im}\, w_4 < \text{Im}\, w_0$

curves. If a non-constant function $w = f(z)$ is analytic in D and continuous on $\overline{D}$, then $|f(z)|$ attains its maximum only on the boundary of D. Under the additional condition that $f(z) \neq 0$, $\forall z \in D$, the function $|f(z)|$ attains its minimum also only on ∂D.

Proof Since $\overline{D}$ is a compact set and the function $w = f(z)$ is continuous on $\overline{D}$, the absolute value $|f(z)|$ is also continuous on $\overline{D}$ and, by the Weierstrass theorem, has a maximum and minimum in $\overline{D}$. However, by Corollary 5.1, $|f(z)|$ does not attain its maximum inside D, then it must attain the maximum only on the boundary of D. If $f(z) \neq 0$, $\forall z \in D$, using the same arguments we can deduce that the minimum of $|f(z)|$ is found also on ∂D. □

Corollary 5.4 (Schwarz's Lemma) *If $f(z)$ is analytic in a disk $|z| < R$, $f(0) = 0$ and $|f(z)| \le M$, $\forall z \in \{|z| < R\}$, then $|f(z)| \le \frac{M}{R}|z|$, $\forall z \in \{|z| < R\}$ and $|f'(0)| \le \frac{M}{R}$. In both estimates the equality is achieved only in the case $f(z) = e^{i\alpha}\frac{M}{R}z$, where α is a real constant.*

Proof Since $f(z)$ is analytic in $|z| < R$ and $f(0) = 0$, the function $f(z)$ can be expanded in a power series $f(z) = \sum_{n=1}^{+\infty} c_n z^n$ in $|z| < R$. Consider the function $g(z) = \frac{f(z)}{z} = \sum_{n=1}^{+\infty} c_n z^{n-1} = \sum_{n=0}^{+\infty} c_{n+1} z^n$, where we set $g(0) = c_1 = f'(0)$. The last expansion converges in $|z| < R$ and $g(z)$ is analytic in this disk. Take any z, $|z| < R$ and fix this point. For any r such that $|z| < r < R$, the function $g(z)$ is analytic in the disk $|z| \le r$. By the Maximum Modulus Principle, the function $|g(z)|$ attains its maximum only on the boundary of the disk $|z| \le r$. Since for $|z| = r$ we have $|g(z)| = \frac{|f(z)|}{|z|} \le \frac{M}{r}$, the same evaluation holds in the entire disk $|z| \le r$. This is true for any r, $|z| < r < R$, and consequently, taking a limit as r approaches R, we get $|g(z)| \le \frac{M}{R}$. Since z is an arbitrary point in $|z| < R$, the last inequality holds for all z in the disk $|z| < R$. From this evaluation we obtain $\left|\frac{f(z)}{z}\right| \le \frac{M}{R}$, that is, $|f(z)| \le \frac{M}{R}|z|$ and $|g(0)| = |f'(0)| \le \frac{M}{R}$.

If at some point z_0, $|z_0| < R$ it happens that $|g(z_0)| = \frac{M}{R}$, then $|g(z)|$ attains its maximum at an interior point of the disk $|z| < R$. By the Maximum Modulus

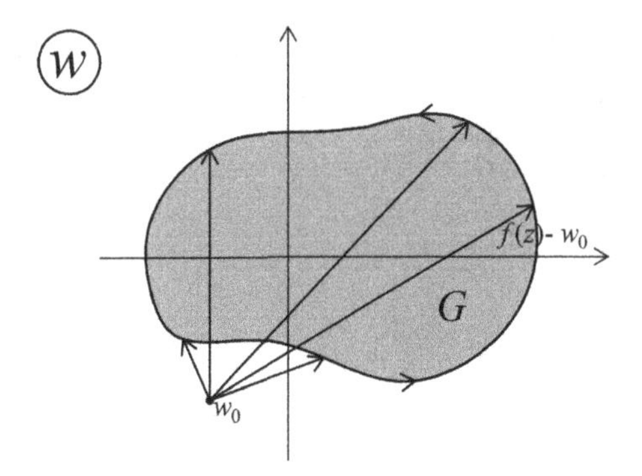

Fig. 5.3 The variation of the vector $f(z) - w_0$ when z traverses once the boundary of D: the case of bounded G and $w_0 \notin \overline{G}$

Principle, it follows that $g(z) = constant = \frac{M}{R}e^{i\alpha}$ (in particular, $|g(0)| = |f'(0)| = \frac{M}{R}$) and consequently $f(z) = \frac{M}{R}e^{i\alpha}z$. □

5.3 The Principle of Boundary Correspondence

Theorem 5.1 (The principle of boundary correspondence) *Suppose D and G are simply connected domains whose boundaries are piecewise smooth curves with folds, the domain G is bounded, and suppose a function $w = f(z)$ is analytic in D and continuous on D up to its boundary. If under the mapping f one positive circuit along ∂D leads to one positive circuit along ∂G, then $w = f(z)$ is a bijective mapping of D onto G.*

Proof We should prove the two properties:
(1) the domain G is the image of the domain D under the transformation f: $G = f(D)$;
(2) the function $w = f(z)$ is univalent on the domain D.

We start with the first property and the second is obtained as a by-product of the used line of reasoning. Take any point w_0 in the plane $\mathbb{C}_w$, which does not belong to the boundary of G, and find the number of zeros of the function $f(z) - w_0$ in the domain D. According the hypothesis, when a point z makes one positive circuit along the boundary of D, the corresponding point $w = f(z)$ also completes one positive circuit along the boundary of G. Therefore, $\underset{z\in\partial D}{var} \arg(f(z) - w_0) = \underset{w\in\partial G}{var} \arg(w - w_0)$. If the point w_0 does not belong to $\overline{G}$, then after one traversal along the boundary of G the vector $w - w_0$ returns to the initial position without a modification of its argument, that is, $\underset{w\in\partial G,\, w_0\notin\bar{G}}{var} \arg(w - w_0) = 0$ (see Fig. 5.3). If the point w_0 belongs to the domain G, then one circuit of w along ∂G leads to one positive circuit of the vector $w - w_0$, and consequently, the argument of $w - w_0$ increases by 2π, that is, $\underset{w\in\partial G,\, w_0\in G}{var} \arg(w - w_0) = 2\pi$ (see Fig. 5.4).

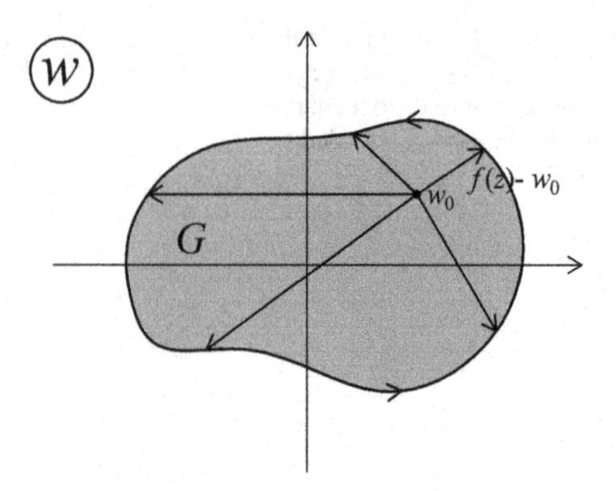

Fig. 5.4 The variation of the vector $f(z) - w_0$ when z traverses once the boundary of D: the case of bounded G and $w_0 \in G$

Hence, using the Argument Principle (see Sect. 3.13), we obtain

$$N_{f(z)-w_0}(D) = \frac{1}{2\pi} \underset{z \in \partial D}{var} \arg(f(z) - w_0) = \frac{1}{2\pi} \underset{w \in \partial G}{var} \arg(w - w_0) = \begin{cases} 1, & w_0 \in G, \\ 0, & w_0 \notin \bar{G}. \end{cases} \tag{5.4}$$

It follows from (5.4) that any point of G is the value of the function $w = f(z)$ at some point in D, ($N_{f(z)-w_0}(D) = 1$, if $w_0 \in G$), which means that every point of G belongs to the image of D under the transformation f: $G \subset f(D)$. At the same time, if some point $w_0 \notin \overline{G}$, then the equality (5.4) implies that w_0 is not a value of $w = f(z)$ at any point of D. Therefore, the image of D is contained in $\overline{G}$: $f(D) \subset \overline{G}$. Thus, $G \subset f(D)$ and $f(D) \subset \overline{G}$.

To show that $f(D) = G$ it remains to prove that no point of the domain D can be carried to a boundary point of G under the transformation $w = f(z)$. Suppose, for contradiction, that there exists a point $z_1 \in D$ such that $f(z_1) = w_1 \in \partial G$. Since $w_1 \in \partial G$, the condition of the Theorem implies that there exists $z_2 \in \partial D$ whose image is also w_1: $f(z_2) = w_1$ (maybe such a point is not unique). Consider a neighborhood of z_1 contained in D and the part of a neighborhood of z_2, which also lies in D, both such small that they do not intersect: $U_{z_1} = \{|z - z_1| < r_1\} \subset D$, $U_{z_2} = \{|z - z_2| < r_2\}$, $U_{z_1} \subset D$, $U_{z_2} \cap D = \tilde{U}_{z_2}$ and $U_{z_1} \cap U_{z_2} = \varnothing$ (see Fig. 5.5).

From the principle of the domain preservation it follows that the function $w = f(z)$ transforms the neighborhood U_{z_1} into a domain U_{w_1} with $w_1 \in U_{w_1}$, and it also transforms the domain $\tilde{U}_{z_2}$ into a domain $\tilde{U}_{w_1}$ with $w_1 \in \partial \tilde{U}_{w_1}$ ($U_{w_1} \subset \overline{G}$ and $\tilde{U}_{w_1} \subset \overline{G}$). Then, the domains U_{w_1} and $\tilde{U}_{w_1}$ have common points (see Fig. 5.5). Pick any point $w^* \in U_{w_1} \cap \tilde{U}_{w_1}$ and $w^* \notin \partial G$ (consequently $w^* \in G$). Since $w^* \in U_{w_1}$ and $U_{w_1} = f\left(U_{z_1}\right)$, there exists $z_1^* \in U_{z_1} \subset D$ such that $f\left(z_1^*\right) = w^*$. On the other hand, since $w^* \in \tilde{U}_{w_1}$ and $\tilde{U}_{w_1} = f\left(\tilde{U}_{z_2}\right)$, there exists $z_2^* \in \tilde{U}_{z_2} \subset D$ such that $f\left(z_2^*\right) = w^*$. Due to the construction, the domains U_{z_1} and $\tilde{U}_{z_2}$ have no common points ($U_{z_1} \cap \tilde{U}_{z_2} = \varnothing$) and consequently $z_1^* \neq z_2^*$ (see Fig. 5.5). In this way, we have found the point w^* in the domain G, which has at least the two inverse images z_1^* and z_2^* in the

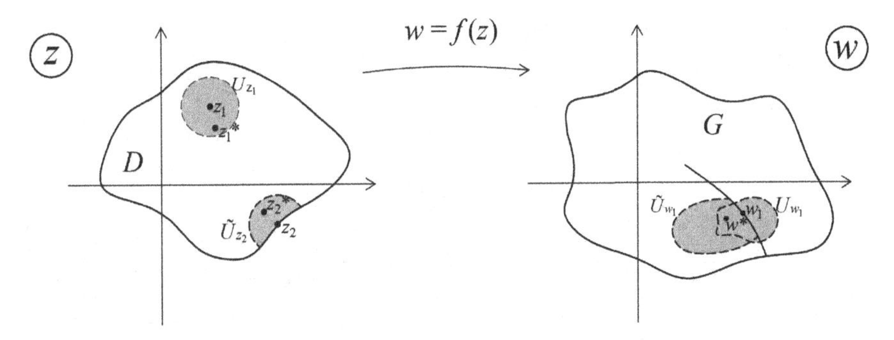

Fig. 5.5 Construction of the points $z_1^*, z_2^* \in D$, $z_1^* \neq z_2^*$ such that $f(z_1^*) = f(z_2^*) = w^* \in G$

domain D under $w = f(z)$, that is, $N_{f(z)-w^*}(D) \geq 2$, $w^* \in G$. Hence, we arrive at the contradiction with the equality (5.4). Therefore, our supposition is false, which means that an interior point of D cannot be mapped into a boundary point of G. Thus, we have that $f(D) \subset G$ and $G \subset f(D)$, that is, $f(D) = G$.

The univalence of the function $w = f(z)$ follows from the equality (5.4), which implies that any point in G has a unique inverse image in D.

The Theorem is proved. □

In the next Theorem, we extend the shown result to the case when the domain G is infinite, which is one of two possible types of unbounded domains.

Theorem 5.2 (The principle of boundary correspondence for infinite domains) *Suppose D and G are simply connected domains whose boundaries are piecewise smooth curves with folds, the domain G is infinite ($\infty \in G$), and suppose a function $w = f(z)$ is analytic in D and continuous on D up to its boundary, except at one simple pole in the interior of D. If under the mapping f a one positive circuit along ∂D causes the one positive circuit along ∂G, then $w = f(z)$ is a bijective mapping of D onto G.*

Proof As in Theorem 5.1, we employ the Argument Principle. Take any point $w_0 \notin \partial G$ and consider the variation of the argument of the vector $f(z) - w_0$ under one positive circuit of the point z along ∂D. Since this circuit induces one positive circuit of the point $w = f(z)$ along ∂G, it follows that $\underset{z\in\partial D}{var} \arg(f(z) - w_0) = \underset{w\in\partial G}{var} \arg(w - w_0)$. If $w_0 \in G$, then after a single circuit of w along ∂G the vector $w - w_0$ returns to its original position without a change of the argument (see Fig. 5.6), that is, $\underset{z\in\partial D}{var} \arg(f(z) - w_0) = 0$. If $w_0 \notin \overline{G}$, then under one positive circuit of z along ∂D, the vector $w - w_0$ makes one loop clockwise along ∂G, which means that $\underset{z\in\partial D}{var} \arg(f(z) - w_0) = -2\pi$ (see Fig. 5.7).

Hence, using the Argument Principle (Sect. 3.13), we obtain

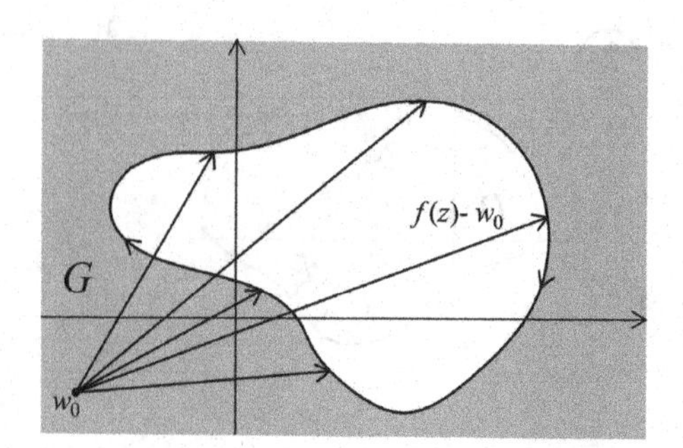

Fig. 5.6 The variation of the vector $f(z) - w_0$ when z traverses once the boundary of D: the case of infinite G and $w_0 \in G$

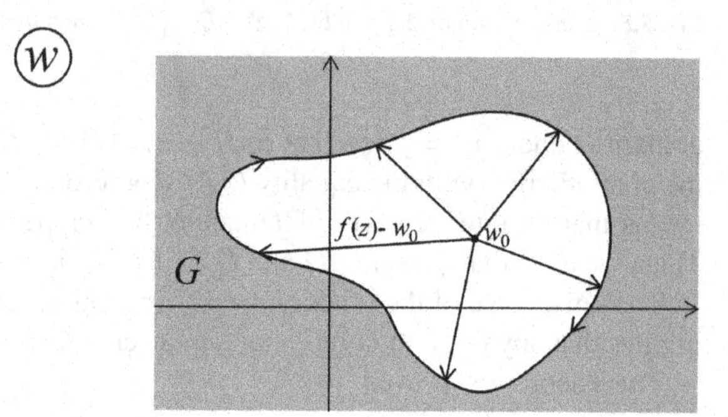

Fig. 5.7 The variation of the vector $f(z) - w_0$ when z traverses once the boundary of D: the case of infinite G and $w_0 \notin \overline{G}$

$$\begin{aligned} N_{f(z)-w_0}(D) - P_{f(z)-w_0}(D) &= \frac{1}{2\pi} \underset{z \in \partial D}{var} \arg\left(f(z) - w_0\right) \\ &= \frac{1}{2\pi} \underset{w \in \partial G}{var} \arg\left(w - w_0\right) = \begin{cases} 0, & w_0 \in G, \\ -1, & w_0 \notin \bar{G}. \end{cases} \end{aligned} \tag{5.5}$$

Recalling that $f(z)$ (and consequently $f(z) - w_0$) has only one simple pole in D, that is, $P_{f(z)-w_0}(D) = 1$ we deduce from (5.5) that

$$N_{f(z)-w_0}(D) = \begin{cases} 1, & w_0 \in G, \\ 0, & w_0 \notin \bar{G}. \end{cases} \tag{5.6}$$

It means that any point $w_0 \in G$ is a function value, that is, $G \subset f(D)$, and, on the other hand, any point $w_0 \notin \overline{G}$ does not belong to the function image, that is, $f(D) \subset \overline{G}$.

The proof that no point of the domain D can be carried to a boundary point of G under the transformation $w = f(z)$ is made in the same way as in Theorem 5.1.

Therefore, $G = f(D)$. Moreover, the first sentence in (5.6) implies that every point $w_0 \in G$ has exactly one inverse image in D, that is, the function $w = f(z)$ is univalent in D, and consequently, it maps bijectively D onto G.

The Theorem is proved. □

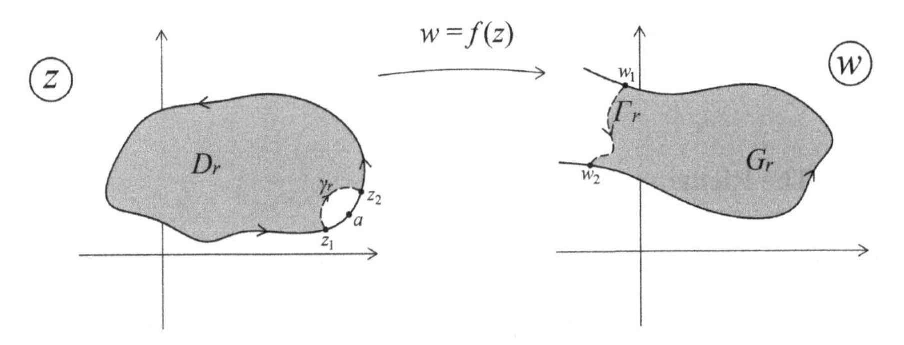

Fig. 5.8 The domain $D_r = D\backslash\{|z-a| \le r\}$ and its image $G_r = f(D_r)$

Remark 5.1 The remaining case of unbounded finite domains, that is, the situation when $\infty \in \partial G$ is more sophisticated. In this case, the principle of argument is not applicable directly. Indeed, consider the function $w = z^5$ in the upper half-plane $D = \{\operatorname{Im} z > 0\}$. The boundary of D is the real axis, and when z passes once along the boundary from $-\infty$ to $+\infty$, the point w also passes once along the real axis from $-\infty$ to $+\infty$. However, the image of the upper half-plane under this transformation is the entire plane $\mathbb{C}_w$. Moreover, under this mapping, the function values fulfill the entire upper half-plane of $\mathbb{C}_w$ thrice and the entire lower half-plane of $\mathbb{C}_w$ twice, because the angle π at the origin passes to 5π under the transformation $w = z^5$.

Due to these reasons, the proof of the principle of boundary correspondence in the case $\infty \in \partial G$ is more intricate and requires the introduction of new concepts and additional results, which are left out of scope of this text.

Remark 5.2 Despite a problem in providing a general proof of the principle of boundary correspondence when $\infty \in \partial G$, in a particular case when D is the domain of univalence or is contained in the domain of univalence of $f(z)$ (which is a common situation in many practical problems) the principle of boundary correspondence is still applicable in the same form as in Theorem 5.1 and can be readily shown even if $\infty \in \partial G$.

In fact, consider, for simplicity, the case when there exists only one point $a \in \partial D$ which is carried to $\infty \in \partial G$. Since $f(z)$ is analytic in D, by the principle of the domain preservation, the image $f(D)$ is also a domain. Take sufficiently small $r > 0$ and remove from D the closed disk $\overline{K}_a = \{|z-a| \le r\}$: $D_r = D\backslash\overline{K}_a$. Denote the arc of the circle $|z-a| = r$, which lies in D, by γ_r and the intersection points of $|z-a| = r$ and ∂D by z_1 and z_2 (see Fig. 5.8). From the hypothesis $f(\partial D) = \partial G$, it follows that $w_1 = f(z_1) \in \partial G$ and $w_2 = f(z_2) \in \partial G$ (since $z_1, z_2 \in \partial D$). Since $w = f(z)$ is analytic in D, the arc γ_r is transformed into a smooth curve $\Gamma_r \subset f(D)$, which links the points w_1 and w_2 on the boundary of G. Besides, due to the univalence of $f(z)$ in D, a single traversal along γ_r (from z_1 to z_2) causes a single traversal along Γ_r (from w_1 to w_2). Notice that the arc Γ_r together with the remaining part of ∂G form the boundary of the new domain G_r such that $\infty \notin \partial G_r$. Therefore, all the conditions of Theorem 5.1 hold, and consequently $f(D_r) = G_r$. Taking the limit as

$r \to 0$, we obtain that $f(D) = G$ and the function $w = f(z)$ maps conformally D onto G.

5.4 The Riemann Mapping Theorem

In this section, we prove one of the fundamental theorems of conformal mappings—the Riemann Mapping Theorem—which says that every simply connected domain with at least two boundary points can be mapped conformally and univalently onto the open unit disk. Evidently, this statement immediately leads to a more general one: any simply connected non-degenerate domain can be mapped conformally and univalently on any other simply connected non-degenerate domain (it was discussed in detail in Sect. 4.1). In other words, any two simply connected non-degenerate domains are *conformally equivalent*. (Recall that a domain is called non-degenerate if it has more than one boundary point.) Although the Mapping Theorem was formulated by Riemann in 1851, its first successful proof was given by Carathéodory in 1912 and simplified by Koebe two years later. The modern approach to the proof based on a solution of an extremal problem was proposed by Fejér and Riesz in 1922.

One of the essential results used commonly in the course of the proof is Montel's Theorem on normality of a family of analytic functions. We start this section by introducing some new concepts required for this theorem and providing one of the standard methods of the proof of Montel's Theorem.

Recall that a sequence of analytic functions defined on a domain D *converges normally on* D (to an analytic in D function) if this sequence *converges uniformly on every compact subset* of D.

Definition 5.1 Let Φ be a family of analytic functions defined on a domain D. This family is said to be *normal on* D if every sequence in Φ has a subsequence which converges normally on D (not necessarily to an element of Φ itself).

Remark *Normal families* are also called *normal sets*, or *relatively compact sets*, or *sequentially compact sets*.

Definition 5.2 A set Φ of analytic functions defined on a domain D is said to be *uniformly bounded on compact sets* in D if for each compact set $K \subset D$ there exists a positive number $B(K)$ such that $|f(z)| \le B(K)$ simultaneously for all $f \in \Phi$ and $z \in K$.

Definition 5.3 A set Φ of analytic functions defined on a domain D is said to be *equicontinuous* on a compact set $K \subset D$ if for any $\epsilon > 0$ there exists $\delta > 0$ such that for any $z, z' \in K$, $|z - z'| < \delta$ it follows that $|f(z) - f(z')| < \epsilon$ for all $f(z) \in \Phi$.

Now, we will prove a version of the well-known Arzelà–Ascoli theorem, which takes the following form suitable for Complex Analysis.

Montel's Theorem *If a set Φ of analytic functions defined on a domain D is uniformly bounded on compact sets in D, then Φ is equicontinuous on every compact set in D and is normal on D.*

Proof We perform the proof in the two steps:
(1) we show the equicontinuity of Φ,
(2) we use a diagonal procedure twice to find a convergent subsequence.

(1) Consider an arbitrary compact set $K \subset D$ and denote by $3r$ the distance between K and the boundary of D: $\operatorname{dist}(K, \partial D) = 3r$ (this distance is positive since both sets K and ∂D are closed, have no common points and at least one of them—K—is bounded). Pick two points $z, z' \in K$ such that $|z - z'| < \delta < r$ and consider the circle C centered at z' of the radius $2r$. By Cauchy's formula, we have

$$f(z) = \frac{1}{2\pi i} \int_C \frac{f(\zeta)}{\zeta - z} d\zeta, \ f(z') = \frac{1}{2\pi i} \int_C \frac{f(\zeta)}{\zeta - z'} d\zeta ,$$

and, using the equality

$$\frac{1}{\zeta - z} - \frac{1}{\zeta - z'} = \frac{z - z'}{(\zeta - z)(\zeta - z')} ,$$

we arrive at the relation

$$f(z) - f(z') = \frac{z - z'}{2\pi i} \int_C \frac{f(\zeta)}{(\zeta - z)(\zeta - z')} d\zeta .$$

In the last integral $|\zeta - z'| = 2r$, $|\zeta - z| > r$, and consequently

$$|f(z) - f(z')| \le \frac{|z - z'|}{2\pi} \int_C \frac{|f(\zeta)|}{|\zeta - z| \cdot |\zeta - z'|} |d\zeta|$$

$$< \frac{|z - z'|}{2\pi} \sup_{z \in C} |f(z)| \cdot \frac{1}{r} \cdot \frac{1}{2r} \cdot 4r\pi = \frac{|z - z'|}{r} \sup_{z \in C} |f(z)| .$$

Let us consider the auxiliary compact K_r consisting of all the points $z \in D$ such that $\operatorname{dist}(z, K) \le 2r$ (geometrically speaking, we attach to the compact K the band around K of the width $2r$). Evidently, K_r lies in D and contains both K and C. Then $\sup\limits_{z \in C} |f(z)| \le \sup\limits_{z \in K_r} |f(z)| \le B(K_r)$, where $B(K_r)$ is the unique constant for all the functions $f(z) \in \Phi$ according to the definition of a uniformly bounded set of functions. Hence, we arrive at the following inequality:

$$|f(z) - f(z')| < \frac{1}{r} B(K_r) \cdot |z - z'| < \frac{1}{r} B(K_r) \cdot \delta ,$$

where r and $B(K_r)$ are the constants depending only on the choice of K, but not depending on z and z'. Therefore, the set Φ is equicontinuous on any compact set $K \subset D$.

(2) Next we prove that the family Φ is normal. First, we take any compact set $K \subset D$ and show that there exists a subsequence of $\{f_n(z)\}$ which converges uniformly on K. Let $\{z_j\}$ be a countable dense subset of K. For each j the numerical sequence $\{f_n(z_j)\}$ is bounded. Therefore, (by Bolzano–Weierstrass theorem of Real Analysis) there exists a subsequence $\{f_{n,1}(z)\}$ of $\{f_n(z)\}$ such that $\{f_{n,1}(z_1)\}$ converges. Next, there exists a subsequence $\{f_{n,2}(z)\}$ of $\{f_{n,1}(z)\}$ such that $\{f_{n,2}(z_2)\}$ converges. Notice that $\{f_{n,2}(z)\}$ also converges at z_1, since $\{f_{n,1}(z)\}$ converges at this point. Proceeding in this manner, at the j-th step we obtain a subsequence $\{f_{n,j}(z)\}$ (of the subsequence $\{f_{n,j-1}(z)\}$), which is convergent at the points $z_1, \dots, z_j$. And we continue this process infinitely. Then, we construct the diagonal subsequence $\{f_{n,n}(z)\}$, whose first element is the first element $f_{1,1}(z)$ of the first subsequence $\{f_{n,1}(z)\}$, the second element is the second element $f_{2,2}(z)$ of the second subsequence $\{f_{n,2}(z)\}$, etc., the n-th element is the n-th element $f_{n,n}(z)$ of the n-th subsequence $\{f_{n,n}(z)\}$, etc. This diagonal subsequence converges at all the points $\{z_j\}$, since starting from j-th element, the diagonal subsequence $\{f_{n,n}(z)\}$ contains only the elements (some elements) of subsequence $\{f_{n,j}(z)\}$, which is convergent at z_j.

Moreover, we can show that $\{f_{n,n}(z)\}$ converges uniformly on K. Indeed, choose any $\epsilon > 0$ and find the corresponding $\delta > 0$ in the definition of equicontinuity of the set Φ. Then consider an open cover of K by all the open disks C_j, centered at z_j with the same radius δ (the disks C_j form a cover of K because $\{z_j\}$ is a countable dense subset of K). Since K is compact, there is a finite subcover containing k disks: $K \subset C_1 \cup \dots C_k$. Since $\{f_{n,n}(z)\}$ converges at z_j, $j = 1, \dots, k$, one can choose (by the Cauchy criterion of Real Analysis) sufficiently large N such that for $\forall m, n > N$ it holds

$$|f_{n,n}(z_j) - f_{m,m}(z_j)| < \epsilon\,, \ \forall j = 1, \dots, k\,. \tag{5.7}$$

If we pick now an arbitrary point $z \in K$, then $z \in C_i$ for some i and we get

$$|f_{n,n}(z)-f_{m,m}(z)| \le |f_{n,n}(z)-f_{n,n}(z_i)| + |f_{n,n}(z_i)-f_{m,m}(z_i)| + |f_{m,m}(z_i)-f_{m,m}(z)| < 3\epsilon\,, \tag{5.8}$$

since the first and third terms are less than ϵ by the definition of equicontinuity and the middle term is less than ϵ by (5.7). Since (5.8) holds for all $z \in K$ simultaneously, it means that the diagonal subsequence $\{f_{n,n}(z)\}$ converges uniformly on K.

Finally, we prove that there exists a subsequence convergent on every compact set. To this end, we construct a sequence of compact sets K_s, $s = 1, 2, \dots$ such that K_s is contained in the interior of K_{s+1} and their union is D. Let $\bar{C}_s$ be the closed disk of radius s centered at the origin and $\bar{D}$ the closure of D. Each K_s is defined as the set of points $z \in \bar{D} \cap \bar{C}_s$ such that $\operatorname{dist}(z, \partial D) \ge \frac{1}{s}$. Then, K_s is compact and contained in the open set of elements $z \in D \cap C_{s+1}$ such that $\operatorname{dist}(z, \partial D) > \frac{1}{s+1}$, that is, K_s is contained in the interior of K_{s+1}. Besides, the union of all K_s is equal to D. It then follows that any compact set K is contained in some K_s because the distance $\operatorname{dist}(K, \partial D) = d > 0$ is greater than $\frac{1}{s}$ for some sufficiently large s.

Let $\{f_n(z)\}$ be an original sequence in Φ. It was already shown that there exists a subsequence $\{f_{n,1}(z)\}$ which converges uniformly on K_1, also a subsequence $\{f_{n,2}(z)\}$ which converges uniformly on K_2, and so on. Then, the diagonal sequence $\{f_{n,n}(z)\}$ converges uniformly on every K_s and consequently on every compact set $K \subset D$.

This completes the proof. □

Remark The converse of Montel's theorem is also true. It can be formulated as follows: if a set Φ of analytic in a domain D functions is normal, then Φ is uniformly bounded on compact sets in D. The proof can be carried out using the contradiction argument. Suppose Φ is normal in D, but is not uniformly bounded on compact sets. Then, there exists a compact subset $S \subset D$ on which $|f(z)|$, $f(z) \in \Phi$ attains as large values as one wants, that is, for any $n \in \mathbb{N}$ one can find such function $f_n(z) \in \Phi$ and such $z_n \in S$ that $|f_n(z_n)| > n$, $\forall n = 1, 2, \ldots$. Since Φ is normal, we can choose from $f_n(z)$ a subsequence $f_{nk}(z)$ which converges uniformly on any compact set in D, in particular, on S. The limit function $f(z)$ is analytic in D and consequently continuous on compact S. Therefore, $f(z)$ is bounded on S and we denote $M = \max\limits_{S} |f(z)| < \infty$. Due to the uniform convergence of $f_{nk}(z)$ on S, the inequalities $|f_{nk}(z) - f(z)| < 1$, $\forall n_k > N$ hold simultaneously for all the points of S, which implies that $|f_{nk}(z)| \le |f(z)| + 1 \le M + 1$ for $\forall n_k > N$ and $\forall z \in S$. However, this contradicts the evaluation $|f_{nk}(z_{nk})| > n_k$, $\forall n_k$, which follows from the made supposition that Φ is not uniformly bounded on compact sets in D. Hence, this assumption is false and Φ is uniformly bounded on compact sets in D.

Frequently Montel's Theorem together with its converse are also referred to as Montel's Theorem.

Now, we are prepared to prove the Riemann Mapping Theorem.

The Riemann Mapping Theorem *For any simply connected non-degenerate domain D, there exists an analytic function $w = f(z)$ defined on D that maps D conformally and univalently onto the open unit disk $|w| < 1$. Moreover, under the normalization conditions $f(z_0) = 0$ and $f'(z_0) > 0$, where $z_0 \in D$, this function is unique.*

Proof We divide the proof into four stages:
(1) transformation of the original problem by simplifying the domain definition;
(2) solution of an extremal problem on the simplified domain in the class of analytic and univalent functions;
(3) verification that the found extremal function defines the required transformation;
(4) proof that this function is unique.

(1) To modify the original domain, we perform some transformations by elementary functions, which map the domain D onto a domain lying inside the unit disk. Consider first the case when there exist an exterior point a with respect to D. It means that there exists r such that the disk $|z - a| \le r$ lies outside D. Then, the fractional linear function $z_1 = \frac{r}{z-a}$ maps the disk $|z - a| \le r$ onto $|z_1| \ge 1$ and the exterior of

the disk $|z-a|>r$ in $|z_1|<1$; in particular, the domain D is mapped conformally onto a domain $D_1 \subset \{|z_1|<1\}$. For a domain D without exterior points, we can take two different finite boundary points a and b (because D is non-degenerate), which belong to the single boundary component $\Gamma = \partial D$ since D is simply connected. Then, the fractional linear function $z_1 = \frac{z-a}{z-b}$ maps D conformally on a domain D_1 with a boundary component Γ_1 such that $0 \in \Gamma_1$ and $\infty \in \Gamma_1$. Since Γ_1 passes through the points 0 and ∞, it is possible to define a single-valued analytic branch of $z_2 = \sqrt{z_1} = \sqrt{\frac{z-a}{z-b}}$, which does not take the same value twice, nor does it take opposite values. Therefore, this function maps D_1 conformally onto a domain D_2 which has exterior points, and consequently, as was shown above, D_2 can be transformed conformally into a domain contained in the unit disk. Hence, instead of an arbitrary domain D in the formulation of the Theorem, we can consider a simply connected domain lying in the unit disk $|z|<1$.

Let us make one more small simplification to carry the point z_0 to the origin. Consider the fractional linear transformation $\zeta = \frac{z-z_0}{1-\bar{z}_0 z}$. This function maps one-to-one the disk $|z|<1$ onto the disk $|\zeta|<1$ in such a way that the point $z_0 \in D$ is carried to $0 \in \zeta(D)$ and $\zeta'(z_0) = \frac{1}{1-|z_0|^2} > 0$. Thus, after the performed simplifications, the original problem can be reformulated in the following manner: for any simply connected domain D, which lies in the unit disk $|z|<1$ and contains the origin, one can find an analytic function $w = f(z)$ that maps D conformally and univalently onto the unit disk $|w|<1$; if, additionally, $f(z)$ satisfies the normalization conditions $f(0)=0$ and $f'(0)>0$, then this function is unique.

(2) Now, we introduce a family Φ of functions $f(z)$, each of which maps conformally the simplified domain D onto a domain inside the unit disk and is such that $f(0)=0$, $f'(0)>0$. This family is not empty, since it contains any linear function $f(z)=az$, $0<a\le 1$. Consider the following extremal problem: among the functions of Φ, find a function that has the maximum value of $f'(0)$. Let us show that there exists a solution to this problem. Since $z=0$ is an interior point of D, there exists a disk $|z|<r$ contained in D. Therefore, all the functions $f(z)\in\Phi$ are analytic and satisfies the conditions of Schwarz's Lemma in this disk (see Corollary 5.4 in Sect. 5.2). Then $f'(0)=|f'(0)|\le \frac{1}{r}$, and consequently, there exists the supremum $\sup\limits_{f\in\Phi} f'(0)=\mu$. Notice that $\mu \ge 1$ because the functions $f(z)=az, 0<a\le 1$ belong to the family Φ. From the definition of the supremum, it follows that there exists a sequence $f_n(z)\in\Phi$ such that $f_n'(0) \underset{n\to\infty}{\to} \mu$. Since the set Φ consists of analytic in D functions and it is uniformly bounded on D (any function $f(z)\in\Phi$ satisfies the inequality $\sup\limits_{z\in D}|f(z)|\le 1$), by Montel's Theorem the set Φ is normal on D, which implies that the sequence $f_n(z)$ has a subsequence $f_{n_k}(z)$ convergent normally on D (that is, uniformly on every compact subset of D) to an analytic limit function $f_0(z)$. It is clear that $f_0(0)=0$, $f_0'(0)=\mu\ge 1$ and $|f_0(z)|\le 1$ in D. Moreover, since $f_0(z)$ is the limit function of the normally convergent on D sequence of univalent analytic functions $f_{n_k}(z)$, by Hurwitz's Theorem (Sect. 3.14) $f_0(z)$ must be constant or univalent on D. The condition $f_0'(0)\ge 1$ guarantees that $f_0(z)$ is not a constant, and consequently, $f_0(z)$ is univalent on D. Notice that there

is no point in D where $|f_0(z)| = 1$, for otherwise the Maximum Modulus Principle would imply that $f_0(z) \equiv 1$, but we have already verified that $f_0(z)$ is not constant. Therefore, $|f_0(z)| < 1$ in D. Thus, $f_0(z) \in \Phi$ and this function solves the posed extremal problem: $f_0'(0) = \mu = \sup\limits_{f \in \Phi} f'(0)$.

(3) Proceeding with our plan, at the third stage we show that the found extremal function $w = f_0(z)$ maps the domain D onto the entire disk $|w| < 1$. Suppose, by contradiction, that the image $G = f_0(D)$ does not cover the entire disk $|w| < 1$. Then there exists a point c, $|c| < 1$ such that $c \notin G$. Since G is a simply connected domain lying inside the disk $|w| < 1$, the point c can be connected with a point on the circle $|w| = 1$ by a curve γ such that $G \cap \gamma = \varnothing$. The fractional linear transformation $w_1 = \frac{w-c}{1-\bar{c}w}$ maps one-to-one the disk $|w| < 1$ onto the disk $|w_1| < 1$. Under this transformation, the domain G is mapped onto a domain G_1, the point $w = c$ is carried to $w_1 = 0$ and the curve γ—to γ_1, where $G_1 \cap \gamma_1 = \varnothing$. Then, in the disk $|w_1| < 1$ with the cut along γ_1, it is possible to define a single-valued analytic branch of $\zeta = \sqrt{w_1}$, that is, in G it is possible to define a single-valued analytic branch of $\zeta = \sqrt{\frac{w-c}{1-\bar{c}w}}$. This function maps univalently the domain G onto a domain G_2 lying in the disk $|\zeta| < 1$. Under this, the point $w = 0$ is carried to $\zeta = b = \sqrt{-c}$. Applying one more fractional linear transformation $t = \frac{b}{|b|} \frac{\zeta - b}{1-\bar{b}\zeta}$, we map univalently the domain G_2 onto a domain $B \subset \{|t| < 1\}$ in such a way that $\zeta = b = \sqrt{-c}$ is carried to $t = 0$. Consider now the following chain of the mappings: $g(z) = \sqrt{\frac{f_0(z)-c}{1-\bar{c}f_0(z)}}$, $F(z) = \frac{b}{|b|} \frac{g(z)-b}{1-\bar{b}g(z)}$, where $f_0(z)$ is the extremal function. Let us show that $F(z) \in \Phi$. In fact, $F(z)$ is analytic and univalent on D as a composition of analytic and univalent functions. The inequality $|F(z)| < 1$ follows from $|f_0(z)| < 1$, and $F(0) = \frac{b}{|b|} \frac{g(0)-b}{1-\bar{b}g(0)} = 0$. It remains to verify that $F'(0) > 0$. To do this, let us calculate the derivatives $g'(z)$ and $F'(z)$:

$$g'(z) = \frac{1}{2} \sqrt{\frac{1 - \bar{c} f_0(z)}{f_0(z) - c}} \cdot \frac{1 - |c|^2}{(1 - \bar{c} f_0(z))^2} \cdot f_0'(z) \,, \ F'(z) = \frac{b}{|b|} \frac{1 - |b|^2}{(1 - \bar{b} g(z))^2} \cdot g'(z) \,.$$

Therefore,

$$F'(0) = \frac{b}{|b|} \frac{1 - |b|^2}{(1 - |b|^2)^2} \cdot g'(0) = \frac{b}{|b|(1 - |b|^2)} \cdot \frac{1 - |c|^2}{2\sqrt{-c}} \cdot f_0'(0) > 0 \,,$$

because $f_0'(0) > 0$ and $b = \sqrt{-c}$. Hence, $F(z) \in \Phi$.

On the other hand, since $|b| < 1$, we get

$$F'(0) = \frac{1 - |c|^2}{2|b|(1 - |b|^2)} \cdot f_0'(0) = \frac{1 + |b|^2}{2|b|} \cdot \mu > \mu \,.$$

This contradicts the fact that $\mu = \sup\limits_{f \in \Phi} f'(0)$. Thus, the made supposition is false and $f_0(z)$ maps D on the entire disk $|w| < 1$.

Fig. 5.9 A simply connected domain conformally equivalent to the unit disk

(4) Finally, we show the uniqueness of the conformal univalent mapping of D onto the unit disk under the imposed normalization conditions. Suppose, by contradiction, that there exist two functions $f_1(z)$ and $f_2(z)$ satisfying the theorem conditions. Then, the function $h(w) = f_1(f_2^{-1}(w))$ defines a one-to-one mapping of $|w| < 1$ onto itself. We know that such a mapping is given by a fractional linear transformation, and, under the given normalization conditions $h(0) = 0$, $h'(0) > 0$, we obtain $h(w) = w$, whence $f_1(z) = f_2(z)$.

The proof of the Theorem is completed. □

Remark Sometimes, in the formulation of the Riemann Mapping Theorem, the normalization condition $\arg f'(z_0) = \alpha$ is used instead of $f'(z_0) > 0$. One can see that these two conditions are equivalent. In fact, if a function $f(z)$ satisfies $\arg f'(z_0) = \alpha$ and all the remaining Theorem conditions, then the function $f_1(z) = e^{-\iota\alpha} f(z)$ also satisfies all the remaining conditions and, additionally, $\arg f_1'(z_0) = 0$, that is, $f_1'(z_0) > 0$.

The Riemann Mapping Theorem establishes the conformal equivalence between two arbitrary simply connected non-degenerate domains. This fact is far to be intuitively clear: it is hardly conceivable that the simply connected domain shown in Fig. 5.9 is conformally equivalent to the open unit disk.

The statement of the Riemann Mapping Theorem appears to be even more remarkable when we move to multiply-connected domains. In what follows we consider a theoretical example of two rings that shows that the situation is different for doubly- (and in general multiply-) connected domains.

Example Recall first that the necessary condition of the conformal equivalence between two domains is the same degree of connectedness of these domains. We consider here doubly-connected domains and will focus our efforts on an attempt to map conformally (and univalently) one circular ring $D = \{0 < r_1 < |z| < r_2 < +\infty\}$ onto another one $G = \{0 < R_1 < |w| < R_2 < +\infty\}$ and will show that this is possible if and only if $\frac{R_2}{R_1} = \frac{r_2}{r_1}$. This statement is sometimes referred to as *Schottky theorem* and the ratio of the two radii $\frac{r_2}{r_1}$ is known as *the modulus of the ring* (frequently the quantity $\frac{1}{2\pi}\ln\frac{r_2}{r_1}$ is also called the modulus of the ring).

In one direction the proof is trivial: if $\frac{R_2}{R_1} = \frac{r_2}{r_1}$, then the linear conformal transformation $w = \frac{R_1}{r_1}z$ maps D onto G. For opposite direction, we suppose that $w = f(z)$ maps conformally D onto G and will analyze the implications of this supposition on the radii of the two rings. Since the function $w = f(z) = u + iv$ is analytic in the ring D, it has the expansion in the Laurent series $f(z) = \sum_{n=-\infty}^{+\infty} c_n z^n$ convergent in D. Consider two circles $|z| = a_1$ and $|z| = a_2$, $r_1 < a_1 < a_2 < r_2$ and denote by Γ_1 and Γ_2 their images under the transformation $f(z)$. Let us find the areas of the domains G_1 and G_2 enclosed by the curves Γ_1 and Γ_2 (using Green's formula of Real Analysis):

$$A(G_1) = \iint_{G_1} dudv = \int_{\Gamma_1} udv = \int_{|z|=a_1} \frac{1}{2}\left(f(z) + \overline{f(z)}\right) \cdot \frac{1}{2i} d\left(f(z) - \overline{f(z)}\right)$$

$$= \frac{1}{4i}\int_0^{2\pi} \left(f(a_1e^{it}) + \overline{f(a_1e^{it})}\right) \cdot \left(f_t(a_1e^{it}) - \overline{f_t(a_1e^{it})}\right) dt$$

$$= \frac{1}{4i}\int_0^{2\pi} \sum_{n=-\infty}^{+\infty} \left(c_n a_1^n e^{int} + \bar{c}_n a_1^n e^{-int}\right) \cdot \sum_{n=-\infty}^{+\infty} i\left(nc_n a_1^n e^{int} + n\bar{c}_n a_1^n e^{-int}\right) dt .$$

Noting that $\int_0^{2\pi} e^{imt}dt = 0$, $\forall m \neq 0$ and $\int_0^{2\pi} e^{i0t}dt = 2\pi$, after multiplication of the series and term-by-term integration (that can be performed due to the uniform convergence) we obtain

$$A(G_1) = \frac{\pi}{2}\sum_{n=-\infty}^{+\infty} \left(nc_nc_{-n} - nc_nc_{-n} + n\bar{c}_n\bar{c}_{-n} - n\bar{c}_n\bar{c}_{-n} + nc_n\bar{c}_n a_1^{2n} + nc_n\bar{c}_n a_1^{2n}\right)$$

$$= \pi \sum_{n=-\infty}^{+\infty} n|c_n|^2 a_1^{2n}.$$

Similarly,

$$A(G_2) = \pi \sum_{n=-\infty}^{+\infty} n|c_n|^2 a_2^{2n}.$$

Since $n(a_2^{2n-2} - a_1^{2n-2}) = 0$ for $n = 0, n = 1$, and $n(a_2^{2n-2} - a_1^{2n-2}) > 0$ for $\forall n \in \mathbb{Z}$, $n \neq 0, n \neq 1$, the obtained expressions for the areas lead to the following evaluation:

$$\frac{A(G_2)}{A(G_1)} - \frac{\pi a_2^2}{\pi a_1^2} = \frac{\pi^2 a_1^2 a_2^2}{\pi a_1^2 A(G_1)} \cdot \sum_{n=-\infty}^{+\infty} n|c_n|^2 \left(a_2^{2n-2} - a_1^{2n-2}\right) \geq 0 .$$

The last inequality is true for $\forall a_1, a_2$ such that $r_1 < a_1 < a_2 < r_2$. Passing to the limit in this inequality as $a_1 \to r_1$ and $a_2 \to r_2$, we get $\frac{\pi R_2^2}{\pi R_1^2} \geq \frac{\pi r_2^2}{\pi r_1^2}$, that is, $\frac{R_2}{R_1} \geq \frac{r_2}{r_1}$. Applying the same reasoning for the interchanged rings D and G (that is, using the inverse conformal mapping from G onto D), we obtain also that $\frac{r_2}{r_1} \geq \frac{R_2}{R_1}$. Hence, $\frac{R_2}{R_1} = \frac{r_2}{r_1}$.

5.5 The Riemann–Schwarz Symmetry Principle

In this section, we formulate and prove two versions of the *Riemann–Schwarz symmetry principle*, also called the *Riemann–Schwarz reflection principle*, or simply the symmetry (reflection) principle. The first version is about an analytic continuation across a segment of the real axis and the second—about an analytic continuation across a circle arc. We also use this symmetry principle to show the uniqueness of a conformal mapping between simply connected domains with a specific relationship between three boundary points.

Theorem 5.1 The Riemann–Schwarz symmetry principle. *Assume D is a domain contained in the upper or lower half-plane whose boundary includes a segment γ of the real axis. Assume also a function $w = f(z)$ is analytic in D, continuous in $D \cup \gamma$ and takes real values at the points of γ (in other words, the image of γ under the transformation $w = f(z)$ is a segment Γ of the real axis: $\Gamma = f(\gamma) \subset \mathbb{R}$). Under these conditions, the function $f(z)$ can be analytically continued across the real axis from the domain D onto the domain D' symmetric to D with respect to the real axis. Moreover, this analytic continuation is defined by the following formula:*

$$F(z) = \begin{cases} f(z), & z \in D \cup \gamma, \\ \overline{f(\overline{z})}, & z \in D'. \end{cases} \tag{5.9}$$

Proof Denote by D^* the domain $D^* = D \cup \gamma \cup D'$. First, we show that the function $F(z)$ defined by (5.9) is continuous on the domain D^*. Take an arbitrary point $z_0 \in D^*$. There are three possibilities: $z_0 \in D$, $z_0 \in D'$ or $z_0 \in \gamma$. If z_0 belongs to D, then according to (5.9) we have $F(z) = f(z)$. By the Theorem conditions, the function $f(z)$ is analytic in D and consequently continuous in D. If z_0 belongs to D', then by formula (5.9) $F(z) = \overline{f(\overline{z})}$, and this function is continuous in D' as the composition of the continuous functions.

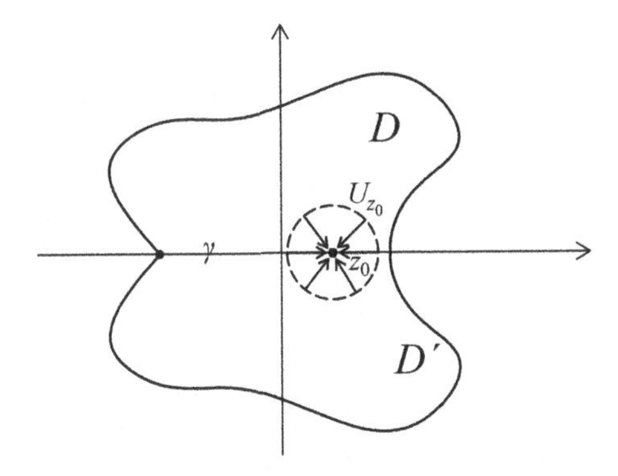

Fig. 5.10 Different paths of z approaching $z_0 \in \gamma$.

Finally, if z_0 lies on γ, then any neighborhood of z_0 contains both the points of D and the points of D'. By formula (5.9), we have $F(z_0) = f(z_0)$. Let us prove the continuity of $F(z)$ at $z_0 \in \gamma$ using the definition of a continuous function, that is, analyzing the limit $\lim\limits_{z\to z_0} F(z)$. Notice that z_0 is an interior point of D^* (and of the segment γ), since $z_0 \in D^*$. Choose a neighborhood $U_{z_0} = \{|z - z_0| < r\}$ of z_0 such that $U_{z_0} \subset D^*$. Since z approaches z_0, we can consider only the points z in this neighborhood. If $z \in D \cup \gamma$, then we have

$$\lim_{z\to z_0,\, z\in D\cup\gamma} F(z) = \lim_{z\to z_0,\, z\in D\cup\gamma} f(z) = f(z_0) = F(z_0), \tag{5.10}$$

because, by the Theorem hypothesis, the function $f(z)$ is continuous in $D \cup \gamma$. If $z \in D'$, then the corresponding point $\bar{z} \in D$ (by the construction of D') and, as $z \to z_0$, we have $\bar{z} \to \bar{z}_0 = z_0$ (see Fig. 5.10). Besides, by the Theorem condition, the function $f(z)$ takes real values at the points of the segment γ, which implies that $\overline{f(z_0)} = f(z_0)$. Therefore, we obtain

$$\lim_{z\to z_0,\, z\in D'} F(z) = \lim_{z\to z_0,\, z\in D'} \overline{f(\bar{z})} = \overline{\lim_{\bar{z}\to z_0,\, \bar{z}\in D} f(\bar{z})} = \overline{f(z_0)} = f(z_0) = F(z_0). \tag{5.11}$$

From (5.10) and (5.11), it follows that $\lim\limits_{z\to z_0} F(z) = F(z_0)$, that is, the function $F(z)$ is continuous at the point z_0. Since z_0 is an arbitrary point of γ, the function $F(z)$ is continuous on γ. Thus, $F(z)$ is continuous at every $z_0 \in D^*$, which means that $F(z)$ is continuous in D^*.

Now, we show that there exists an analytic element of the function $F(z)$ at any point $z_0 \in D^*$. Again we consider the three cases: $z_0 \in D$, $z_0 \in D'$, or $z_0 \in \gamma$. If $z \in D$, then $F(z) = f(z)$ and, by the Theorem condition, $f(z)$ is analytic in D. If z_0 is a point of D', then by formula (5.9) we have $F(z_0) = \overline{f(\bar{z}_0)}$. Let us prove that $F(z)$ is differentiable in D'. Indeed, since D' is a domain, the point z_0 belongs to D' togeteher with a neighborhood: there exists $r > 0$ such that $U'_{z_0} = \{|z - z_0| < r\} \subset D'$. From the construction of D', we have that $U_{\bar{z}_0} = \{|\bar{z} - \bar{z}_0| < r\} \subset D$. To find the derivadive of $F(z)$ at z_0, we appeal to the definition. Since $z \to z_0$, we can consider

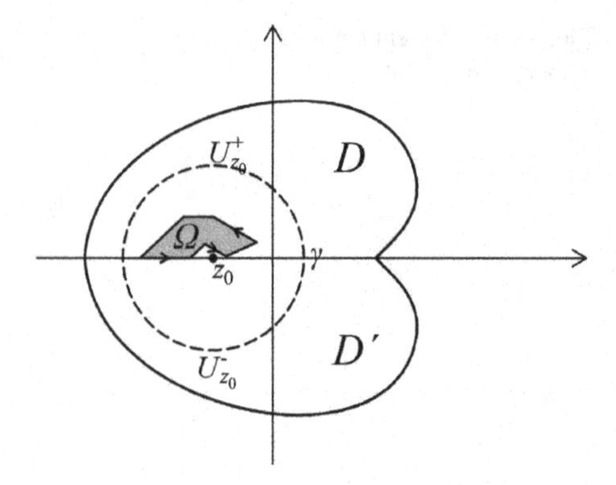

Fig. 5.11 Construction of the polygon Ω such that $\bar{\Omega} \subset U^+_{z_0}$.

only such points z that lie in U'_{z_0}. Notice that if z and z_0 belong to $U'_{z_0} \subset D'$ then $\bar{z}$ and $\bar{z}_0$ belong to $U_{\bar{z}_0} \subset D$, and the approach $z \to z_0$ is equivalent to $\bar{z} \to \bar{z}_0$. Using the condition of analyticity of $f(z)$ in D, which implies the differentiability of $f(z)$ at the point $\bar{z}_0 \in D$, we obtain

$$
\begin{aligned}
F'(z_0) &= \lim_{z \to z_0,\, z_0 \in D'} \frac{F(z) - F(z_0)}{z - z_0} = \lim_{z \to z_0} \frac{\overline{f(\bar{z})} - \overline{f(\bar{z}_0)}}{z - z_0} \\
&= \lim_{z \to z_0} \overline{\left(\frac{f(\bar{z}) - f(\bar{z}_0)}{\bar{z} - \bar{z}_0} \right)} = \overline{\left(\lim_{\bar{z} \to \bar{z}_0,\, \bar{z}_0 \in D} \frac{f(\bar{z}) - f(\bar{z}_0)}{\bar{z} - \bar{z}_0} \right)} = \overline{f'(\bar{z}_0)} .
\end{aligned}
\tag{5.12}
$$

Formula (5.12) shows that $F(z)$ is differentiable at any point $z_0 \in D'$, that is, it is differentiable in D', and consequently, $F(z)$ is analytic in D'.

Finally, we prove that $F(z)$ has an analytic element at any point z_0 of γ (notice that z_0 is an interior point of γ). The point z_0 is interior in the domain D^*, and consequently, it belongs to D^* together with a neighborhood: there exists $r > 0$ such that $U_{z_0} = \{|z - z_0| < r\} \subset D^*$. Denote by $U^+_{z_0}$ and $U^-_{z_0}$ the parts of the neighborhood U_{z_0} contained in the sets $D \cup \gamma$ and $D' \cup \gamma$, respectively. Take any polygon Ω whose closure is contained in the disk U_{z_0} ($\bar{\Omega} \subset U_{z_0}$) and calculate the integral $\int_{\partial\Omega} F(z)\,dz$. If $\bar{\Omega} \subset U^+_{z_0}$, then $F(z) = f(z)$ and, using the analyticity of $f(z)$ in D and continuity in $D \cup \gamma$, we obtain from Goursat's Theorem that $\int_{\partial\Omega} F(z)\,dz = \int_{\partial\Omega} f(z)\,dz = 0$ (see Fig. 5.11). If $\bar{\Omega} \subset U^-_{z_0}$, it was already proved that $F(z)$ is analytic in D' and continuous in $D' \cup \gamma$, and, due to Goursat's Theorem, we have again that $\int_{\partial\Omega} F(z)\,dz = 0$ (see Fig. 5.12).

It remains to consider the case when the polygon Ω is not contained either in $U^+_{z_0}$ or in $U^-_{z_0}$. Then, the segment γ divides this polygon in a finite number of polygons Ω_k, $k = 1, \ldots, p$, each of which is completely contained either in $U^+_{z_0}$ or in $U^-_{z_0}$ (see Fig. 5.13). Therefore, the equality $\int_{\partial\Omega_k} F(z)\,dz = 0$, $k = 1, \ldots, p$ is true for each polygon Ω_k. Since $\int_{\partial\Omega} F(z)\,dz = \sum_{k=1}^{p} \int_{\partial\Omega_k} F(z)\,dz = 0$, we conclude that the function $F(z)$ is continuous in the disk U_{z_0}, and the integral along the boundary of any polygon Ω contained in this disk is equal to zero. Then, by Morera's Theorem,

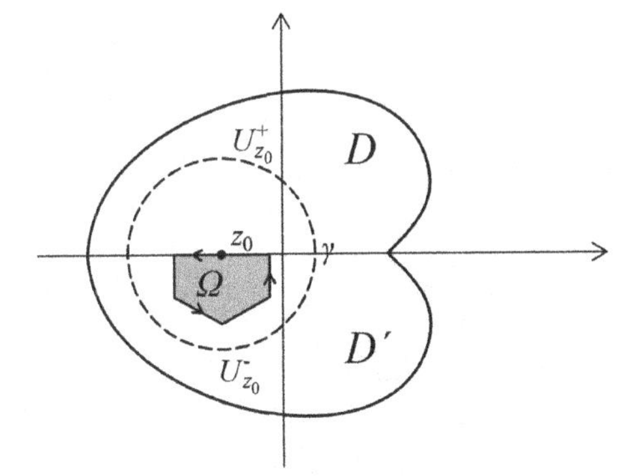

Fig. 5.12 Construction of the polygon Ω such that $\bar{\Omega} \subset U^-_{z_0}$.

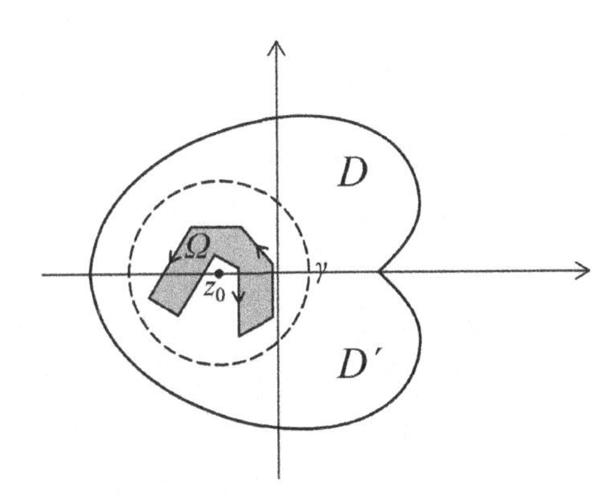

Fig. 5.13 Construction of the polygon Ω such that $\bar{\Omega} \subset U_{z_0}$ and $\Omega \cap U^+_{z_0} \neq \varnothing$, $\Omega \cap U^-_{z_0} \neq \varnothing$.

the function $F(z)$ is analytic in the disk U_{z_0}, which means that it has an analytic element at every point $z_0 \in \gamma$.

Thus, the function $F(z)$ has an analytic element at every point in the domain D^*. By the Theorem condition, the domain D is entirely contained in the upper or lower half-plane, which means that $D \cap D' = \varnothing$. Therefore, $F(z)$ is defined in a unique way in D^* by formula (5.9), which, together with the existence of an analytic element of $F(z)$, $\forall z \in D^*$, implies that $F(z)$ is analytic in D^*. Moreover, for $\forall z \in D$, we have $F(z) = f(z)$, that is, $F(z)$ is analytic continuation of $f(z)$ from D onto the domain D^*.

This terminates the proof of the Theorem. □

Remark 5.1 Let G and G' be the images of the domains D and D', respectively, under the transformation $F(z)$. Formula (5.9) shows immediately that not only D and D' are symmetric with respect to the real axis, but their images G and G' are symmetric with respect to the real axis as well (see Fig. 5.14).

Remark 5.2 If the domain D is not contained entirely in the upper or lower half-plane, then the domains D and D' have common points: $D \cap D' = E \neq \varnothing$. It was shown in Theorem 5.1 that the function $F(z)$ defined by (5.9) has an analytic element

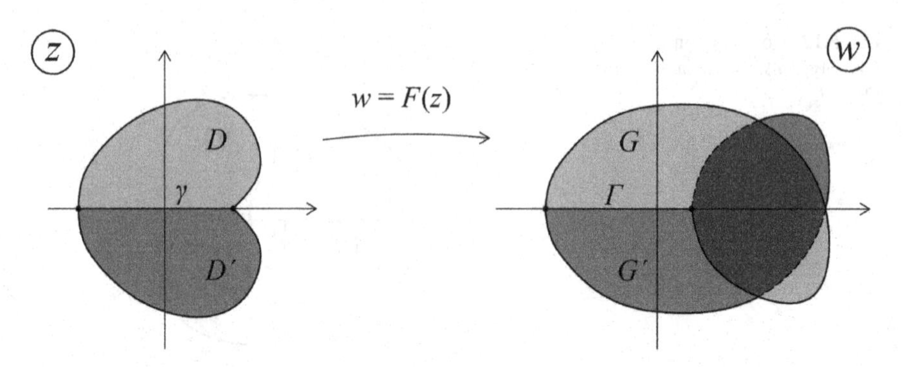

Fig. 5.14 Domains D and D' symmetric with respect to the real axis and their images $G = F(D)$ and $G' = F(D')$ also symmetric with respect to the real axis

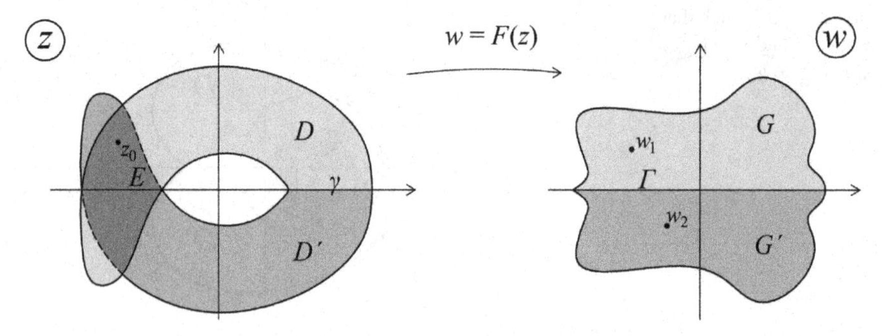

Fig. 5.15 Different images of $z_0 \in E = D \cap D'$ under $w = F(z)$: $w_1 = f(z_0)$, $w_2 = \overline{f(\bar{z_0})}$.

at every point $z \in D^* = D \cup \gamma \cup D'$. However, if $z \in E = D \cap D'$, then considering this point z in D we have analytic element $F(z) = f(z)$, while treating the same point z as a point of D' we get $F(z) = \overline{f(\bar{z})}$ and these analytic elements can be different (see Fig. 5.15). Therefore, in this case, the function $F(z)$ is analytic continuation of $f(z)$ from D onto D^*, but, in general, $F(z)$ is a multi-valued function in D^*.

Remark 5.3 If, under the conditions of Theorem 5.1, the function $w = f(z)$ maps conformally D onto G and, additionally, $G \cap G' = \varnothing$, where $G' = F\left(D'\right)$, then the continuation $F(z)$ is analytic in the domain $D^* = D \cup \gamma \cup D'$ and maps conformally D^* onto $G^* = G \cup \Gamma \cup G'$ (see Fig. 5.16).

Corollary *Suppose D is a domain whose boundary contains an arc γ of a circle Γ and a function $w = f(z)$ is analytic in D, continuous in $D \cup \gamma$ and transforms the arc γ into arc l of the circle L: $l = f(\gamma)$. In this case, the function $f(z)$ can be analytically continued from the domain D onto the domain D' symmetric to D with respect to Γ. Moreover, this analytic continuation $F(z)$ is defined by the following rule: if the points z and z' are symmetric with respect to the circle Γ, then the points*

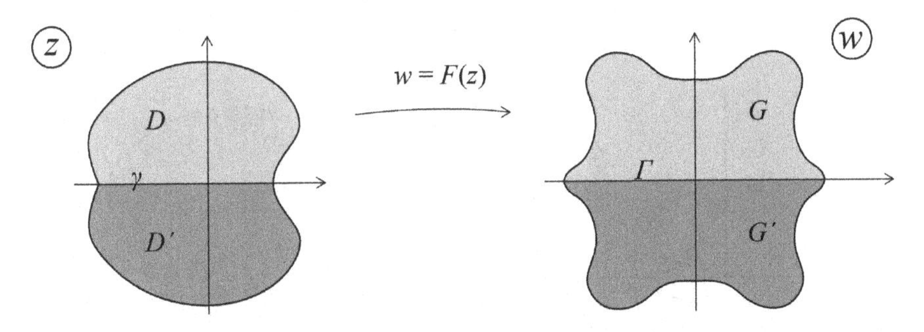

Fig. 5.16 Conformal mapping of $D^* = D \cup \gamma \cup D'$ onto $G^* = G \cup \Gamma \cup G'$ under the function $w = F(z)$: $D \cap D' = \varnothing$ and $G \cap G' = \varnothing$.

$F(z)$ and $F(z')$ are symmetric with respect to the circle L. This means that the domains $B = F(D)$ and $B' = F(D')$ are symmetric with respect to the circle L.

Proof Let us consider a few auxiliary fractional linear transformations. The first, $t = \varphi(z)$, is a fractional linear function that transforms the circle Γ of the plane $\mathbb{C}_z$ into the real axis of the plane $\mathbb{C}_t$ (see Fig. 5.17). For simplicity, we can use such a fractional linear function that carries the point $t = \infty$ to a point of the circle Γ, which does not belong to the arc γ. (The existence of such a mapping is shown in Property 8 of Sect. 4.5.) Then, the domain D is mapped onto a domain D_1 and the arc γ is carried to a segment γ_1 of the real axis. Similarly, construct a fractional linear function $\zeta = h(w)$ that transforms the circle L of the plane $\mathbb{C}_w$ into the real axis of the plane $\mathbb{C}_\zeta$ in such a manner that the point $\zeta = \infty$ corresponds to some point of L, which is not located on the arc l. Under this transformation, the domain B is mapped onto a domain B_1 in the plane $\mathbb{C}_\zeta$ and the arc l is carried to a segment l_1 of the real axis (see Fig. 5.17).

Consider the function $z = \varphi^{-1}(t)$, which is fractional linear as the inverse to a fractional linear function $t = \varphi(z)$ (see Property 2 in Sect. 4.5) and the composition of the functions $\zeta = g(t) = h\left(f\left(\varphi^{-1}(t)\right)\right)$. Let us show that the last function satisfies the conditions of Theorem 5.1. Indeed, $\zeta = g(t)$ is analytic in the domain D_1 as a composition of analytic functions and is continuous in $D_1 \cup \gamma_1$ as a composition of continuous functions. Under the transformation $\zeta = g(t)$ the image of the segment γ_1 of the real axis is the segment l_1 of the real axis, and consequently, by Theorem 5.1, the function $g(t)$ can be analytically continued from the domain D_1 onto the domain D_1', which is symmetric to D_1 with respect to the real axis. The resulting function $G(t)$ is defined by the formula

$$G(t) = \begin{cases} g(t), & t \in D_1 \cup \gamma_1 \\ \overline{g(\bar{t})}, & t \in D_1' \end{cases}, \tag{5.13}$$

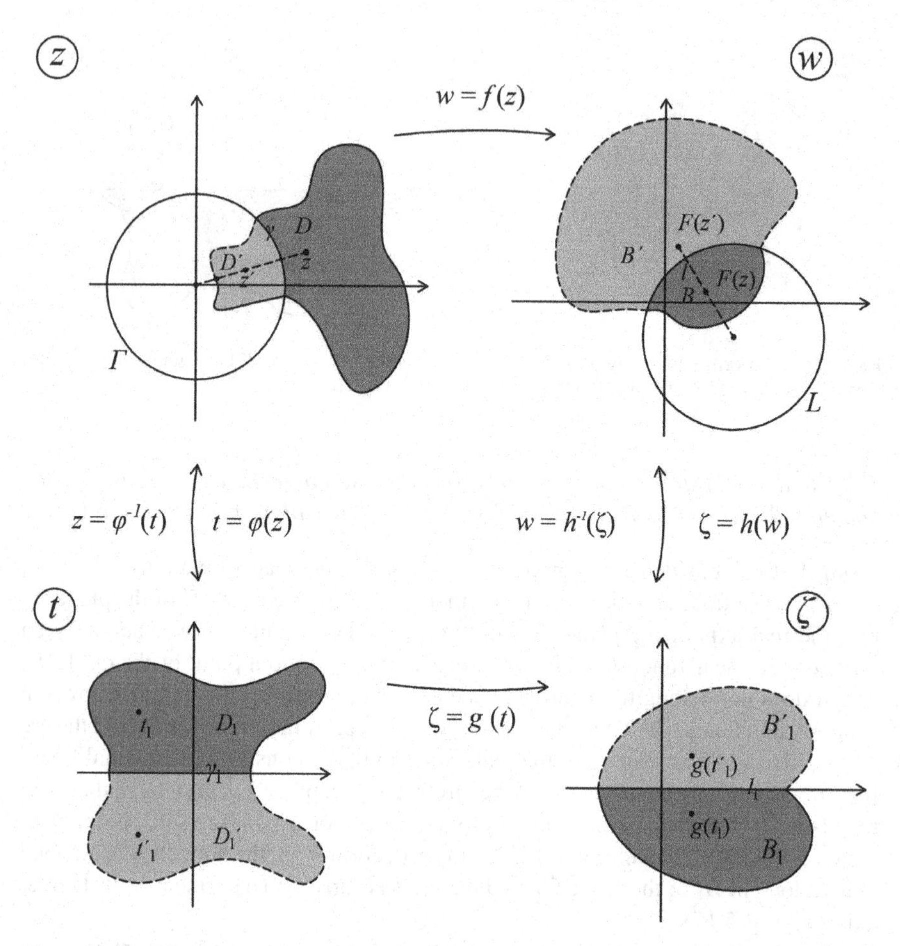

Fig. 5.17 The domains D and B in the planes $\mathbb{C}_z$ and $\mathbb{C}_w$, the auxiliary domains D_1 and B_1 in the planes $\mathbb{C}_t$ and $\mathbb{C}_\zeta$ and their analytic continuations

Therefore, at the points t and t' symmetric with respect to the real axis in the plane $\mathbb{C}_t$ the function $G(t)$ takes the values $\zeta = G(t)$ and $\zeta' = G(t')$ symmetric with respect to the real axis in the plane $\mathbb{C}_\zeta$ (see Fig. 5.17).

Return now to the planes $\mathbb{C}_z$ and $\mathbb{C}_w$, that is, to the domains D and B. Notice that the function $w = f(z)$ can be considered as the composition of the fractional linear functions and the function $g(t)$: $w = f(z) = h^{-1}(g(\varphi(z)))$, where $w = h^{-1}(\zeta)$ is the inverse to the fractional linear function $\zeta = h(w)$. Since the fractional linear functions $\varphi(z)$ and $h^{-1}(\zeta)$ can be analytically continued to the entire complex plane, and the function $g(t)$ can be analytically continued from the domain D_1 onto the domain D_1' symmetric to D_1 with respect to the real axis, the function $w = f(z) = h^{-1}(g(\varphi(z)))$ can be also analytically continued from D onto a domain D'. Recall that under a fractional linear transformation the points symmetric with respect to a circle are mapped to the points symmetric with respect to the image of this circle (see Property 6 in Sect. 4.5). Since the domains D_1 and D_1' are symmetric with respect to the real axis, their images D and D' under the fractional linear transformation $z = \varphi^{-1}(t)$ are symmetric with respect to the circle Γ. Thus, the function $f(z)$ can be analytically continued from the domain D onto the domain D' symmetric to D with respect to the circle Γ.

Let us check the properties of the obtained analytic continuation. Denote this continuation by $F(z)$ and notice that this function is defined in the domain $D^* = D \cup \gamma \cup D'$ as follows: $F(z) = h^{-1}(G(\varphi(z)))$. Take any two points z and z' in the domain D^* such that they are symmetric with respect to the circle Γ. These two points are carried by the fractional linear transformation $t = \varphi(z)$ to the points $t = \varphi(z)$ and $t' = \varphi(z')$ symmetric with respect to the segment γ_1 of the real axis. By formula (5.13), the analytic continuation $\zeta = G(t)$ carries the points t and t', symmetric with respect to the real axis in the plane $\mathbb{C}_t$ to the points $\zeta = G(t)$ and $\zeta' = G(t')$ symmetric with respect to the real axis in the plane $\mathbb{C}_\zeta$. Then, we use again the property of the invariance of the symmetric points under a fractional linear transformation. Since the points $\zeta = G(t)$ and $\zeta' = G(t')$ are symmetric with respect to the real axis, they are mapped by the fractional linear transformation $w = h^{-1}(\zeta)$ to the points w and w' symmetric with respect to the circle L. Thus, we have performed the analytic continuation of $f(z)$ from the domain D onto the domain D' symmetric to D with respect to the circle Γ, and, additionally, at the points $z \in D$ and $z' \in D'$ symmetric with respect to Γ, the function $F(z)$ takes the values, which are symmetric with respect to the circle L.

Hence, the Corollary is proved. □

Remark 5.2 This Corollary is also frequently called the *Riemann–Schwarz symmetry/reflection principle* or simply the *symmetry/reflection principle*.

As an application of the symmetry principle, we prove the result on the uniqueness of a conformal mapping of a simply connected domain onto another simply connected domain with a specified correspondence between three boundary points.

Theorem 5.2 *Let D and G be two simply connected domains whose boundaries are piecewise smooth curves. There exists one and only one conformal mapping of D onto G such that arbitrary three points $z_k \in \partial D$, $k = 1, 2, 3$, which appear in the order z_1, z_2, z_3 under the positive traversal along ∂D, are carried to the respective points $w_k \in \partial G$, $k = 1, 2, 3$, which have the corresponding order w_1, w_2, w_3 under the positive traversal along ∂G.*

Proof First, we consider the special case when both D and G are the unit disks. There exists a fractional linear function $w = f(z)$ that transforms the unit disk $|z| < 1$ into the unit disk $|w| < 1$ and that satisfies the condition $f(z_k) = w_k$, $|z_k| = 1$, $|w_k| = 1$, $k = 1, 2, 3$ (see Property 7 in Sect. 4.5). Let us show the uniqueness of such a function. Suppose, by contradiction, that there exists another function $w = g(z)$, which satisfies the Theorem hypotheses: it maps conformally $D = \{|z| < 1\}$ onto $G = \{|w| < 1\}$ with the indicated correspondence between the three pairs of the

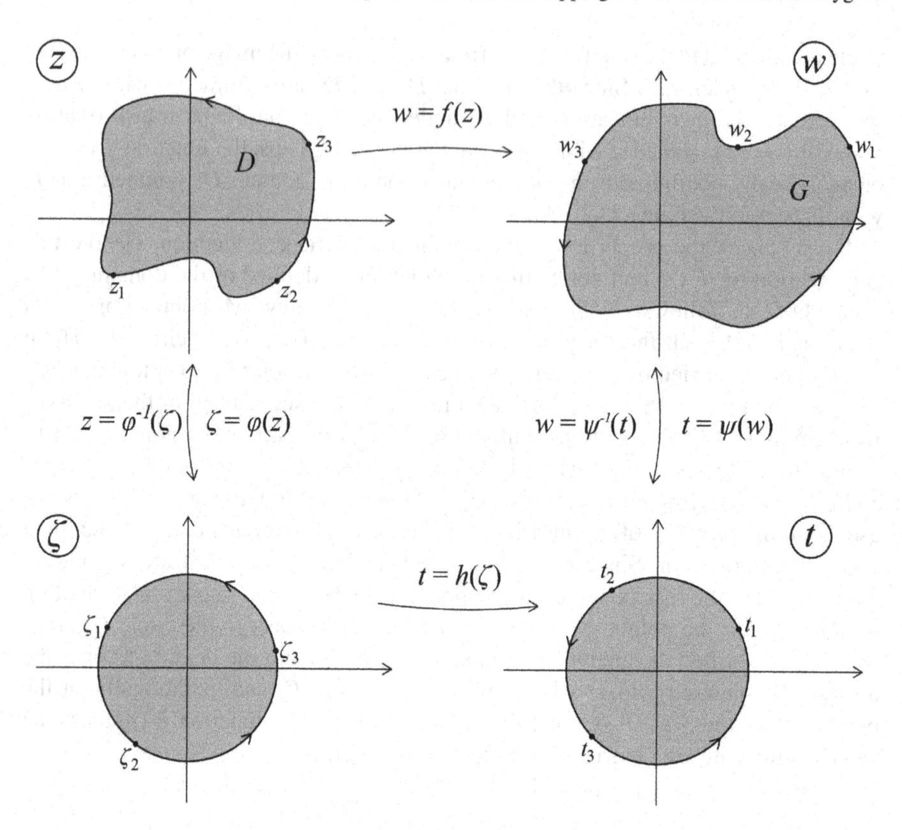

Fig. 5.18 Construction of the mapping of D onto G through the auxiliary mappings of D and G onto the unit disks

points. This function and the domain D satisfy the conditions of the symmetry principle formulated in the above Corollary. Then, the function $w = g(z)$ can be analytically continued from D onto the domain D' symmetric to D with respect to the unit circle $|z| = 1$, that is, $D' = \{|z| > 1\}$. Denote the extended function by $w = G(z)$. By the symmetry principle, $G(D) = g(D) = B = \{|w| < 1\}$ and $G(D') = B' = \{|w| > 1\}$. Besides, $D \cap D' = \varnothing$ and $B \cap B' = \varnothing$, which implies that $w = G(z)$ conformally maps the entire extended plane $\overline{\mathbb{C}}_z$ onto the entire extended plane $\overline{\mathbb{C}}_w$ (since $w = g(z)$ is a conformal transformation of D into $G = B$). Then Corollary 5.3 of Sect. 5.1 guarantees that $G(z)$ is a fractional linear function (including a linear function as a singular case). Since, by the Theorem conditions, this function carries the chosen points z_k, $|z_k| = 1$, $k = 1, 2, 3$ to the corresponding points w_k, $|w_k| = 1$, $k = 1, 2, 3$, Property 7 of fractional linear functions (Sect. 4.5) ensures that such a function is unique. However, $w = f(z)$ satisfies the same conditions in the domain $D = \{|z| < 1\}$ as the function $w = g(z)$. Therefore, $G(z) = f(z)$, for $\forall z \in D$, which implies the uniqueness of the function $f(z)$.

Let us consider now the general case. From the Riemann Mapping Theorem, it follows that there exist conformal transformations $\zeta = \varphi(z)$ and $t = \psi(w)$ of the domains D and G in the unit disks $|\zeta| < 1$ and $|t| < 1$, respectively (see Fig. 5.18). Under such transformations $\zeta_k = \varphi(z_k)$, $|\zeta_k| = 1$ and $t_k = \psi(w_k)$, $|t_k| = 1$, $k = 1, 2, 3$. By the first part of the current theorem, which is already proved, there exists a unique (fractional linear) function $t = h(\zeta)$ that maps conformally $|\zeta| < 1$ onto $|t| < 1$ and satisfies the condition $h(\zeta_k) = t_k$, $k = 1, 2, 3$. Construct now the function

$$w = f(z) = \psi^{-1}(h(\varphi(z))), \tag{5.14}$$

where ψ^{-1} is the inverse to the transformation ψ. The function (5.14) carries the domain D onto G in such a way that $f(z_k) = w_k$, $k = 1, 2, 3$. We show the uniqueness of this function by contradiction. If there exists another function $w = g(z)$ that performs a conformal mapping of D onto G with the condition $g(z_k) = w_k$, $k = 1, 2, 3$, then there exists another function $t = \psi\left(g\left(\varphi^{-1}(\zeta)\right)\right) = H(\zeta)$, different from $t = h(\zeta)$, that carries $|\zeta| < 1$ onto $|t| < 1$ in the way that $H(\zeta_k) = t_k$, $k = 1, 2, 3$ (see Fig. 5.18). However, this contradicts the uniqueness of the function $t = h(\zeta)$.

This completes the proof of the Theorem. □

5.6 Transformation of the Upper Half-Plane into a Rectangle

Consider the following function defined on the upper half-plane D:

$$F(k, z) = \int_0^z \frac{dt}{\sqrt{\left(1 - t^2\right)\left(1 - k^2 t^2\right)}}, \quad 0 < k < 1, \ \operatorname{Im} z > 0. \tag{5.15}$$

This function is called *the Legendre elliptic integral of the first type*. The integrand in (5.15), $g(z) = \frac{1}{\sqrt{(1-z^2)(1-k^2z^2)}}$, has the following singular points: $z = \pm 1$, $z = \pm\frac{1}{k}$ and $z = \infty$. All these points belong to the real axis, which is the boundary of the domain D. Since the function $g(z)$ has the square root in the denominator, it implies that $g(z)$ is a multi-valued function. The singularities $z = \pm 1$, $z = \pm\frac{1}{k}$ and $z = \infty$ are the branch points of a finite order of $g(z)$ (later we will see that $z = \infty$ is not a branch point, but it is a removable singularity). At the same time, the upper half-plane (the domain D) does not contain singular points of $g(z)$, and consequently, a traversal along any closed curve contained in D does not represent a circuit around any singular point. Therefore, we can choose an analytic single-valued branch of the function $g(z)$ in the domain D. To this end, we fix the value of $g(z)$ at some arbitrary point, for instance, we take $z = 0$ and choose the value of the root such that $g(0) = 1$. In what follows we consider the chosen branch of the function $g(z)$, which is an analytic function in the domain $D = \{\forall z : \operatorname{Im} z > 0\}$ and is continuous in $\bar{D}$, except for the points $z = \pm 1$ and $z = \pm\frac{1}{k}$ belonging to ∂D. Since D is a simply connected domain and the function $g(z)$ (or, more precisely, the chosen branch of $g(z)$) is analytic in D, the Theorem on the primitive of analytic function (Sect. 2.6) implies that the function $F(k, z)$ is also analytic in the upper half-plane D.

Let us analyze what kind of transformation is performed by the function $F(k, z)$. The core part of this study is an application of the principle of boundary correspondence. However, before this principle can be used, we have to verify some properties of the integral (5.15). Therefore, we divide the analysis into the following three steps:
(1) first, we study the behavior of the integral (5.15) at the singular points $z = \pm 1$, $z = \pm\frac{1}{k}$ and $z = \infty$;
(2) then, we find the image of the boundary of D (the real axis) under the transformation $w = F(k, z)$;
(3) finally, we use the principle of boundary correspondence to make conclusions about the image of the upper half-plane D under $F(k, z)$.

(1) We start the investigation of the integral (5.15) with a consideration of the singular point $z = 1$ and the corresponding integral $\int_0^1 \frac{dt}{\sqrt{(1-t^2)(1-k^2t^2)}}$. Since

$$\lim_{t\to 1} \frac{g(t)}{\frac{1}{(1-t)^{1/2}}} = \lim_{t\to 1} \frac{(1-t)^{1/2}}{\sqrt{(1-t^2)(1-k^2t^2)}} = \frac{1}{\sqrt{2(1-k^2)}} = m\,,$$

where $m = constant \neq 0$, the integral $\int_0^1 \frac{dt}{(1-t)^{1/2}}$ converges, because $\frac{1}{2} < 1$, and, therefore, the integral $\int_0^1 g(t)\,dt$ also converges. Applying the same arguments, we deduce that the integral $\int_1^a g(t)\,dt\,,\ 1 < a < \frac{1}{k}$ also converges.

Switch now to the singular point $\frac{1}{k}$ and the corresponding integral

$$\int_a^{1/k} \frac{dt}{\sqrt{(t^2-1)(1-k^2t^2)}} \quad (\forall a,\ 1 < a < \frac{1}{k})\,. \tag{5.16}$$

Notice that

$$\lim_{t\to\frac{1}{k}} \frac{1/\sqrt{(t^2-1)(1-k^2t^2)}}{1/\sqrt{1-kt}} = \lim_{t\to\frac{1}{k}} \frac{\sqrt{1-kt}}{\sqrt{(t^2-1)(1-k^2t^2)}} = \frac{k}{\sqrt{2(1-k^2)}} = m_1$$

($m_1 = constant \neq 0$), and the integral $\int_a^{1/k} \frac{dt}{(1-kt)^{1/2}}$ converges because $\frac{1}{2} < 1$. Therefore, the integral (5.16) also converges. In the same manner, the integral $\int_{1/k}^b g(t)dt$ converges for $\forall b > \frac{1}{k}$.

Similarly, we can show the convergence of the integral (5.15) at the points $z = -1$ and $z = -\frac{1}{k}$.

Turning to the point ∞, we need to consider the improper integral $\int_b^{\infty} \frac{dt}{\sqrt{(1-t^2)(1-k^2t^2)}}$ $(\forall b > \frac{1}{k})$. The evaluation of the auxiliary limit

$$\lim_{t\to+\infty} \frac{g(t)}{1/t^2} = \lim_{t\to+\infty} \frac{t^2}{\sqrt{(t^2-1)(k^2t^2-1)}} = \frac{1}{k} \tag{5.17}$$

shows that the integral $\int_b^{+\infty} g(t)\,dt$ converges due to the convergence of the integral $\int_b^{+\infty} \frac{1}{t^2} dt$ (the last converges since the power in the denominator is greater then 1). Similarly, the integral $\int_c^{-\infty} g(t)\,dt$ converges for $\forall c,\ c < -\frac{1}{k}$.

Hence, the integral (5.15) converges at all the singular points. Besides, from the representation of the function $g(t)$ in a neighborhood of ∞ in the form $g(t) = \frac{1}{t^2\sqrt{\left(1-\frac{1}{t^2}\right)\left(k^2-\frac{1}{t^2}\right)}}$ it follows that $t = \infty$ is a single-valued isolated singularity, because the function $\sqrt{\left(1-\frac{1}{t^2}\right)\left(k^2-\frac{1}{t^2}\right)}$ is analytic at ∞. Since $\lim_{t\to\infty} g(t) = 0$, the point $t = \infty$ is a removable singularity of $g(t)$.

(2) Now we are ready to find the image of the real axis under the transformation $w = F(k, z)$. We start by moving from the point $z = 0$ in the positive direction of the real axis. First consider $z = x$ varying along the segment $[0, 1)$. If $z = 0$ then $w = F(k, 0) = 0$. For $0 < x < 1$, the variable of integration t is also real: $0 \le t \le x$. The chosen analytic branch of $g(t)$, which satisfies $g(0) = 1$, has real positive values in the segment $t \in [0, x]$ for any $0 < x < 1$. Then, the function $F(k, x) = \int_0^x g(t)\,dt$ is also real and positive for $x \in [0, 1)$. Besides, $F(k, x)$ is increasing in the segment $x \in [0, 1)$ because $g(t) > 0$. Therefore, when the variable $z = x$ varies from 0 to 1, the function $w = F(k, x)$ takes positive values and increases from $w = F(k, 0) = 0$ to $F(k, 1) = \int_0^1 \frac{dt}{\sqrt{(1-t^2)(1-k^2t^2)}}$ (see the segment l_1 and its image γ_1 in Fig. 5.19). As it was shown, the last improper integral is convergent. It is referred to as the *complete Legendre integral of the first type* and denoted by $K(k) = \int_0^1 \frac{dt}{\sqrt{(1-t^2)(1-k^2t^2)}}$. The values of this integral are usually determined through the tables of pre-calculated values for k, $0 < k < 1$.

Let us investigate what happens with the function $g(z)$, and consequently with the function $w = F(k, z)$ when z moves through the point $z = 1$. Since $z = 1$ is a singular point of $g(z)$, instead of moving along the segment $[1-\delta, 1+\delta]$, where $0 < \delta < \inf\left\{1, \frac{1}{k} - 1\right\}$, we use the traversal along the upper half-circle $|z - 1| = \delta$, $\operatorname{Im} z > 0$. Notice that when z moves along $|z - 1| = \delta$, $\operatorname{Im} z > 0$, the vector $1 - z$ (with the initial point at 1) changes its argument by $-\pi$ (see Fig. 5.20) and, therefore, the function $(1 - z)^{1/2}$ changes its argument by $-\frac{\pi}{2}$. In other words, after passing through the point $z = 1$, the function $(1 - z)^{1/2}$ takes the form

$$(1-z)^{1/2}\Big|_{z=x>1} = |1-z|^{1/2} \cdot e^{i\arg(1-z)^{1/2}}\Big|_{z=x>1} = (x-1)^{1/2} \cdot e^{-i\frac{\pi}{2}}.$$

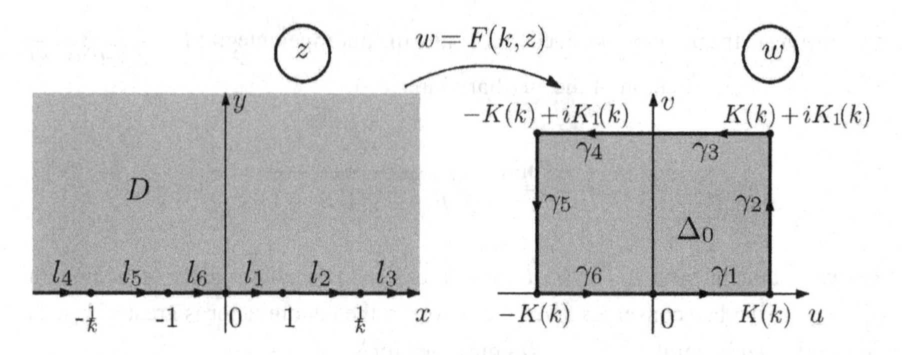

Fig. 5.19 The real axis $z = x$ and its image under the transformation $F(k, z)$

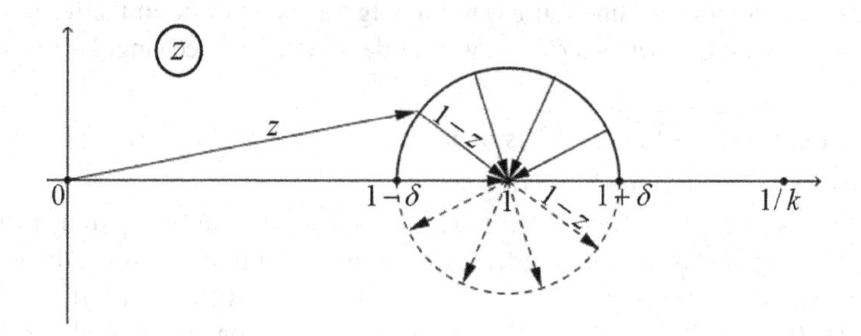

Fig. 5.20 Traversal of the vector $1 - z$ along $|z - 1| = \delta$, $\operatorname{Im} z > 0$.

At the same time, the vectors $(1 + z)$, $(1 - kz)$ and $(1 + kz)$ do not change their arguments when z passes through $z = 1$. Therefore, after passing the point $z = 1$ (that is, on the segment $\left(1, \frac{1}{k}\right)$), the function $g(t)$ takes the form

$$g(t) = \frac{1}{\sqrt{(1-t^2)(1-k^2t^2)}} = \frac{1}{e^{-i\frac{\pi}{2}}\sqrt{(t+1)(t-1)(1-k^2t^2)}} = \frac{e^{i\frac{\pi}{2}}}{\sqrt{(t^2-1)(1-k^2t^2)}}.$$

Besides, $|g(t)| = \frac{1}{\sqrt{(t^2-1)(1-k^2t^2)}}$ has positive real values on the interval $\left(1, \frac{1}{k}\right)$, and consequently, for $z = x \in \left(1, \frac{1}{k}\right)$ we have

$$F(k, x) = \int_0^x \frac{dt}{\sqrt{(1-t^2)(1-k^2t^2)}} = \int_0^1 \frac{dt}{\sqrt{(1-t^2)(1-k^2t^2)}} + \int_1^x \frac{dt}{\sqrt{(1-t^2)(1-k^2t^2)}}$$

$$= K(k) + e^{i\frac{\pi}{2}} \int_1^x \frac{dt}{\sqrt{(t^2-1)(1-k^2t^2)}}.$$

The function inside the integral $\int_1^x \frac{dt}{\sqrt{(t^2-1)(1-k^2t^2)}}$ on the right-hand side is positive on the segment $\left(1, \frac{1}{k}\right)$, which implies that this integral is positive on $\left(1, \frac{1}{k}\right)$ and increases from 0 to $K_1(k) = \int_1^{1/k} \frac{dt}{\sqrt{(t^2-1)(1-k^2t^2)}}$. The last integral is referred to as the *Legendre elliptic (convergent) integral* and its values are also tabulated.

Let us return to the function $w = F(k, x)$. If $x \in \left[1, \frac{1}{k}\right]$, then the values of the function $F(k, x)$ fill the segment perpendicular to the real axis: $w = F(k, x) = K(k) + e^{i\frac{\pi}{2}} v = K(k) + iv$, where v varies from 0 to $K_1(k)$ (see the segments l_2 and γ_2 in Fig. 5.19). Consider now what happens with the functions $g(z)$ and $F(k, z)$ when z passes through the point $z = \frac{1}{k}$. Notice that $g(z)$ has a singularity at this point only due to the factor $(1 - kz)$: the point $z = \frac{1}{k}$ is not singular for the function $(z^2 - 1)^{1/2} \cdot (1 + kz)^{1/2}$. Therefore, it suffices to evaluate the behavior of $(1 - kz)$ under the passage through the point $z = \frac{1}{k}$. Take any $\delta > 0$ ($\delta < \frac{1}{k} - 1$) and substitute moving the point along the segment $\left[\frac{1}{k} - \delta, \frac{1}{k} + \delta\right]$ by the traversal along the half-circle $\left|z - \frac{1}{k}\right| = \delta$, $\operatorname{Im} z > 0$. The function $(1 - kz)$ can be written in the form $(1 - kz) = k\left(\frac{1}{k} - z\right)$. The vector $\left(\frac{1}{k} - z\right)$ changes its argument by $-\pi$ when z moves along the half-circle $\left|z - \frac{1}{k}\right| = \delta$, $\operatorname{Im} z > 0$, and consequently, the function $(1 - kz)^{1/2} = k^{1/2}\left(\frac{1}{k} - z\right)^{1/2}$ changes its argument by $-\frac{\pi}{2}$. Then, after the traversal along the half-circle $\left|z - \frac{1}{k}\right| = \delta$, $\operatorname{Im} z > 0$, the factor $(1 - kz)^{1/2}$ takes the form:

$$(1 - kz)^{1/2}\Big|_{z=x>1/k} = \left(\left|k^{1/2}\left(\tfrac{1}{k} - z\right)\right| \cdot e^{i \arg\left(\frac{1}{k} - z\right)^{1/2}}\right)\Big|_{z=x>1/k}$$

$$= e^{-i\frac{\pi}{2}} \cdot k^{1/2} \cdot \left(x - \frac{1}{k}\right)^{1/2} = e^{-i\frac{\pi}{2}} (kx - 1)^{1/2},$$

which is valid for all the points $z = x \in \left(\frac{1}{k}, +\infty\right)$, because there are no other singular points of the function in this interval. Therefore, $g(t)$ has the following form on the interval $\left(\frac{1}{k}, +\infty\right)$:

$$g(t) = \frac{e^{i\frac{\pi}{2}}}{\sqrt{\left(t^2 - 1\right)\left(1 - k^2t^2\right)}} = \frac{e^{i\frac{\pi}{2}}}{(1 - kt)^{1/2}\sqrt{\left(t^2 - 1\right)(1 + kt)}}$$

$$= \frac{e^{i\frac{\pi}{2}}}{e^{-i\frac{\pi}{2}} (kt - 1)^{1/2}\sqrt{\left(t^2 - 1\right)(1 + kt)}} = \frac{e^{i\frac{\pi}{2}} \cdot e^{i\frac{\pi}{2}}}{\sqrt{\left(t^2 - 1\right)\left(k^2t^2 - 1\right)}} = \frac{-1}{\sqrt{\left(t^2 - 1\right)\left(k^2t^2 - 1\right)}}.$$

Notice that the function $\sqrt{\left(t^2 - 1\right)\left(k^2t^2 - 1\right)}$ takes positive real values on the interval $\left(\frac{1}{k}, +\infty\right)$.

Analyze now the function $F(k, z)$ on the same interval $\left(\frac{1}{k}, +\infty\right)$:

$$F(k, x) = \int_0^x \frac{dt}{\sqrt{\left(1-t^2\right)\left(1-k^2t^2\right)}} = \int_0^1 \frac{dt}{\sqrt{\left(1-t^2\right)\left(1-k^2t^2\right)}} + \int_1^x \frac{dt}{\sqrt{\left(1-t^2\right)\left(1-k^2t^2\right)}}$$

$$= K(k) + e^{i\frac{\pi}{2}} \int_1^x \frac{dt}{\sqrt{(t^2-1)(1-k^2t^2)}} = K(k) + e^{i\frac{\pi}{2}} \int_1^{1/k} \frac{dt}{\sqrt{(t^2-1)(1-k^2t^2)}}$$

$$+ e^{i\frac{\pi}{2}} \cdot e^{i\frac{\pi}{2}} \int_{1/k}^x \frac{dt}{\sqrt{(t^2-1)(k^2t^2-1)}} = K(k) + iK_1(k) - \int_{1/k}^x \frac{dt}{\sqrt{(t^2-1)(k^2t^2-1)}} .$$

If $x \in \left(\frac{1}{k}, +\infty\right)$, then (as was seen above) the function $\frac{1}{\sqrt{(t^2-1)(k^2t^2-1)}}$ takes positive real values, implying that the integral $\int_{1/k}^x \frac{dt}{\sqrt{(t^2-1)(k^2t^2-1)}}$ has positive real values and increases on the interval $\left(\frac{1}{k}, +\infty\right)$. Therefore, since $\operatorname{Im} F(k, x) = K_1(k) = constant$, $\forall x \in \left(\frac{1}{k}, +\infty\right)$, the function $F(k, x)$ takes the values that belong to the segment parallel to the real axis: $F(k, x) = K(k) - u(x) + iK_1(k)$ (see the segments l_3 and γ_3 in Fig. 5.19).

It was already shown that the integral $\int_{1/k}^{+\infty} \frac{dt}{\sqrt{(t^2-1)(k^2t^2-1)}}$ converges. Let us calculate its value. Using the change of variable $\frac{1}{kt} = y$, we have $t = \frac{1}{ky}$, $dt = -\frac{dy}{ky^2}$, the lower $t = \frac{1}{k}$ and upper $t = \infty$ limits of the integration are changed to $y = 1$ and $y = 0$, respectively, and we obtain

$$\int_{1/k}^{+\infty} \frac{dt}{\sqrt{(t^2-1)(k^2t^2-1)}} = -\int_1^0 \frac{dy}{ky^2\sqrt{\left(\frac{1}{k^2y^2}-1\right)\left(\frac{1}{y^2}-1\right)}}$$

$$= \int_0^1 \frac{dy}{\sqrt{(1-y^2)(1-k^2y^2)}} = K(k) .$$

Then, when the variable $z = x$ moves along the interval $\left(\frac{1}{k}, +\infty\right)$, the point $w = F(k, x)$ moves along the straight line $\operatorname{Im} w = K_1(k)$ from the point $K(k) + iK_1(k)$ to the point $K(k) + iK_1(k) - K(k) = iK_1(k)$.

Due to the symmetry of the function $F(k, x)$: $F(k, -x) = -F(k, x)$, the traversal along the negative part of the real axis in the z-plane results in the contour in the w-plane, which is symmetric with respect to the imaginary axis to the already obtained curve. More specifically, when z moves along the x-axis from $-\infty$ to 0 (in the positive direction), we obtain the following images under $F(k, z)$:

The interval $(-\infty, -\frac{1}{k}]$ is carried to the segment $iK_1(k) + u$, $u \in [-K(k), 0]$ parallel to the real axis and described from right to left (the segments l_4 and γ_4 in Fig. 5.19);

the segment $[-\frac{1}{k}, -1]$ is mapped to the segment $-K(k) + iv$, $v \in [0, K_1(k)]$ parallel to the imaginary axis and traced downward (the segments l_5 and γ_5 in Fig. 5.19);

the segment $[-1, 0]$ is carried to the segment of the real axis $[-K(k), 0]$ described from left to right (the segments l_6 and γ_6 in Fig. 5.19).

Joining the results obtained for all the parts of the real axis $z = x$, we conclude that the real axis of the z-plane is mapped to the four line segments: the two of them—the segments $[-K(k), K(k)]$ and $[-K(k) + iK_1(k), K(k) + iK_1(k)]$—are parallel to the real axis (the first one is a part of the real axis), and other two—the segments $[-K(k), -K(k) + iK_1(k)]$ and $[K(k), K(k) + iK_1(k)]$—are parallel to the imaginary axis. These four segments form the boundary of the rectangle Δ_0, which is traversed counterclockwise (see Fig. 5.19).

(3) Finally, using the principle of boundary correspondence, we conclude that the function $F(k, z)$ is univalent on the upper half-plane $\text{Im}\, z > 0$ and maps bijectively this half-plane onto the rectangle Δ_0. Thus, the function $w = F(k, z)$ defined by formula (5.15) is analytic in the upper half-plane and maps it conformally on the rectangle Δ_0 (see Fig. 5.19).

Since the function $w = F(k, z)$ is analytic and univalent in the upper half-plane, there exists the inverse function $z = F^{-1}(k, w)$, which is analytic and univalent in the rectangle Δ_0 and maps bijectively Δ_0 onto the upper half-plane. This function is called the *elliptic sine* and denoted by $z = F^{-1}(k, w) = \text{sn}\,(k, w)$. Let us analyze some properties of this function. In what follows we use slightly different notations of the boundary components (more convenient in this case): the intervals of the real axis in the z-plane are denoted by

$$l_1 = (-1, 1)\ , \ l_2 = \left(1, \frac{1}{k}\right)\ , \ l_3 = \left(-\frac{1}{k}, -1\right)\ , \ l_4 = \left(\frac{1}{k}, +\infty\right) \cup \left(-\infty, -\frac{1}{k}\right),$$

and the corresponding sides of the rectangle Δ_0 are denoted by $\gamma_1, \gamma_2, \gamma_3, \gamma_4$ (see Fig. 5.21). We can apply the Riemann–Schwarz symmetry principle to the function $z = \text{sn}\,(k, w)$ (equivalently, we can use this principle for $w = F(k, z)$). The function $z = \text{sn}\,(k, w)$ is analytic in the rectangle Δ_0, continuous on $\Delta_0 \cup \gamma_1$ and takes the values that fill the interval l_1 when the point w moves along the interval γ_1. Then, according to the symmetry principle, we can analytically continue the function $z = \text{sn}\,(k, w)$ from the domain Δ_0 to the domain Δ_1, which is symmetric to Δ_0 with respect to the real axis; the values of the function onto the rectangle Δ_1 cover the entire lower half-plane in $\mathbb{C}_z$ (see Fig. 5.21). Since the rectangles Δ_0 and Δ_1 do not intercept, the extended function is single-valued (and consequently analytic) in the domain $\Delta_0 \cup \gamma_1 \cup \Delta_1$. This function is denoted in the same way as its original branch—$z = \text{sn}\,(k, w)$.

Notice that the values of the analytically continued function cover the entire plane $\mathbb{C}_z$ with the cuts along the intervals $(-\infty, -1]$ and $[1, +\infty)$ (see Fig. 5.21). Similarly, we can analytically continue the function $z = \text{sn}\,(k, w)$ from the domain Δ_0 onto the domain Δ_2, which is symmetric to Δ_0 with respect to the interval γ_2. The values, that the extended function $z = \text{sn}\,(k, w)$ takes on the rectangle Δ_2, cover again the entire lower half-plane. For this reason, we take another exemplar of the lower half-plane and connect it with the upper half-plane along the corresponding interval l_2. In the same manner, we can analytically continue the function $z = \text{sn}\,(k, w)$ from the domain Δ_0 onto the domains Δ_3 and Δ_4 symmetric to Δ_0 with respect to the intervals γ_3 and γ_4, respectively. Then, we obtain two more copies of the lower half-plane in the plane $\mathbb{C}_z$, which are attached to the upper half-plane along the corresponding intervals l_3 and l_4.

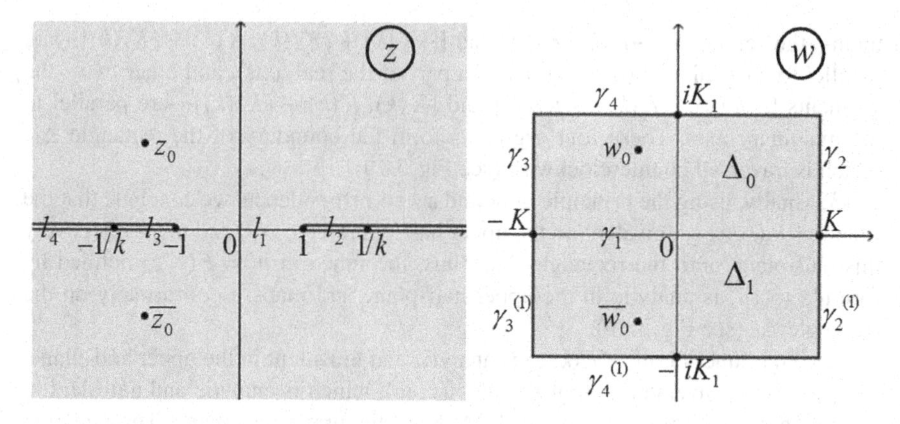

Fig. 5.21 Analytic continuation of the function $z = \text{sn}\,(k, w)$ from Δ_0 to $\Delta_0 \cup \gamma_1 \cup \Delta_1$

Thus, we have analytically continued the function $z = \text{sn}\,(k, w)$ from the domain Δ_0 onto the domain $\Delta_0 \cup \gamma_1 \cup \Delta_1 \cup \gamma_2 \cup \Delta_2 \cup \gamma_3 \cup \Delta_3 \cup \gamma_4 \cup \Delta_4$. Besides, we have obtained on the plane $\mathbb{C}_z$ four copies of the lower half-plane connected with the upper half-plane through the intervals l_1, l_2, l_3, or l_4, respectively. Notice that the rectangles $\Delta_k, k = 0, 1, 2, 3, 4$ have no intersection in the plane $\mathbb{C}_w$, which implies that the extended function $z = \text{sn}\,(k, w)$ remains to be single-valued (and consequently analytic) in the entire amplified domain except at the point $w = iK_1(k)$. This point is a pole of the function $z = \text{sn}\,(k, w)$, because $\text{sn}\,(k, iK_1\,(k)) = \infty$ (recall that the function $w = F(k, z)$ carries the point $z = \infty$ to the point $w = iK_1(k)$). Since the transformation is bijective in a neighborhood of the point $w = iK_1(k)$, the function $z = \text{sn}\,(k, w)$ is univalent at this point. Therefore, by the criterion for local univalence, $w = iK_1(k))$ is a simple pole of the function $z = \text{sn}\,(k, w)$.

Consider now the domain Δ_1. Its boundary is composed of intervals, that is, the arcs of circles with infinite radius. The function $z = \text{sn}\,(k, w)$ is analytic in the domain Δ_1, continuous on $\overline{\Delta_1}$, except at the point $-iK_1(w) \in \partial\Delta_1$, and takes on the boundary intervals the values that belong to the intervals l_1, l_2, l_3, or l_4 of the real axis. Then, the function $z = \text{sn}\,(k, w)$ can be analytically continued from the domain Δ_1 onto the domains symmetric to Δ_1 with respect to the intervals $\gamma_2^{(1)}, \gamma_3^{(1)}, \gamma_4^{(1)}$. Besides, the values of the extended function cover all the upper half-plane, that is, we have to consider three copies of the upper half-plane in $\mathbb{C}_z$ and connect them to the lower half-plane (which corresponds to the domain Δ_1) along the corresponding intervals l_2, l_3, and l_4 of the real axis. Similarly, the function $z = \text{sn}\,(k, w)$ can be analytically continued from the rectangle Δ_2 onto a larger domain, and so on. As a result of these analytic continuations, we obtain the function $z = \text{sn}\,(k, w)$ extended analytically onto the entire plane $\mathbb{C}_w$. Correspondingly, on the plane $\mathbb{C}_z$ we have the surface composed of an infinite number of copies of the upper and lower half-planes, connected along the intervals l_1, l_2, l_3, and l_4. Since the rectangles on the plane $\mathbb{C}_w$ have no intersection, the extended function $z = \text{sn}\,(k, w)$ is analytic on the entire complex plane $\mathbb{C}_w$, except at the simple poles

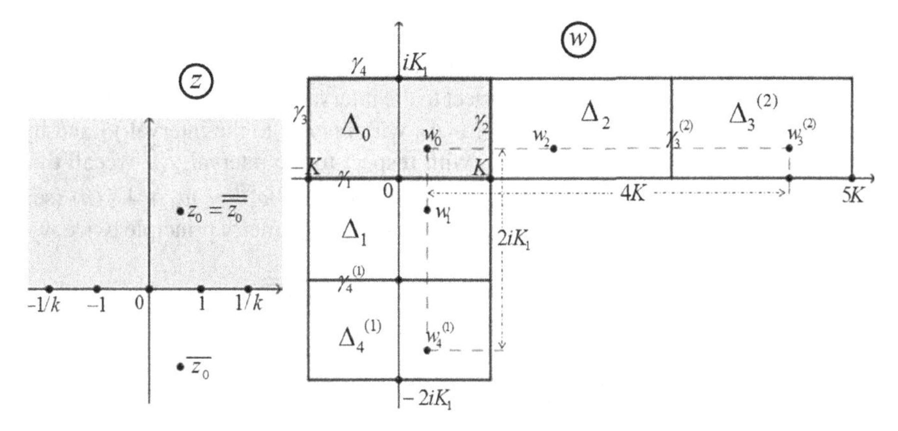

Fig. 5.22 Periods $4K$ and $2iK_1$ of the function $z = \operatorname{sn} w$

$$w_{n,m} = 2nK(k) + i(2m+1)K_1(k), \ \forall n, m \in \mathbb{Z}.$$

However, $z = \operatorname{sn}(k, w)$ is not univalent on the plane $\mathbb{C}_w$, because it carries different points of $\mathbb{C}_w$ to the same points in $\mathbb{C}_z$. Therefore, the inverse function (denoted by $F(k, z)$ as before) is extended onto the complex plane $\mathbb{C}_z$ with "holes" at the points $z = \pm 1, \pm\frac{1}{k}$. Although this function is not single-valued, it is analytic multi-valued. The surface on the plane $\mathbb{C}_z$, composed on an infinite number of the upper and lower half-planes linked by the intervals l_1, l_2, l_3, and l_4, is called the Riemann surface of the multi-valued analytic function $F(k, z)$.

We also point out the following interesting feature of the function $z = \operatorname{sn}(k, w)$. Let us pick an arbitrary rectangle in the plane $\mathbb{C}_w$, which corresponds to a single copy of the upper or lower half-plane in the plane $\mathbb{C}_z$. Choose, for instance, the rectangle Δ_0. Fix any point of this rectangle—$\forall w_0 \in \Delta_0$—and denote $z_0 = \operatorname{sn}(k, w_0)$, where $\operatorname{Im} z_0 > 0$. By the Riemann–Schwarz symmetry principle, the function $z = \operatorname{sn}(k, w)$ can be analytically continued from the domain Δ_0 onto the domain Δ_1, symmetric to Δ_0 with respect to the interval γ_1. Denote by $w_1 \in \Delta_1$ the point of the rectangle Δ_1, which is symmetric to w_0 with respect to γ_1. According to the symmetry principle, $\operatorname{sn}(k, w_1) = \overline{z}_0$. Now, we analytically continue the function $\operatorname{sn}(k, w)$ from Δ_1 onto the rectangle $\Delta_4^{(1)}$, symmetric to Δ_1 with respect to the interval $\gamma_4^{(1)}$ (see Fig. 5.22). Denote by $w_4^{(1)} \in \Delta_4^{(1)}$ the point, which is symmetric to w_1 with respect to the interval $\gamma_4^{(1)}$. Then, using again the symmetry principle, we obtain $\operatorname{sn}(k, w_4^{(1)}) = \overline{\operatorname{sn}(k, w_1)} = \overline{(\overline{z_0})} = z_0$ (see Fig. 5.22), that is, $\operatorname{sn}(k, w_4^{(1)}) = \operatorname{sn}(k, w_0)$. Notice that $w_4^{(1)} = w_0 - 2iK_1(k)$ (see Fig. 5.22), or equivalently, $\operatorname{sn}(w_0 - 2iK_1(k)) = \operatorname{sn}(w_0)$ (we temporarily drop the parameter k). This means that the number $2iK_1(k)$ is a period of the function $z = \operatorname{sn}(k, w) = \operatorname{sn} w$, because the point $w_0 \in \Delta_0$ was chosen arbitrarily.

Now, we analytically continue the function $z = \mathrm{sn}(k, w) = \mathrm{sn}w$ from Δ_0 onto the domain Δ_2, symmetric to Δ_0 with respect to the interval γ_2, and then onto the domain $\Delta_3^{(2)}$, symmetric to Δ_2 with respect to the interval $\gamma_3^{(2)}$ (see Fig. 5.22). Denote by $w_2 \in \Delta_2$ the point symmetric to $w_0 \in \Delta_0$ with respect to the interval γ_2 and by $w_3^{(2)} \in \Delta_3^{(2)}$ the point symmetric to w_2 with respect to the interval $\gamma_3^{(2)}$ (recall that w_0 is an arbitrary point of the rectangle Δ_0). Notice that $w_3^{(2)} = w_0 + 4K(k)$ (see Fig. 5.22). Therefore, applying the Riemann–Schwarz symmetry principle twice, we obtain

$$\mathrm{sn}\,(w_0 + 4K\,(k)) = \mathrm{sn}\,w_3^{(2)} = \overline{\mathrm{sn}\,w_2} = \overline{(\overline{\mathrm{sn}\,w_0})} = \mathrm{sn}\,w_0\,.$$

Hence, the number $4K(k)$ is also a period of the function $z = \mathrm{sn}(k, w) = \mathrm{sn}w$. Therefore, the following equality is satisfied $\forall m, n \in \mathbb{Z}$:

$$\mathrm{sn}\,(w + 4nK(k) + 2imK_1(k)) = \mathrm{sn}w, \;\; \forall w \in \mathbb{C}, \;\; \forall m, n \in \mathbb{Z}\,,$$

that is, the function $z = \mathrm{sn}w$ is bi-periodic, because it has two different periods $4K(k)$ and $2iK_1(k)$. In this way, as a result of analytic continuation we obtain a bi-periodic function $z = \mathrm{sn}(k, w)$, which is analytic in the entire complex plane $\mathbb{C}_w$, except at the simple poles $w_{n,m} = 2nK(k) + i(2m+1)K_1(k)$, $\forall n, m \in \mathbb{Z}$.

Notice that the function

$$F\,(k, z) = \int_0^z \frac{dt}{\sqrt{\left(1 - t^2\right)\left(1 - k^2t^2\right)}}$$

is a particular case of the *Schwarz–Christoffel mapping (integral)*, which has the form

$$f\,(z) = c \int_0^z (t - a_1)^{\alpha_1 - 1}\,(t - a_2)^{\alpha_2 - 1} \cdots (t - a_n)^{\alpha_n - 1}\,dt + c_1\,, \;\; \mathrm{Im}\,z > 0\,,$$

and transforms the upper half-plane into a polygon in such a way that the points a_k, $k = 1, 2, \ldots, n$ belonging to the real axis are carried to the vertices A_k of this polygon and the angle at the vertex A_k is equal to $\pi\alpha_k$. The Schwarz-Christoffel transformation will be studied in the next section.

An example of the application of the Legendre elliptic integral of the first type to conformal mapping of double-connected domains is presented below.

Example Let us construct a conformal mapping of the entire plane z with the cuts along the real intervals $[-4\,,\,0]$ and $[2\,,\,4]$ onto the ring $1 < |w| < R$.

We start with the conformal mapping of the z-plane on itself, which carries the two original cuts into two cuts on the real axis symmetric with respect to the origin. To do this, we draw two auxiliary circles passing through the endpoints of the given cuts: $C_1 = \{z : |z| = 4\}$ and $C_2 = \{z : |z - 1| = 1\}$ (see Fig. 5.23). The eccentric ring formed by the circles C_1 and C_2 can be transformed conformally into a concentric ring centered at the origin. This kind of problem was solved in Sect. 4.5 (see Example 4.2),

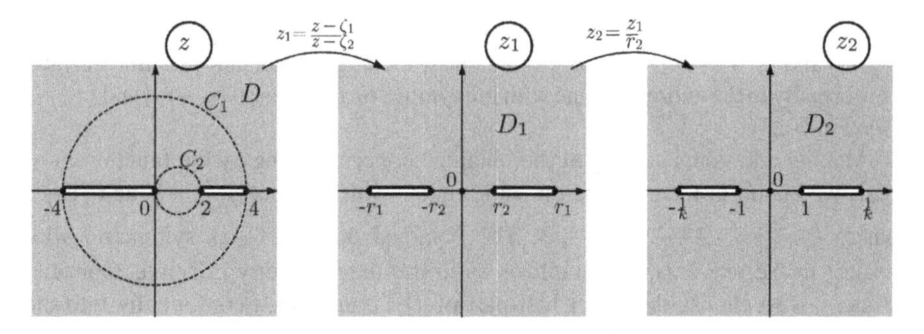

Fig. 5.23 Example: original domain D and symmetrization of its cuts through a fractional linear function

and it was shown that the solution consists in finding the points ζ_1 and ζ_2, which are symmetric with respect to both circles C_1 and C_2, and constructing a fractional linear function that sends one of these points to 0 and another one—to ∞.

Since the centerpoints of the circles C_1 and C_2 lie on the real axis, the points ζ_1 and ζ_2 are also real: $\zeta_1 = x_1$ and $\zeta_2 = x_2$. The condition of symmetry of two points (Eq. (4.13) in Sect. 4.3) applied for the circles C_1 and C_2 leads to the following system:

$$\begin{cases} x_1 = \dfrac{16}{x_2} \\ x_1 - 1 = \dfrac{1}{x_2 - 1} \end{cases} .$$

This system can be reduced to the quadratic equation for any of two unknowns $x^2 - 16x + 16 = 0$, whose roots are

$$\zeta_1 = x_1 = 8 - 4\sqrt{3}\,, \ \zeta_2 = x_2 = 8 + 4\sqrt{3}\,.$$

Therefore, the fractional linear function

$$z_1 = \frac{z - \zeta_1}{z - \zeta_2} = \frac{z - (8 - 4\sqrt{3})}{z - (8 + 4\sqrt{3})}$$

transforms the given eccentric ring into a concentric ring, and consequently, the cuts $[-4, 0]$ and $[2, 4]$ are carried to real segments symmetric with respect to the origin. To specify the location of the concentric ring and the cuts in the z_1-plane, let us calculate the images of the two points:

$$z_1(-4) = \frac{-4 - 8 + 4\sqrt{3}}{-4 - 8 - 4\sqrt{3}} = \frac{3 - \sqrt{3}}{3 + \sqrt{3}} = 2 - \sqrt{3}\,,\ z_1(0) = \frac{-8 + 4\sqrt{3}}{-8 - 4\sqrt{3}} = \frac{2 - \sqrt{3}}{2 + \sqrt{3}} = 7 - 4\sqrt{3}\,.$$

Therefore, the radius of the first circle in the z_1-plane is $r_1 = |z_1(-4)| = 2 - \sqrt{3}$ and that of the second—$r_2 = |z_1(0)| = 7 - 4\sqrt{3}$. Since $r_2 < r_1$, the ring in the z_1-

plane has the form $r_2 < |z_1| < r_1$, and consequently, the segment $[-4, 0]$ is sent to $[r_2, r_1]$ and $[2, 4]$—to $[-r_1, -r_2]$. In this way, the original domain D is transformed conformally to the entire z_1-plane with the symmetric real cuts $[-r_1, -r_2]$ and $[r_2, r_1]$ (see Fig. 5.23).

Making additional scaling of the obtained concentric ring by the function $z_2 = \frac{1}{r_2} z_1$, we obtain in the z_2 plane the entire plane with the cuts $[-\frac{1}{k}, -1]$ and $[1, \frac{1}{k}]$, where $k = \frac{r_2}{r_1} = \frac{7-4\sqrt{3}}{2-\sqrt{3}} = 2 - \sqrt{3}$. The obtained domain D_2 is symmetric with respect to the real axis, which allows us to use the symmetry principle. Our next strategy is to choose the upper half-part of D_2, transform it conformally into the upper half-ring and then return to the original domain and its image according to the symmetry principle.

To perform these stages, we first make the additional cuts along the intervals $(-\infty, -\frac{1}{k}]$, $[-1, 1]$ and $[\frac{1}{k}, +\infty)$ of the real axis and choose the upper half-plane $\tilde{D}_2$ (the upper half-part of D_2). Now our goal is to map conformally the domain $\tilde{D}_2$ onto a half-ring in such a way that the boundary elements of $\tilde{D}_2$, which are the parts of the original boundary, are carried to the half-circles of the ring, while the auxiliary cuts—to the radial segments. To this end, first we transform $\tilde{D}_2$ into a rectangle by employing the Legendre elliptic integral of the first type (5.15):

$$z_3 = F(k, z_2) = \int_0^{z_2} \frac{dt}{\sqrt{(1-t^2)(1-k^2t^2)}}, \quad \operatorname{Im} z_2 > 0 .$$

It was shown in this section that $F(k, z)$ maps the upper half-plane onto the rectangle D_3, whose base is the segment $[-K, K]$ located on the real axis and the height is K_1, where

$$K = \int_0^1 \frac{dt}{\sqrt{(1-t^2)(1-k^2t^2)}}, \quad K_1 = \int_1^{1/k} \frac{dt}{\sqrt{(t^2-1)(1-k^2t^2)}}, \quad k = 2 - \sqrt{3} .$$

Under this transformation, the auxiliary cut $[-1, 1]$ is carried to the real segment $[-K, K]$, the two remaining auxiliary cuts $(-\infty, -\frac{1}{k}]$ and $[\frac{1}{k}, +\infty)$—to the horizontal segments $[-K + iK_1, iK_1]$ and $[iK_1, K + iK_1]$, respectively, while the original cuts $[-\frac{1}{k}, -1]$ and $[1, \frac{1}{k}]$ are sent to the vertical segments $[-K, -K + iK_1]$ and $[K, K + iK_1]$, respectively (see Fig. 5.24).

Shifting the rectangle D_3 to the first quadrant and stretching it by the linear function $z_4 = (z_3 + K)\frac{\pi}{K_1}$, we get the rectangle D_4 with the base $[0, 2\pi \frac{K}{K_1}]$ and the height π. Then, we apply the exponential function $w = e^{z_4}$, which maps conformally the rectangle D_4 onto the upper half-ring G (see Fig. 5.24). It remains to use the symmetry principle, which restores the domain D_2 in the z_2-plane and leads to the ring $1 < |w| < R$, $R = e^{2K\pi/K_1}$ in the w-plane. Thus, the chain of the used conformal mappings transforms the original doubly-connected domain into the ring.

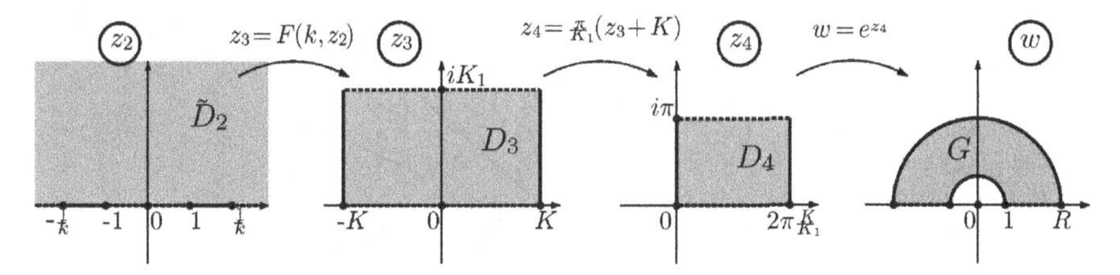

Fig. 5.24 Example: conformal mapping of the domain $\tilde{D}_2$ onto the upper half-ring

5.7 Transformation of Polygons

In this section, we will find the function that transforms conformally the upper half-plane $\operatorname{Im} z > 0$ into a given polygon P on the plane $\mathbb{C}_w$ with the vertices $A_k,\ k = 1, \ldots, n$ and the corresponding angles $\pi\alpha_k$, where $\sum_{k=1}^{n} \alpha_k = n - 2$. By the Riemann Mapping Theorem, such transformation exists. Our purpose is to specify the formula of the function that performs the desired transformation. Initially, we will consider bounded polygons, but later on we will withdraw this restriction.

Schwarz–Christoffel Theorem *Let P be a bounded polygon with the successive vertices A_k and the angles $\pi\alpha_k$, $0 < \alpha_k \le 2$, $k = 1, \ldots, n$ at the corresponding vertices. If a function $w = f(z)$ transforms conformally the upper half-plane* $\operatorname{Im} z >$ 0 *into this polygon P in such a way that the points a_k, $-\infty < a_1 < a_2 < \ldots < a_n < +\infty$ are carried to the corresponding vertices $A_k,\ k = 1, \ldots, n$, then the function $f(z)$ has the form*

$$f(z) = c_0 \int_{z_0}^{z} (t - a_1)^{\alpha_1 - 1} (t - a_2)^{\alpha_2 - 1} \cdots (t - a_n)^{\alpha_n - 1}\, dt + c_1\,, \quad \operatorname{Im} z > 0\,, \tag{5.18}$$

where z_0, c_0, c_1 are some constants, $\operatorname{Im} z_0 \ge 0$.

Proof According to the Riemann Mapping Theorem such a mapping $w = f(z)$ exists and our goal is to show that it is defined by formula (5.18). For now, we suppose that the points a_k on the real axis, which correspond to the vertices A_k, $k = 1, \ldots, n$ of the polygon P, are known and $a_k \ne \infty$, $\forall k = 1, \ldots, n$. We split the proof of the theorem in a few steps.

1. Since $w = f(z)$ transforms bijectively the domain $\operatorname{Im} z > 0$ into the polygon P, on each of the intervals $\gamma_k = (a_k, a_{k+1})$, $k = 1, \ldots, n$ of the real axis this function has the values that belong to the straight-line segment $\Gamma_k = (A_k, A_{k+1}) \subset \partial P$ (we assume here that $a_{n+1} = a_1$ and $A_{n+1} = A_1$). Using the symmetry principle, we can analytically continue the function $w = f(z)$ from the upper half-plane onto the lower half-plane with respect to the interval γ_k. The extended function maps conformally the domain $\operatorname{Im} z < 0$ onto the polygon P'_k, symmetric to P with respect to the segment Γ_k (see Fig. 5.25), and the straight-line segments $\Gamma_j \subset \partial P$ are reflected (mapped) to the corresponding straight-line segments $\Gamma'_j \subset \partial P'_k$, $j = 1, \ldots, n$, $j \ne k$. This

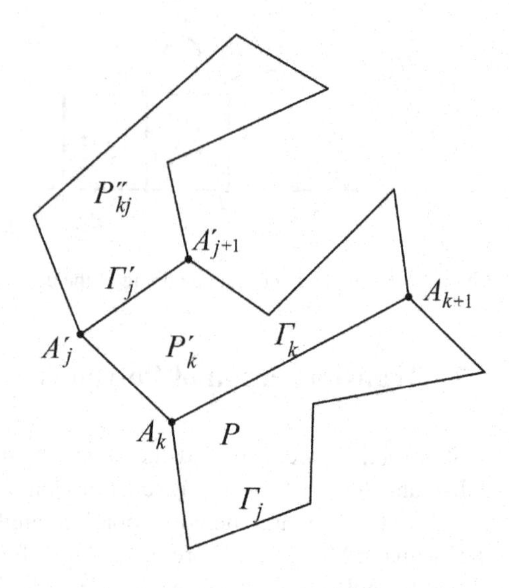

Fig. 5.25 Analytic continuation of $w = f(z)$: P and P''_{kj} are the images of $\operatorname{Im} z > 0$ under the analytic branches $f(z)$ and $f_{kj}(z)$, and P'_k is the image of $\operatorname{Im} z < 0$ under the branch $f_k(z)$

analytic continuation $w = f_k(z)$ can be again continued onto the upper half-plane using the symmetry with respect to any interval $\gamma_j = (a_j, a_{j+1})$, $j \neq k$, and the extended function $w = f_{kj}(z)$ transforms conformally the upper half-plane $\operatorname{Im} z > 0$ into the polygon P''_{kj}, symmetric to P'_k with respect to the straight-line segment Γ'_j (see Fig. 5.25).

Suppose that we have made all possible analytic continuations of this type. As a result, we obtain an analytic multi-valued function $w = F(z)$ on the entire plane $\mathbb{C}_z$, except at the points $z = a_k$, $k = 1, \ldots, n$, for which the function $f(z)$ is one of analytic branches on the upper half-plane (original analytic branch).

Let us analyze the relation between the analytic branches $f(z)$ and $f_{kj}(z)$ of the function $F(z)$, $\operatorname{Im} z > 0$. It follows from the symmetry principle that the functions $f(z)$ and $f_k(z)$ map symmetric points z and $\bar{z}$ of the plane $\mathbb{C}_z$ in symmetric points with respect to the straight line containing the segment Γ_k. If $\varphi_k = \arg(A_{k+1} - A_k)$, then the linear function $\zeta = (w - A_k)\, e^{-i\varphi_k}$ maps this straight line onto the real axis $\operatorname{Im} \zeta = 0$, and consequently, the composite functions $\zeta = h(z) = (f(z) - A_k)\, e^{-i\varphi_k}$ and $\zeta = h_k(z) = (f_k(z) - A_k)\, e^{-i\varphi_k}$ carry z and $\bar{z}$ to the points $\zeta = h(z)$ and $\bar{\zeta} = \overline{h_k(\bar{z})}$, that is, $\overline{(f(\bar{z}) - A_k)} e^{i\varphi_k} = (f_k(z) - A_k)\, e^{-i\varphi_k}$, from which we have

$$f_k(z) = e^{2i\varphi_k} \overline{f(\bar{z})} + B_k\,, \quad \operatorname{Im} z < 0\,. \tag{5.19}$$

Using the same reasoning for the branches $f_k(z)$ and $f_{kj}(z)$ of the function $F(z)$, we get

$$f_{kj}(z) = e^{2i\varphi'_j} \overline{f_k(\bar{z})} + B'_j\,, \quad \operatorname{Im} z > 0\,. \tag{5.20}$$

From (5.19) and (5.20), it follows that

$$f_{kj}(z) = e^{i\beta_{kj}} f(z) + B_{kj}, \quad \operatorname{Im} z > 0. \tag{5.21}$$

The last formula shows that successive application of two symmetries with respect to any two straight lines results in a linear function.

Using the same arguments for any two analytic branches $f_*(z)$ and $f_{**}(z)$ of the function $F(z)$ on the upper (or lower) half-plane, we arrive at the formula similar to (5.21):

$$f_{**}(z) = e^{i\beta} f_*(z) + B, \quad \operatorname{Im} z > 0 \text{ (or } \operatorname{Im} z < 0). \tag{5.22}$$

2. Introduce the function

$$g(z) = \frac{F''(z)}{F'(z)}. \tag{5.23}$$

Since $F(z)$ is analytic on the entire plane $\mathbb{C}_z$, except at the points $a_k,\ k = 1, \ldots, n$ (and, possibly, at infinity), and any analytic branch $f_*(z)$ of $F(z)$ maps conformally the upper (or lower) half-plane onto a polygon P^*, that is, $f_*(z)$ is univalent in $\operatorname{Im} z > 0$ (or in $\operatorname{Im} z < 0$), then $f_*'(z) \neq 0$, which implies that $F'(z) \neq 0,\ \forall z \in \{\operatorname{Im} z > 0\} \cup \{\operatorname{Im} z < 0\}$. Due to the same reasons, $F'(z) \neq 0$ on the real axis, except at the points $z = a_k,\ k = 1, \ldots, n$ and, possibly, $z = \infty$. From the equality (5.22), it follows that

$$\frac{f_{**}''(z)}{f_{**}'(z)} = \frac{f_*''(z)}{f_*'(z)}$$

for any two analytic branches of the function $F(z)$ on the upper (or lower) half-plane. This means that the function $g(z)$ defined by (5.23) is analytic and single-valued in the entire complex plane $\mathbb{C}_z$, except at the points a_k, which correspond to the vertices of the polygon P. Notice that $g(z)$ is also analytic at the point $z = \infty$, because any analytic branch $f_*(z)$ of $F(z)$ carries $z = \infty$ to the point belonging to one of the sides of the polygon P^*, but not to one of its vertices.

Let us show that $g(z)$ is also univalent at ∞. Consider, for simplicity, the original analytic branch $f(z)$ of the function $F(z)$. From the symmetry principle it follows that $f(z)$ is analytic and univalent in $\{\operatorname{Im} z > 0\} \cup \gamma_n$, in particular, it is analytic and univalent at the point $z = \infty \in \gamma_n$. Therefore, it can be represented in the Laurent series in a neighborhood of ∞:

$$f(z) = d_0 + \frac{d_{-1}}{z} + \frac{d_{-2}}{z^2} + \ldots, \quad |z| > R, \tag{5.24}$$

where $d_0 = f(\infty) \in \Gamma_n$ e $d_{-1} = -\operatorname*{res}_{z=\infty} f(z) \neq 0$, due to the univalence of $f(z)$ at ∞. Substituting the series (5.24) in formula (5.23) and using the already proved result that the function $g(z)$ is single-valued, that is, the values of this function do not depend on the choice of an analytic branch of $F(z)$, we have

$$g(z) = \frac{F''(z)}{F'(z)} = \frac{f''(z)}{f'(z)} = \frac{2d_{-1}z^{-3} + 2 \cdot 3d_{-2}z^{-4} + \ldots}{-d_{-1}z^{-2} - 2d_{-2}z^{-3} - \ldots} = -\frac{2}{z} + \frac{c_{-2}}{z^2} + \ldots, \quad |z| > R. \tag{5.25}$$

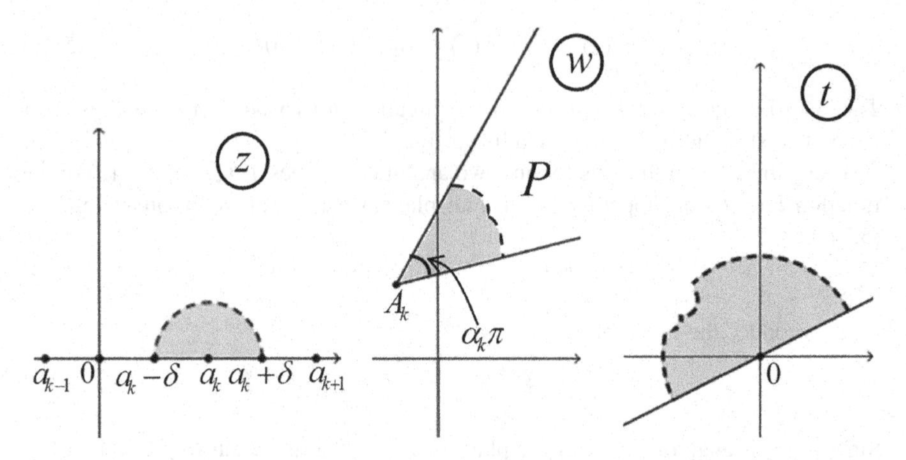

Fig. 5.26 Transformation of the half-disk $|z - a_k| < \delta, \operatorname{Im} z > 0$ by the function $t = h(z) = (f(z) - A_k)^{\frac{1}{\alpha_k}}$

This expansion shows that the function $g(z)$ is analytic and univalent at the point ∞.

3. Let us analyze the behavior of the function $f(z)$ at the point $z = a_k$. Recall that $f(z)$ transforms conformally the upper half-plane $\operatorname{Im} z > 0$ into the polygon P and, under this mapping, the point a_k is carried to the vertex A_k of this polygon: $f(a_k) = A_k$. Introduce the auxiliary variable $t = (w - A_k)^{\frac{1}{\alpha_k}}$, which increases $\frac{1}{\alpha_k}$ times the angles at the point A_k. Then, the composite function

$$t = (f(z) - A_k)^{\frac{1}{\alpha_k}} = h(z) \tag{5.26}$$

represents a conformal mapping of the upper part of a neighborhood of the point a_k ($|z - a_k| < \delta$, $\operatorname{Im} z > 0$, where $\delta > 0$ is sufficiently small) onto a part of a neighborhood of the point $t = 0$ contained in one of the half-planes. Besides, the interval $(a_k - \delta, a_k + \delta)$ of the real axis is mapped onto an interval (see Fig. 5.26). By the symmetry principle, the function $t = h(z)$ can be analytically continued onto the entire neighborhood of a_k and the extended function is analytic at the point a_k, which implies that

$$h(z) = b_0 + b_1 (z - a_k) + b_2 (z - a_k)^2 + \dots, \quad |z - a_k| < \delta .$$

Notice that $b_0 = h(a_k) = 0$ because $f(a_k) = A_k$, and that $b_1 = h'(a_k) \neq 0$ due to univalence of the function $h(z)$ in the neighborhood $|z - a_k| < \delta$ (see Corollary 5.3 in Sect. 5.4).

Using these considerations and relation (5.26), we can express $f(z)$ in the neighborhood $|z - a_k| < \delta$ in the form

$$f(z) = A_k + (h(z))^{\alpha_k} = A_k + (z-a_k)^{\alpha_k} (h_1(z))^{\alpha_k} = A_k + (z-a_k)^{\alpha_k} (b_1 + b_2(z-a_k) + \ldots)^{\alpha_k}. \tag{5.27}$$

Since $b_1 = h_1(a_k) = h'(a_k) \neq 0$, we can choose an analytic single-valued branch of the multi-valued analytic function $(h_1(z))^{\alpha_k}$ in a neighborhood of a_k, and expand this branch in a power series in the neighborhood of a_k:

$$(h_1(z))^{\alpha_k} = c_0 + c_1(z-a_k) + c_2(z-a_k)^2 + \ldots, \; c_0 \neq 0.$$

Substituting the last expression in formula (5.27), we obtain

$$f(z) = A_k + (z-a_k)^{\alpha_k} \left(c_0 + c_1(z-a_k) + c_2(z-a_k)^2 + \ldots\right). \tag{5.28}$$

It is seen from (5.28) that $f'(a_k) = 0$ if $\alpha_k > 1$ and $f'(a_k) = \infty$ if $\alpha_k < 1$. If we consider the inverse transformation $z = \varphi(w)$, then we obtain the opposite result: $\varphi'(A_k) = \infty$ if $\alpha_k > 1$ and $\varphi'(A_k) = 0$ if $\alpha_k < 1$. It also follows from relation (5.28) that in the case $\alpha_k \neq 1$, $\alpha_k \neq 2$ the point a_k is a branch point of the function $f(z)$.

4. Now, we come back to the function $g(z)$ and analyze its behavior at the point $z = a_k$. At the second stage of the proof we have shown that $g(z)$ is single-valued, and consequently, we can choose any analytic branch of the function $F(z)$ in (5.23), in particular, the original analytic branch $f(z)$. Using formula (5.28), we obtain the Laurent series expansion of $g(z)$ in a neighborhood of $z = a_k$:

$$g(z) = \frac{F''(z)}{F'(z)} = \frac{f''(z)}{f'(z)} = \frac{(\alpha_k - 1)\alpha_k c_0 (z-a_k)^{\alpha_k - 2} + \alpha_k(\alpha_k + 1) c_1 (z-a_k)^{\alpha_k - 1} + \ldots}{\alpha_k c_0 (z-a_k)^{\alpha_k - 1} + (\alpha_k + 1) c_1 (z-a_k)^{\alpha_k} + \ldots}$$

$$= \frac{\alpha_k - 1}{z - a_k} + c_0' + c_1'(z-a_k) + \ldots.$$

This expansion reveals that the point $z = a_k$ is a simple pole of the function $g(z)$ (if $\alpha_k \neq 1$) and $\operatorname*{res}_{z=a_k} g(z) = \alpha_k - 1$.

Thus, the function $g(z)$ possesses exactly n singular points, all of which are simple poles, in the entire complex plane $\mathbb{C}_z$. Therefore, the function

$$G(z) = g(z) - \sum_{k=1}^{n} \frac{\alpha_k - 1}{z - a_k} \tag{5.29}$$

is analytic in the entire extended complex plane. Besides, formula (5.25) implies that $g(\infty) = 0$, and consequently $G(\infty) = 0$ also according to (5.29). Using Liouville's Theorem (Sect. 3.6), we conclude that $G(z) \equiv 0$, for $\forall z \in \overline{\mathbb{C}}_z$, which means that

$$g(z) = \sum_{k=1}^{n} \frac{\alpha_k - 1}{z - a_k}, \quad \forall z \in \overline{\mathbb{C}}_z.$$

Applying the last formula to the points of the upper half-plane, we obtain

$$\frac{f''(z)}{f'(z)} = \sum_{k=1}^{n} \frac{\alpha_k - 1}{z - a_k}, \quad \operatorname{Im} z > 0. \tag{5.30}$$

5. At the last stage of the proof, we integrate the expression (5.30) along any curve contained in the upper half-plane and obtain

$$\ln f'(z) = \sum_{k=1}^{n} (\alpha_k - 1) \ln (z - a_k) + \tilde{c} = \ln \prod_{k=1}^{n} (z - a_k)^{\alpha_k - 1} + \tilde{c},$$

that is,

$$f'(z) = c_0 (z - a_1)^{\alpha_1 - 1} (z - a_2)^{\alpha_2 - 1} \dots (z - a_n)^{\alpha_n - 1}. \tag{5.31}$$

Applying one more integration to the last expression, we arrive at the Schwarz–Christoffel formula (5.18).

This completes the proof of the Theorem. □

Now, we proceed to different remarks, which extend the application of the Schwarz–Christoffel transformation (5.18).

Remark 5.1 Consider the case when one of the vertices of the polygon P is the image of the point $z = \infty$. Without a loss of generality we can assume that $a_n = \infty$. This case can be reduced to that considered in the above Theorem by using the fractional linear function

$$\zeta = -\frac{1}{z} + a'_n, \tag{5.32}$$

which maps the upper half-plane $\operatorname{Im} z > 0$ onto the upper half-plane $\operatorname{Im} \zeta > 0$ in such a way that the points $a_1, \dots, a_{n-1}$, $a_n = \infty$ of the real axis are carried to the finite points $a'_1, a'_2, \dots, a'_n$ of the real axis. If one of the points $a_k = 0$, then instead of (5.32) we can employ the function $\zeta = -\frac{1}{z-a} + a'_n$, where $a \neq a_k$, $\forall k = 1, \dots, n-1$. Applying the Schwarz–Christoffel integral to the variable ζ, making the change of the variable (5.32) in this integral and recalling that $\alpha_1 + \dots + \alpha_n = n - 2$, we obtain

$$w = c'_0 \int_{\zeta_0}^{\zeta} (\zeta - a'_1)^{\alpha_1 - 1} \cdots (\zeta - a'_{n-1})^{\alpha_{n-1} - 1} (\zeta - a'_n)^{\alpha_n - 1} \, d\zeta + c'_1$$

$$= c'_0 \int_{z_0}^{z} \left((a'_n - a'_1) - \frac{1}{z} \right)^{\alpha_1 - 1} \cdots \left((a'_n - a'_{n-1}) - \frac{1}{z} \right)^{\alpha_{n-1} - 1} \left(-\frac{1}{z} \right)^{\alpha_n - 1} \frac{dz}{z^2} + c'_1$$

$$= (-1)^{\alpha_n - 1} c'_0 (a'_n - a'_1)^{\alpha_1 - 1} \cdots (a'_n - a'_{n-1})^{\alpha_{n-1} - 1}$$

$$\times \int_{z_0}^{z} (z - a_1)^{\alpha_1 - 1} \cdots (z - a_{n-1})^{\alpha_{n-1} - 1} \frac{dz}{z^{\alpha_1 + \dots + \alpha_n - n + 2}} + c'_1$$

Fig. 5.27 Polygon P with $A_k = \infty$

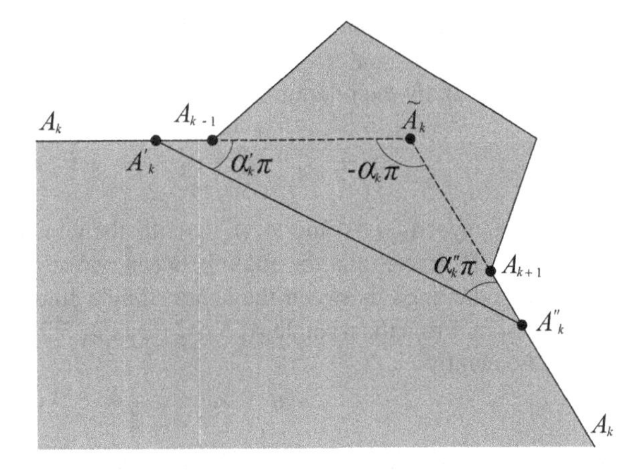

$$= c_0 \int_{z_0}^{z} (z - a_1)^{\alpha_1 - 1} \cdots (z - a_{n-1})^{\alpha_{n-1} - 1} \, dz + c_1 ,$$

where

$$c_0 = c'_0 \, (-1)^{\alpha_n - 1} \left(a'_n - a'_1\right)^{\alpha_1 - 1} \cdots \left(a'_n - a'_{n-1}\right)^{\alpha_{n-1} - 1} , \; a_k = \frac{1}{a'_n - a'_k} , \; k = 1, 2, \ldots, n - 1 ,$$

and $c_1 = c'_1$.

Thus, we arrive at the following conclusion: if one of the vertices of the polygon P, for instance A_n, corresponds to the point $z = \infty$, then the Schwarz–Christoffel formula does not contain the factor related to this vertex and the Schwarz–Christoffel transformation takes the form

$$w = f(z) = c_0 \int_{z_0}^{z} (z - a_1)^{\alpha_1 - 1} (z - a_2)^{\alpha_2 - 1} \cdots (z - a_{n-1})^{\alpha_{n-1} - 1} \, dz + c_1 . \tag{5.33}$$

Remark 5.2 Consider now the case when one of the vertices of the polygon P, say A_k, is ∞ (a similar analysis can be made if various vertices of P are infinite points). Take any points A'_k and A''_k on the corresponding sides (rays) $A_{k-1}A_k$ and $A_k A_{k+1}$, and connect them by the straight-line segment $A'_k A''_k$. Then, we obtain an auxiliary bounded polygon P' with $n + 1$ vertices (see Fig. 5.27).

Using the Schwarz–Christoffel formula (5.18) for conformal mapping of the upper half-plane $\text{Im}\, z > 0$ onto the auxiliary polygon P', we have

$$w = c_0 \int_{z_0}^{z} (z - a_1)^{\alpha_1 - 1} \cdots \left(z - a'_k\right)^{\alpha'_k - 1} \left(z - a''_k\right)^{\alpha''_k - 1} \cdots (z - a_n)^{\alpha_n - 1} \, dz + c_1 , \tag{5.34}$$

where a_k', a_k'' are the points on the real axis of the plane $\mathbb{C}_z$, which correspond to the vertices A_k', A_k'', and $\pi\alpha_k'$, $\pi\alpha_k''$ are the angles of P' at the vertices A_k', A_k''. The angles of P' satisfy the relation

$$\alpha_1 + \ldots + \alpha_k' + \alpha_k'' + \ldots + \alpha_n = n - 1\,. \tag{5.35}$$

Extend the rays $A_{k-1}A_k$ and A_kA_{k+1} until their intersection at a finite point $\tilde{A}_k$ (see Fig. 5.27). Recall that the angle between two straight lines at the infinity point is equal to the angle between the same straight lines at the finite point, but with opposite sign. From the triangle $A_k'\tilde{A}_kA_k''$ (see Fig. 5.27), we have $\alpha_k' + \alpha_k'' - \alpha_k = 1$ and consequently

$$\alpha_k' + \alpha_k'' - 2 = \alpha_k - 1\,, \tag{5.36}$$

which is true for $\forall A_k' \in A_{k-1}A_k$ and $\forall A_k'' \in A_kA_{k+1}$.

Let us change the position of the segment $A_k'A_k''$ of the polygon P' in such a way that it keeps the same direction (all the segments $A_k'A_k''$ are parallel to each other) and goes away from the intersection point $\tilde{A}_k$ approaching the infinity point. In the limit of this process, the points A_k' and A_k'' in the plane $\mathbb{C}_w$ turn into the same point ∞, and the corresponding points a_k' and a_k'' on the real axis of the plane $\mathbb{C}_z$ end up at the same point a_k, which corresponds to the vertex $A_k = \infty$. Under this, the product $\left(z - a_k'\right)^{\alpha_k'-1}\left(z - a_k''\right)^{\alpha_k''-1}$ in (5.34) is transformed to the factor $(z - a_k)^{\alpha_k'+\alpha_k''-2}$. Applying this limit to formula (5.34) and using relation (5.36), we obtain

$$w = c_0 \int_{z_0}^{z} (z - a_1)^{\alpha_1 - 1} \cdots (z - a_k)^{\alpha_k - 1} \cdots (z - a_n)^{\alpha_n - 1}\, dz + c_1\,.$$

Hence, formula (5.18) is true even if one of the vertices of the polygon P is the infinity point. (The same result can be obtained if various vertices of P are ∞.)

Notice that the relation between the angles of the infinite polygon P is the same as for a bounded polygon. Indeed, substituting (5.36) in (5.35), we get

$$\alpha_1 + \ldots + \alpha_k + \ldots + \alpha_n = \sum_{k=1}^{n} \alpha_k = n - 2\,.$$

(Once more recall that the angle between two straight lines at infinity is defined as the negative value of the angle between these lines at their finite point of the intersection.)

Thus, we conclude that the Schwarz–Christoffel transformation (5.18) keeps the same form if one (or some) of the vertices of the polygon P is (are) the infinity point. Notice that in the case when one of the vertices (finite or infinite) of P corresponds to the point $z = \infty$, formula (5.33) should be used instead of (5.18).

Remark 5.3 Consider a conformal transformation of the upper half-plane $\operatorname{Im} z > 0$ into the exterior of the bounded polygon P. This situation has the only one substantial difference with the analyzed case of the interior part of a polygon in the Schwarz–Christoffel Theorem: at some point $z = a$, which corresponds to $w = \infty$, the function $w = f(z)$ has a pole. Notice that this pole is simple due to the univalence of $f(z)$ in $\mathrm{Im}\, z > 0$. Therefore, there exists a deleted neighborhood of the point $z = a$ in which the Laurent expansion of $f(z)$ has the form

$$f(z) = \frac{b_{-1}}{z-a} + b_0 + b_1(z-a) + b_2(z-a)^2 + b_3(z-a)^3 + \dots, \quad 0 < |z-a| < \delta.$$

Substituting this expression in formula (5.23), we obtain the Laurent expansion of the function $g(z)$ in a neighborhood of $z = a$:

$$\begin{aligned} g(z) &= \frac{F''(z)}{F'(z)} = \frac{f''(z)}{f'(z)} = \frac{2b_{-1}(z-a)^{-3} + 2b_2 + 2\cdot 3b_3(z-a) + \dots}{-b_{-1}(z-a)^{-2} + b_1 + 2b_2(z-a) + 3b_3(z-a)^2 + \dots} \\ &= -\frac{2}{z-a} + c_0 + c_1(z-a) + \dots, \quad 0 < |z-a| < \delta, \ \mathrm{Im}\, a > 0. \end{aligned} \tag{5.37}$$

From (5.37), it follows that $z = a$ is also a simple pole of $g(z)$ with $\operatorname*{res}_{z=a} g(z) = -2$. The situation is similar at the point $z = \bar{a} \in \{\mathrm{Im}\, z < 0\}$, because $z = \bar{a}$ is a simple pole of the analytic branch $f_k(z)$, which is an analytic continuation of the original branch $f(z)$ onto the lower half-plane with respect to the interval $\gamma_k = (a_k, a_{k+1})$, $\forall k = 1, \dots, n$. Therefore, the function

$$G(z) = g(z) - \sum_{k=1}^{n} \frac{\alpha_k - 1}{z - a_k} + \frac{2}{z-a} + \frac{2}{z-\bar{a}}$$

is analytic in the entire extended complex plane $\overline{\mathbb{C}}_z$. Like in the Schwarz–Christoffel Theorem, using Liouville's Theorem and the fact that $g(\infty) = 0$, we conclude that $G(z) \equiv 0$, $\forall z \in \overline{\mathbb{C}}_z$, which implies that

$$g(z) = \sum_{k=1}^{n} \frac{\alpha_k - 1}{z - a_k} - \frac{2}{z-a} - \frac{2}{z-\bar{a}}. \tag{5.38}$$

Again like in the Schwarz–Christoffel Theorem, integrating the last formula twice along a curve contained in the upper half-plane with the initial point z_0 and final point z, we obtain

$$f(z) = c_0 \int_{z_0}^{z} (z-a_1)^{\alpha_1 - 1} \cdots (z-a_n)^{\alpha_n - 1} \frac{dz}{(z-a)^2 (z-\bar{a})^2} + c_1, \quad \mathrm{Im}\, z > 0. \tag{5.39}$$

Notice that in formula (5.39) the angles $\pi\alpha_k$, $k = 1, \dots, n$ are external angles of the bounded polygon P.

Remark 5.4 Now, we return to the analysis of the Schwarz–Christoffel formula (5.18). This formula was considered under the supposition that the points a_k of the real axis, which correspond to the vertices of a polygon P, are known. In practice, in the majority of cases, only the elements of a polygon P—the vertices A_k and the corresponding angles $\pi\alpha_k$—are known, whereas the points a_k are not given. In this situation, the essential difficulty in the use of the Schwarz-Christoffel integral is the finding of the points a_k and evaluation of the constants $z_0,\ c_0,\ c_1$.

Since z_0 is an arbitrary fixed point such that $\operatorname{Im} z_0 \geq 0$, a usual choice is $z_0 = 0$ or $z_0 = a_1$ (or $z_0 = a_k,\ k = 1, \dots, n$). If, for instance, we take $z_0 = a_1$ in formula (5.18), then

$$A_1 = f(a_1) = c_0 \int_{a_1}^{a_1} f'(z)\,dz + c_1 = c_1\,,$$

that is, $c_1 = A_1$. Besides, from Theorem 5.2 of Sect. 5.4 it follows that the three points a_k, say $a_1,\ a_2,\ a_3$, we can choose arbitrarily. Then the problem is to find the remaining points $a_k,\ k = 4, \dots, n$ and the constant c_0. We will consider different ways to solve this problem in the examples analyzed at the end of this section. For now, we present one possible solution.

From formula (5.31), it is seen that for $z{=}x > a_n$ the condition $\arg\,(x - a_k)^{\alpha_k - 1} = 0,\ \forall k = 1, \dots, n$ holds, which implies that $\arg f'(x) = \arg c_0,\ \ x > a_n$. In addition, the interval $\gamma_n = (a_n, a_{n+1}) = (a_n, a_1)$, which contains the infinity point, is carried into the segment $\Gamma_n = (A_n, A_1)$. Denoting by θ the angle between this segment Γ_n and the real axis $u = \operatorname{Re} w$, we have $\arg c_0 = \arg f'(x) = \theta$. Hence, $c_0 = \tilde{c}e^{i\theta}$, where $\tilde{c} = |c_0| > 0$, and formula (5.18) takes the form

$$f(z) = \tilde{c}e^{i\theta} \int_{a_1}^{z} (t - a_1)^{\alpha_1 - 1} \cdots (t - a_n)^{\alpha_n - 1}\,dt + A_1\,. \tag{5.40}$$

Notice that when $z \in \gamma_k = (a_k, a_{k+1})$ then $w = f(z) \in \Gamma_k = (A_k, A_{k+1})$, that is, $\arg f(z) = constant$ and $\arg\left[(z - a_1)^{\alpha_1 - 1} \cdots (z - a_n)^{\alpha_n - 1}\right] = const.,\ \ z \in \gamma_k$. Therefore, it follows from (5.40) that

$$|A_{k+1} - A_k| = |f(a_{k+1}) - f(a_k)| = \tilde{c} \int_{a_k}^{a_{k+1}} \left|(t - a_1)^{\alpha_1 - 1} \cdots (t - a_n)^{\alpha_n - 1}\right| dt\,,\ k = 1, \dots, n-1\,. \tag{5.41}$$

According to the Schwarz–Christoffel Theorem, the system (5.41) has a unique solution, which means that this system defines the real parameters $\tilde{c},\ a_4, \dots, a_n$ in a unique mode. However, in the application problems, an analytic solution to (5.41) is usually hard or even impossible to find. Nevertheless, for some specific polygons of a simpler form there are other techniques to find the parameters in the Schwarz–Christoffel formula, which we will see in the solved examples at the end of this section.

Remark 5.5 It is straightforward to note that formula (5.15) in the previous section is a particular case of the Schwarz–Christoffel integral (5.18) when P is a rectangle and $a_1 = -\frac{1}{k}\,,\ a_2 = -1\,,\ a_3 = 1\,,\ a_4 = \frac{1}{k}\,,\ 0 < k < 1$.

Remark 5.6 A conformal transformation of the unit disk $|\zeta| < 1$ into the interior of a bounded polygon P can be constructed as the composition of two conformal mappings: first we transform the unit disk into the upper half-plane and then we employ the Schwarz–Christoffel transformation. Recall that fractional linear function

$$z = i\frac{1+\zeta}{1-\zeta}\,, \quad \zeta = \frac{z-i}{z+i}\,, \tag{5.42}$$

is a one-to-one correspondence between the unit disk $|\zeta| < 1$ and the upper half-plane $\operatorname{Im} z > 0$. Using the change of variable (5.42) in the Schwarz–Christoffel integral (5.18), we rewrite the expressions in this integral in the form

$$dz = \frac{2i}{(1-\zeta)^2}d\zeta\,, \tag{5.43}$$

$$(z-a_k)^{\alpha_k-1} = \left(i\frac{1+\zeta}{1-\zeta} - a_k\right)^{\alpha_k-1} = \left(\frac{\zeta\,(a_k+i)-(a_k-i)}{1-\zeta}\right)^{\alpha_k-1}$$
$$= \left(\frac{a_k+i}{1-\zeta}\right)^{\alpha_k-1}\left(\zeta - \frac{a_k-i}{a_k+i}\right)^{\alpha_k-1} = \left(\frac{a_k+i}{1-\zeta}\right)^{\alpha_k-1}(\zeta-b_k)^{\alpha_k-1}\,,\ b_k = \frac{a_k-i}{a_k+i}\,. \tag{5.44}$$

Bringing (5.43) and (5.44) in (5.18) and using the relation $\alpha_1 + \ldots + \alpha_n = n-2$, we obtain

$$\begin{aligned} f(\zeta) &= d_0\int_{\zeta_0}^{\zeta}(\zeta-b_1)^{\alpha_1-1}\cdots(\zeta-b_n)^{\alpha_n-1}\frac{d\zeta}{(1-\zeta)^{\alpha_1+\ldots+\alpha_n-n+2}} + d_1 \\ &= d_0\int_{\zeta_0}^{\zeta}(\zeta-b_1)^{\alpha_1-1}\cdots(\zeta-b_n)^{\alpha_n-1}\,d\zeta + d_1\,, \end{aligned} \tag{5.45}$$

where $d_0 = 2ic_0\,(a_1+i)^{\alpha_1-1}\cdots(a_n+i)^{\alpha_n-1}$, $d_1 = c_1$.

Thus, formula (5.45) of the conformal mapping of the unit disk $|\zeta| < 1$ onto a polygon P has the identical form to that of the integral (5.18). For the conformal transformation of the exterior of the unit disk $|\zeta| > 1$ into the the exterior of a polygon P with the additional condition that the points $\zeta = \infty$ and $w = \infty$ correspond each other, we obtain the same formula (5.45). Notice that $\pi\alpha_k,\ k = 1,\ldots,n$ are the interior angles of the polygon P in the conformal mapping of the interior of the unit disk onto the interior of P, while the same angles are exterior under the conformal mapping between the exteriors of the disk and polygon. The points $b_k,\ k = 1,\ldots,n$ are located on the unit circle and correspond to the vertices of the polygon P.

Remark 5.7 Let us find a conformal mapping of the unit disk $|\zeta| < 1$ onto the exterior of a bounded polygon P under the supposition that the point $\zeta = 0$ is carried to $w = \infty$: $f(0) = \infty$. Since the transformation $w = f(\zeta)$ is conformal, the point $\zeta = 0$ is a simple pole of $f(\zeta)$, that is, the Laurent expansion of $f(\zeta)$ in a deleted neighborhood of $\zeta = 0$ has the form

$$f(\zeta) = \frac{p_{-1}}{\zeta} + p_0 + p_1\zeta + p_2\zeta^2 + p_3\zeta^3 + \ldots\,,$$

and consequently

$$\frac{f''(\zeta)}{f'(\zeta)} = \frac{2p_{-1}\zeta^{-3} + 2p_2 + 6p_3\zeta + \ldots}{-p_{-1}\zeta^{-2} + p_1 + 2p_2\zeta + 3p_3\zeta^2 + \ldots} = -\frac{2}{\zeta} - \frac{2p_1}{p_{-1}}\zeta + \ldots .$$

From the last relation, it is seen that the function $\frac{f''(\zeta)}{f'(\zeta)} + \frac{2}{\zeta}$ is analytic in the entire disk $|\zeta| < 1$ and vanishes at the origin.

Like in the proof of the Theorem, we can continue analytically the function $g(\zeta) = \frac{f''(\zeta)}{f'(\zeta)}$ from the unit disk onto the entire extended complex plane and, using the same reasoning as in part 4 of the proof and in Remark 5.3 and applying also Liouville's Theorem, we obtain that the function

$$G(\zeta) = g(\zeta) - \sum_{k=1}^{n} \frac{\alpha_k - 1}{\zeta - b_k} + \frac{2}{\zeta}$$

is a constant. The condition $G(\infty) = 0$ implies then that $G(\zeta) \equiv 0\,,\ \forall\zeta \in \overline{\mathbb{C}}_\zeta$ and, conseuquently,

$$\frac{f''(\zeta)}{f'(\zeta)} = \sum_{k=1}^{n} \frac{\alpha_k - 1}{\zeta - b_k} - \frac{2}{\zeta}\,, \quad |\zeta| < 1\,. \tag{5.46}$$

Following the part 5 of the proof, we integrate the equality (5.46) twice along a curve contained in the unit disk with the initial point ζ_0 and final point ζ, and arrive at the following result:

$$w = f(\zeta) = d_0 \int_{\zeta_0}^{\zeta} (\zeta - b_1)^{\alpha_1 - 1} \cdots (\zeta - b_n)^{\alpha_n - 1} \frac{d\zeta}{\zeta^2} + d_1\,. \tag{5.47}$$

Notice that in the last expression $\pi\alpha_1, \ldots, \pi\alpha_n$ are the exterior angles of the polygon P and $b_1, \ldots, b_n$ are the points of the unit circle, which correspond to the vertices of P.

Consider now a few examples.

Example 5.1 Let us construct a conformal transformation of the upper half-plane $\operatorname{Im} z > 0$ into the interior of the triangle T with the vertices $A = 0,\ B = 1$ and C ($\operatorname{Im} C > 0$) and the corresponding angles $\pi\alpha, \pi\beta$ and $\pi\gamma, 0 < \alpha, \beta, \gamma < 1\,,\ \alpha + \beta + \gamma = 1$ (see Fig. 5.28).

In Remark 5.4, we have already noted that the three points $a_1,\ a_2,\ a_3$ of the real axis, corresponding to the vertices of the triangle, can be chosen arbitrarily. Pick, for instance, $a_1 = 0\,,\ a_2 = 1\,,\ a_3 = \infty$. In this case, using the condition $f(0) = 0$ (that is, $c_1 = 0$), the Schwarz–Christoffel formula (5.33) takes the form

$$f(z) = c_0 \int_0^z z^{\alpha-1}(z-1)^{\beta-1}\,dz = \tilde{c}_0 \int_0^z z^{\alpha-1}(1-z)^{\beta-1}\,dz\,,$$

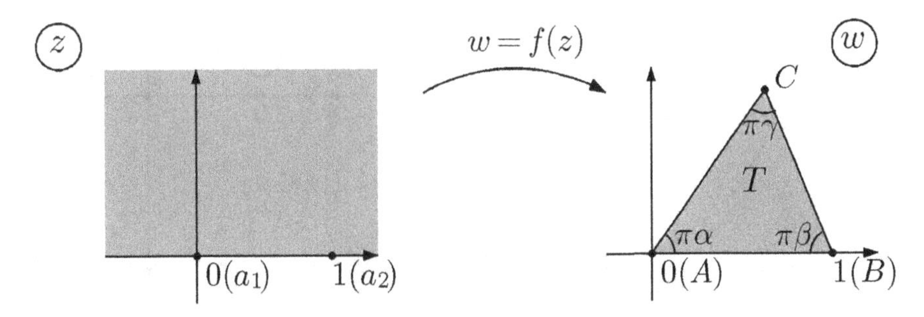

Fig. 5.28 Conformal mapping of the upper half-plane onto the triangle T

where $\tilde{c}_0 = c_0 e^{i\pi(\beta-1)}$. Since $z^{\alpha-1}(1-z)^{\beta-1} > 0$ for $z = x \in (0,1)$ and $f(x) \in (0,1)$, that is, $f(x) > 0$ if $z = x \in (0,1)$, we have that $\tilde{c}_0 > 0$. To evaluate this parameter, we apply the condition that $f(1) = 1$:

$$1 = f(1) = \tilde{c}_0 \int_0^1 z^{\alpha-1}(1-z)^{\beta-1}\,dz = \tilde{c}_0 B(\alpha,\beta) = \tilde{c}_0 \frac{\Gamma(\alpha)\Gamma(\beta)}{\Gamma(\alpha+\beta)}. \tag{5.48}$$

In the last formula, $B(\alpha,\beta)$ is the beta function and $\Gamma(t)$ is the gamma function. Recalling that $\alpha+\beta+\gamma = 1$ and applying the known property of the gamma function $\Gamma(t)\Gamma(1-t) = \frac{\pi}{\sin \pi t}$, $0 < t < 1$ to (5.48), we obtain

$$\tilde{c}_0 = \frac{\Gamma(\alpha+\beta)}{\Gamma(\alpha)\Gamma(\beta)} = \frac{\Gamma(1-\gamma)}{\Gamma(\alpha)\Gamma(\beta)} = \frac{\pi}{\sin\pi\gamma}\frac{1}{\Gamma(\alpha)\Gamma(\beta)\Gamma(\gamma)}. \tag{5.49}$$

Then, the function

$$f(z) = \frac{\pi}{\sin\pi\gamma}\frac{1}{\Gamma(\alpha)\Gamma(\beta)\Gamma(\gamma)}\int_0^z z^{\alpha-1}(1-z)^{\beta-1}\,dz \tag{5.50}$$

transforms conformally the upper half-plane in the considered triangle.

To find the vertex C of this triangle, we use the condition $C = f(\infty)$. From (5.48), (5.49), (5.50) and the inequality $\operatorname{Im} C > 0$ it follows that

$$C = f(\infty) = \tilde{c}_0 \int_0^{+\infty} z^{\alpha-1}(1-z)^{\beta-1}\,dz = \tilde{c}_0 \int_0^1 z^{\alpha-1}(1-z)^{\beta-1}\,dz$$

$$+\tilde{c}_0 \int_1^{+\infty} z^{\alpha-1}(1-z)^{\beta-1}\,dz = 1 + \tilde{c}_0 e^{-i\pi(\beta-1)} \int_1^{+\infty} z^{\alpha-1}(z-1)^{\beta-1}\,dz.$$

Changing variable in the last integral—$z = \frac{1}{t}$, $dz = -\frac{dt}{t^2}$, using formula (5.49) and the condition $\alpha+\beta+\gamma = 1$, we arrive at the following result:

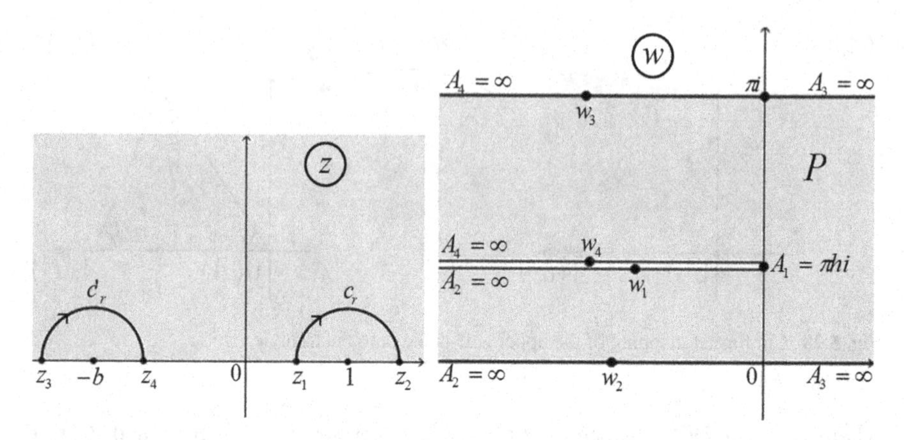

Fig. 5.29 Conformal transformation of the upper half-plane into the strip (quadrilateral) P

$$C = 1 + \tilde{c}_0 e^{i\pi(1-\beta)} \int_0^1 \frac{1}{t^{\alpha-1}} \frac{(1-t)^{\beta-1}}{t^{\beta-1}} \frac{dt}{t^2} = 1 + \tilde{c}_0 e^{i\pi(1-\beta)} \int_0^1 t^{\gamma-1} (1-t)^{\beta-1} dt$$

$$= 1 + \tilde{c}_0 e^{i\pi(1-\beta)} \frac{\Gamma(\gamma)\Gamma(\beta)}{\Gamma(\gamma+\beta)} = 1 + \tilde{c}_0 e^{i\pi(1-\beta)} \frac{\sin\pi\alpha}{\pi} \Gamma(\alpha)\Gamma(\beta)\Gamma(\gamma) = 1 + \frac{\sin\pi\alpha}{\sin\pi\gamma} e^{i\pi(1-\beta)} .$$

Example 5.2 Let us find a conformal mapping of the upper half-plane $\operatorname{Im} z > 0$ onto the infinite strip $0 < \operatorname{Im} w < \pi$ with the cut along the ray $(-\infty + \pi hi,\ \pi hi]$, $0 < h < 1$ (see Fig. 5.29). This strip can be considered as an unbounded quadrilateral P with the vertices $A_1 = \pi hi$, $A_2 = \infty$, $A_3 = \infty$, $A_4 = \infty$, $\alpha_1 = 2$, $\alpha_2 = \alpha_3 = \alpha_4 = 0$. Choose three points of the real axis in the plane $\mathbb{C}_z$ in the following way: $a_1 = 0$, $a_2 = 1$, $a_3 = \infty$. Then, $a_4 = -b$, where $0 < b < +\infty$.

Using formula (5.33) of Remark 5.1 with $z_0 = 0$ and using also the condition $f(0) = \pi hi$, we obtain the function that transforms conformally the upper half-plane $\operatorname{Im} z > 0$ into the strip P in the following form:

$$w = f(z) = c_0 \int_0^z \frac{t\,dt}{(t-1)(t+b)} + \pi hi . \tag{5.51}$$

To find $\arg c_0$, consider a point $z_1 \in (0, 1):\ z_1 = x_1,\ 0 < x_1 < 1$. The image of this point, $w_1 = f(z_1)$, belongs to the side (A_1, A_2) of the quadrilateral P, that is, $w_1 = u_1 + iv_1,\ u_1 < 0,\ v_1 = \pi h$ (see Fig. 5.29). From (5.51), we have that

$$w_1 - \pi hi = c_0 \int_0^{x_1} \frac{t\,dt}{(t-1)(t+b)} .$$

Since the function inside the last integral is real negative for $0 \le t \le x_1 < 1$ and $w_1 - \pi hi = u_1 < 0$, it implies that $c_0 > 0$.

It remains to find the two real positive parameters in formula (5.51): $c_0 > 0$ and $b > 0$. Choosing a sufficiently small $r > 0$ and considering the points $z_1 = x_1 = 1 - r$ and $z_2 = x_2 = 1 + r$ on the real axis, we have $w_1 = f(z_1) \in (A_1, A_2)$ and $w_2 = f(z_2) \in (A_2, A_3)$, which implies that

$$\operatorname{Im}(w_2 - w_1) = -\pi h. \tag{5.52}$$

Denote by C_r the upper half-circle $C_r = \{|z-1| < r,\ \operatorname{Im} z \geq 0\}$ traversed in the clockwise direction (see Fig. 5.29). From (5.51), we have

$$w_2 - w_1 = f(z_2) - f(z_1) = c_0 \int_{C_r} \frac{t\,dt}{(t-1)(t+b)}. \tag{5.53}$$

Notice that $t = 1$ is a simple pole of the function inside the last integral, and consequently $\operatorname*{res}_{t=1} \frac{t}{(t-1)(t+b)} = \frac{1}{1+b}$. Therefore, in a deleted neighborhood of $t = 1$, this function can be represented in the following form

$$\frac{t}{(t-1)(t+b)} = \frac{1}{1+b} \cdot \frac{1}{t-1} + g(t), \tag{5.54}$$

where $g(t)$ is an analytic function at $t = 1$, which implies that $g(t)$ is bounded in a neighborhood of $t = 1$, that is, there exists $M > 0$ such that $|g(t)| \leq M,\ |t-1| < r$.

Using (5.54), we can rewrite the integral in (5.53) in the following manner:

$$\int_{C_r} \frac{t\,dt}{(t-1)(t+b)} = \frac{1}{1+b} \int_{C_r} \frac{dt}{t-1} + \int_{C_r} g(t)\,dt.$$

Taking into account that

$$\int_{C_r} \frac{dt}{t-1} = \int_{\pi}^{0} \frac{re^{i\varphi} i\,d\varphi}{re^{i\varphi}} = -i\pi, \quad \left| \int_{C_r} g(t)\,dt \right| \leq \int_{C_r} |g(t)|\,|dt| \leq M \cdot r\pi \underset{r\to 0}{\to} 0,$$

and that the integral in (5.53) does not depend on r, we can pass to the limit in (5.53) as $r \to 0$ and obtain

$$w_2 - w_1 = c_0 \left(\frac{1}{1+b} \int_{C_r} \frac{dt}{t-1} + \int_{C_r} g(t)\,dt \right) = -c_0 \frac{i\pi}{1+b}. \tag{5.55}$$

Comparing the relations (5.52) and (5.55), we conclude that $\pi h = \frac{\pi c_0}{1+b}$, that is,

$$h = \frac{c_0}{1+b}. \tag{5.56}$$

Similar reasoning in a neighborhood of A_4 provides one more relation between c_0 and b. To this end, take the points $z_3 = x_3 = -b - r,\ z_4 = x_4 = -b + r$, where $r > 0$ is sufficiently small and consider the upper half-circle $C_r' = \{|z+b| = r,\ \operatorname{Im} z \geq 0\}$, traversed in the clockwise direction. Notice that $w_3 = f(z_3) \in (A_3, A_4)$, $w_4 = f(z_4) \in (A_4, A_1)$ (see Fig. 5.29), which implies that

$$\operatorname{Im}(w_4 - w_3) = \pi h - \pi = \pi(h-1). \tag{5.57}$$

On the other hand, the function $\frac{t}{(t-1)(t+b)}$ has a simple pole at the point $t = -b$ with $\operatorname*{res}_{t=-b} \frac{t}{(t-1)(t+b)} = \frac{b}{b+1}$, which means that this function can be represented in the form $\frac{t}{(t-1)(t+b)} = \frac{b}{b+1} \cdot \frac{1}{t+b} + \tilde{g}(t)$, where $\tilde{g}(t)$ is an analytic (and consequently bounded) function in a neighborhood of $t = -b$. Bringing this representation to (5.51), we obtain

$$w_4 - w_3 = f(z_4) - f(z_3) = c_0 \int_{C'_r} \frac{tdt}{(t-1)(t+b)} = \frac{bc_0}{1+b} \int_{C'_r} \frac{dt}{t+b} + c_0 \int_{C'_r} \tilde{g}(t)\,dt. \tag{5.58}$$

Taking into account that

$$\int_{C'_r} \frac{dt}{t+b} = \int_{\pi}^{0} \frac{re^{i\varphi} id\varphi}{re^{i\varphi}} = -i\pi, \quad \left| \int_{C'_r} \tilde{g}(t)\,dt \right| \le \tilde{M} \cdot r\pi \underset{r\to 0}{\to} 0$$

and also that the integral in (5.58) does not depend on r, we can take the limit in (5.58) as $r \to 0$ and obtain

$$w_4 - w_3 = -\frac{c_0 b i\pi}{1+b}. \tag{5.59}$$

Comparing (5.57) and (5.59), we arrive at the equation

$$\frac{bc_0}{1+b} = 1 - h. \tag{5.60}$$

Solving the system (5.56) and (5.60) with respect to b and c_0, we get

$$b = \frac{1-h}{h}, \quad c_0 = 1.$$

Substituting these values in (5.51) and calculating the integral, we obtain

$$w = f(z) = \int_0^z \frac{tdt}{(t-1)(t+b)} + \pi hi = h \int_0^z \frac{dt}{t-1} + (1-h) \int_0^z \frac{dt}{t + (1-h)/h} + \pi hi$$

$$= h \ln(t-1)|_0^z + (1-h) \ln\left(t + \frac{1-h}{h}\right)\Big|_0^z + \pi hi$$

$$= h \ln(z-1) - h \ln(-1) + (1-h) \ln\left(z + \frac{1-h}{h}\right) - (1-h) \ln \frac{1-h}{h} + \pi hi$$

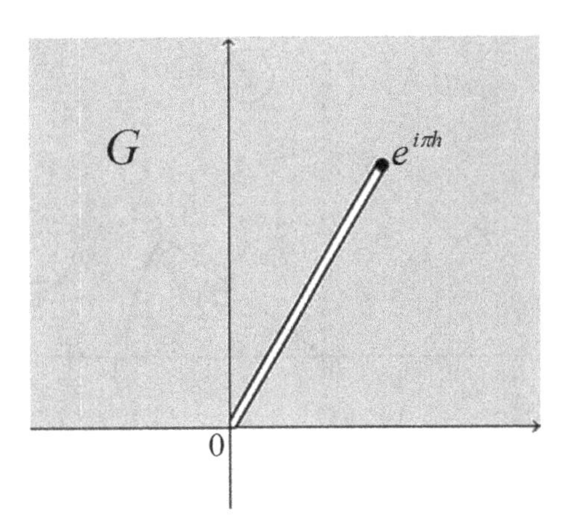

Fig. 5.30 The upper half-plane with the cut along the interval $\left[0\,,\ e^{i\pi h}\right]$

$$= h\ln(z-1) + (1-h)\ln\left(1+\frac{zh}{1-h}\right) = \ln\left[(z-1)^h\left(1+\frac{zh}{1-h}\right)^{1-h}\right].$$

Notice that the function $\zeta = e^w$ transforms conformally the quadrilateral P into the domain G, which is the upper half-plane with the cut along the interval $\left[0\,,\ e^{ih\pi}\right]$ (see Fig. 5.30). Therefore, the function

$$\zeta = (z-1)^h\left(1+\frac{zh}{1-h}\right)^{1-h}$$

transforms conformally the upper half-plane $\operatorname{Im} z > 0$ into the domain G.

Example 5.3 Let us find the conformal transformation of the unit disk $|z| < 1$ into a pentagram (a regular five-point star) (see Fig. 5.31). The last domain is a decagon with five obtuse angles equal to $\pi\alpha = \frac{7}{5}\pi$ and five acute angles equal to $\pi\beta = \frac{\pi}{5}$. Since this domain possesses symmetries, we detach the tenth part of the star—the triangle OA_1B_1, which we denote by G. Our purpose is to map bijectively the sector $D = \left\{|z| < 1\,,\ 0 < \arg z < \frac{\pi}{5}\right\}$ onto the domain G in such a way that the interval $[0,\ 1]$ of the plane $\mathbb{C}_z$ is carried to the interval $[0,\ A_1]$ of the plane $\mathbb{C}_w$, and $\left[0,\ e^{i\pi/5}\right]$ is carried to $[0,\ B_1]$.

By the Riemann Mapping Theorem, the transformation we are looking for exists and if three points on the boundary of D and three corresponding points on the boundary of G are fixed, then this transformation is unique. At the same time, using the symmetry principle ten times, we obtain the mapping of the disk $|z| < 1$ onto the star in such a manner that the points $a_k = e^{\frac{2(k-1)\pi}{5}i}$ and $b_k = e^{\frac{(2k-1)\pi}{5}i}$, $k = 1, 2, 3, 4, 5$ are carried to the star points A_k and B_k, respectively (see Fig. 5.31). Since the transformation is unique, we can determine it by formula (5.45) (see Remark 5.6):

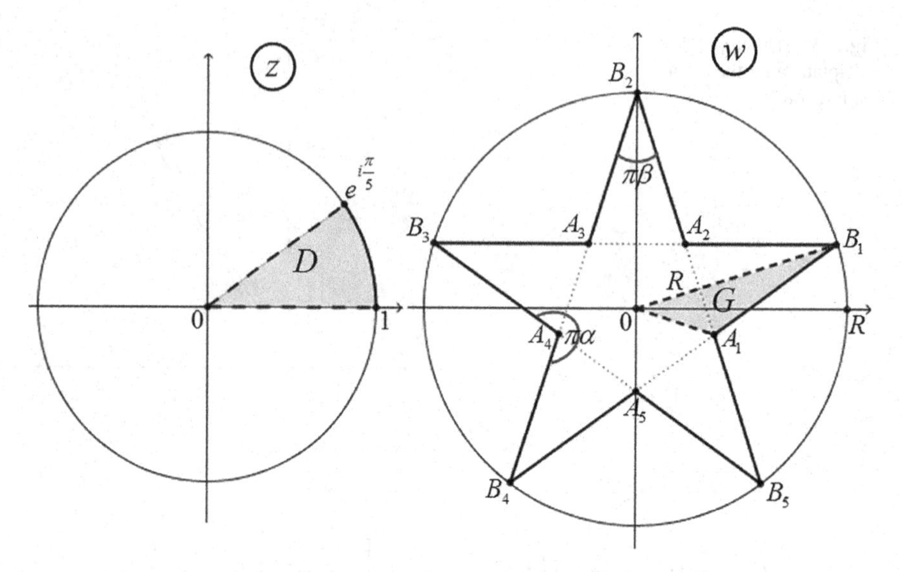

Fig. 5.31 Conformal transformation of the unit disk into pentagram

$$w = f(z) = c \int_0^z \frac{\prod_{k=1}^{5} (z - a_k)^{\frac{2}{5}}}{\prod_{k=1}^{5} (z - b_k)^{\frac{4}{5}}} dz\,.$$

Noting that

$$\prod_{k=1}^{5} (z - a_k) = \prod_{k=1}^{5} \left(z - e^{\frac{2(k-1)\pi}{5} i}\right) = z^5 - 1\,, \quad \prod_{k=1}^{5} (z - b_k) = \prod_{k=1}^{5} \left(z - e^{\frac{(2k-1)\pi}{5} i}\right) = z^5 + 1\,,$$

we can represent the mapping in the form

$$w = f(z) = c \int_0^z \frac{\left(z^5 - 1\right)^{2/5}}{\left(z^5 + 1\right)^{4/5}} dz\,. \tag{5.61}$$

Denote the star radius (the radius of the circle which the star is inscribed in) by R. Since the point $z = 0$ corresponds to the star center and the point $z = -1$ to one of the rays $OB_k = R$, we can find the relation for constant c in (5.61), assuming that it is real (the last follows from the star rotation):

$$R = c \int_{-1}^{0} \frac{\left(1 - x^5\right)^{2/5}}{\left(1 + x^5\right)^{4/5}} dx\,.$$

Using the change of the variable in the last integral

$$t = \left(\frac{1+x^5}{1-x^5}\right)^2; \quad x^5 = -\frac{1-t^{1/2}}{1+t^{1/2}}; \quad x = -\left(\frac{1-t^{1/2}}{1+t^{1/2}}\right)^{1/5};$$

$$dx = -\frac{1}{5}\left(\frac{1-t^{1/2}}{1+t^{1/2}}\right)^{-4/5} \frac{-\frac{1}{2}t^{-1/2}\left(1+t^{1/2}\right) - \frac{1}{2}t^{-1/2}\left(1-t^{1/2}\right)}{\left(1+t^{1/2}\right)^2}dt$$

$$= \frac{1}{5}t^{-1/2}\left(1-t^{1/2}\right)^{-4/5}\left(1+t^{1/2}\right)^{-6/5}dt,$$

we obtain

$$R = c\int_0^1 \left(1-\frac{t^{1/2}-1}{t^{1/2}+1}\right)^{2/5}\left(1+\frac{t^{1/2}-1}{t^{1/2}+1}\right)^{-4/5} \cdot \frac{1}{5}t^{-1/2}\left(1-t^{1/2}\right)^{-4/5}\left(1+t^{1/2}\right)^{-6/5}dt$$

$$= \frac{c}{5\cdot 2^{2/5}}\int_0^1 t^{-2/5}\left(1+t^{1/2}\right)^{2/5}t^{-1/2}\left(1-t^{1/2}\right)^{-4/5}\left(1+t^{1/2}\right)^{-6/5}dt$$

$$= \frac{c}{5\cdot 2^{2/5}}\int_0^1 t^{-9/10}(1-t)^{-4/5}dt = \frac{c}{5\cdot 2^{2/5}}B\left(\frac{1}{10},\frac{1}{5}\right) = \frac{c}{5\cdot 2^{2/5}}\frac{\Gamma\left(\frac{1}{10}\right)\Gamma\left(\frac{1}{5}\right)}{\Gamma\left(\frac{3}{10}\right)}.$$

Solving for c, we get

$$c = 2^{2/5}\cdot 5R\frac{\Gamma\left(\frac{3}{10}\right)}{\Gamma\left(\frac{1}{10}\right)\Gamma\left(\frac{1}{5}\right)}. \tag{5.62}$$

Thus, the function (5.61) with the constant c defined by (5.62), transforms conformally the unit disk $|z| < 1$ into the star of the radius R with the angles $\pi\alpha_k = \frac{7}{5}\pi$ and $\pi\beta_k = \frac{\pi}{5}$, $k = 1, 2, 3, 4, 5$.

Exercises

1. Show that the following functions are not univalent in the given domains D, although they are univalent at every point of D:

   (1) $f(z) = z^2$, $D = \{1 < |z| < 2\}$;
   (2) $f(z) = z^3$, $D = \{\mathrm{Im}\, z > 0\}$;
   (3) $f(z) = e^z$, $D = \{|z| < 4\}$.

2. Prove that the function $f(z) = z^2 + az$ is univalent in the upper half-plane $\mathrm{Im}\, z > 0$ if and only if $\mathrm{Im}\, a \geq 0$.

3. Let $f(z) = z^n + ne^{i\alpha} z$, where $\forall \alpha \in \mathbb{R}$, $n \in \mathbb{N}$, $n \geq 2$. Show that $f(z)$ is univalent in the unit disk $|z| < 1$.
4. Prove that the function $f(z) = z + e^z$ is univalent in the left half-plane $\operatorname{Re} z < 0$.
5. Show that an analytic in a convex domain D function $f(z)$, that satisfies the condition $\operatorname{Re} f'(z) > 0$, is univalent in D. Give an example that shows that this result fails for non-convex domains.
6. Transform the following domains into the upper half-plane using the Riemann–Schwarz symmetry principle:

(1) the entire plane z $(z = x + iy)$ with the cuts $\begin{cases} -1 \leq x \leq 1 \\ y = 0 \end{cases}$ and $\begin{cases} x = 0 \\ -b \leq y \leq a \end{cases}$, $a, b > 0$;

(2*) the entire plane z $(z = x + iy)$ with the cuts $\begin{cases} -1 \leq x \leq 1 \\ y = 0 \end{cases}$ and $\begin{cases} 0 \leq |z| \leq \sqrt{2} \\ \arg z = \pi/4 + k\pi/2 \end{cases}$, $k = 0, 1, 2, 3$;

(3) the strip $-2 < y < 2$ with the cuts $\begin{cases} x \geq 1 \\ y = \pm 1 \end{cases}$, $\begin{cases} x \leq -1 \\ y = \pm 1 \end{cases}$, $\begin{cases} x \geq 3 \\ y = 0 \end{cases}$ and $\begin{cases} x \leq 0 \\ y = 0 \end{cases}$;

(4) the entire plane z with the cuts $\begin{cases} -4 \leq x \leq 2 \\ y = 0 \end{cases}$, $\begin{cases} |z| = 2 \\ -\frac{\pi}{3} \leq \arg z \leq \frac{\pi}{3} \end{cases}$ and $\begin{cases} |z| = 2 \\ \frac{2\pi}{3} \leq \arg z \leq \frac{4\pi}{3} \end{cases}$;

(5*) the entire plane z with the cuts $\begin{cases} x \leq 1 \\ y = 0 \end{cases}$ and $\begin{cases} 0 \leq |z| \leq 1 \\ \arg z = \pm\pi/3 \end{cases}$;

(6*) the upper half-plane with the cuts $\begin{cases} x = k \\ 0 \leq y \leq 1 \end{cases}$, $\forall k \in \mathbb{Z}$;

(7) the exterior of the unit disk (centered at the origin) with the cuts $\begin{cases} 1 \leq |z| \leq 2 \\ \arg z = 2k\pi/n \end{cases}$, $k = 0, 1, \ldots, n-1$, $\forall n \in \mathbb{N}$;

(8*) the entire plane z with the cuts $\begin{cases} k - a \leq x \leq k + a \\ y = 0 \end{cases}$ and $\begin{cases} x = k \\ y \leq b \end{cases}$, $\forall k \in \mathbb{Z}$, $b > 0$, $0 < 2a < 1$;

(9) the disk $|z| < 2$ with the cuts $\begin{cases} -1 \leq x \leq 2 \\ y = 0 \end{cases}$ and $\begin{cases} x = 0 \\ -1 \leq y \leq 1 \end{cases}$;

(10) the upper half-plane with the cuts $\begin{cases} x = 0 \\ 1/2 \leq y \leq 2 \end{cases}$ and $\begin{cases} |z| = 1 \\ \pi/2 \leq \arg z \leq \pi \end{cases}$.

7. Show that using an appropriate change of variable the integral

$$K_1(k) = \int_1^{1/k} \frac{dt}{\sqrt{(t^2-1)(1-k^2t^2)}}\,, \quad 0 < k < 1\,,$$

considered in Sect. 5.6, can be transformed to the form

$$K_1(k) = K(k_1) = \int_0^1 \frac{dt}{\sqrt{(1-t^2)(1-k_1^2t^2)}}\,, \quad 0 < k_1 < 1\,,$$

where k_1 is an additional parameter such that $k^2 + k_1^2 = 1$.

8. Let T be an infinite triangle with the vertices $A = 0\,,\ B = 1\,,\ C = \infty$ and the corresponding angles $\pi\alpha,\ \pi\beta,\ \pi\gamma$, where $0 < \alpha \le 2\,,\ 0 < \beta \le 2\,,\ -2 \le \gamma \le 0\,,\ \alpha + \beta + \gamma = 1$. Find the function that transforms conformally the upper half-plane $\operatorname{Im} z > 0$ into this triangle in the following cases:

(1) $\alpha = \frac{3}{4}\,,\ \beta = \frac{1}{2}\,,\ \gamma = -\frac{1}{4}\,;$
(2) $\alpha = 2\,,\ \beta = \frac{1}{2}\,,\ \gamma = -\frac{3}{2}\,;$
(3) $\alpha = \frac{3}{2}\,,\ \beta = \frac{3}{2}\,,\ \gamma = -2\,;$
(4) $\alpha = \frac{2}{3}\,,\ \beta = \frac{2}{3}\,,\ \gamma = -\frac{1}{3}\,.$

9. Construct conformal mappings of the upper half-plane $\operatorname{Im} z > 0$ onto the domains in the plane w shown in Figs. 5.32, 5.33, 5.34, 5.35, 5.36, 5.37, 5.38, and 5.39, using the indicated correspondence between the points. Find the values of a and b for these mappings:

(1) $z = -1;\ b;\ 1;\ \infty \leftrightarrow w = A;\ B;\ C;\ D;$
(2) $z = -1;\ -b;\ 0;\ b;\ 1;\ \infty \leftrightarrow w = A;\ B;\ C;\ D;\ E;\ K;$
(3*) $z = -1;\ 1;\ \infty \leftrightarrow w = E;\ O;\ D;$
(4) $z = \infty;\ 0;\ 1 \leftrightarrow w = C;\ B;\ A;$

(5) $z = 0;\ 1;\ \infty \leftrightarrow w = O;\ C;\ A;$

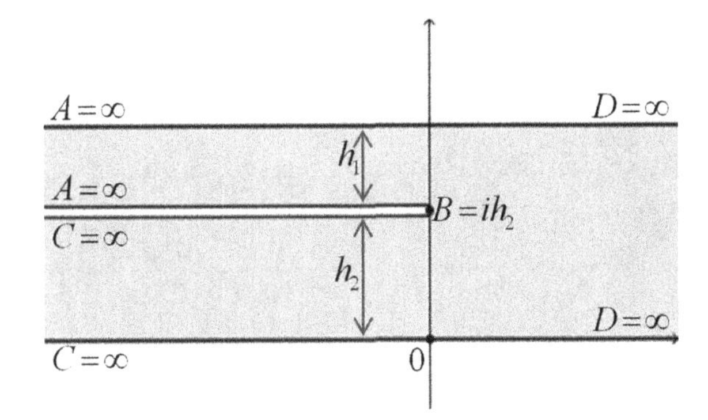

Fig. 5.32 Domain of exercise (9.1)

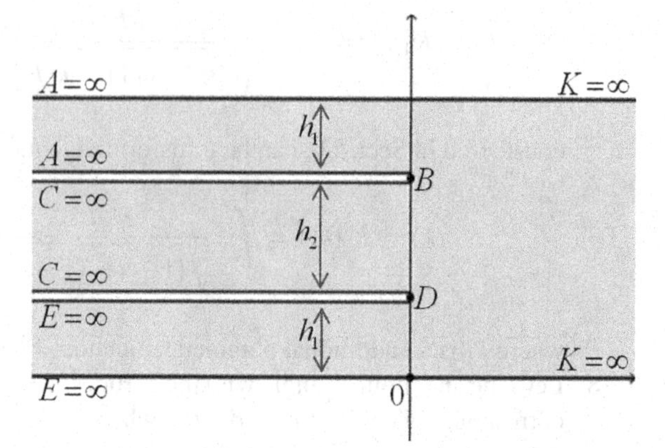

Fig. 5.33 Domain of exercise (9.2)

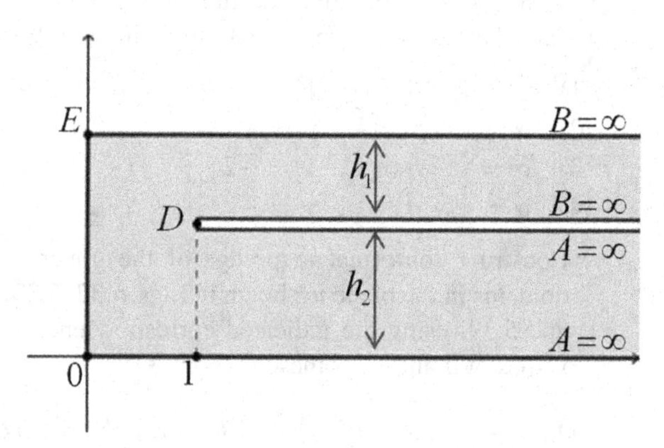

Fig. 5.34 Domain of exercise (9.3)

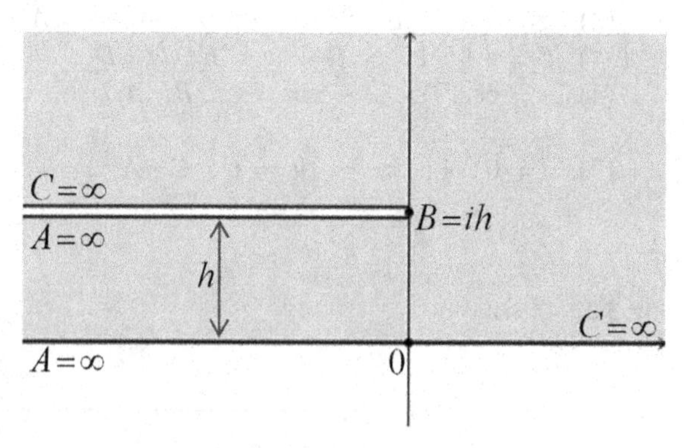

Fig. 5.35 Domain of exercise (9.4)

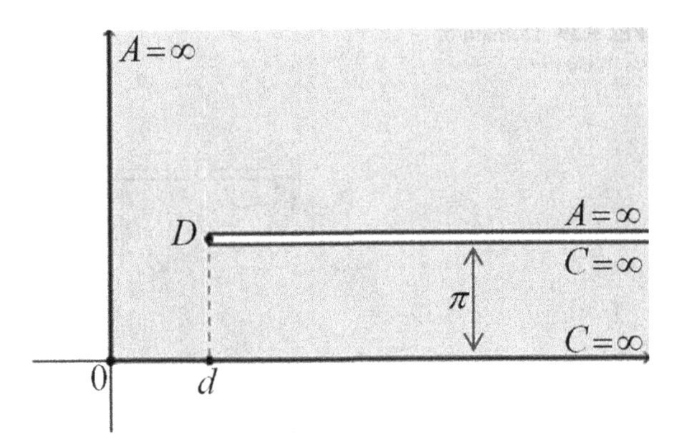

Fig. 5.36 Domain of exercise (9.5)

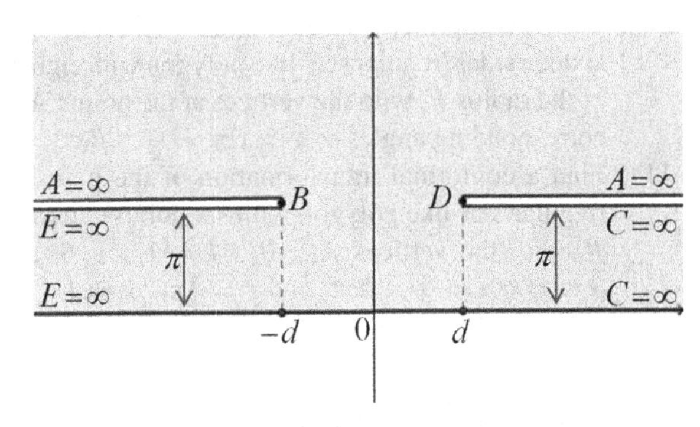

Fig. 5.37 Domain of exercise (9.6)

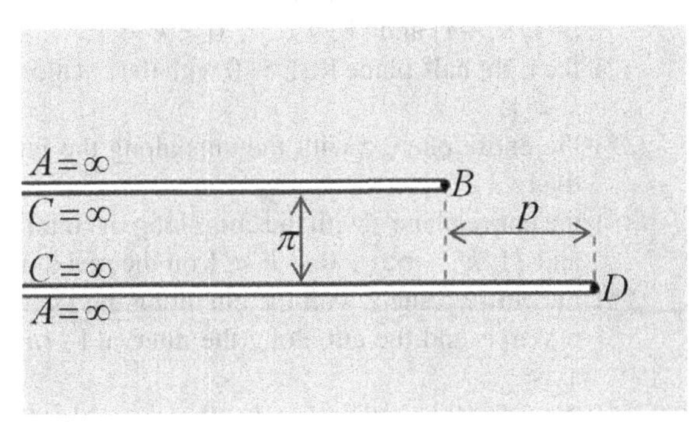

Fig. 5.38 Domain of exercise (9.7)

(6) $z = -1\,;\ 0\,;\ 1 \leftrightarrow w = E;\ O;\ C;$

(7) $z = \infty\,;\ -(1+a)\,;\ -1\,;\ 0 \leftrightarrow w = A\,;\ B;\ C;\ D;\ p = \text{Re}\ (D - B)\,;$

(8) $z = \infty\,;\ -1\,;\ a\,;\ 1 \leftrightarrow w = A\,;\ B\,;\ C;\ D;\quad p{=}\text{Re}\ (D - B)\ ,$
$h = \text{Im}\ (B - D)\,.$

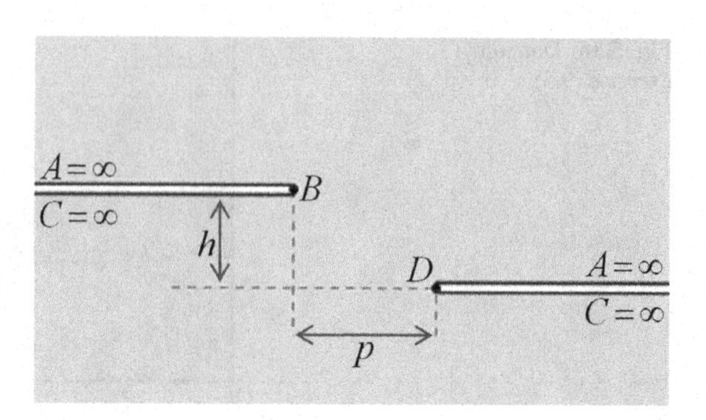

Fig. 5.39 Domain of exercise (9.8)

10*. Find a conformal transformation of the unit disk $|z| < 1$ into a polygon of sixteen sides (regular star-like polygon with eight points), inscribed in the circle of the radius R, with the vertices at the points $A_k,\ B_k,\ k = 1, \ldots, 8$, and the corresponding angles $\alpha_k\pi = \alpha\pi = \frac{\pi}{2},\ \beta_k\pi = \beta\pi = \frac{5}{4}\pi,\ k = 1, \ldots, 8$.

11*. Find a conformal transformation of the unit disk $|z| < 1$ into a dodecagon (regular star-like polygon with six points), inscribed in the circle of the radius R, with the vertices $A_k,\ B_k,\ k = 1, \ldots, 6$, and the corresponding angles $\alpha_k\pi = \alpha\pi = \frac{\pi}{3},\ \beta_k\pi = \beta\pi = \frac{4}{3}\pi,\ k = 1, \ldots, 6$.

12. Using the Legendre elliptic integral of the first type, construct conformal mappings of the following doubly-connected domains onto the ring $1 < |w| < R$:

(1) the entire plane z with the cuts along the two intervals of the real axis $[-1/k, -1]$ and $[1\,,\ 1/k]\ ,\ 0 < k < 1$;
(2) the right half-plane $\operatorname{Re} z > 0$ with the cut along the interval $[1\,,\ 1/k]\ ,\ 0 < k < 1$;
(3*) the entire plane z with the cuts along the intervals $[-4\,,\ 0]$ and $[1\,,\ 2]$ on the real axis;
(4) the entire plane z with the cuts along the intervals $(-\infty\,,\ -1/k]\ ,\ [-1\,,\ 1]$ and $[1/k\,,\ +\infty)\ ,\ 0 < k < 1$ on the real axis;
(5*) the entire plane z with the cut along the interval $[1\,,\ 1+d]\ ,\ d > 0$ of the real axis and the cut along the interval $[-ih\,,\ ih]$, $h > 0$ of the imaginary axis;
(6) the disk of the radius $1/\sqrt{k}\ ,\ 0 < k < 1$ (centered at the origin) with the cut along the interval $[-1, 1]$;
(7) the exterior of the disk of the radius $1/\sqrt{k},\ 0 < k < 1$ (centered at the origin) with the cuts $(-\infty\,,\ -1/k]$ and $[1/k,\ +\infty)$ on the real axis;
(8) the entire plane with the cuts $[0\,,\ k]$ and $[1/k,\ +\infty)\ ,\ 0 < k < 1$ on the real axis;
(9) the unit disk $|z| < 1$ with the cut along the interval $[0\,,\ k]\ ,\ 0 < k < 1$ of the real axis;
(10) the exterior of the unit disk $|z| > 1$ with the cut along the interval $[1/k, +\infty)$, $0 < k < 1$ of the real axis;

(11) the entire plane with the cuts $[-1, 0]$ and $[p, +\infty)$, $p > 0$ on the real axis;

(12) the entire plane with one cut along the interval $(-\infty, 0]$ on the real axis and another cut along the arc of the unit circle: $|z| = 1$, $|\arg z| \leq \alpha$, $0 < \alpha < \pi/2$;

(13) the entire plane with the cuts $(-\infty, -\sin\alpha]$, $[\sin\alpha, +\infty)$ on the real axis and the cut $[-i\cos\alpha, i\cos\alpha]$ on the imaginary axis, where $0 < \alpha < \pi/2$;

(14) the strip $-\frac{\pi}{2} < \operatorname{Re} z < \frac{\pi}{2}$ with the cut along the interval $[-iH, iH]$, $H > 0$ on the imaginary axis;

(15*) the strip $-\frac{\pi}{2} < \operatorname{Re} z < \frac{\pi}{2}$ with the cut along the interval $[\alpha, \beta]$, $-\frac{\pi}{2} < \alpha < \beta < \frac{\pi}{2}$ on the real axis;

(16) the half-strip $\begin{cases} -\pi/2 < \operatorname{Re} z < \pi/2 \\ \operatorname{Im} z > 0 \end{cases}$ with the cut along the interval $[iH_1, iH_2]$, $0 < H_1 < H_2 < +\infty$ on the imaginary axis.

13. Suppose $f(z)$ is analytic in the rectangle $D = \{z = x + iy : 0 < x < 1, |y| < h\}$, continuous in $\overline{D}$ and satisfies the conditions $\operatorname{Im} f(z) = 0$, for $\forall z : x = 0, |y| < h$ and $\operatorname{Im} f(z) = 1$, for $\forall z : x = 1, |y| < h$. Demonstrate that $f(z)$ can be analytically extended to the strip $|y| < h$ $(\forall x)$ and the corresponding function $F(z)$ has the form $F(z) = iz + F_1(z)$, where $F_1(z)$ is analytic in the strip $|y| < h$ and has the period 2.
Hint: use the Riemann–Schwarz symmetry principle.

14 . Suppose that $f(z) \neq constant$ is an analytic function in a domain D and $\Gamma = \{\forall z \in D : |f(z)| = c\}$, $c = constant$, is a simple closed curve such that $\Gamma = \partial G$, $G \cup \Gamma = \overline{G} \subset D$. Prove that there exists at least one zero of $f(z)$ in G.

15. Suppose that $f(z)$ is analytic in the disk $|z| < R$, $f(a) = 0$, $|a| < R$, and $|f(z)| \leq M$, $\forall z \in \{|z| < R\}$. Demonstrate that $|f(z)| \leq MR\left|\frac{z-a}{R^2 - z\bar{a}}\right|$, $\forall z \in \{|z| < R\}$, and $|f'(a)| \leq \frac{MR}{R^2 - |a|^2}$. For what functions the equality sign in these evaluations is achieved.
Hint: use the fractional linear transformation of a disk into another disk.

16. Suppose that $f(z)$ is analytic in the disk $|z| < R$, $|f(z)| \leq M$, $\forall z \in \{|z| < R\}$ and $f(0) = w_0$, $|w_0| < M$. Prove the following inequalities:

(1) $\left|\frac{f(z) - w_0}{M^2 - f(z)\overline{w}_0}\right| \leq \frac{|z|}{MR}$, $\forall z \in \{|z| < R\}$;

(2) $|f'(0)| \leq \frac{M^2 - |w_0|^2}{MR}$.
Investigate if there exist functions for which the equality in the above evaluation is achieved.

17. Suppose that $f(z)$ is analytic in the right half-plane $\operatorname{Re} z > 0$, continuous in $\operatorname{Re} z \geq 0$ and $|f(z)| < 1$, $\forall z \in \{\operatorname{Re} z > 0\}$. Suppose also that $f(z_k) = 0$ at the points z_k, $k = 1, \ldots, m$ located in the right half-plane. Show that $|f(z)| \leq \frac{|z - z_1| \cdot |z - z_2| \cdot \ldots \cdot |z - z_m|}{|z + \bar{z}_1| \cdot |z + \bar{z}_2| \cdot \ldots \cdot |z + \bar{z}_m|}$ for any point z in the right half-plane.

18*. Let $P(z)$ be a polynomial of degree n and $M(r) = \max_{|z|=r} |P(z)|$. Demonstrate that the inequality $\frac{M(r_1)}{r_1^n} \geq \frac{M(r_2)}{r_2^n}$ holds for $\forall r_1, r_2, 0 < r_1 < r_2$. Additionally, show that $P(z) = az^n$ if the equality in the last evaluation occurs at least for one pair of r_1, r_2.
Hint: consider the auxiliary function $g(t) = \frac{1}{a_n} P(\frac{1}{t}) t^n$ and apply the maximum modulus principle.

19. Let $P(z) = z^n + a_{n-1} z^{n-1} + \ldots + a_0$. Prove that either $|P(z)| > 1$ at least at one point of the circle $|z| = 1$ or $P(z) = z^n$.
Hint: make use of the result of Exercise 18.

20*. Suppose $g(z) = z + \sum_{n=1}^{+\infty} \frac{a_n}{z^n}$ is analytic in the domain $|z| > 1$ except a simple pole at $z = \infty$, and univalent in $|z| > 1$. Show that:

(1) $\sum_{n=1}^{+\infty} n |a_n|^2 \leq 1$;
(2) $|a_n| \leq \frac{1}{\sqrt{n}}, \forall n \in \mathbb{N}$; investigate for which function this inequality turns into equality $|a_m| = \frac{1}{\sqrt{m}}$ for some $m \in \mathbb{N}$.
Find the functions for which $|a_1| = 1$, and the image of the domain $|z| > 1$ under transformation by these functions.
Hint: use the fact that the area of a domain bounded by the image of the circle $|z| = \rho > 1$ is positive and calculate this area using Green's formula.

21*. Let function $f(z) = z + \sum_{n=2}^{+\infty} c_n z^n$ be analytic and univalent in the disk $|z| < 1$. Prove that $|c_2| \leq 2$. Investigate when the inequality turns into equality and find the image of the unit disk under such functions.
Hint: apply the result of Exercise 20 to the function $g(z) = \frac{1}{f_2(\frac{1}{z})}$, where $f_2(z) = \sqrt{f(z^2)}$ is one of the functions considered in Exercise 28 in Chap. 4.

Bibliography

Textbooks on Complex Analysis

1. Ahlfors, L.: Complex Analysis. McGraw-Hill, New York (1979)
2. Boas, R.P.: Invitation to Complex Analysis. Random House, New York (1987)
3. Evgrafov, M.A.: Analytic Functions. Dover Publication, Mineola (2019)
4. Freitag, E., Busam, R.: Complex Analysis. Springer, Berlin (2009)
5. Gamelin, T.W.: Complex Analysis. Springer, Berlin (2003)
6. Goluzin, G.M.: Geometric Theory of Functions of a Complex Variable. AMS, New York (1969)
7. Henrici, P.: Applied and Computational Complex Analysis. Wiley, New York (1993)
8. Jenkins, J.: Univalent Functions and Conformal Mapping. Springer, Berlin (1958)
9. Lang, S.: Complex Analysis. Springer, Berlin (2003)
10. Markushevich, A.I.: Theory of Functions of a Complex Variable, Vol.1-3. AMS, New York (2005)
11. Nehari, Z.: Conformal Mapping. Dover Publications, Mineola (2011)
12. Remmert, R.: Theory of Complex Functions. Springer, Berlin (1998)
13. Rodriguez, R.E., Kra, I., Gilman, J.P.: Complex Analysis: In the Spirit of Lipman Bers. Springer, Berlin (2012)
14. Rudin, W.: Real and Complex Analysis. McGraw-Hill, New York (1986)
15. Saff, E., Snider, A.D.: Fundamentals of Complex Analysis: With Applications to Engineering and Science. Pearson, London (2017)
16. Sidorov, Y.V., Fedoryuk, M.V., Shabunin, M.I.: Lectures on the Theory of Functions of a Complex Variable. Mir Publication, Moscow (1985)
17. Stein, E.M., Shakarchi, R.: Complex Analysis. Princeton University Press, Princeton (2003)
18. Sveshnikov, A., Tikhonov, A.: The Theory of Functions of a Complex Variable. Mir Publication, Moscow (1973)
19. Zill, D.G., Shanahan, P.D.: Complex Analysis. Jones & Bartlett, Burlington (2015)

A. Bourchtein and L. Bourchtein, *Complex Analysis*, Hindustan Publishing Corporation,
https://doi.org/10.1007/978-981-15-9219-5

History of Complex Analysis and Function Theory

1. Belhoste, B.: Augustin-Louis Cauchy: A Biography. Springer, Berlin (1991)
2. Bottazzini, U.: The Higher Calculus: A History of Real and Complex Analysis from Euler to Weierstrass. Springer, Berlin (1986)
3. Bottazzini, U., Gray, J.: Hidden Harmony - Geometric Fantasies: The Rise of Complex Function Theory. Springer, Berlin (2013)
4. Grabiner, J.V.: The Origins of Cauchy's Rigorous Calculus. MIT Press, Cambridge (1981)
5. Gray, J.: The Real and the Complex: A History of the Analysis in the 19th Century. Springer, Berlin (2015)
6. Klein, F.: Development of Mathematics in the 19th Century. Mathematical Sciences Press (1979)
7. Nahin, P.J.: An Imaginary Tale: The Story of $\sqrt{-1}$. Princeton University Press, Princeton (2016)
8. Smithies, F.: Cauchy and the Creation of Complex Function Theory. Cambridge University Press, Cambridge (1997)

Problems and Exercises of Complex Analysis

1. Alpay, D.: A Complex Analysis Problem Book. Birkhauser, Basel (2016)
2. Knopp, K.: Problem Book in the Theory of Functions, Vol. 1, 2. Dover Publications, Mineola (2000)
3. Krzyz, J.G.: Problems in Complex Variable Theory. Elsevier, Amsterdam (1971)
4. Milewski, E.G.: The Complex Variables Problem Solver. Research & Education Association, Piscataway (1987)
5. Mitrinovic, D.S., Keckic, J.D.: The Cauchy Method of Residues: Theory and Applications. Reidel, Dordrecht (1984)
6. Pap, E.: Complex Analysis through Examples and Exercises. Kluwer, Amsterdam (1999)
7. Shakarchi, R.: Problems and Solutions for Complex Analysis. Springer, Berlin (1999)
8. Volkovyskii, L.I., Lunts, G.L., Aramanovich, I.G.: A Collection of Problems on Complex Analysis. Dover Publications, Mineola (2011)

Index

A. Bourchtein and L. Bourchtein, *Complex Analysis*, Hindustan Publishing Corporation,
https://doi.org/10.1007/978-981-15-9219-5

Printed in the United States
by Baker & Taylor Publisher Services